A GUIDE
TO
PRACTICAL
TECHNOLOGICAL
FORECASTING

Contributions by

CLARK C. ABT
President
Abt Associates, Inc.

A. WADE BLACKMAN, JR.
Principal Systems Analyst
United Aircraft Research Laboratories

JAMES R. BRIGHT
Professor of Management
University of Texas at Austin

DAVID W. BROWN
President
Technical Marketing Associates, Inc.

JAMES L. BURKHARDT
Senior Associate
Technical Marketing Associates, Inc.

MARVIN J. CETRON
President
Forecasting International, Ltd.

WILLIAM H. CLINGMAN, JR.
President
W. H. Clingman & Co.

RICHARD C. DAVIS
Manager of Technological Forecasting
Whirlpool Corporation Research Laboratories

DONALD N. DICK
Advanced Planning and Analysis Staff
U.S. Naval Ordnance Laboratory

DAVID V. EDWARDS
Associate Professor of Government
University of Texas at Austin

JOHN R. EHRENFELD
President
Walden Research Corporation

SELWYN ENZER
Director of Operations—East Coast Office
Institute for the Future

MAURICE E. ESCH
Director, Military and Space Sciences Department
Honeywell, Inc.

RICHARD N. FOSTER
Senior Engineering Scientist
Abt Associates, Inc.

ALAN R. FUSFELD
Research Assistant
Sloan School of Management, M.I.T.

STEPHEN J. GAGE
Associate Professor of Mechanical Engineering
University of Texas at Austin

LUCIEN GERARDIN
Director of Look-out Studies
Thomson—CSF (France)

FREDERICK P. GLAZIER
Senior Staff Engineer
Sun Oil Company

THEODORE J. GORDON
President
The Futures Group

ROBERT L. HANEY
Director of Marketing Coordination
Transamerica Corporation

JARED E. HAZELTON
Associate Professor of Economics
University of Texas at Austin

ERICH JANTSCH
Research Associate
Sloan School of Management, M.I.T.

STEPHEN JECKOVICH
Vice President
PPG Industries

DAVID G. JOPLING
Administrative Assistant
Florida Power and Light Company

KENNETH E. KNIGHT
Associate Professor of Management
University of Texas at Austin

HAROLD A. LINSTONE
Professor and Director, Systems Science Institute
Portland State University

LAURENCE D. MCGLAUCHLIN
Staff Scientist
Honeywell, Inc.

JOSEPH P. MARTINO
Department of the Air Force
Air War College

DENNIS L. MEADOWS
Thayer School of Engineering
Dartmouth

M. EUGENE MERCHANT
Director of Research Planning
Cincinnati Milacron Inc.

KARLE S. PACKARD
Director of Long Range Planning
AIL Division of Cutler Hammer

DONALD L. PYKE
Planning Coordinator
University of Southern California

ELISABETH K. RABITSCH
Corporate Economist
Celanese Corporation

RICHARD ROCHBERG
Assistant Professor of Mathematics
Washington University

M. J. RAFFENSPERGER
The Futures Group

ROBERT H. REA
Vice President
Abt Associates, Inc.

NICHOLAS RESCHER
University Professor of Philosophy
University of Pittsburgh

MILTON E. F. SCHOEMAN
Assistant Professor of Management
University of Texas at Austin

WILLIAM L. SWAGER
Battelle Memorial Institute
Columbus Laboratories

W. H. CLIVE SIMMONDS
Program Planning and Analysis Group
National Research Council of Canada

A GUIDE
TO
PRACTICAL
TECHNOLOGICAL
FORECASTING

Editors

JAMES R. BRIGHT
MILTON E. F. SCHOEMAN

Graduate School of Business Administration
University of Texas

Prentice-Hall Inc., Englewood Cliffs, New Jersey

Library of Congress Cataloging in Publication Data

A GUIDE TO PRACTICAL TECHNOLOGICAL FORECASTING.
    Lectures from the Industrial Management Center courses, 1968 through 1971.

    Includes bibliographical references.
    1. Technological forecasting. I. Bright, James
Rieser, ed. II. Schoeman, Milton E., ed.
III. Industrial Management Center. IV. Title.
T174.G84   658.4′01   72-399
ISBN 0-13-370536-6

Printed in the United States of America

10 9 8 7 6 5 4 3 2 1

PRENTICE-HALL INTERNATIONAL, *London*
PRENTICE-HALL OF AUSTRALIA, PTY. LTD., *Sydney*
PRENTICE-HALL OF CANADA, LTD., *Toronto*
PRENTICE-HALL OF INDIA PRIVATE LTD., *New Delhi*
PRENTICE-HALL OF JAPAN, INC., *Tokyo*

Contents

Preface

The goal of this book is to advance practical technological forecasting by presenting new materials emerging from industry, government, and academia in the last five years. It is specifically designed to supplement the first United States book on technology forecasting.

In May, 1967, the Industrial Management Center, Inc. conducted what may have been the first conference for industry on formal methods of forecasting technological developments.[1] Recognition of the need had grown out of the development of a course on technological innovation at Harvard

---

[1] The office of Aerospace Research sponsored a symposium on Long-Range Forecasting and Planning at the U.S. Air Force Academy, August 16–17, 1966. This was the first assembly of analysts and planners directed specifically at technology forecasting. As might be expected, it was slanted to military application.

A symposium "Long-Range Forecasting Methodology" was conducted by the Department of Defense at Alamagordo, New Mexico, October 11–12, 1967. This symposium included a number of technological forecasting discussions. The Military Operations Research Society held several classified programs on technological forecasting about this time. Apparently these Department of Defence efforts were predecessors to civilian symposiums and conferences.

Business School. Several points were strikingly evident in that 1960–65 period:

1. Managers in industry and government were facing more and more decisions in which technological prospects were a major part of the issues. It was increasingly necessary, therefore, to anticipate technological changes and possibilities.
2. There was virtually no published methodology for forecasting technology. Past practice had been to use expert opinion. Opinion was not a very satisfying predictive device, as was evidenced by the highly uneven record of scientists and engineers with unassailable records of technical competence.
3. It seemed inconceivable that there could not be some predictive methodology that would be better, at least in some instances, than just opinion.

That first conference was attended by about 130 persons from six countries. The proceedings were edited and combined to form a major book, *Technological Forecasting For Industry and Government*.[2] After 1967, many other organizations, professional societies, management training firms, universities, and government agencies also conducted seminars and short courses on technological forecasting. These courses often were intertwined with topics such as research and development planning, corporate planning, and similar subjects. Three-journals heavily involving technological forecasting were introduced by 1969.[3] Technological forecasting became very popular, in both the best and the worst senses. On one hand was the promising search for useful methodology. On the other was faddism of the overenthusiastic and naive hopes of some that here was a crystal ball yielding wise decisions that would identify the next Xerox or Polaroid.

Nevertheless, a study of the literature and, in particular, the actual forecasts which speakers described at numerous conferences here and abroad showed definite progress.

This volume includes materials from the IMC courses 1968 through 1971, specifically selected for the industrial reader. They encompass:

1. More than half a dozen new forecasting theories and concepts, and in two cases improved explanations of old concepts.
2. Actual forecasts, as examples of what others have done. (We have deliberately omitted the many descriptions of Delphi studies which have been widely published and so are readily available.)
3. Approaches to improving technological forecasting through insight on areas affecting technology developments—the environment, political forces, and social change.
4. Descriptions of how successful technological forecasting efforts have been organized.

2 J.R. Bright, ed., *Technological Forecasting for Industry and Government: Methods and Applications*. (Englewood Cliffs, N.J.: Prentice-Hall, Inc., 1968).
3 *Futures, Technological Forecasting and Social Change, Long Range Planning*.

Thus this volume supplements the first book by bringing the field reasonably up to date as of the fall of 1971. Exercises for teaching technological forecasting methods have been a major feature of IMC courses. They are too voluminous to include here and are more properly directed to the academician. Because they are frequently revised and expanded, they have been put into workbook form.[4]

ACKNOWLEDGMENTS

Each chapter is identified with its author as listed, and we are deeply appreciative of every contribution. Our thanks go to these practitioners and researchers who have been willing to share their ideas. Unhappily, we have been unable to include the work of all our lecturers, but we appreciate their contributions. Although it is impractical to list them all, we are equally indebted to the 1,000 or so industrialists, government officials, and academicians who have participated in the short courses of the Industrial Management Center and the Management Development programs at the University of Texas. They have provided invaluable criticism, suggestions, and new industrial data.

The reader should know that, with three exceptions, all these papers were first presented before these IMC classes. IMC's policy has been to encourage its staff to publish their papers wherever possible, in order to speed the testing and improvement of TF concepts. Interim publications are acknowledged herewith:

"Mapping—A System Concept for Displaying Alternatives" by Donald L. Pyke appeared in *Technological Forecasting and Social Change,* Vol. 2, 1971.

"The Technological Progress Function: A New Technique for Forecasting" by Alan R. Fusfeld appeared in *Technological Forecasting,* Vol. 1, 1970.

"The Relevance Tree Method for Planning Basic Research" by Theodore J. Gordon and M.J. Raffensperger appeared as Chapter IV in *Technological Forecasting,* J.R. Bright and M.E.F. Schoeman, eds. (Canoga Park, California: Xyzyx Information Corp., 1970).

"Honeywell's PATTERN: Planning Assistance Through Technical Evaluation of Relevance Numbers" by Maurice E. Esch appeared as Chapter VI in *Technological Forecasting,* J.R. Bright and M.E.F. Schoeman, eds. (Canoga Park, California: Xyzyx Information Corp., 1970).

"An Application of Technological Forecasting to the Computer Industry" by Kenneth E. Knight appeared as Chapter VIII in *Technological Forecasting,* J.R. Bright and M.E.F. Schoeman, eds. (Canoga Park, California: Xyzyx Information Corp., 1970).

"Normex Forecasting of Jet Engine Characteristics" by A. Wade Blackman, Jr. appeared in *Technological Forecasting and Social Change,* Vol. 2, 1970.

[4] The workbook is in use at the Graduate School of Business, the University of Texas at Austin. Current information can be obtained by writing to Professor James R. Bright at that address.

"Characteristics and Implications of Forecasting Errors in the Selection of R&D Projects" by Dennis L. Meadows appeared as Chapter X in *Technological Forecasting,* J.R. Bright and M.E.F. Schoeman, eds. (Canoga Park, California: Xyzyx Information Corp., 1970).

"Political Forecasting" by David V. Edwards appeared as Chapter XV in *Technological Forecasting,* J.R. Bright and M.E.F. Schoeman, eds. (Canoga Park, California: Xyzyx Information Corp., 1970).

"New Organizational Forms for Forecasting" by Erich Jantsch appeared in *Technological Forecasting,* Vol. 1, 1969.

Each author has updated his original paper for this summary of five years progress in practical technological forecasting.

Our special thanks is due to Dean George Kozmetsky, of the Graduate School of Business, the University of Texas at Austin. He has provided both encouragement and opportunity to pursue this interactive learning experience and has facilitated the materials development and interchange between our school and industry.

We are grateful and express our very special thanks to Ralph Alpher and the General Electric Company for their provision of a grant for the support of doctoral work in this field.

The critical academician and keen industrialist may very properly challenge the degree of proof behind some of these concepts and gaps in the coverage. We accept this criticism but do not apologize for making the effort to sum up progress. A comparison with the predecessor volume will show that substantial gains have been made in five years. In another twenty years, perhaps, we shall begin to catch up with economic forecasting. Our basic position remains that the anticipation of technology is too vital to be ignored. And so we invite anyone to contact us with ideas, data, hypotheses, or experiences that may help to advance understanding of technological forecasting.

JAMES R. BRIGHT
MILTON E.F. SCHOEMAN

Austin, Texas
May 1972

A GUIDE
TO
PRACTICAL
TECHNOLOGICAL
FORECASTING

TOWARD THE IMPROVEMENT OF TECHNOLOGICAL FORECASTING

PART ONE

What concepts can help to improve our basic approach to a forecasting problem? Four authors offer specific ideas. The first chapter describes the process by which technology emerges from technical idea into widespread usage. Seven stages or "milestones" in this process provide measuring points which can be related to time and so provide a sense of rate of emergence. The emerging technology also is identified by different types of things over this period—papers, laboratory demonstrations, prototypes, commercial models, and so on—and so suggests the need for the forecaster to measure different things depending on his problem.

The second chapter springs from the experience of a vice-president of a major firm. He analyzes the pitfalls in trying to draw out economic significance from technology trend statistics. The dilemma of the manager is how to get effective planning guidance from technology forecasts. Some concepts are provided.

Now do we know that we have a good forecast in hand? How can we test its validity and relevance? In a unique and widely sought after analysis, Lt. Col. Joseph Martino describes his "interrogation" model for evaluating the validity of a technological forecast.

The final chapter deals with a growing aspect of technology—social resistance to technological change. It is painfully apparent that the old predictive procedures do not adequately reflect today's societal attitudes toward technology. Therefore, the forecaster must begin to include the social climate in his analysis. This chapter offers an explicit example and model for forecasting resistance to new technology.

The Process of Technological Innovation– An Aid to Understanding Technological Forecasting

JAMES R. BRIGHT

Success in forecasting technology can be improved if we understand the process by which technology emerges into social use. If we know the sources of technical concepts and those factors that support or inhibit its progress as it grows into physical reality and diffuses throughout society, we will have a better idea of what things to measure and what data to consider. And if we know the patterns of the time needed to carry out the parts of this process in the past, we might have the basis for better time estimates in the future. This chapter offers some views about the process of technological innovation against which technological forecasting concepts can be considered and, hopefully, data selection can be improved.

The *process of technological innovation* is a phrase intended to embrace those activities by which *technical knowledge is translated into a physical reality and becomes used on a scale having substantial societal impact.* This definition includes more than the act of invention. It includes initiation of the technical idea and acquisition of necessary knowledge, its transformation into usable hardware (or a process), its introduction into society, and its diffusion and adoption to the point where its impact is significant.[1] This full process will take time measured in years, and, more often, decades.

The technological forecaster then must recognize that the parts of this process do not take place in a vacuum but as activities in a very complex social system involving, at times, as many as five major environments. Exhibit 1 schematically suggests these environments and the institution's interaction and outlook on them. The significance of these environments to technological forecasting is enormous and, I believe, often substantially neglected or undervalued. Consider that many major technological innovations such as the jet engine, numeric control, the computer, the space program, water desalting, the SST, urban renewal, and the high speed passenger train were initiated, supported, and controlled to varying degrees by governments. Other technical innovations such as DDT, cyclamates, bacteriological warfare, and automobile safety are controlled, limited, or banned by government action. Some of these formal political steps were substantially inspired in the social environment. Recently some phenomena occurring in the ecologic environment itself (notably

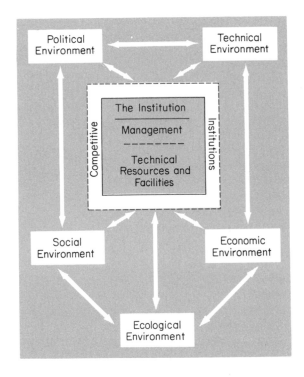

Exhibit 1. Technology and Its Environmental Interactions.

[1] The definition recognizes that technological innovations are not restricted to physical things but also include intellectual concepts such as operations research and decision theory as well as procedures and standards, such as the FAA certification testing required of new aircraft or the SAE standards for specification of screw threads.

population pressures and water and air pollution) have triggered social concern, which is resulting in political action affecting the traditional economic value of present and proposed technologies.

Therefore, the forecaster and the forecast user should establish their first operating principle: The economic viability of many existing technologies and the emergence and economic worth of new technologies often will be triggered and controlled by activities and events in the nontechnological environments. Furthermore, these forces and value systems are not necessarily constant. Indeed, values seem to be changing more rapidly in recent times, possibly due to improved communications, heightened social awareness, and more organization of social groups to resist or institute changes.

As a case in point, consider Ford Motor Company's 1957 effort to introduce automobile safety devices in their product line. Seat belts, padded dashes, recessed knobs, and similar items were designed and offered on the 1957 Ford cars. Although literally millions were spent by the company to promote these automobile safety features and their usage, the new technology did little for sales. Indeed, some industry specialists believe that, if anything, Ford's "safety" campaign hurt sales. But look at this same safety technology a decade later. It is required by law. Now these same safety devices (and more) *must* be provided! What happened? The 1957–1967 technologies and their economics are virtually identical. Something in our other environments has shifted. The emergence of this technology (automobile safety features), hence its value, was influenced in environments outside those traditional areas of business analysis—economics and technology. Forecasts, in this instance, should have somehow included the influential factors in the social and political environments.

As another example consider the fate of DDT. Because of perceptions, concerns, and pressures in the ecologic and social environments, political action has terminated the use of DDT in some countries and states. It may soon be banned internationally. It needs no crystal ball to see that other technology for controlling insects is certain to receive support, and the more promising will emerge into test and adoption. Again, we are taught the lesson that technical progress is not autonomous but is triggered and paced in part by forces outside of the businessman's and the scientist's traditional areas of analysis.

Our national SST decision of May, 1971, is a striking example of a current major technical advance that has been halted, or at least delayed, by a value change in the socio-political environments. Thus forecasting the emergence of technology and its social diffusion obviously must embrace more than the technological environment.

The process of technological innovation is extremely complex, different from case to case, and has been only partially investigated.[2] In the following

2 For an international survey of studies of technological innovation see *The Conditions For Success In Technological Innovation* (Paris: OECD, 1971).

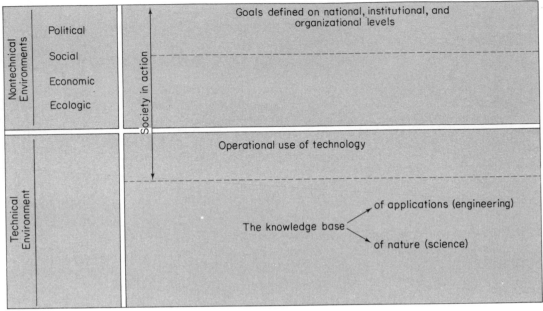

Goals defined on national, institutional, and organizational levels

Society in action

Nontechnical Environment

Political

Social

Economic

Ecologic

Operational use of technology

Technical Environment

The knowledge base — of applications (engineering)

of nature (science)

Exhibit 2. Schematic of Stages in the Process of T/I.

discussion I shall start with an oversimplified model of this process and then add recognition of some major complexities. Consider Exhibit 2 as a schematic representation. T/I (technological innovation) takes place in large environment, and this environment is composed of many subenvironments. For simplicity these are grouped in five categories. Four nontechnical aspects of the environment are shown on the upper left of the chart to convey that they are omnipresent, and their influence is diffused throughout the time and activities under consideration. The technical element of the environment is shown as the base from which the technical aspects of T/I are drawn. Now, what are the stages or milestones that we can use to measure the rate of progress of the emergence of the innovation? While many variations are possible, I have defined eight stages in Exhibit 3.

Stage 1 is the starting point for technological innovations, which seem to emerge in one of three ways: by *scientific suggestion,* meaning the speculations, hypotheses, and inferences of the scientist and engineer arising out of his search for new knowledge; by *discovery,* meaning the identification of a new phenomena in the course of pursuing scientific and engineering activities; and by *recognition of need or opportunity.* Exhibit 4 is intended to convey that the latter source can occur in any of the environments. Indeed, one technological forecasting concept specifically rests upon the hypothesis that by

Page	Identified By	Comment
1	Scientific suggestion, discovery, recognition of need or opportunity	Latter source seems to be origin of majority of contemporary innovations
2	Proposal of theory or design concept	Implying crystalization of theory or design concept that is ultimately successful; usually culmination of much trial and error
3	Laboratory verification of theory or design concept	Demonstration of existence or operational validity of concepts suggested in previous stage; may be difficult for manager to assess since thing demonstrated usually is phenomenon rather than application
4	Laboratory demonstration of application	Principle is embodied in laboratory "bread board" model of device (or sample material or its process equivalent) showing theory of Stage 2 reduced to (hopefully) useful form
5	Full scale or field trial	Concept has moved from laboratory bench into its first trial on large scale; succession of prototypes follows, leading eventually to saleable model.
6	Commercial introduction or first operational use	First sale of an operational system; may be deliberate or unconscious premature application of previous stage and so be replete with debugging problems and subsequent changes in technology
7	Widespread adoption as indicated by substantial profits, common usage, and significant impact	Admittedly, not sharply defined; individual firm might choose to classify as recovering its R&D investment through profits on sale of innovation or simply achievement of profitability; nationally, one might define this as given percent displacement, or as given percent of adoptors.
8	Proliferation	Technical device is applied to other uses; or principle is adapted to different purposes; this stage may begin much earlier.

Exhibit 3. Stages in the Process of Technological Innovation.

identifying future needs we can predict the most desirable, hence most likely, technology to merge. This may become a much more important source of forecasts in the future, as society organizes to choose the goals and means of achievement that it decides are desirable.

Stage 2 is the *proposal of theory* or, for many innovations, the *proposal of design concept* (combination of existing techniques and knowledge that will yield the desired technical effect) and is the next stage we can usually identify. Much trial and error generally takes place in reaching this stage. In this scheme, Stage 2 means the technical concept that is ultimately workable enough to become the basis of the T/I as first introduced into operational use. We must

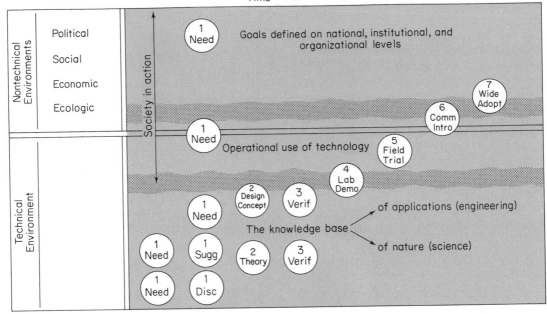

Exhibit 4. Stages in the Process of T/I. Related to Times and Environment.

realize that sometimes an earlier concept or a new concept may be refined and may then replace this original Stage 2 concept.

Stage 3 is the *verification of theory or design concept*. This stage is marked by the accomplishment of the experiment that confirms the validity of the proposed theory or design concept. It implies demonstration of an effect or a phenomenon as distinct from its application to a useful purpose.

Stage 4 is the *laboratory demonstration of application*. This is the first primitive model of the technology concept in a useful form. This is the laboratory "bread board" model and its equivalent for new processes and materials. Between this and the next stage there are numerous trials of alternative configurations, materials, and variations of scale. Eventually the T/I concept leaves the laboratory and development continues on a "life-size" scale and under field conditions.

Stage 5 is the *field or full-scale trial*. Again, we must recognize that there are very likely to be failures in the field trials or that results are so imperfect as to require a return to the laboratory. For forecasting purposes, we are concerned with identifying the full-scale approach that ultimately becomes a prototype of the T/I as introduced into everyday use.

Stage 6 is the *commercial introduction* (first sale) or *first operational use* (for some military or government adoptions) that marks the time when the T/I is believed to be ready for application as acceptable practice. This stage may be confused because some introductions are conducted in this operational mode but undertaken with the understanding that they are further trials. Also, sometimes the T/I may be deliberately or unwittingly introduced prematurely, before the technology is adequately refined or supported.

Stage 7 is *wide adoption,* meaning that time when the innovation has achieved usage on a scale great enough to have societal impact as measured by profits to producing firms or by major reaction in society (such as the alteration of military and foreign policies due to the explosion of two atomic bombs or by the mass immunization of the United States population for polio).

An eighth stage, *proliferation,* is listed, but it is not necessarily in proper sequence. The potential application of new technology to other purposes sometimes begins in Stage 1 and almost invariably in Stage 4. However, the sponsor usually tends to choose to pursue only one or a few applications initially. So most proliferation seems to take place after the first T/I reaches Stage 6.

This brief conceptual scheme is overly simplified and needs some major qualifications. First, the entire process or any of its parts may be filled with frustrating delays, false starts, and apparent progress that eventually run into technical, economic, or social blind alleys. It is not a simple linear process. Therefore, the "stages" must be thought of simply as marking the achievement of a particular state of growth. No equivalence of time, effort, importance, or value is implied between these "milestones."

Second, technological devices comprise a number of components or subsystems. These often are at different stages in the process of T/I. It is important to consider the status of each of these elements as we attempt to assess the present condition and future prospects of technological innovation.

Let us return to Exhibit 4 where we imposed symbols, representing the achievement of each stage, on a schematic chart. Several other things become clearer.

The activities associated with achieving the different stages take place in different physical environments and locations. The skills and resources necessary for success are very different. Therefore, the factors we must identify and measure in order to predict technological progress probably are quite different in at least some of the stages.

For the T/I to move from one stage to the next eventually means emerging in new environments. Here further progress will be influenced by new forces. We can surmise that the rate of progress will be delayed and speeded by a variety of events and activities in the particular environment, and some of these may have little direct relationship to the technical activities. Hence our ability to forecast must somehow involve perception and allowances for these forces.

What we mean in using the term "technological forecasting" needs clarification and more precision in use. If technology emerges from idea to widespread reality, it is identified and represented by different things. For instance, in Stage 2 it may be a scientific paper and in Stage 4 a crude laboratory model. The purpose and use of the forecast differs. A forecast might be directed at

Exhibit 5. Levels of Emergence or Impact in Technological Forecasting.

1. That certain *knowledge* of nature or *scientific understanding* will be acquired.
2. That it will be possible to *demonstrate* a new *technical capability* (on a laboratory basis).
3. That the new technical capability will be *applied to a full-scale prototype* (field trial).
4. That the new technology will be put to first *operational use* (commercial introduction).
5. That the new technology will be *widely adopted* (as measured by such things as number of units in use, output of the new technology, dollars of sales generated, or percentage measured relative to competitive technology).*
6. That certain *social and economic consequences* will result from the use of the new technology.†
7. That future economic, political, social, and technical conditions will *require the creation* of certain new technical capability.‡

Note: Levels 1, 2, 3, and 4 are simple predictions of technical achievement.
* Implies a prediction of economic desirability and implies a prediction about *all other competing* technology, as well as economic conditions.
† May or may not include the assumptions of Level 5, depending on whether it concerns micro or macro usage; but it adds a prediction of the effect on human behavior and institutions.
‡ Implies a synthesis of forecasts about *many attributes* of future society plus some judgments about the value of various goals of that future society. Obviously, Levels 5, 6, and 7 involve *more* than technology and probably should not be thought of simply as "technological" forecasts

any of several levels of emergence or degrees of impact. Therefore, the apparently clearcut term technological forecasting is not clear at all. We must agree upon a level of emergence significant to what we are considering. Exhibit 5 lists these "levels of emergence or impact."

Note that Levels 5, 6, and 7 are the levels of popular economic and social concern, yet the technical forecast is only one input to predictions on these levels.

These levels, as well as the stages, can help the forecaster by suggesting what he should be looking for and where data might be found. Indicators of coming technology, therefore, can be found in speeches, journals, professional meetings, political statements, laboratories, experimental installations, announcements in the public and trade press, technical data in catalogs, the historical statistics of an industry, laboratory activities, universities, census data, and informal professional exchanges. The positions of government officials on social problems, agency programs, funding, and appointments, all may provide impetus, direction, delay, or insight.

Consideration of the process of technological innovation offers the forecaster help in other ways. What does history tell him about the time, the influential factors, and other considerations to use in making his prediction? Since 1960 I have been collecting data on recent and current technical innovations. While the data are highly uneven, there are enough commonalities to suggest patterns of experience. Based on some thirty innovations (in great detail) and some key events in thirty more, the following propositions are offered for the forecaster's consideration:

Proposition 1 states that *the full process of T/I usually takes upward of ten years, with twenty to twenty-five years more likely.* This may seem contrary to the common statement that the time to innovate is shortening. Parts of the process may be shorter today, but, in general, Stages 1 through 3 involve many years. Moving from Stage 6 to Stage 7 is almost certain to take upwards of five years. To test this proposition I urge the reader to examine the record of innovation in his own company or industry. He will find that the original concept of many major innovations goes back very far. As a few examples consider: the computer prior to 1840, the fuel cell prior to 1830, photo type-setting prior to 1860, and so on. Even xerography took about eighteen years to go from Stage 1 to Stage 7, and another five years were required to achieve Carlson's original goal of a successful office copier. The NSF TRACES study, Project Hindsight, Arthur D. Little studies, and case studies in Jewkes, Sawers and Stillerman, *The Sources of Invention*[3], confirm these long periods of the total innovation process. Integrated circuits, however, took only twelve years from Stage 1 through 7, and the laser reached Stage 6 in perhaps five years. I leave it to the reader's judgment as to whether it has yet reached Stage 7. At the best, the laser took close to a decade to go through the process.

A fair argument is that this time pattern may be generally true but is not really pertinent. After all, most firms tend to encounter innovations in Stage 3 or 4. They are concerned with taking the innovation to Stage 7. Therefore, decision to commit major resources is really the significant start to the process of T/I to the firm. Whatever one's opinion on this argument, for the forecaster the point is that it is likely to take one to two decades or more to turn a technical *suggestion* into a widespread reality. The process of diffusion from Stage 6 to Stage 7 alone is a five- to ten-year job (depending upon the innovation). The forecaster must be rigorous in defining the stages he is considering. And he must make his forecast definite in its relationship to these stages.

Proposition 2 states that *many factors will influence the progress and direction of the technology.* In addition to the items that might arise in any of the five environments in Exhibit 1, at times such things as *leadership, policy, organization, funding,* and *chance* will be influential.

Proposition 3 states that *government actions to choose directions, goals, or amount of support or to control technology are becoming far more significant in T/I.* Therefore, the technological forecaster must begin to add political inputs to his predictive processes.

Proposition 4 maintains that *technological capabilities such as improvements in speed, power, abrasion resistance, and so on, grow in an exponential manner.* The rate may be very slow at first but eventually becomes explosive. History shows that the technologist working in the field usually finds it hard

[3] John Jewkes, David Sawers, Richard Stillerman, *The Sources of Invention*, 2nd ed. (New York: W. W. Horton and Co., Inc., 1969).

to believe that this rate of improvement can be maintained. However, case after case seems to verify this exponential phenomenon. The forecaster must respect exponential improvement as virtually a law of nature. He may not know the exponents but he must expect this characteristic in the rate of progress. Ultimately, this progress will be limited by technological, social, environmental, economic, or political factors.

Proposition 5 states that *forecasts of technological capabilities are not the equivalents or indicators of profits or amounts of usage.* There often are social, economic, and political barriers or restraints that prevent or delay usage. We must *not* equate rates of technical improvement with rates of usage, adoption, or with profits.

Proposition 6 states that *accelerated and unsuspected progress often comes when one technology impinges upon another.* These mergers create new forms or break previous barriers to progress or to the adoption of technology. The computer and solid state physics exemplify the point.

Proposition 7 states that is *the mode of financing usage of T/I's is of utmost significance to* (a) its speed of diffusion and (b) the financial returns to the firm. Few technological forecasters have paid much attention to the design of the system by which users or consumers will pay for the use of their innovation. The importance of this concept can be illustrated by the case of xerography. From 1960 to 1965 Xerox Corporation installed about 65,000 Model 914 Copiers. If these units had been sold for twice manufacturing cost, the gross revenue would have been about $330 million. If they had been leased on a flat monthly charge of $300, the income would have been about $360 million, with the advantage that further annual lease fees would continue to accrue and the company would gain the cash flow from depreciation. However, using the charge-per-copy system that was actually installed, Xerox's gross revenue was about $660 million with the same advantages of cash flow and further income from usage. Roughly, the additional revenue over these five years was at least $300 million more than the first two plans—a decision equivalent to doubling the value of the innovation. Furthermore, there is no doubt that the very low cost of the charge-per-copy plan greatly encouraged adoption and speeded diffusion. Consider how slowly the computer would have spread had IBM only sold their machines.

Forecasts of usage, therefore, must be made with the realization that the mode of providing the technology to the user will influence adoption and economic returns.

CONCLUSION

Although the foregoing is incomplete, it does point out to the forecaster that he should become a student of technological innovation. The patterns and events in technological history will guide him toward a better understanding of the factors that will shape future technology.

Managerial Planning
Beyond
Statistical Forecasting

STEPHEN JECKOVICH

There is an important distinction between forecasting technological parameters and forecasting the economic or social usage of that technology.

Trend extrapolation concepts have proven useful in projecting the progress of certain performance parameters in some technologies. The author forcefully demonstrates the dangers that can befall the forecast user who blindly applies trend extrapolation techniques for forecasting the economic output of an industry or individual firm. Despite these possible pitfalls Jeckovich does not dismiss the utility of trend extrapolation. He argues that such projections can indeed serve as a useful baseline for managerial planning. He offers several guidelines for modifying the raw projections and transforming them into more useful data.

Many efforts to plan for the future, including an increasing concern with generating innovations, rest heavily on the projection of past statistical trends into the future. In order to enhance the "scientific basis" for such forecasts as well as to avoid charges of sheer speculation, there is frequent resort to two kinds of statistical underpinning: analysis of the industry's past growth pattern and analysis of the growth patterns of its major products. And perhaps the most widely prevailing expectation with respect to each of these is the "life

13

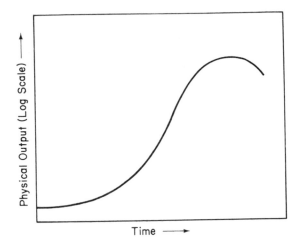

Exhibit 1. S-Shaped Retardation Curve.

cycle" concept, suggested by Arthur F. Burns[1] along with others[2,3] and shown in Exhibit 1, with physical output on the ordinate in log scale and time on the abscissa in arithmetic scale: progressive retardation as a result of the seemingly inevitable gradual pushing aside of old products and industries after a period of lively growth. Logical as it seems, however, extensive testing fails to support this widely accepted view.[4,5] Moreover, detailed analysis reveals other serious limitations on the usefulness of such generalized guides for managerial planning. These may be illustrated by reviewing experience in a major industry with which the author is familiar: glass.[6,7]

INDUSTRY GROWTH PATTERN

Exhibit 2 shows that United States glass industry's physical output increased nearly thirtyfold between 1899 and 1966, for an average annual rate of about five percent or almost double that for total manufacturing. However,

1 A. F. Burns, *Production Trends in the United States Since 1870* (New York: National Bureau of Economic Research, 1934).

2 Simon S. Kuznets, *Secular Movements on Production and Prices* (New York: Houghton Mifflin, 1930).

3 J. F. Gaston, *Growth Patterns in Industry: A Reexamination* (New York: The Conference Board, 1961).

4 Bela Gold, "Industry Growth Patterns Theory and Empirical Results," *Journal of Industrial Economics* (November, 1964).

5 Bela Gold, D. Huettner, R. Mitchell, and R. Skeddle, "Long-Term Growth Patterns of Industries, Firms, and Products" (Working Paper No. 8, Research Program in Industrial Economics, School of Management, Case Western Reserve University, August, 1968).

6 Bureau of the Census, U. S. Department of Commerce, *Census of Manufactures* (Washington, D. C.: Government Printing Office, various dates.).

7 Stephen Jeckovich, *An Economic Study of the United States Glass Industry, 1899–1947,* (Ph. D. dissertation, University of Pittsburgh, 1961.)

there was no characteristic S-shaped pattern of retardation. On the contrary, the long-term growth pattern of the glass industry could be represented by a straight line on a logarithmic scale, reflecting growth at a constant average rate. But what are the implications of this industry pattern for the management of individual firms and for planning innovations?

Closer examination of Exhibit 2 shows that the industry's growth was far from steady. In 1923 output was only slightly higher than 1914 and the intervening period was characterized by severe adjustments from the managerial viewpoint. Further sharp fluctuations ensued over the next sixteen years as well, only to be followed by the turbulence of World War II and postwar adjustments. Hence, although a statistician could easily look back and discern a comfortably steady average rate of growth, it is apparent that this was not the experience of the industry's management. Moreover, it might even be argued that none of the resumptions of the interrupted growth were inevitable, for example, bound to occur had management been passive about products, processes, and markets. Accordingly a mere continuation of past rates of growth would require as much innovational effort and success, relative to competitive industries, as in the past and with the emergence of more innovative competi-

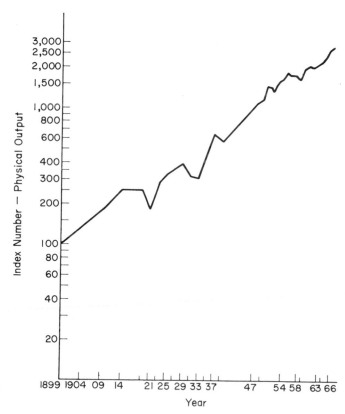

Exhibit 2. Glass Industry Physical Output (1899–1966).

tors, a progressive increase in the innovative effort and success of the glass industry would be necessary in order for it to simply maintain its relative position.

PRODUCT GROWTH PATTERNS

Further perspectives on forecasting as well as an appreciation of the pressures for innovation may be derived by examining the growth patterns of the major products comprising the output of the glass industry. Available data restrict such efforts to analyzing the growth of certain major product groups, as shown in Exhibit 3. It is apparent that there were widely disparate rates and patterns of growth among these products. Between 1899 and 1947, average annual rates ranged from seven percent for polished plate glass to three percent for window glass and included also a sharp decline for lamp chimneys. Polished plate output increased at more than the glass industry rate while window glass expanded at about one half that rate. Milk bottle output grew about eleven percent per year from 1899 to 1925 and at about three percent

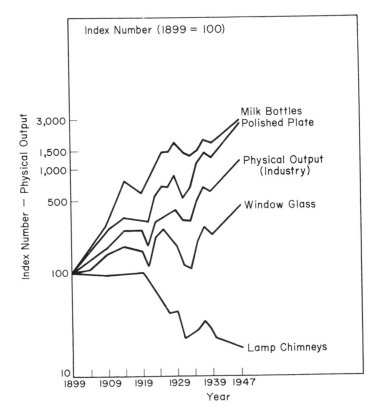

Exhibit 3. Growth Patterns of Glass Products (1899–1947).

per year from 1925 to 1947. Thus milk bottle output gave evidence of early stages of retardation; lamp chimney output evidenced the final phases of retardation from 1919 onward; and polished plate and window glass experienced no apparent retardation between 1899 and 1947.

Closer examination of these series again reveals a variety of short-term fluctuations which render attempts to plan for R&D programs, new facilities, market development, and operating programs on the basis of long-term average growth rates, a hazardous undertaking. And the reality is even more formidable than has been suggested both because Exhibit 3 does not contain annual data and because reliance on statistics of product groups ignores further differences among individual products. Thus one may gain further insight into the usefulness of statistical guides for planning an innovational program for a given product by considering three examples, window glass, milk bottles, and polished plate glass, in greater detail.

Window Glass[8]

A more detailed study of window glass output during the 1899 to 1968 interval, shown on the lower panel of Exhibit 4, highlights several points of interest. One of the major problems in forecasting concerns the determination of position. Setting aside the advantages of hindsight, a managerial assessment at time A (1914), if based on an extension of the 1899 to 1914 trend, would have suggested the need for additional capacity which, in turn, would have been unused for the next nine years. However, an assessment at time B (1921), based on the 1899 to 1921 pattern, would have suggested that the product was undergoing retardation, and hence neither additional capacity nor further technological improvements were required. An assessment at time C (1925) gave the impression of rapid recovery and growth, calling for more capacity which would have been essentially idle for the next eleven years. An assessment made at time D (1933) again might have suggested retardation arguing against both capacity increases as well as further technological improvements. Since 1935 there has been recovery and major growth with some suggestion of topping off between 1953 and 1968.

The foregoing analysis suggests that determination of position could be seriously in error if the forecast is simply an extrapolation of the then apparent trend. With the clearer vision of hindsight, the dips in 1921 and 1933 (and 1958) seem to relate to recessions and/or depressions and to that extent could be rationalized on an economic basis. It is therefore of interest next to analyze next this period relative to the major technological developments which took place and are shown on the upper panel of Exhibit 4.

As late as 1899 the Bureau of the Census reported that all of the 100

[8] Jeckovich, *An Economic Study of the United States Glass Industry,* op. cit. pp. 162–68.

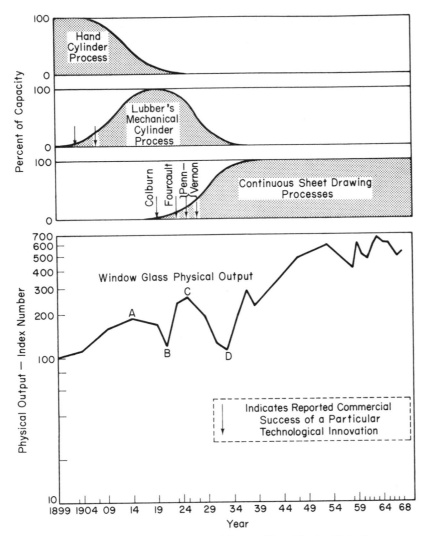

Exhibit 4. Growth Pattern of United States Window Glass. Physical Output and Major Processing Technological Innovations (1899–1968).

window glass plants used the hand-cylinder method of blowing window glass. The first major technological development was the Lubber's cylinder blowing machine, which was developed under the auspices of the then American Window Glass Company, and even though the first machine was installed in 1900 it was not until 1903 or 1907 that the process was reported to be commercially successful. Basically, the Lubber's process simply substituted scaled-up (for example forty-inch diameter by forty-foot long cylinders as compared to twenty-inch diameter by six-foot long cylinders) mechanical methods for the hand gathering and manual blowing operations, while capping, splitting, and flattening continued to be hand operations. Advantages resulting were reduced production costs (especially by eliminating the need for highly skilled and highly paid gatherers and blowers so that by 1925 all hand operations were practically nonexistent) and larger marketable size glass of improved optical quality and more uniform thickness. The Lubber's process dominated the scene from about 1903 until about 1927.

However it was still necessary to cut the cylinder and flatten and anneal the glass in separate operations. Breakage during the blowing operation tended on occasion to be high and the flattening operation created surface distortions. There was also an apparent reluctance on the part of the owners of the Lubber's patent to license that technology; hence competitors were motivated to develop an innovation to break out of the bind.

As early as 1860 there were recorded experiments of the continuous vertical drawing, in sheet form, of window glass. Efforts intensified following the turn of the century, stimulated by the Lubber's process. After many plant failures as well as personal financial failure, Colburn's process became commercially successful in 1919. The glass sheet was drawn vertically about twenty-four inches, and then turned at right angles over a roll, and carried through an annealing lehr. Some authorities believed that a large part of Colburn's difficulties was due to his lack of practical experience in glass manufacturing and ignorance of the physical properties of glass. The commercial success of the Colburn process culminated in the formation of the Libbey-Owens Sheet Glass Company (predecessor to the present Libbey-Owens-Ford Company).

In the meantime, Emile Fourcault of Belgium was busy perfecting his method. The Fourcault process drew a flat sheet vertically from between a slot in a refractory member floating in molten glass. The glass was annealed during its upward travel through a short vertical lehr above the forming operation. Even though certain patents were allowed as early as 1902, its first reported commercial use was not until 1923. Interestingly the owner of the Lubber's machine, attempting to regain its market position, picked up the Fourcault process in 1928, and after about ten years of difficult and costly research, was finally able to produce an acceptable glass, superior to anything produced on the Fourcault machine.

Parallel developments were undertaken by the then Pittsburgh Plate Glass Company which had patents issued in 1918, and subsequent development work, which was both difficult and at times discouraging, resulted in commercial successes in 1925 and 1927. The Pittsburgh, or Pennvernon process as it was called, differed from both the Colburn and Fourcault processes in that the continuous sheet was drawn vertically from an open bath of glass over a submerged refractory bar and annealed in a relatively short vertical lehr.

The result of these three continuous sheet drawing innovations was a substantial improvement in glass quality and increase in market availability of larger sizes in a greater range of thicknesses, thus enhancing window glass market appeal. Machine capacity increased considerably and by 1929, although the number of plants had shrunk from 100 to 19, annual output per plant increased from about 2 million square feet to about 21 million square feet. The market position of some of the producers was drastically affected by the technological path they followed. The technological approaches were different and required considerable time and money to perfect. However the end products were roughly comparable and apparently competitive in the market, thus illustrating the notion of technological equivalence.

Subsequent developments focused on capacity additions, scale-up, through-put increases, and in the early 1950s an especially intensive effort to improve quality by reducing distortion.

Milk Bottles

As shown in Exhibit 5, the sharp trend over the prior fifteen years would have led an analyst in 1914 (A) to urge major increases in capacity which would then have remained unused for the next seven years. A similar interpretation in the early 1920s (B) would have led to an even longer period of underutilization. Again in 1947 (C) a projection of the 1933 to 1947 trend would have encouraged a program to increase through-put on existing tech-

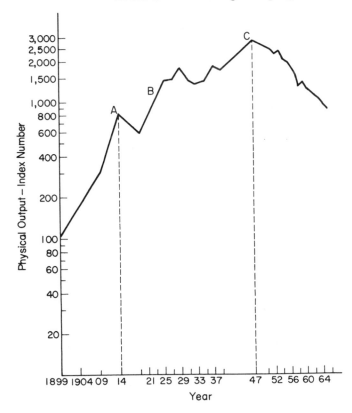

Exhibit 5. Growth Pattern of Milk Bottle Output (1899–1965).

nology or to expand capacity by developing new technology with greater productivity and profitability. But, as shown in Exhibit 5, milk bottle output reached a peak in 1947 and thereafter underwent a steady decline, which amounted to sixty-eight percent between 1947 and 1965, under the competitive pressures of paperboard and plastic containers. This illustrates the danger resulting from marketing myopia, as discussed by Levitt,[9] which fails to distinguish between product market function and product features as major guidance for an innovational program. Clearly, a milk bottle program oriented in the historical tradition of decreased weight, increased strength, better quality, and possibly lower cost did not appear to meet the needs of the market. While the sheet glass growth curve illustrated to an extent the effect of innovations within the industry, the milk bottle curve illustrates the impact of innovations outside the industry.

Polished Plate Glass

Exhibit 6 shows the output of polished plate glass for the period 1899 through 1968. However, to a manager looking ahead at time A (in the late 1940s), the consideration of innovational potentials could not be avoided, for additional capacity suggested by an extrapolation of the 1899 to 1947 trend might embody new technology as well as prevailing practices, that is, present state of the art. The technology reflected in the latter at that time embodied a large continuous melting tank furnace equipped with a continuous rolling facility for producing a continuous "rough" ribbon, which after continuous annealing was converted to polished plate glass by two basic methods. In the conventional process which accounted for the larger portion of capacity the rough ribbon was cut into individual stock sheets, laid on adjoining rectangular tables which moved continuously on tracks or ways through a multiplicity of grinding heads, and fed with a slurry of progressively finer abrasives (sand, garnet) followed by similar processing through a series of rouge-fed felt polishing heads. After completion of the first side, the glass plate was turned over and processed through a second line where the second side received similar treatment. Individual stock sheets were then stripped from the tables, washed, inspected, and cut for different fabricating orders. In the twin process the continuous rough ribbon, after continuous annealing, was processed through a twin grinder where both sides were continuously and simultaneously ground. The twin ground ribbon was cut into stock sheets which were laid on tables and each side polished separately as in the conventional process.

Innovational possibilities had from time to time speculated about the prospects for combining the continuous twin grinding with some form of continuous twin polishing, or even more optimistically, of eliminating the grinding and

9 Theodore Levitt, "Marketing Myopia," *Harvard Business Review,* July-August, 1960, pp. 45–56.

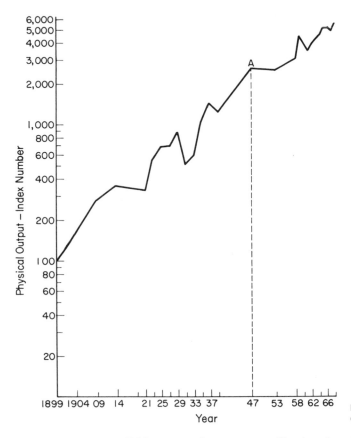

Exhibit 6. Growth Pattern of Polished Plate Glass (1899–1968).

polishing steps by a more effective forming operation. Accordingly, at least three companies at that point in time embarked on what turned out to be major innovational programs. After a decade of intensive efforts and R&D investments of approximately equal amounts, the Pilkington Float process emerged in 1960 as the apparent "winner" thereby tending to make obsolete both conventional as well as twin grinding and twin polishing processes for the manufacturer of plate glass. In this case the projections of future industry market demand did materialize, but individual company forecasts and planning regarding innovational approaches to be pursued differed markedly. Inasmuch as there did not appear to be technological equivalence among the competing approaches, the float process appears to be the surviving process with the individual company consequences, especially with regard to royalties and technological position, differing sginificantly.

LIMITATIONS OF STATISTICAL FORECASTING

In brief, the contributions of statistical extrapolations to the definition of managerial planning targets are likely to prove far more limited than is commonly recognized. Such shortcomings may be discerned at several levels.

For example, over the short run the projection of past averages tends to miss both the extraordinary increases and the extraordinary decreases from year to year, the correct anticipation of which is a primary objective and a major potential contribution of operational planning. And over the longer run, the projection of past averages tends to concentrate on a level of aggregation which ignores the changing nature of products, the changing technology of production processes and facilities, and the changing characteristics of markets, all of which should be basic foci of strategic planning.

More serious difficulties arise from the unstated assumption underlying most statistical extrapolations that the future will be like the past. The trends which are projected invariably represent "backcasts" from the known end point of a statistical series so as to provide a reasonable approximation to the data for prior years rather than a true forecast made five or ten years earlier and still proved relevant as the unknown future was gradually revealed. But the economic system, unlike the physical system which is closed (as a rocket in flight), is an open or reactive system, subject to changes in environmental conditions and to the introduction of new variables. Hence, forecasting on the basis of hindsight has been successful enough—in the limited sense of roughly approximating past complex adjustment patterns with some simpler average trend—but forecasting into the future has quite generally been a dismal failure.[10]

Even more important, close analysis of past patterns almost invariably reveals downturns which were arrested and reversed not by passively awaiting for the "natural forces" to restore equilibrium but rather by energetic if not frantic efforts to offset unfavorable pressures. But the statistical analyses associated with most attempts to project past patterns do not even attempt to identify the kind of managerial counter efforts underlying reversals of expected declines, much less to determine their relative effectiveness under various conditions.

It is also worth recalling the sources of the relative stability reflected by product and industry growth trends and projections. In part, these are attributable to the casual ignoring of intermittent fluctuations on the basis of hindsight which may have represented frightening threats at the time to the managers seeking to cope with them. Additionally, reliance on totals for entire product lines and even larger output groupings means that the introduction of a superior competitive product was only gradually reflected in output categories dominated by other established commodities, relegating such series to providing only belated confirmation of major shifts such as the displacement of milk bottles by paper and plastic containers.

SUGGESTIONS FOR ADDITIONAL GUIDANCE

Although attention has been called to some limitations of statistical forecasts as guides to managerial planning, there can be no denying their useful-

[10] Bela Gold, *Journal of Industrial Economics,* November, 1964, op. cit.

ness for certain purposes. In particular estimates of the probable results of a continuation of established trends and relationships constitute an important *base line* for policy and program planning. And within such undertakings, it may prove possible to strengthen contributions by collecting and analyzing more detailed information by product and market subcategories, by experimenting with forecasting methods and time periods more clearly differentiated according to the purposes to be served by the results, and by more critical evaluation of the managerial uses made as well as the accuracy of the various forecasts provided. Continuing studies are underway in this area.

In addition, however, effective planning requires more intensive efforts to supplement the foregoing with estimates of the effects of factors tending to produce significant deviations from established trends and relationships. And activities, especially during the past decade, reveal a growing array of forecasting techniques with increasing emphasis on technological forecasting as a means of uncovering alternative considerations for the R&D portfolio. While it still may be premature to judge effectiveness, it appears that these techniques are more successful in predicting evolutionary improvements rather than in predicting breakthroughs or innovations.

One suggestion for providing additional guidance is embodied in the four categories of the Forecast Adjustment Matrix, shown in Exhibit 7, which includes the major determinants of change. Externally generated sources of deviations from market trends and relationships would center around important changes in consumer purchasing power, needs, and tastes as well as the availability and prices of competitive products. Internally generated sources of deviations from such trends and relationships would involve planned modifications in marketing, distribution, and pricing.

Whether generated internally or externally, the technological factors likely to cause significant deviations from past output and cost trends and relationships would relate to products, processes, and production facilities. One rich source of stimuli for assessing or predicting deviations from the technological base line is the so-called "boney pile" of literature and issued patents. On the assumption that a certain number of ideas are articulated before their time, one should expect that a search through relevant literature and patents might uncover technological approaches which for a variety of reasons are now both pertinent and feasible. Ideally a search for such "sleeping beauties" should be

	Externally generated	Internally generated
Market factors		
Technological factors		

Exhibit 7. Externally/Internally Generated Sources of Deviations.

conducted by personnel experienced in the industry and sensitive to its needs and opportunities.

Accordingly, it is suggested by way of hypothesis that a next major advance in strengthening managerial planning may be obtained by *formally* and *regularly* calling for separate estimates of each of these four determinants of deviations from past growth patterns instead of allowing partial reflections of them to sidle into planning perspectives on an informal, helter-skelter basis. Such formalization offers several potential benefits. First, it would clearly place responsibility for making each estimate on the particular group with the highest relevant expertise and would require the group to present such assessments on a regularized basis along with supporting data and arguments. Second, regularized auditing of such estimates and supporting material, in comparison with actual developments, may stimulate organized efforts to improve such estimates. And, third, direct confrontation of the problems of integrating these estimates with one another and with estimates of the effects of continuing trends may encourage increasing exploration of the interactions among these different sources of change as the basis for rearranging managerial planning and control arrangements to increase their effectiveness in the face of such complexities.

With regard to organizational implications within individual corporations, it is apparent that estimating the nature and effects of both externally and internally generated changes in market factors might readily be concentrated in one organizational unit in order to facilitate effective responses by the latter to the former and in order to provide inputs to planning which seek to take account of both. The desirability of such an arrangement has already been recognized in many progressive corporations through the development of market research or market planning groups. As yet, however, few seem to have effectively separated the processes and personnel involved in estimating the effects of prevailing trends and relationships from the quite different methods and expertise involved in identifying and estimating the effects of developments tending to generate significant deviations from such trends.

It may also be argued that estimating the nature and effects of both externally and internally generated technological factors in products, processes, and facilities ought to be similarly concentrated in one organizational unit. But this is still extremely uncommon, most corporations continuing to have such responsibilities fragmented among research and development, engineering, production, and even sales or marketing. Both the interaction of the relevant technologies, however, and their common bearing on the planning of basic corporate development as well as the attendant formulation of a soundly balanced R&D portfolio for various innovational purposes suggest that the planning and evaluation of technical innovations is likely to be increasingly integrated and centralized. And one of the valuable by-products of such integration may well be the development of far more effective estimates of the potential effects of alternative innovational programs, with consequent strengthening of the corporate planinng to which such estimates constitute increasingly important inputs.

Evaluating Forecast Validity

JOSEPH P. MARTINO

Does the value of a forecast lie in its ultimate correctness? The author says "not necessarily." The value of any forecast must be based upon its contribution in providing information to a decision making process. In many instances, the decision maker can exert some degree of control over the outcome of a particular forecasted condition. Hence, he may take actions tending to make the forecast incorrect whenever a discrepancy exists between the particular outcome forecasted and his desired outcome.

Colonel Martino presents an interrogation model as a rational and explicit method for evaluating the utility of a forecast for decision making. This model is based upon a series of iterative questions and the construction of a relevance tree. He illustrates the model through a hypothetical situation involving an aircraft manufacturer concerned with evaluating a specific forecast of domestic revenue passenger miles flown by United States commercial airlines.

What is a good forecast? The immediate answer might seem to be: one that turns out to be correct. After all, one might ask, what good is a forecast that turns out to be wrong? Consider the person seeking a weather forecast. While he might obtain a weather forecast out of simple curiosity, more likely he wants to make some plans for some activity which will be affected by the weather. Naturally, he wants to plan correctly, either to take advantage of good weather or to prepare for bad weather. If he bases his plans on a forecast of good weather and the weather turns out to be bad, the forecast has been useless to him. Worse yet, it may have misled him into foregoing his routine preparations for bad weather. It appears, then, as though a weather forecast has to be correct to be useful.

But the story is not always that simple. The weather is an example of one class of situations, those over which the persons affected have absolutely no control. At the other extreme are situations which are completely controlled by a single person or group. In between are situations where several persons or groups each exercise partial control over the outcome.

In the case of a completely controlled situation, a forecast is superfluous. The person in control does not need a forecast of what is going to happen because the outcome will be whatever he wishes it to be. Other persons who may be affected by his actions might find a forecast of his actions useful, but he himself does not need one.

When a decision maker has no control over the outcome of a particular situation, he wants to tailor his actions to what the situation is going to be like, either to maximize the benefit from a favorable outcome, or to minimize the impact of an unfavorable outcome. As with the case of the weather, it appears that an incorrect forecast is useless, or worse.

In the inbetween cases the decision maker is interested in more than just taking advantage of a good situation or avoiding as much of a bad situation as possible. He is interested in tailoring his actions so as to make maximum use of his influence over the situation. In order to do this, he needs to know the outcome of various courses of action open to him. In short he needs one or more forecasts.

Thus, in either the partial control or no control cases, we are primarily concerned with forecasts as decision information, that is, as inputs to decisions. Throughout this discussion we will take the viewpoint of a decision maker who has been given a forecast as part of the information on which to base his decision. Normally he will not prepare the forecast himself. Instead it will be prepared for him by someone else. This someone else may be a subordinate who has been requested to provide the forecast. Or the forecast may be provided by someone outside his organization who is advocating a specific course of action and attempting to influence the decision maker in a particular direction. Or the forecast may have been prepared for some other purpose and "borrowed" for the occasion because it appeared to be useful.

Furthermore, it must be recognized that the forecast is not the only factor to be considered in making a decision; it may not even be the most important. Other factors such as personal and organizational goals, guidance or direction from higher authority, or the amount and kinds of resources available must also be taken into account. Our viewpoint, then, is that of the rational decision maker who has analyzed the decision problem facing him, decided what he needs to know, and determined how these pieces of information are to be fitted together in order to arrive at his decision. The decision may be as simple as whether or not to take a raincoat to work, or it may be a decision having a major impact on the future of some corporate or government enterprise. In any case, we assume his plans are consciously and deliberately based on rational factors, including a forecast of the appropriate portion of the environment.

As pointed out earlier in the case of no control by the decision maker, it appears that a forecast must be correct to be of value. However, in the situation where the decision maker has partial control over the outcome, forecasting then becomes more complicated. If the decision maker is given a forecast of a particular event which he considers undesirable, he may use what control he has to forestall the event. To the extent that he is successful, he invalidates the forecast. Conversely, if the forecast calls for an event which he considers desirable, he may use his partial control to increase the likelihood of that event. To the extent that he is successful, he validates the forecast.

This leads to the subject of self-fulfilling and self-defeating forecasts. A self-fulfilling forecast is one which will come true if people believe it and act rationally on their belief. A self-defeating forecast is one that will not come true if people believe it and act rationally on their belief.

As an example of a self-fulfilling forecast, let us suppose the President of the United States announces that his economic advisers have informed him the country is going to have a depression. If I believe the forecast, there are certain actions which follow logically from my belief: I should reduce my spending, pay off my debts, and save my money. This will put me in a better position to withstand the depression when it comes. However, if enough people do this, the depression will certainly come. Unlike preparing for bad weather, preparing for a depression can actually bring it about. This is an example of a self-fulfilling forecast, since if people believe it and act rationally on their belief, they will bring about the event which was forecast.

As an example of a self-defeating forecast, let us suppose that the Department of Agriculture forecasts for next year a shortage of some crop on which there are no production controls. As a result of the forecast farmers, processors, and others will expect high prices for that crop. Farmers will respond by increasing the acreage they plant in that crop and thereby produce more than they would have without the forecast. The amount of the shortage will be reduced as a result; and if the response has been too vigorous, there may be a surplus. Here again, we see people believing a forecast and acting rationally on their belief and, in this case, forestalling the event which was forecast.

The possibility of self-fulfilling and self-defeating forecasts complicates the issue of evaluating a forecast. Even if a given forecast turned out to be true, it may have been self-fulfilling. Had the forecast not been made, events might well have turned out some other way. Conversely, if a forecast fails to come true, it might have been self-defeating. Had the forecast not been made, the event forecast might well have come about instead of being forestalled. This casts some doubt on the notion that a good forecast is one that turns out to be correct.

Consider yet another case, that of an orange grower who is given a forecast of freezing weather tonight. If he believes the forecast, he will take steps to protect his orange crop from freezing. Suppose the forecast is in error and it does not freeze. Was the forecast therefore useless? One argument might go as follows. He has the option of preparing for freezing weather on every night on which his trees and their crop might be damaged. This will cost him a certain amount of money which must be deducted from his profits. He also has the option of preparing for freezing weather only on those nights on which frost is forecast. This will reduce his cost of preparing for frost, since he will not make preparations every night. However, he runs the risk of losing the entire crop and therefore all his profit if a forecast of good weather turns out to be in error. We can thus conclude the following. If the likelihood of error in a forecast of good weather is small enough, a forecast of bad weather is still useful even when it is wrong. The reason is that it causes the farmer to make preparations for frost on all those nights when the likelihood of frost is high enough that the expected cost of a lost crop exceeds the cost of preparation for frost.

We can generalize this conclusion to all kinds of forecasts. A forecast was useful if, as seen by hindsight, it led the decision maker to make a correct and timely decision even though the forecast itself may not have been valid. Put another way, as viewed from the standpoint of considering forecasts as decision information, a forecast was useful if it led to a good decision at the time, even if events did not turn out to agree with the forecast.

But this is all in retrospect. It enables one to tell after the fact whether or not a given forecast was useful at the time it was made. But this is not the usual problem. The decision maker is not usually concerned with evaluating the utility of forecasts made and used at some time in the past for decisions whose correctness is already known. His concern, on the contrary, is with a given forecast whose outcome is still unknown which is to be an input to a decision as yet unmade. How does he evaluate the forecast?

The essential point is that his evaluation is still in terms of the forecast's utility for decision making. He is not really concerned with whether or not the forecast will come true, especially since that may involve the paradox of a self-fulfilling or self-defeating forecast. He is really concerned with whether or not the forecast he has been given as information for a decision is in fact useful for making that decision and whether or not it is likely to lead to a correct decision.

Is it possible for the decision maker to evaluate a forecast at the time he needs to use it as to its utility for the decision he is going to make? In one sense, the answer is yes. Decision makers since time immemorial have been evaluating and using forecasts. The question really is not whether or not it can be done but whether or not there is a formal and rational scheme available which is teachable to and usable by the decision maker. One such scheme, known as an interrogation model, has been developed and is presented here.

INTERROGATION MODEL

The goal of the research which led to this scheme was to provide a rational and explicit method for systematically evaluating forecasts as to their utility for decision making. The scheme is designated *interrogation* rather than *analysis* because it makes use of a structured series of questions rather than statements. This approach was chosen deliberately in an attempt to avoid a priori assumptions about what is and what is not of concern in evaluating a forecast. The model is deliberately designed to maximize the likelihood that all relevant considerations are taken into account.

The structure of the model, a particular series of questions in a given order, has the following effects: (1) the questions asked imply the criteria used in the evaluation; (2) the order acts to identify the problem initially; (3) once the problem is identified, the order allows a systematic identification of key problem elements; (4) the key problem elements then serve as criteria for seeking additional information; and (5) once the relevant and available knowledge has thus been identified, it can be tested for reliability. In short the model allows an efficient search for those elements which turn out to be important in each specific case.

The model involves the following steps, which are described in more detail below.

1. Interrogation for need.
2. Interrogation for underlying cause.
3. Interrogation for relevance.
4. Interrogation for reliability.

These steps are taken in the order given, as described in the following paragraphs.

Interrogation for Need

This step is intended to identify those factors which it is necessary to know about the future. Questions which follow from this objective are directed at the purpose or goal of the forecast. The following specific questions get at the need for the forecast.

Who needs this forecast? The objective here is to obtain a specific identification of the individual, organization, agency, company, or nation by whom or for whom the decision in question is to be made. The decision must take into account the values of this actor, since human values are not universal. All subsequent questions must be relevant to this specifically identified actor or entity who will be referred to below as "we."

What are we interested in knowing about the future? This depends not only on the actor "we" but on the specific decision to be made. This subject area should initially be as broad as possible in order to avoid the danger of excluding elements which would subsequently turn out to be important but which might be missed by an initial overly narrow definition of "our" area of concern.

How critical is the need for this forecast? This question is intended to get at the value structure of the actor. The answer will depend on what we are seeking to avoid or obtain and how much that is worth to us. It will also depend on the criticality of the decision to be made, for example, whether it affects the survival of the actor, only the continued efficient operation of the actor, or merely the convenience and comfort of the actor. The answer to this question will establish the criteria which determine how accurate the forecast has to be and how much effort should be expended in evaluating it.

Through what future time range do we need to forecast? The answer here will depend on the activities we are engaged in and the decision to be made. Regarding the activities, attention must be given to the rates of change, sequences, lags, and delays. We must consider the total time elapsed from the first availability of information through the making of the decision to its ultimate impact on reality. This must include the implementation of decisions through activities such as programming, construction, training, debugging, and the shakedown period to full operational status. In short, how far we need to look into the future depends on what we are trying to do and how long it takes us to do it. Among other things, the answer to this question will be used to eliminate factors which cannot have any influence, or whose influence is essentially constant during the relevant time range.

Interrogation for Underlying Cause

This step is intended to identify those factors which can cause changes in those things we want to know about the future during the time range of interest. This step is carried out iteratively in that each time a factor capable of causing change is identified, we next ask what can cause that factor to change. The end result of this step, after iteration, is a relevance tree. In particular, it is a relevance tree specific to the area of our concern over the time range of interest to us. Unlike most relevance trees, however, the bottom level of "fundamental causes" should contain a few elements repeated many times. These are the key elements which will be examined in detail in later steps of the model. The following specific questions can be used to assist in carrying out this step.

In what ways can this general subject area change? Here we examine the subject area which we have determined is of interest to us and attempt to identify its modes of change. In particular, we should look for changes in size, either larger or smaller. We would also look for changes in composition or structure. Will the relative importance of the present components change? Will some of the present components be eliminated? Will new components enter? Once we have identified the possible modes of change, we are ready for the next step.

What can cause these changes? Now we examine each of the possible modes of change in order to identify possible causes of change. It should be noted that many possible causes of change may in themselves not be sufficient, or even necessary, for change to take place. They may be only contributory. Nevertheless, they must be included. As an aid to identifying possible causes of change, we take into account the fact that a technology can be considered as existing in a complex environment. Each of the elements of that environment may be a cause of change or an inhibitor of change. Hence we can more easily identify causes of change by examining the components of the environment. The following list of components has been found to be fruitful as an aid to this interrogation for causes of change: economic, technological, intellectual, managerial, political, social, religious/ethical, ecological. We will take up each component separately and identify the way it can cause or inhibit change.

The *economic* component of the environment deals with costs and returns in monetary or economic terms. Some useful questions are the following: Will costs of production, use, or operation change? Will the general economic climate change? Will financing be available on the scale required by a specific potential change? Will government subsidies be initiated, increased, or decreased? Will there be a change in the size of the potential market?

A *technology* does not exist in isolation but in context with other technologies—some competing, some supporting, some complementary. Some useful questions to ask about the technological component of the environment are the following: Will there be a change in a competing technology? If so, how will this affect the competitive standing of the competing technology and the technology being forecast? Is there a chance that a technology currently "in embryo" may be brought to perfection? Can a combination of advances in several fields have an impact? Does the technology being forecast draw on other technologies for support (for example, production, maintenance, energy supply, transportation)? If so, what are the possible changes in the supporting technologies? Must the technology being forecast be compatible with existing technologies (for example, power supply frequency, physical dimensions)? If so, what changes might occur in these complementary technologies?

The current *intellectual* environment has an impact on technology, and changes in this environment are reflected by changes in technology. Some useful questions to ask about the intellectual component of the environment are as follows: Is a change likely in the general climate of opinion? Is change being hindered by biases or preconceptions which might change? What are the policies and attitudes of decision makers and their institutions? What are the policies and attitudes of opinion leaders? What changes are possible in these policies and attitudes? What are the policies and attitudes of private and governmental decision makers, of opinion leaders, and of the general public toward specific sources of possible change? What changes are possible in these policies and attitudes?

It is frequently overlooked that enterprises must be *managed*. At any given time, there is a state of the art of managerial capability which limits the size or complexity of the enterprise which can be managed. Changes in the state of the managerial art can either force or permit changes in technology. Some useful questions to ask about the managerial component of the environment are the following: Would a possible technological change require a scale of enterprise beyond the capabilities of current managerial technique or beyond the capabilities of the specific managerial group responsible for bringing it about? Will there be changes in managerial capability which will force the abandonment of current technological practices? Will there be changes in managerial capability which will enable the adoption of technologies previously unmanageable?

Man is by nature a *political* animal. Thus the political component of the environment inevitably has an impact on technological change. The two basic questions about the political environment are: Who benefits and who gets hurt? However, these can usefully be broken down into detailed questions. Will there be changes in the missions and responsibilities of existing institutions? Will new institutions be created with missions and responsibilities currently not assigned to anyone? Will there be changes in the attitudes of the political leadership? Is there the possibility of change of political leadership, especially through death or retirement of long-term office holders? What is the possibility of the new leadership having different attitudes? Are there likely to be changes in government incentives or restrictions on technological change? Are there vested interests in either a continuation of the current technological status or in a change of the current technological status? If so, who benefits by continuation or by change? Who gets hurt by continuation or by change? Would certain political objectives, unrelated to technology per se, be supported by change or be threatened by change? Would any political objectives be supported or threatened by the acceptance of change from a particular source?

Ultimately technology is used by society and will be affected by society. Hence it is necessary to examine the *social* component of the environment to identify possible causes of change in technology. The following questions can be useful in this regard. Will social problems and pressures force a change from current practice? Is there any evidence for a possible large-scale change in the value systems of society or any major components of society? What is the likely change in population size? What is the likely change in age distribution of the population? What is the likely change in geographical distribution of the population? What technological changes might arise from these changes? What has been the general social reaction to change? How might this change?

Both individuals and groups tend to judge events and situations in terms of right and wrong. This judgment is based on *religious* or *ethical* considerations. Hence these considerations must be taken into account in assessing the possibility of change. The following questions are useful in identifying these considerations. Is there likely to be a change in a religious or ethical attitude which has supported or tolerated current practice? What is the likely response of religious or ethical groups to a specific possible change? Is there likely to be a change in religious or ethical attitudes which have heretofore impeded change?

Until recently, considerations of an *ecological* nature have not weighed heavily on the possibility of technological change. However, it is likely that these considerations will become more important in the future, perhaps becoming more significant than purely technological considerations. Some useful questions are the following: Is the ecological impact of current practice likely to force a change? What would be the ecological impact of a specific possible change initially and over the long term if exploited on a large scale?

Having carried out the analysis of each mode of change in terms of these components of the environment, we have identified several factors which appear to have the potential of causing a change in the technology we are forecasting. Next we carry the process a step further.

For each of the factors identified, we iterate the process. We ask, what can cause this factor to change? Each factor must then be examined in terms of the appropriate components of the environment to determine what factors might cause it to change. This iteration is repeated at as many levels as necessary until the interrogation reaches a set of "fundamental" causes which is not deemed worthy of further investigation (for example, it is probably not worthwhile investigating to decide whether or not the sun will continue to shine throughout the period of interest).

What are the key elements of this subject area? Once the iteration of interrogation for underlying cause has reached termination in a set of "fundamental" causes, it will normally be the case that these causes appear several times at the bottom layer of the relevance tree. These repeatedly appearing causes are then the key elements of the subject area. Of course, a factor which enters only once may also be a key element if its influence is large enough.

Interrogation for Relevance

In the previous step we identified those items which appeared to be key elements in the subject area. We now, for the first time, look at the actual forecast we are attempting to evaluate. We determine those key elements which are found to be present in the forecast. In general, there will be three categories of elements: (1) relevant and included, for example those which were found to be key elements and which were taken into account by the forecast; (2) relevant but excluded, for example those which were found to be key elements which the forecast omits; and (3) irrelevant but included, for example those elements which were found not to be relevant but which are included in the forecast. Those elements found to be relevant and included are then evaluated for reliability in the next step. It should be noted that the presence of irrelevant material may not necessarily indicate a fault in the forecast. The forecast may have been prepared originally for some other purpose or for some subject area which only partially overlaps our area of interest.

Interrogation for Reliability

All information which is found to be both relevant and included in the forecast must now be evaluated for its reliability. This includes not only the purported facts but the assumptions and the methods used as well. None of the forecast should be accepted at face value.

Are the methods replicable? A replicable method is one in which the steps can be repeated by someone other than the original forecaster. If any steps in the methodology involve intuition or expert judgment, the method is not replicable. This statement is not intended to imply that intuition should not be used, but it is intended to make the point that the reliability of intuition is not subject to external evaluation. The decision maker using a forecast has a right to know which portions of the methodology are nonreplicable.

Are the logics used formally consistent? Once a method has been determined to be replicable, the only useful question which can be asked about it is whether it is formally consistent. This is because the prime criterion of formal consistency is that the conclusions must be a necessary consequence of the premises or assumptions. If a replicable method is used, it should be examined using the rules of formal logic. Logical fallacies such as circular reasoning, begging the question, non sequitur, post hoc, ergo propter hoc, and others may be found in the methodology or in its specific application. A logical fallacy, if present, will render unreliable all conclusions from that point in the methodology onward. This question does not apply to empirical logics such as induction since their conclusions do not follow from the data of necessity but rather are only suggested by the data.

What are the assumptions? The term assumption is intended to include all laws, principles, axioms, hypotheses, theories, and premises underlying the forecast. Especially important are any assumptions only implicit in the forecast but not stated explicity. It is necessary to unearth these and make them explicit.

Are these assumptions adequately defined? In a subsequent step the assumptions will be tested for validity. This can be done only for assumptions which are adequately defined. In particular, an assumption must specify the conditions to which it does and does not apply and must specify the evidence necessary for affirmation or denial. If an assumption is inadequately defined, it is not possible to determine the reliability of conclusions deduced from it. Some secondary questions which can assist in determining whether definitions are adequate are as follows.

Is this a static or dynamic assumption? Static assumptions refer to states or conditions and must specify a particular point in time at which they are valid. Dynamic assumptions involve a rate, acceleration, sequence, delay, or causal linkage. All dynamic assumptions must specify a dimension of change and an interval of time during which the change takes place. A static assumption, valid for one point in time, may not be reliable if applied to some other point in time. A dynamic assumption, valid over some interval, may not be reliable if applied to some other interval.

Is this a realistic or humanistic assumption? Realistic assumptions concern phenomena essentially independent of human opinions. These may include facts about the physical universe or past actions of human beings. Humanistic assumptions refer to the values, perceptions, future actions, intentions, or goals of human beings. A humanistic assumption must specify the group or individual to which it applies since it may not be true of some other group or individual. Assumptions of value, worth, cost, or risk must specify to whom they apply.

How valid are the assumptions? Having identified the assumptions underlying the forecast and defined them properly, we are now in a position to test their validity. There are two ways of doing this: submission to authority and submission to evidence.

Does this assumption necessarily follow from some laws, principles, or axioms, or is it supported by some expert in the field? A no answer would mean that the assumption was denied and therefore invalid. A yes answer would mean only that we have pushed the question one step back. The laws, principles, and axioms or the expertise of the purported expert must be questioned. While the issue of how far back through a sequence of laws or experts to pursue the investigation is a matter for judgment, one either reaches a point where the validity of the law or expert must be accepted on faith, or one reaches a point where the validity can be determined empirically. In the latter case, submission to authority is transformed into submission to evidence which is considered next.

What evidence is available to support or deny this assumption? We have rendered explicit the assumptions underlying the forecast and assured that they are adequately defined. They are now subject to test by empirical means. Note, however, that we are attempting to evaluate a forecast. Complete confirmation is possible only in the future and may not even be possible there (for example, self-fulfilling or self-defeating paradox). Hence, whether the evidence available is sufficient to validate an assumption becomes a matter of judgment. Generally accepted scientific laws, for instance, usually can be assumed valid. Assumed changes in the values held by other people may, however, be of extremely low reliability. Judgment about the sufficiency of evidence may be assisted by some additional questions given below.

Is the evidence relevant to this assumption? That is, one must raise the question of whether the evidence available is of the type demanded by the nature of the assumption to adequately test the assumption. The following criteria are applicable. Evidence to support or deny a static assumption must be obtained from the particular point in time specified by that assumption (this may be a specific stage in a dynamic process). Evidence to support or deny a dynamic assumption must be gathered sequentially in time or process. The frequency of observation and the time span over which observations are taken must be consistent with the nature of the assumption. Evidence to support or deny a realistic assumption (a fact) must be gathered, processed, and presented in a manner as free from human evaluation and prejudices as possible. This is important, since what people do or do not believe is irrelevant to matters of fact. Inclusion of human evaluations only serves to reduce the reliability of the purported facts. Evidence to support or deny a humanistic assumption (for example, about the values or perceptions of some person or group) must be obtained from physical, written, or verbal behavior of the person or group to whom the assumption refers. Different groups of people may interpret the same reality in different ways and may have different ideas about what is desirable or undesirable. The only valid means for determining what other people believe or perceive is through examination of their behavior. In particular, it is necessary to avoid the error of attributing to others our own values or of assuming they perceive reality in the same way we do.

How accurate is the evidence? This question gets at the issue of how well we know what we think we know. There are two sources of error: uncertainties of observation or measurement and conscious or unconscious distortion. Uncertainties in observation are inherent in any sensor. However, in the physical sciences they can usually be evaluated. In the social sciences they may be harder to evaluate since the act of measurement may change the thing (for example, people's values or perceptions) being measured. In either case it is necessary to estimate these uncertainties for the specific data at hand. Uncertainties in measurement may arise even with precisely observed data. Sources are rounding, aggregation, sampling from a large population, and so on. Conscious distortion is the situation where observations are distorted to fit preexisting human values. The person or group which gathered the data must be identified and his past actions examined to identify possible biases which may then be applied to the observations. Unconscious distortion arises from the philosophy, culture, or ideology of the observer. It originates in such factors as "self-evident truths" of a culture, unchallenged axioms of a scientific discipline, or the implicit assumptions of a common philosophy. This unconscious distortion shows up in both selection and classification of data. It is extremely difficult to identify, especially when the forecast evaluator shares the unconscious assumptions of the data gatherers. However, it is necessary to try, especially when the forecast involves the actions of people who may not share these unconscious assumptions.

Does the evidence available tend to confirm or deny this assumption? Assumptions about future conditions can never be denied or confirmed absolutely on the basis of available evidence. The evaluator must make use of judgment to determine whether to accept or reject an assumption on the basis of the evidence available. Assumptions about past or present conditions can, in principle, be confirmed absolutely, but it may not be worthwhile to gather "all" the necessary evidence. Furthermore, evidence about the past may no longer be available (through death of witnesses, or destruction of records, for example). Here again, judgment must be applied to the question of whether or not the available evidence is sufficient to make a decision about the assumption. In some cases, all the evidence which can be obtained within the time or cost contraints imposed may still be insufficient to determine the reliability of an assumption. In that case, the assumption must be evaluated as being of indeterminable reliability, and all consequences derived from it must also be evaluated as of indeterminable reliability.

This completes the description of the interrogation model. In summary, the model leads the evaluator through four steps, summarized as follows.

1. In interrogation for need, the decision is analyzed to determine what kind of forecast is required, including what should be forecast, how accurately it need be forecast, and over what time range it should be forecast.

2. In interrogation for underlying cause, those elements which have been identified in the previous step as required parts of the forecast are analyzed to identify causes of change.

3. In interrogation for relevance, the forecast itself is analyzed to determine how much information it contains about those factors previously identified as affecting the things to be forecast.

4. In interrogation for reliability, the items of information previously determined to be both necessary and actually present in the given forecast are evaluated for reliability.

By asking this structured sequence of questions about the subject area and about the given forecast, we are led through an evaluation of the relevance and the reliability of the forecast. A forecast which is highly relevant to the subject area in question and highly reliable (for example, draws sound conclusions from the best data available) is very useful as an input to a decision. Note that this is not an assertion that the forecast is in any sense valid. It is an assertion that the decision maker would be well advised to make use of the forecast in his decision. Put another way, he is likely to do better if he uses the forecast than he is if he rejects or ignores it. If the forecast turns out to be irrelevant or unreliable, or both, then the decision maker is well advised to reject it. If the matter is sufficiently important and the available forecast is not sufficiently relevant or reliable, then he is well advised to initiate the preparation of a useful forecast.

The use of the interrogation model will be shown by a specific example below. In the usual case, of course, we find a decision maker who is embedded in a situation and who is faced with the problem of evaluating a forecast which he has been given. In the example the situation is reversed. A forecast has been selected for evaluation, primarily on the basis of availability and convenience, and it is necessary to postulate a hypothetical decision maker and a hypothetical situation to allow the evaluation to be carried through. Thus there is a certain amount of artificiality to the analysis. This shows especially in the inability to generate the amount and kind of historical data which would in fact be available to a real life decision maker. On the other hand, the very artificiality of the example allows us to cast it in a form which highlights the evaluation rather than the decision; hence, the example can have considerable utility.

We assume the following scenario. An executive of the Hifly Airplane Company is responsible for making a decision in the near future on the number and timing of new models of aircraft which the Hifly Airplane Company will develop and produce for the United States domestic airline market. Alternatively, his decision may be to postpone making a decision for a specified time until additional information becomes available. He must estimate the market for different models of aircraft which might be developed, the time when they should be introduced on the market, and the production rate and length of production run the company should plan for. As one input to his decision, he is interested in knowing the market for air travel, in particular, the domestic revenue passenger miles to be flown by United States commercial airlines. A forecast of this market is issued annually by the FAA (Federal Aviation Administration), and a recent (Fiscal Years 1969–1980) edition is to be evaluated. This FAA forecast is to be evaluated strictly from the standpoint of its utility to the Hifly Airplane Company for this specific decision. Its general validity or its utility to other decision makers for other decisions is not in question.

In addition, it should be stated at the outset that none of the comments below are intended to be derogatory towards either the forecast or the FAA. The FAA is one of the few government agencies with the courage to put out a single point "fearless" forecast instead of a uselessly wide band of forecasts or a range of forecasts, each one of which is based on a different set of assumptions. The FAA forecast is clearcut and unambiguous in its predictions of future events and is unquestionably one of the best forecasts issued by the United States Government.

The evaluation will be carried out by going through the interrogation model step by step determining what is needed, what the causes of change are, how much relevant information is contained in the forecast, and how reliable this information is. This will lead to a conclusion about the utility of the forecast to the decision maker in the Hifly Airplane Company.

Interrogation for Need

Here we attempt to determine exactly what is required in the nature of a forecast so that we can subsequently compare these requirements with the actual forecast being evaluated.

Who needs this forecast? This forecast is required by the Hifly Airplane Company.

What are we interested in knowing about the future? We want to know what the airline requirements for new aircraft will be so that we can schedule our introduction and production of new models accordingly. We assume these airline requirements will be based, at least in part, on the demand for air travel by the traveling public. Thus we are interested in knowing the domestic revenue passenger miles to be flown by United States commercial airlines.

How badly do we need this forecast? This requires evaluation, from the viewpoint of the Hifly Airplane Company, of the costs of both underestimating and overestimating the demand for new aircraft by the airlines. Currently we are tooled to produce a particular line of aircraft models. A gross underestimate of the market for new aircraft would probably mean that we would be far too late in introducing new models to replace those we are presently producing. In the face of competition from newer models from other manufacturers, we would be unable to sell the aircraft we are presently tooled to produce. Not only would this mean a loss of sales and profits, it would also mean laying off skilled production workers since we would not have work for them on newer models. This would disrupt our production team as well as our facilities. It could be a threat to the long-term survival of the company. A moderate underestimate would mean that we would have new models ready but perhaps not quite on time. We would be tooled for a lower production rate than would actually be required. This would result in some loss of sales to competitors able to meet earlier delivery dates. An attempt to accelerate production would raise costs, reducing the profit margin on those aircraft we sold. Thus, a moderate underestimate would result in reduced profits as compared with a correct estimate.

A moderate overestimate would mean that new models might be ready too early and thus be slightly inferior technologically to those of our competitors who made more timely introduction of their new models. This would mean some loss of sales and a reduced profit margin on those aircraft sold since we would have to cut our price somewhat to meet the competition of technologically superior aircraft. In addition, we would be tooled for greater production than actually required, which would mean operating at less than capacity with increased overhead. This would also cut into our profit margin on each aircraft sold. A gross overestimate would probably mean that we would develop too many new models and be tooled for production far in excess of that actually required. This would involve writing off the development costs of some of the models. It would also mean scrapping much of the tooling well before it was amortized. Either would hurt the company's profit picture badly. In addition, we would have built up an excessively large work force which we would have to lay off. Finally, the losses would probably make it more difficult for us to generate funds for development of the next generation of commercial transports, thus damaging our long-term prospects.

In summary, either an underestimate or an overestimate would be bad but an underestimate probably more so. In any case, considerable effort is justified in attempting to get an accurate forecast, since the potential additional profits from an accurate forecast are quite large.

To step outside the analysis for a moment, the reader should note that in an actual case the executives of the Hifly Airplane Company would have data available which could be used to obtain quantitative estimates of the loss to be expected from a given error in the forecast. In an actual case these quantitative estimates should be obtained. Enough effort should be expended on obtaining them so that their accuracy is compatible with the accuracy required of the forecast.

Through what time range do we need to forecast? The time range will be determined by two considerations. The first is the time required to bring an aircraft from the design stage through prototype, FAA certification, and production. The second is the length of time during which the aircraft will still be competitive, which will in part determine the length of the production run and the production rate to be planned for. Historically, the Hifly Airplane Company has taken an average of four years from design to commercial introduction, and the production lifetime of a new model has averaged about five years after commercial introduction. Thus, we are concerned with a time range of at least nine years if a decision is to be made about starting a new design right now. If a decision to start design is to be postponed for any length of time, this time must be added to the time range of the forecast. (The reader should note that while the time figures given above are not unreasonable for the aircraft industry as a whole, they may not be valid for specific companies. A specific company should make use of its own historical data as a starting point for these estimates.)

Interrogation for Underlying Cause

Here we attempt to identify the causes of change in the item to be forecast; namely, the domestic revenue passenger miles flown by United States commercial airlines.

We start by identifying two types of change possible in the variable, namely, a change in amount and a change in composition. The total number of revenue passenger miles may change in absolute magnitude, and it may also change in the distribution of types of travel within a given total. We next attempt to identify factors which might influence each of these types of change.

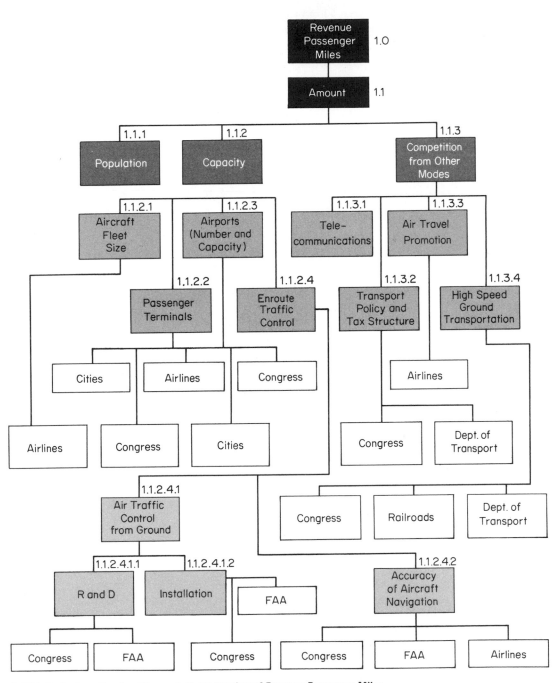

Exhibit 1. Factors Causing Change in Total Number of Revenue Passenger Miles.

We, in turn, identify the causes of change of those factors and continue this process until we reach some factor which we can consider fundamental in that it is not worth pursuing any lower level causes of change.

The result of this sequence of iterations through interrogation for underlying cause is a pair of relevance trees (Exhibit 1 and Exhibit 2). Exhibit 1 shows the sequence of factors causing change in total number of revenue passenger miles. Exhibit 2 shows the sequence of factors causing change in composition of the total.

As an example, we will trace through one branch of the relevance tree. The amount of revenue passenger miles is affected by the total population (since the more people there are, the more there are who are likely to fly), by the capacity of the system (since inadequate capacity will place constraints on the amount of flying done), and by the amount and type of competition from other modes of transportation and communication (since these may be used instead of air travel). The competition from other modes is affected by

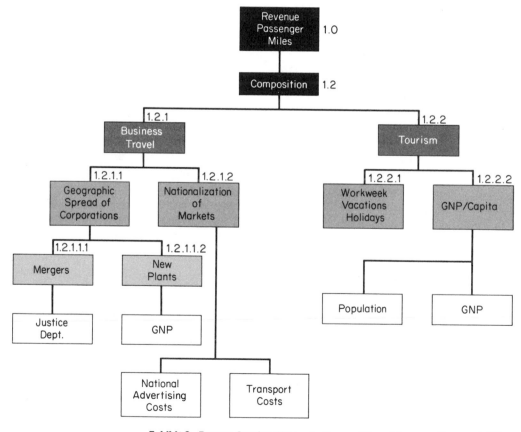

Exhibit 2. Factors Causing Change in Composition of Revenue Passenger Miles.

the state-of-the-art of telecommunications (which to some extent may serve as a substitute for travel), by national transportation policy and the tax structure (since subsidies and taxes, e.g., the property tax on railroad right-of-way, will have different impacts on different modes), by the extent of air travel promotion through advertising and other methods, and by the state-of-the-art of high speed ground transportation (since this may be a substitute for air travel in some cases). High speed ground transportation is in turn affected by actions of the U.S. Congress, the Department of Transportation, and the railroads.

All the other branches shown in the two figures may be traced in the same way. At each step one asks: What factors can cause change in this element? At the bottom or "fundamental" level of the relevance trees, we find the same items appearing repeatedly. The airlines, the U.S. Congress, the FAA, the GNP (gross national product), and the population appear several times. These appear to be highly significant factors in the total picture.

A complete analysis of the relevance tree of causation is carried out in the Appendix. The reader should note that in a hypothetical example such as this, there is little point in carrying out the analysis to an excessive level of detail. Thus the analysis in the Appendix is intended more to illustrate the method and approach than it is to tell the reader anything about the airline industry. In a real case, however, a much more exhaustive analysis should be undertaken to assure that all significant factors have been uncovered at all levels of causation. The analysis given in the Appendix would probably be totally inadequate for a case where a decision involving several hundred million dollars and the possible continued existence of a major corporation is to be made.

If we were intending to construct our own forecasting model, we would next identify the exact nature of the causal relationships at each level of the relevance tree and obtain data about the size and direction of the various influences of the different factors. However, we are instead evaluating a given forecast. Our interest, then, is in determining the extent to which it takes into account the various factors we have found to be involved in determining the amount and composition of airline revenue passenger miles.

Interrogation for Relevance

Here we examine the forecast to determine those items which are (1) relevant and included, (2) relevant but excluded, and (3) irrelevant but included.

The forecast itself is given in Table 1. The far right column is the FAA Fiscal Year 1969 forecast prepared in January, 1969, and covers Fiscal Years 1969 to 1980. We observe that it does in fact make a forecast of the item of interest, the domestic revenue passenger miles flown by airlines, and that it covers the time range we have earlier determined to be necessary.

The methodology for preparing the forecast, as given by the FAA, is as follows. First, the GNP of the United States is forecast. This is assumed to grow at a constant rate of 4.25 percent per year in constant dollars. Next, the

Year for Which Forecast	Fiscal Year in Which Forecast Was Made						
	1962	1964	1965	1966	1967	1968	1969
1957	23.8						
1958	25.5						
1959	26.8						
1960	30.0						
1961	30.5						
1962	32.0	33.0					
1963	35.0	35.1					
1964	38.0	38.2	41.3				
1965	41.5	40.0	46.0	47.3			
1966	44.0	42.5	49.0	54.2	57.9		
1967	45.0	45.0	52.0	59.6	66.1	65.7	
1968		47.0	55.0	64.6	74.0	80.2	81.6
1969		49.0	58.0	70.0	81.0	89.5	91.2
1970		51.0	61.0	75.9	89.4	100.0	100.5
1971				82.3	100.0	110.0	111.8
1972					112.0	123.0	125.0
1973					126.0	136.0	139.0
1974						151.0	155.0
1975		61.0					170.0
1976							
1977					200.0		
1978							
1979						258.0	260.0
1980							288.0

Note: Data above the lines in each column is the actual historical data; the forecasts appear below the lines.

proportion of GNP spent on air travel is estimated. This is developed from a trend extrapolation of historical data and is assumed to grow from 0.6 percent in Fiscal Year 1969 to 0.85 percent in 1980. The product of the forecast GNP and the percent of GNP spent on air travel gives total passenger revenue by year. The yield per passenger mile is then estimated, based on some assumptions about the fare structure, shifts from first class to coach travel, and so forth. Specifically, for this forecast the fare per passenger mile in constant dollars is assumed to decrease at 3 percent per year during the period of the forecast. The total passenger revenue is then divided by the yield per passenger mile for each year to obtain the total revenue passenger miles for each year.

Thus we find that the forecast takes into account growth in the GNP, but it does not specifically account for population growth nor attitudes of agencies responsible for providing the necessary funds. Competition from other modes is taken into account only implicitly in the assumption that air travel captures a growing portion of the GNP. The whole issue of composition is only implicitly addressed by the assumptions about shifts in revenue per passenger mile as a result of shifts in traveler preferences in accommodations. We find, then, that the forecast contains no irrelevant information but only part of the relevant information.

Having determined the relevant information obtained in the forecast, we next examine this information for reliability.

1. Are the methods replicable? The methodology as given is completely replicable. All information needed to repeat the computations is available either from the written forecast itself or from generally available sources.

2. Are the logics formally consistent? The answer to this is trivially, yes. The only logic employed in the formal, stated model is that of arithmetic which can be assumed consistent.

3. What are the assumptions? The methodology contains a number of assumptions, some explicit and some implicit. These are listed here.
   a. The GNP will continue to grow at 4.25 percent per year in constant dollars.
   b. Domestic fares per revenue passenger mile in constant dollars will decrease at 3 percent per year during the forecast period.
   c. The proportion of GNP spent on air travel will grow steadily from 0.6 percent in 1969 to 0.85 percent in 1980.
   d. Growth will not be slowed by airport congestion or air traffic control limitations.
   e. The shift from other modes of transportation to air travel will not be reversed during the forecast period.

4. Are the assumptions adequately defined? The five assumptions listed above appear to be defined adequately for purposes of testing.

5. How valid are the assumptions? While each assumption could be treated in considerable detail, we will make only a brief examination of each one.
   a. From the information given in the forecast, it appears that the FAA has taken advantage of the best information available in choosing this figure for GNP growth rate. This is probably the most reliable figure one can obtain.
   b. Fares per passenger mile cannot be evaluated on the basis of the information in the forecast. Presumably more details could be available from the FAA and could be used to determine whether the assumption seems supportable or some other change in fares should be assumed.
   c. Proportion of GNP spent on air travel in turn is based on several other assumptions about the affluence of potential tourists, the need for business travel, and competition from other modes of transportation. It is not possible to determine its validity from the data given in the forecast.
   d. Growth limitations from facilities constraints is one of the most critical assumptions underlying the whole forecast, especially for the longer term, but it is not addressed at all. The forecast gives no grounds for evaluating it as to validity.
   e. Competition from other modes, although actually subsumed under c above, is called out specifically because of its high current interest. There appears to be a concentrated effort to improve ground transportation, and success would have considerable impact on air travel.

In this example we have considered each of the assumptions of the forecast only briefly. In an actual case considerable effort should be expended on verifying these assumptions since we have seen above that a high degree of accuracy in the forecast is justifiable from the standpoint of avoiding significant losses. Assumption a is probably the only one whose validity cannot be improved by additional effort. Assumptions b and c should be examined carefully. Industry data and data from the FAA itself should be gathered and examined to the maximum extent feasible since errors in the numbers assumed for these factors are directly translatable into errors in the forecast. Assumptions d and e present a somewhat different problem. These assumptions, in effect, say that certain restrictions on air travel growth will not exist or will have negligible effect. If these assumptions are valid, then they do not actually enter into the computation of the forecast of air travel. If the assumptions are not valid, however, then their impact must be evaluated carefully. They were essentially ignored in the preparation of the FAA forecast and could have a major impact on the growth of air travel.

As the above analysis shows, two assumptions were made which were not adequately taken into account in the forecast. Failure of these assumptions to hold would tend to reduce the amount of air travel and thereby make the forecast an overestimate. However, in the past there have been no significant facilities limitations on the growth of air travel. In addition, the railroads, bus lines, and private automobiles have not given the airlines any significant competition for those types of travel for which air was commonly used. An increased level of affluence has led to increased travel, and much of this increase went to airlines rather than to competing modes. Since these potential restrictions on air travel did not exist, we thus have an opportunity to check the historical validity of the first three assumptions (taken in combination) and the predictions based on them. We know that the GNP did in fact grow, that average fares did in fact decline, and that an increasing proportion of the growing GNP was in fact spent on air travel. How accurate were past forecasts in which these validated assumptions were used? Some of the data of Table 1 is plotted in Exhibit 3, which compares historical performance with some past forecasts. It can be seen that past forecasts have tended to underestimate the future growth of air travel, sometimes by quite significant factors. However, this comparison is not entirely fair since the methodology used by the FAA has been refined considerably since the first forecasts were made, and presumably the more recent forecasts should be better. Nevertheless, it appears that if assumptions d and e are valid, the forecast will probably provide an underestimate of the actual amount of air travel.

Thus we see that the interrogation model has led us to the critical assumptions underlying the forecast and has allowed us to evaluate the relevance and reliability of the forecast with respect to the requirements of a specific decision. In particular, it has identified a number of restrictive factors which might become operative and which, therefore, deserve additional evaluation. It has also identified several assumptions whose reliability, on the basis of data available in the forecast, is indeterminable. The historical check showed that the reliability of these assumptions was lower than that required for the decision

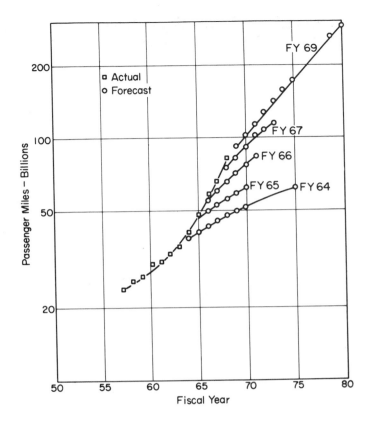

Exhibit 3. Actual and Forecast Revenue Passenger Miles.

Chart labels: Passenger Miles – Billions (y-axis); Fiscal Year (x-axis); □ Actual; ○ Forecast; FY 69, FY 67, FY 66, FY 65, FY 64.

to be made and that acceptance of them without further check would probably lead to an underestimate of the market size. (Note that in actual practice this historical check would have been included under Step 5 of the interrogation for reliability. However, this would have left us unable, in this example, to check the effectiveness of the interrogation model.)

It is also worth mentioning, in conclusion, that the series of questions embodied in the model are not a substitute for thought or for knowledge of the subject area. The interrogation model is not a mechanical procedure which one can go through by rote and succeed in evaluating the utility of a forecast. Instead, the model is intended to be of assistance to the evaluator by directing his evaluation in an efficient and orderly pattern and by suggesting specific questions which he might overlook in a less explicit or more intuitive approach to forecast evaluation. Of particular benefit to the evaluator would be some knowledge of how forecasts in his area of interest have failed in the past. Use of the interrogation model to codify and render explicit this knowledge of past forecasting errors can permit much more effective and explicit application of past experience to the evaluation of a current forecast.

Finally, of course, it should be clear that the interrogation model, as it now stands, is not a perfected piece of work. It is a very useful and helpful tool for the analysis of forecasts. However, it still requires improvement in a number of areas; and additional research towards improving it is vitally necessary.

ACKNOWLEDGEMENTS

The interrogation model on which this paper is based is the work of Lieutenant Colonel Ben H. Swett, formerly assigned to the Office of Research Analyses, U.S. Air Force.

APPENDIX: RELEVANCE TREE OF CAUSATION

This appendix describes, in more detail than that presented in the main portion of the paper, the development of the relevance tree of causes of change in air travel. This explanation is intended more as an example of the reasoning which must go into development of the relevance tree than it is an exposition of the inner dynamics of the airline industry. The explanation is on a level-by-level basis.

The total domestic revenue passenger miles flown by United States commercial airlines can change in two ways. The absolute amount of travel can change and the composition of that travel can change.

The amount of travel taking place can change as a a result of changes in the factors of population, capacity of the system, and the level of competition from other modes. All other things being equal, the more people there are, the more will be the absolute number of people who travel by all means, including air. The capacity of the system has only a restrictive influence. Adding capacity to an unsaturated system will not increase the amount of travel, but failure to increase capacity will limit travel to the saturation level. The degree of competition from other modes will affect the absolute amount of air travel since other modes are at least potentially capable of providing services partially or completely equivalent to air travel.

The composition of the air travel market can alter in the proportions which represent business travel and tourism.

The population is, for the purposes of this analysis, a fundamental cause of change. We will not investigate further the causes of change in population.

The capacity of the system can in turn be changed by changes in the following factors: size of the commercial air fleet available, passenger terminal size and efficiency, number and capacity of airports, and effectiveness of en route traffic control. Each of these factors is a potential bottleneck in the total system. Inadequacy in any of these factors can set an upper limit on the absolute amount of air travel which can actually be achieved.

Competition from other modes can be changed by changes in the following factors: telecommunications, transportation policy and tax structure, air travel promotion, and high speed ground transportation. Telecommunications, to some extent, provides a substitute for travel. This is true of both business travel and travel for pleasure. Frequent telephone calls to distant relatives may be a substitute for periodic visits. Similarly, telephone conference calls, transmission of documents by electronic photocopying, and so on may be a partial substitute for business travel. As these capabilities are improved, need for travel may diminish. Transportation policy and tax structure affects competition from other modes to the extent that different modes of transportation are treated differently. Taxes on railroad right-of-way, for instance, impose costs on railroads which are not felt by airlines, buses, and so on. Subsidies to airlines may to some extent offset disadvantageous economic factors. Policies which foster or hinder the growth of other modes of transportation affect the degree to which these can compete with air travel. Changes in policies and in taxing practices can therefore change the amount of competition which air travel meets from other modes of transportation. High speed ground transportation is treated separately because it is being heavily funded by the Federal Government and because it is consciously conceived as a substitute for air travel in such heavily traveled regions as the Northeast Corridor. Success of the effort to develop a rapid and convenient system of high speed ground transportation will attract some portion of the people who currently travel by air.

Business travel can be altered by the degree of geographic spread of corporations and by the "nationalization" of markets. If corporations tend to increase the geographic spread of their divisions, subsidiaries, and so on, this will tend to increase the amount of travel among these elements of the corporations. Similarly, if geographically limited corporations find they can sell to a national market, this will tend to increase the travel of sales and marketing personnel, field engineers, and so on.

Tourism can be changed by changes in the length of the workweek and the time available for vacations and holidays and the degree of affluence of the population, which can be approximated by the ratio of GNP per capita.

The size of the commercial air fleet can be changed by actions taken by the airlines with regard to acquisition of new aircraft and the retention or disposition of old aircraft. For the purposes of this analysis, the airlines will be considered a fundamental cause, and the causes of change in their activities will not be pursued.

The size and efficiency of passenger terminals can be changed by actions of the cities which they serve, by Congressional appropriations for construction and expansion of passenger terminals, and by the actions of the airlines, not only with regard to expenditures for physical facilities but with regard to policies and procedures for ticket selling, baggage handling, and so on. These factors will be considered fundamental, and further causes of change will not be pursued.

The number and capacity of airports can be changed by actions on the part of the cities which the airports serve and by Congressional appropriations for construction and expansion of new airports. These causes of change will be considered fundamental and will not be pursued further.

The capacity of the en route traffic control system is affected by changes in the system's ability to control aircraft from the ground and by the navigation accuracy of aircraft. Direct control of aircraft from the ground imposes a workload on the traffic control system. There will be some saturation level for a given number of air traffic controllers provided with a given type of equipment. Increased accuracy of aircraft navigation will reduce the work load on traffic control personnel and will also permit more aircraft to use the same volume of airspace. Both results tend to increase the capacity of the traffic control system.

Transportation policy and the tax structure can be changed by actions of the Congress and the Department of Transportation. These will be considered fundamental causes and will not be pursued further.

The level of promotion of air travel can be changed by actions on the part of the airlines to increase or decrease their advertising, adoption of promotional fares, and so on. This will be considered a fundamental cause and not pursued further.

The impact of high speed ground transportation can be affected by actions of the Congress, the railroads, and the Department of Transportation. These will be considered fundamental causes and will not be pursued further.

The geographic spread of corporations is affected both by mergers of smaller corporations in dispersed locations and by the construction of new plants at locations nearer to markets, nearer to raw materials, or in other advantageous sites away from the current plant locations.

The degree to which a corporation with one manufacturing plant or several plants confined to a single region can sell to a national market will be affected by the costs of reaching a national market with advertising and costs of transportation. Reductions in these costs will make reaching a national market easier; increases will make it more difficult. For the purpose of this analysis, these will be considered fundamental causes and will not be pursued further.

For the purposes of this analysis, the free time available to workers through setting the length of the workweek and the extent to which productivity increases are taken as additional vacations and holidays will be considered a fundamental cause and will not be pursued further.

The GNP per capita can be altered by changes in both the GNP itself and by changes in the population. For the purposes of this analysis, these factors will be considered fundamental causes and will not be pursued further.

The capability to control an increased amount of air traffic from the ground will be affected by the amount of R&D done to devise means for increasing the productivity of air traffic controllers and by the installation of this productivity increasing equipment on a large scale.

The accuracy of aircraft navigation will be affected by the willingness of the airlines to purchase and install more accurate navigation equipment and by the willingness of the FAA and Congress to fund the installation of the ground-based portions of improved navigation systems. For the purposes of this analysis, these causes will be considered fundamental and will not be pursued further.

Mergers of corporations to produce more geographically dispersed organizations will be affected by the antitrust policies of the Justice Department. For the purpose of this analysis, this will be considered a fundamental cause of change and will not be pursued further.

The extent to which new plants are built will depend on the levels of anticipated business and the availability of capital to finance new construction. Both of these will be affected by changes in the GNP. This will be considered a fundamental cause and will not be pursued further.

R&D on means for increasing the productivity of ground-based air traffic controllers will be affected by the willingness of the FAA and Congress to expend funds for this purpose. These factors will be considered fundamental causes of change and will not be pursued further.

Installation of improved equipment for ground-based air traffic controllers will be affected by the willingness of the FAA and Congress to expend funds for this purpose. These factors will be considered fundamental causes of change and will not be pursued further.

As illustrated here, it is possible to carry out the interrogation for underlying cause, analyzing the phenomenon of interest. The interrogation can be carried out iteratively, with each identified cause of change in turn subjected to the same interrogation. Eventually, the analysis reaches a fundamental level of causation which does not justify further analysis. In general, also, the same fundamental causes appear several times. That is, a few fundamental causes affect the factor of interest through a multitude of branches of the tree. In many cases, it may well turn out that the effects arising from the same cause but traveling through separate branches tend to offset each other.

If this example were being carried out for practical rather than expository purposes, it would probably be desirable to carry out the analysis to even lower levels, investigating the probable attitudes of Congress, the FAA, the airlines, and the Department of Transportation. It might also be desirable to include other factors at various levels or to split some of the factors used above into finer subelements at each level. In any case it should be clear that carrying out an interrogation for underlying cause and developing the appropriate relevance tree are in no sense automatic procedures which can be carried out without thought and knowledge. The formal interrogation for underlying cause, with its series of suggestive questions, is not intended as a substitute for thought but only as an aid to the analysis.

Forecasting Public Resistance to Technology: The Example of Nuclear Power Reactor Siting

DAVID G. JOPLING
STEPHEN J. GAGE
MILTON E. F. SCHOEMAN

If we are to predict coming technology with any accuracy, we must understand forces that affect its emergence and diffusion. Here is the first proposal for systematically analyzing public resistance to new technology. A pattern of resistance is identified with phases and time scales. While there is no presumption that this pattern applies to all technologies and societies, similar studies could be made for other types of technology. If such patterns should be found to exist, we could do a much better job of predicting. More significantly, management could handle the introduction of new technology more wisely.

Public concern over the perceived deterioration of the environment by new technological activities is growing, and even threatens to halt them. This public resistance is, in some instances, a constraint or limit to the advance of technology. It is analogous, perhaps, to the physical limitations of science and engineering. For this reason, a forecast of public resistance may become a primary requisite to a comprehensive assessment of technological proposals.

We shall demonstrate a methodology for anticipating public resistance to technology. Our seven stage model of resistance was developed originally in the context of the siting of nuclear power reactors. The general model, however, seems potentially applicable to many other technology-based activities. We shall first describe the key features of the general model, followed by a specific application of the concepts to the situation surrounding the siting of nuclear reactors. Our discussion concludes with some observations of potential significance to anticipating resistance and some recommendations for actions that such forecasts may prompt.

MODEL OF PUBLIC RESISTANCE

An initial observation is that public resistance to technology develops over time in an orderly progression of actions. These actions are distinguishable both by the individuals who perform them and by the outcome towards which they are directed. Although alternative groupings of these actions are possible, it seems useful to isolate the seven stages which are shown in Exhibit 1.

Stage	1	2	3	4	5	6	7
Description	Public Disclosure	Expert Inquiry ↓	Information Distribution	Citizens Organize ↓	Technical Disagreements	Uncompromising Conflict ↓	Legal Confrontation
Time	Months	↑ Days	Months	↑ Month or Two	Months	↑ Year or More	Unlimited

Exhibit 1. Seven-Stage Model of Public Resistance to Technology.

Stage One

The first stage takes the technology into the public domain through various governmental institutions. Prior to this stage the technology and its manifestations may be effectively shielded from public view by its producer. Usually this stage involves the satisfaction of some highly technical requirements established by legislation. Primary interaction is between the producer of the technology-based activity or product and several regulatory agencies in the local, state, or federal government. The time which is spent in satisfying previously established statutory requirements may vary depending upon the nature of the activity or product and the specific governmental agency involved. This process, however, does seem to be generally a lengthy one, consuming several months, or even years, as in the case of commercial aircraft satisfying FAA regulations.

Stage Two

This stage begins as a natural outgrowth of the first. Prompted by public disclosure of a new technological activity or product that is seeking legal sanction, an individual or a group considered technically competent initiates a direct inquiry to the producer about technical specifications. The central issue appears to be the lack of credibility by scientific experts-at-large in the information provided publicly by the producers of the activity or product. Credibility is weakened further by the tendency of the producers to ignore these inquiries outright or to summarily dismiss them as irrelevant. Relative to the first stage, the time spent here is short, perhaps spanning only a few days. The swiftness with which this stage transpires and its position as an adjunct to the first sometimes result in the two stages occurring virtually simultaneously. Nevertheless, they do constitute distinct actions. They may require different remedial actions to overcome the resistance present.

Stage Three

This is characterized by attempts on the part of the experts previously ignored or spurned to inform the general public and arouse it. Many of these experts orient their information to the technical or semitechnical factors which they perceive as a matter of serious public concern. Naturally, public response is variable in scope and is highly dependent upon the manner in which the information is disseminated. Mass communications media accomplish this stage in a matter of a few weeks; more conventional methods such as public meetings and literature distribution may cause this stage to continue for several months.

Stage Four

The fourth stage is entered as citizens groups organize and broad public activism develops. Protest letters proliferate and protest meetings are well attended. Public pressure on legislators mounts and may establish the basis for new or renewed legislative investigation. This stage follows the previous one in a manner similar to the progress from stage one to stage two. However, efforts to organize usually require a moderate time to yield results. The stage is likely to be a few months rather than a few days.

Stage Five

The fifth stage is the most critical in the development of the resistance. It is within this stage that the technology producers finally attempt to respond to the public criticism and quell it. Technical arguments by the producers are countered by those of the experts-at-large. Disagreement between equally qualified experts contributes to a continued public uneasiness about the proposed activity or product. Several months may elapse in this confrontation between opposing scientific viewpoints.

Stage Six

The politics that characterize the preceding stages are the politics of compromise which are necessary to work out solutions when no obvious solution to a problem is self-evident. However, the politics of the sixth stage are the most unsophisticated, mudslinging operations available. Any means for getting publicity are employed. Actions indicate the nearly total collapse of communication channels between the public and the technology producer. Time spent in this stage is usually long, sometimes encompassing more than a year.

Stage Seven

This stage is the final legal confrontation to decide the fate of the technology-based activity or product. The location may be a courtroom or a legislative chamber. Compromise is rarely possible at this stage. If the producer does not choose to voluntarily withdraw his activity or product, the time spent may be considerable. Court or legislative proceedings may drag on for years. It is even possible that new statutory requirements on the technology involved may be an output of this stage.

Naturally, the precise characteristics of these seven stages will vary somewhat with the technology involved. Several stages may overlap to the extent that they appear to occur simultaneously. Timing of each stage may be altered as there is an increase in the number of attempts to introduce an activity or product based on a single technology. The case of siting nuclear power reactors will draw attention to the manner in which characteristics of the general model must be adapted to fit specific cirumstances.

PUBLIC RESISTANCE TO NUCLEAR POWER REACTORS

In response to the growing nuclear power industry and the prospects of a future proliferation of nuclear power plants, numerous members of the general public and some atomic energy specialists have expressed doubts about the adequacy of the standards that regulate the emissions of radioisotopes and thermal wastes from nuclear power reactors.

The radioactive emissions and thermal discharge problems of nuclear utilities have entangled the atomic energy industry in a national controversy over environmental pollution. The public no longer regards nuclear power as a perfect solution to urban air pollution from conventional power plants but considers atomic energy to be an air and water polluter itself.

The public's apprehension over radioactive and thermal discharges has become crucially important for the cost competitiveness of nuclear power over fossil fuels. The installation of additional equipment for complete containment of radiation and for heat dispersal adds significant costs to a nuclear plant's capital investment, operating expenses, maintenance, safety procedures, and efficiency.

The presence of the U.S. AEC (Atomic Energy Commission) in its dual role of promoter and regulator of atomic energy has served to precipitate an additional controversy, a reevaluation of the regulatory jurisdiction which the AEC alone has exercised in all matters concerning nuclear power. Critics of standards regulating radioactive exposure, charging a conflict of interest exists between the AEC's promotional and regulatory functions, have called for a transfer of all environmental monitoring and surveillance programs as well as

MPC (maximum permissible concentrations) standards enforcements to another national government agency like the new Environmental Protection Agency. Other critics favor a sharing of regulatory functions by the federal and state agencies.

The new controversy over thermal effects on aquatic life has further complicated the reevaluation of the AEC's jurisdiction. While claiming regulatory authority over radiation dangers, the Commission has refused until recently (*Nuclear News*, January, 1971) to accept jurisdiction over thermal effects problems. Critics have argued that if radioactive emissions are within the Commission regulatory purview, thermal pollution must also fall within that purview.

In an effort to learn more of the nature of the public resistance to major proposals for nuclear power plants, a study was initiated that involved the careful analysis of the following four specific cases:

1. Pacific Gas and Electric Proposal, Bodega Bay, California.
2. Consolidated Edison Proposal, Ravenswood, Queensboro, New York City.
3. New York State Electric and Gas Proposal, Cayuga Lake, New York.
4. Northern States Power Proposal, Monticello, Minnesota.

The development of public resistance in each case followed the definite pattern suggested in the previous section. Exhibit 2 shows the timing of each of the seven stages for these specific case studies. Exhibit 3 shows the stages of resistance in a flow model specifically adapted to nuclear reactor siting. A brief description of these stages as they actually occurred in each of the four cases follows.

Exhibit 2. Timing of Seven Stages for Four Specific Nuclear Reactor Siting Proposals.

| Stage | Siting Proposal | | | |
	Pacific G&E; Bodega Bay	Con. Ed., Ravenswood	N.Y. State E&G, Cayuga Lake	N. States Power, Monticello
1	8 months	2 months	9 months	18 months
2	5 days	1 day*	2–3 months	2 months
3	1 month	1 day*	2 months	12 months
4	6 months	1 month	5 months	3 months†
5	6 months	1 month	8 months	2 months
6	18 months	8 months	‡	§
7	Proposal withdrawn	Proposal withdrawn	Proposal withdrawn	Litigation in process

* It appears that stages two and three occurred in the same day for this case. This highly compressed time (as well as that of the following stages) may be due to the fact that the first inquiry about the reactor proposal's safety came from David Lilienthal, the frst AEC Commissioner. His prestige gave credibility to his statements and widespread publicity to the siting proposal.

† Organization was reported to have been completed in only one month, but evidence suggests firm action did not begin for at least another couple of months.

‡ Due to insufficient information, stages five through seven for this case were virtually indistinguishable. However, stage six, almost certainly lasted no longer than a very few months (perhaps less than one month).

§ This stage seems to be shortening in duration as the number of reactor siting cases increase. In Monticello, it took only a few months (two to four) to intensify the public utility confrontation.

Exhibit 3. Seven-Stage Model of Public Resistance to Nuclear Reactor Siting.

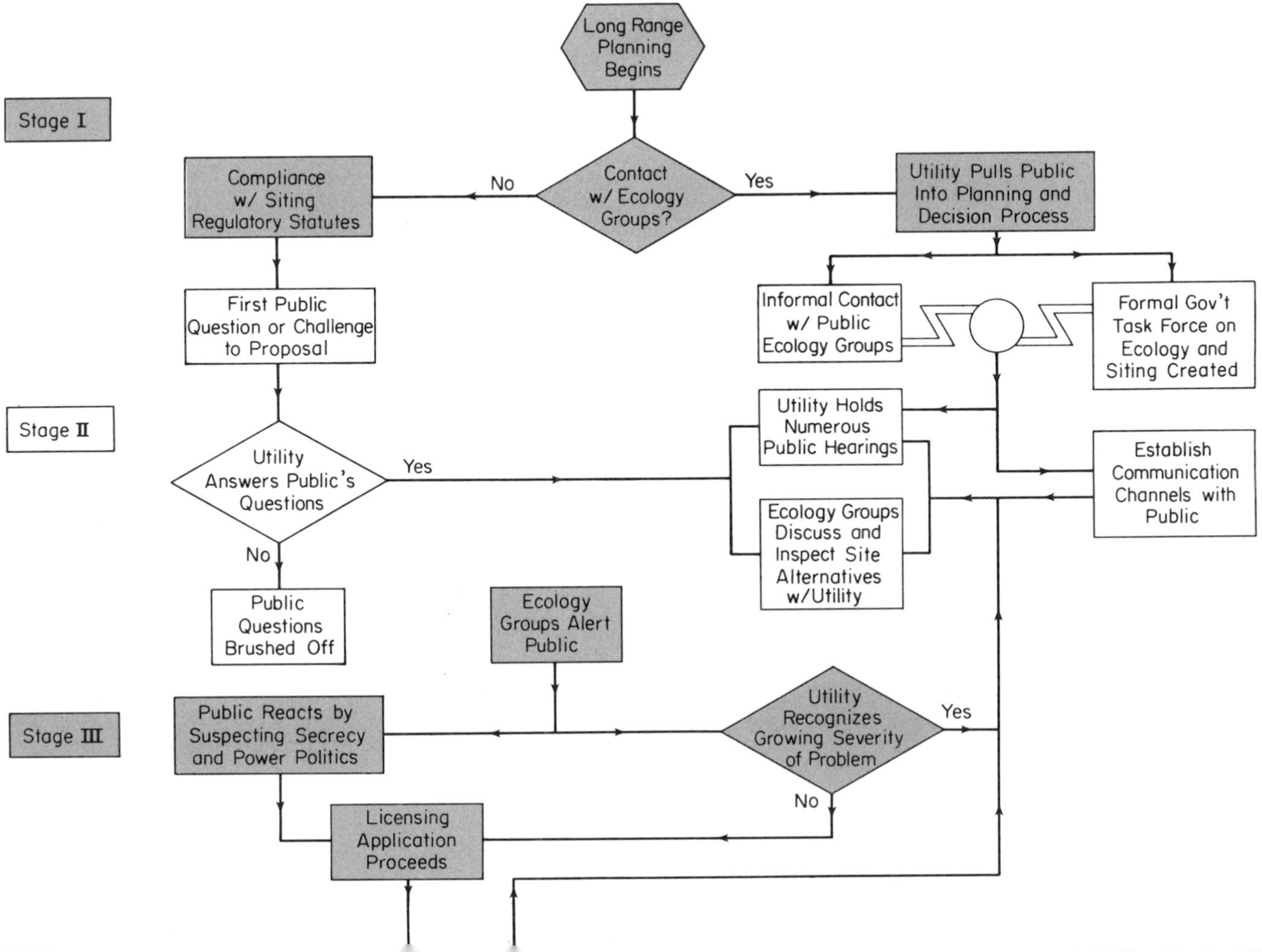

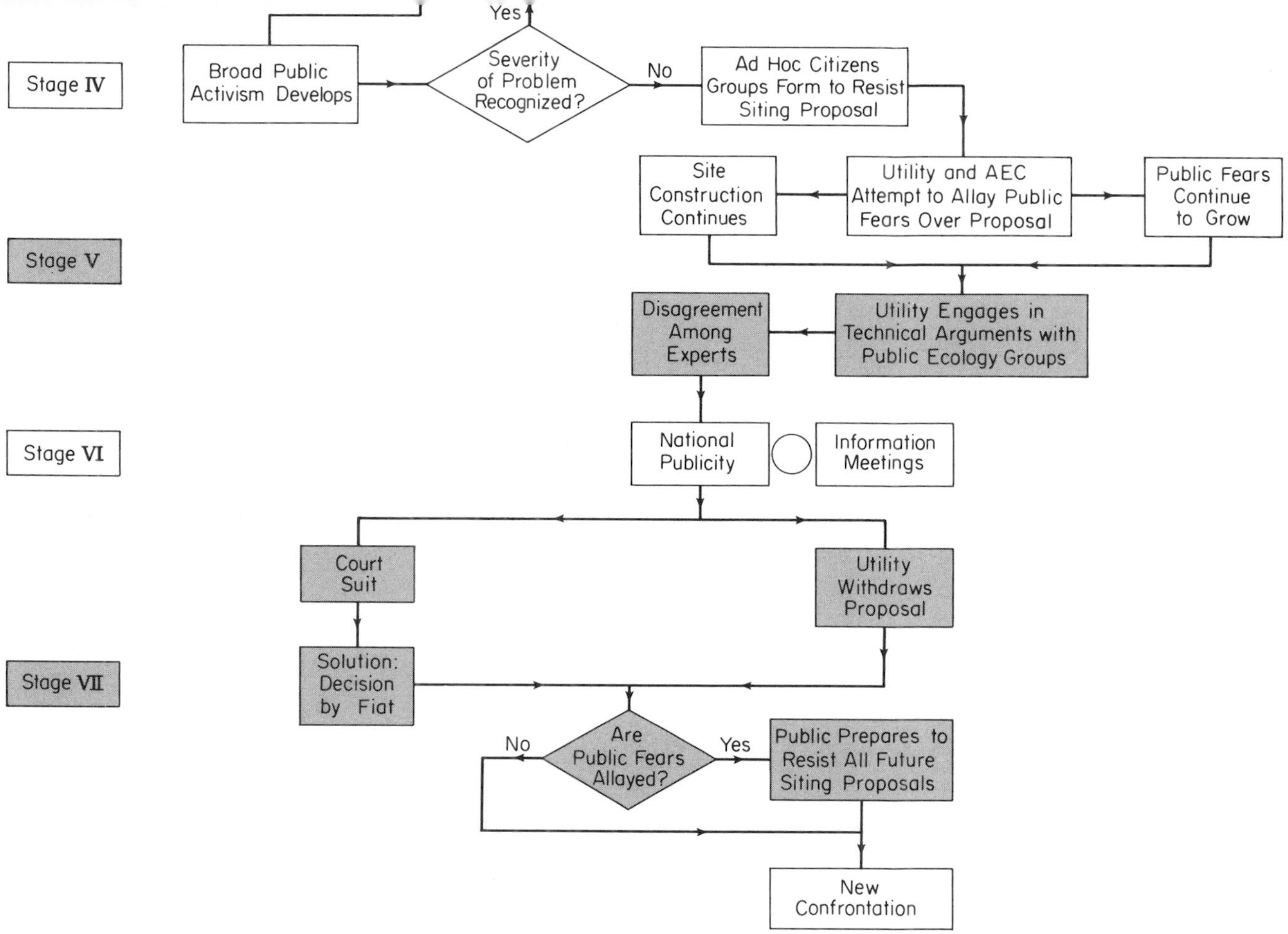

Stage IV

Stage V

Stage VI

Stage VII

Exhibit 3 (cont.)

Stage One

During the first stage, the utility company proceeded to satisfy the existing statutory requirements of the AEC and the state to obtain the necessary licenses to construct and operate a nuclear power plant. The company met with the appropriate divisions of the AEC and with other national government agencies such as the U.S. Public Health Service and the Federal Water Pollution Control Administration. The utility also contacted state regulatory bodies such as the departments of health, water pollution, and conservation.

The utility did *not* make an active effort, however, to contact public groups and individuals who were interested in conservation or who were active in anti-pollution efforts. In the Bodega Bay case, the Pacific Gas and Electric Company of California made no effort to inform the powerful Sierra Club of the company's plans to site the reactor on Bodega Head. Consolidated Edison allowed a long lead time for the Ravenswood project in order to enable interested parties to voice any fears over the reactor's urban location; Consolidated Edison, however, did not actively take their siting proposal and the arguments in its favor to the public.

Stage Two

The second stage began with either a first inquiry for information on the reactor siting proposal or by a challenge to the proposal's safety from some individual or group regarded by the public as a scientific expert. These first inquiries were brushed off as unimportant or rebuffed—for example, the telephone inquiry of Dean Arnold, a Cornell research fellow, to New York State Electric and Gas Company concerning the Cayuga Lake siting drew the response from a company official: "We are not in the habit of discussing our plans with the public." In all cases when the first show of public concern took the form of a challenge to the siting proposal's safety, the challenger was condemned as uninformed, misled, or called nothing more than a political activist. The Joint Committee on Atomic Energy even accused former AEC Commissioner Lilienthal of "loose talk" for expressing doubts about the safety of the urban site in Ravenswood, New York City.

Increasingly, however, the first inquiry or challenge to a reactor proposal has begun to come from standing groups organized to make such inquiries. The Committee for Nuclear Information, the Sierra Club, and the Scientists Committee for Radiation Information have become highly regarded by the public as expert bodies who will actively represent the public's interest in health and safety. When these groups have challenged a reactor proposal, countercharges of irresponsibility have carried little weight with the public.

Stage Three

The failure of the utility to answer the first inquiry or to rebut the first challenge to the proposal caused the developing controversy to enter its third stage. During this stage those individuals or groups who first questioned the proposal set out to inform the general public of the proposal's danger. Several means were used to inform the public. In the Bodega case, for example, Karl Kortum wrote a newspaper article for the *San Francisco Chronicle*. The Eipper group resorted to pamphleteering about New York State Electric and Gas Company's plans for Cayuga. Papers by Abrahamson, Pogue, and Olson appeared in the published proceedings of the Minnesota Academy of Science and drew widespread public attention to the Monticello site.

Upon learning of the siting proposal, many members of the public suspected that the utility had desired to keep the public unaware of the possible dangers from the reactor. These suspicions of secrecy would have heightened if the utility had not held numerous public hearings on the proposed plant. The fact that the utility was a monopoly and may have exercised condemnation proceedings to acquire title to the proposed site almost certainly turned the suspicions of secrecy into charges of power politics, as the Bodega case clearly exemplified.

Throughout the time that the public was learning of the proposal, the utility's licensing application proceeded relentlessly forward. This continuation of company efforts in apparent disregard to snowballing public questions propelled the growing controversy into its fourth stage.

Stage Four

During the fourth stage, public distress over the siting application became so great that a broad base of public activism against the proposal appeared. The rapid increase in protest letters, from 100 to 2,500 received by the California Public Utilities Commission holding its first three hearings on the Bodega proposal, illustrated very well the appearance of a broad base of public concern. The appearance of 300 observers and important political leaders at a routine monthly meeting of the Minnesota Pollution Control Agency to consider the Monticello site was equally important evidence of widespread public concern. In a somewhat different way, the gradual change in the attitude of the Cornell Water Resources Center from one of cynical complacency over the thermal pollution of Cayuga Lake to one of sincere concern was evidence of growing misgivings over the Cayuga proposal.

The utility company often continued to fail to recognize the potential seriousness of the growing public alarm. In the Bodega case, Pacific Gas and Electric believed that most ordinary people were disinterested. In the Cayuga Lake controversy, the AEC, the utility, and the industry media discounted thermal emissions as a potential threat to the lake's aquatic life.

In response to the general unwillingness of the industry or the utility company to acknowledge the legitimacy of public fears, ad hoc citizens groups began to form for the purpose of fighting the reactor proposal. These groups, as exemplified by the Citizens to Save Cayuga Lake, often were well managed and possessed numerous expert advisors although none were as well financed as the utility or industry groups which supported the siting proposals.

Stage Five

The fifth stage was characterized by a general refusal of organized resistance to die down despite all of the arguments which utility, industry, and the AEC could provide to allay public fears. Several factors combined to sustain each resistance case through this, the longest stage. Public fears continued to gain a public forum. Newspapers and magazines gave each siting controversy extensive coverage. National, state, and local governmental agencies held hearings on the cases. Several political figures began to involve themselves, although these figures were usually of minor significance. In addition, however, the ad hoc citizens groups became firmly established and began to prosecute their cases against the siting proposals by drawing as much attention to themselves as possible. Some threats of special legislation also appeared during this period and served to draw additional news and opinion commentary.

Another factor which kept public fears alive was the oftentimes unceasing effort of the utility company to continue to construct the reactor and the balance of plant regardless of the growing uproar. Such construction was made possible by a "certificate of convenience and necessity." By this time, however, the company had definitely begun to engage in technical arguments with the reactor's critics. In some cases the utility even funded special studies in an effort to rebut the arguments of the alarmed citizens groups.

The technical arguments between the utility and its critics led immediately to a very important development—disagreement among the experts. The involvement of government experts from numerous agencies at both national and state levels and the involvement of independent experts from consulting firms and research organizations resulted in a barrage of expert opinions, no one of which entirely agreed with the conclusion of any other expert opinion on the extent of potential dangers from the proposed reactor. The inability of expert opinion to allay public fears caused the siting controversy to become amenable only to a political solution.

Stage Six

The sixth stage saw public political activity rise to its greatest frenzy. Public groups conducted demonstrations and public relations campaigns against the utility and its proposal. For example, the Northern California Association to Preserve Bodega Head and Harbor held a public relations campaign at "Earthquake McGoon's" to draw attention to the seismic nature of the Bodega site. Such tactics always garnered national news coverage.

During the sixth stage, fairly well-organized public information meetings, like the University of Minnesota's Symposium on Nuclear Power and the Public, were likely to occur. By the sixth stage, however, these information meetings could do little or nothing to lessen the public uproar over the reactor proposal; these meetings served only to recapitulate all of the issues and questions which had remained unresolved in the public's mind. Thus, they drew additional attention to the controversy.

The political furor of the sixth stage resulted in two types of outcome. In three cases, the company withdrew its siting proposal in the face of bad visibility on a nationwide scale. In Monticello, however, the dispute was taken to the courts. Northern States Power has gained a court order halting the state of Minnesota's bid to exercise regulatory jurisdiction over the Monticello reactor. The citizens groups have vowed to appeal this decision. Even if the higher courts sustain the recent ruling in Northern States Power's favor, this confrontation between the utility and the Minnesota public will become only another Pyrrhic victory for the nuclear power industry.

CONCLUDING OBSERVATIONS ON PUBLIC RESISTANCE TO TECHNOLOGY

Public resistance to the siting of nuclear power reactors exhibits several characteristics which are, or may be significant, for public resistance to the development of other technologies.

A Long Construction Lead Time

Increasingly, high technology industry requires long lead times for the construction of equipment installations. This lead time has lengthened so much that discussion of construction as a set of short-run problems is largely impossible. In the case of nuclear power plants, lead times of five years are average, and with any delay in licensing, adjudication of labor problems, public resistance cases, or delays in equipment deliveries, this lead time can easily extend to seven, eight, or nine years. The time frame for these high technology construction projects is most appropriately considered the intermediate to long term.

The long lead times require any utility or any high technology industry to look far in advance of the present if they are to anticipate the relevant problem dimensions of their business. A company that continues to believe that it can live solely by solving short-run problems, whose attitude toward long-range planning is, "We'll cross those bridges when we come to them," is leaving itself vulnerable to very costly delays, costly in terms of money and political health.

These high technology industries should realize that in some cases historical forces, technical problems, and so forth are drawn together in a confluence which can be anticipated in some general way. In the case of public resistance to reactor siting, resistance is only part of much larger national attitudes and problems, but utility companies should formulate some strategy in anticipation of the distinct possibility of political problems.

Spillover

The occurrence of a political crisis in one geographic area in high technology industry is almost certain to generate a very similar crisis over the same type of industry in an adjacent geographic area. For example, the Monticello case and the Cayuga Lake case very definitely have had strong influences on a controversy developing over a reactor which is being sited at Palisades, Michigan. It is no accident that the same citizens groups which have been politically active in New York and Minnesota have been in communication with citizens groups becoming active in Michigan.

In addition a spillover phenomenon may also occur between industries. For example, the national attention to environmental pollution began initially around the emissions of devices utilizing fossil fuels, such as automobiles, power stations, oil tankers, and runaway oil and gas wells at sea. But almost immediately the nuclear power industry, which utilizes a vastly different technology from fossil fuels, was caught up in the same pollution controversy. If this type of spillover continues, legislative actions in the problems of fossil fuel emissions, while supposedly dealing with completely different problems, may come to have very great relevance to the regulation of the nuclear power industry. A very specific case in point is a recent decision by a federal court to allow the State of California to impose stricter emission requirements for automobiles than those set by federal agencies. Such a ruling may also have important implications for state regulation of radioactive and thermal emissions from nuclear power plants.

Fear of Potential Catastrophes

The nuclear power controversy has been characterized repeatedly by the incidence of certain critical events, accidents in particular, which have had a profound influence over the intensity of the confrontations. For example, in the case of a proposal for the Enrico Fermi breeder reactor (Lagoona Beach, Michigan), the critics referred repeatedly to a core melt that the AEC's experimental breeder reactor EBR-1 had experienced about six months prior to the Fermi proposal. The Fermi proposal received another setback when the AEC's stationary low power reactor SLPR-1 exploded and killed three technicians. During the Bodega Bay case, the Alaska earthquake of 1964 underscored the public fear over possible damage to the Bodega reactor from earthquakes and possible tidal waves. In the Monticello case, critics repeatedly enumerated every major reactor accident in their arguments against the construction of this facility. While these accidents have not changed the whole course of development for nuclear technology, they have slowed tremendously the rate at which this technology is being commercialized. This is in marked contrast to a trend which is generally reducing the time between creation of a new technology and its widespread introduction in the marketplace.

Increasingly, industry is being subjected to some kind of licensing and regulation process. This is especially true for high technology industries because their impact, both economic and physical, is so widespread. A generalization which may be possible from the nuclear power cases is that whenever an industry is subjected to some king of licensing process, extensive public involvement lengthens the licensing process, the operations of hearings, and almost certainly leads to an involvement in judicial appeals and reviews if public resistance to the industry is severe.

Speed of Intervention

The first cases of public activism against a particular technology will probably be somewhat inept and badly managed and, therefore, unsuccessful. In time, however, the public is capable of attracting sufficient expertise to make its organized resistance efforts much more competent. The result is that the public learns very quickly how to fight and stop the development of a technology.

Acceleration in Confrontation

Sufficient data on the nuclear power controversy exists to observe that as more and more confrontations occur, they seem to develop much more quickly, especially in the latter stages of the controversy. The period for licensing and meeting statutory requirements, of course, is lengthening, but once intervenors have filed their initial petition to halt construction, the controversy develops very rapidly, so rapidly in fact that if a company has not formulated some strategy for dealing with the possibility of political resistance, they will be unable to assess the issues and marshall their arguments quickly enough to escape the confrontation.

These six areas of observation suggest that there will be a continuing need for technology producing companies to become aware of potential public resistance as soon as possible. Using a seven stage model such as that exhibited here may permit a company to assess its position and anticipate resistance. Judiciously timed action such as public education programs, not traditional public relations, at the onset of public inquiries into the technology may avert the future failure to commercialize the technology. Naturally, a constantly changing societal value system presents a moving target of public expectations. But recognizing what stage of resistance is being entered should provide some guidance for resolving disagreements and keep the technology producer from escalating the resistance to its activities.

ACKNOWLEDGMENT

The foregoing paper is based upon work more fully described in "The Politics of Nuclear Reactor Siting," an unpublished Master's Thesis by D.G. Jopling for the Graduate School of Business at the University of Texas at Austin (December, 1970); a complete bibliography is contained in this thesis. Some of the material also appeared in *Nuclear News* (March, 1971), pp. 32–49.

ADVANCES IN TECHNOLOGICAL FORECASTING TECHNIQUES

PART TWO

Since 1967 the vigorous efforts of industrial and academic researchers have produced some very promising new forecasting methods. It was generally assumed that forecasting scientific progress was impossible. William Clingman has been the first to outline a way of forecasting *some* scientific and exploratory development prospects of the near term time horizon. The first industrial Delphi study by North and Pyke at TRW evolved into a second effort, PROBE II, that has now been developed by Pyke into a most promising kind of "map" of technological possibilities in relationship to sociopolitical developments.

Is some technological progress basically a function of learning, hence practice? If so, technological advances should be related to cumulative production, which is not always regularly distributed over time. This novel idea of a *technological progress function* has been tested and shows some substantiation.

Trend extrapolation is not new, but some guidance on selecting parameters is new and is included in a following chapter.

Goal-oriented (normative) forecasting is a fashionable intellectual concept, but who has done it and how? Two chapters describe practical applications of relevance tree theory to (1) a scientific experiment for NASA and (2) exploratory development planning for a major corporation. A variation, *perspective trees,* by Battelle consultants suggests matching technology and the environment through relevance analysis.

A popular term in planning circles today is *scenario.* The idea of considering possible futures to find the factors that are truly critical and so to better prepare for a "suprprise free" future is unquestionably useful. But how does one prepare scenarios? Researchers from Abt Associates show a technique for scenario generation that has been applied to a major government problem.

In another chapter, a keen observer of industry examines step-functions in some major technical facilities built by the chemical industry. He suggests that we are looking at *patterns of behavior* and that these patterns provide a guide to the technical steps that industry will take. Another extrapolation concept has been developed by two General Electric researchers. Their *substitution theory,* backed up by some forty examples, holds that the displacement of an existing technology by a newcomer proceeds in a highly orderly manner. The rate of displacement can be predicted with considerable accuracy after the initial rate is determined.

Monitoring to predict technical progress has been done implicitly or explicitly in many cases. This chapter formalizes the concept and specifically suggests searching for signals of technological change in five environments.

Despite much academic literature on the desirability of dynamic modeling, one looks long to find any applications to the forecasting of technology. The closing chapter shows one major attempt at *dynamic modeling* in the research laboratory of a large firm.

Predicting Scientific and
Exploratory Development

WILLIAM H. CLINGMAN, JR.

Until recently, no one has offered a methodology for predicting advances in science (as distinct from technology). Clingman points out that advances in many fields of science (and exploratory development) are the direct result of research funding by the Federal Government. Currently, the Federal Government funds some $17 billion of such work, and this funding is done largely against proposals that describe specific goals, technical means, and often include time tables for accomplishment. Therefore, it follows that a critical analysis of Federally funded research projects would describe, at least to some extent, the specific goals, the amount of effort, fields of work, and some likely scientific accomplishments in the next few years.

Using some NASA examples, the author transforms some research proposals into a set of verbal descriptors. These descriptors are further checked against progress reports and final research reports. Analysis of the nature and frequency of use of these descriptors in a body of research activity clearly suggests the thrust of efforts and probable accomplishment.

While this forecasting idea in no way pretends to embrace all scientific progress, it certainly suggests a way of anticipating some results from the massive Federal funding of research. It is, we believe, the first specific suggestion at forecasting at least part of the progress at the "science" end of the R&D spectrum.

INTRODUCTION

About two thirds of the research and development in the United States is funded by the Federal Government. Although the prime purpose of these programs concerns military, space, and other government objectives, the programs contribute to the advancement of many technological areas important to industry. Computer sciences, electronic miniaturization, high temperature alloys and ceramics, and light structural materials such as titanium are a few examples. Government funded programs have made significant contributions in these fields. What are the future technological areas in which such programs will have impact? This paper concerns one technique that is proposed to answer this question.

As a part of the NASA Technology Utilization Program a methodology was developed by the writer for predicting descriptive terms applicable to technology resulting from present NASA programs.[1] These descriptors identify the technological areas in which the most progress will be made over the next two to five years. The method should be generally applicable to any program. The starting point is a brief description of technical objectives for each project in the program. Such descriptions are available for a large fraction of government funded research. The forecasting procedure discussed in this paper analyses technical objectives to provide a set of descriptors for the technology anticipated from a group of projects.

The predicted technology can be defined to the extent that such a set of descriptors can be chosen. The descriptors are individual subject terms selected from a fixed vocabulary. There will be several descriptors applying to each project. If the same descriptor is used for a large number of projects of a given program, one can expect that program to make significant impact in the technological area corresponding to the descriptor. The method by which these descriptors are selected constitutes the main part of the forecasting methodology discussed in this paper.

In the next section the testing of the methodology as it was developed for the NASA Technology Utilization Program is discussed. The accuracy and completeness with which forecasting can be done can be derived from the results achieved in this NASA study. The forecasting technique and examples are then presented.

BACKGROUND

To meet the needs of the space program scientists and engineers have been producing technology at a far more rapid rate than in the past. Information is being generated faster than it can be disseminated by normal means, such as

1. This work was sponsored by the National Aeronautics and Space Administration under Contract NASw-1812. The author wishes to acknowledge the guidance and assistance provided by Leonard A. Ault of the NASA Technology Utilization Division.

technical journal publications. Much of the technology being developed is applicable to nonaerospace problems in industry. The NASA Technology Utilization Program was established to meet the challenge of transferring this technology. Its purpose is fourfold: to increase the return on the national investment in aerospace research by encouraging additional use of the results; to shorten the time gap between the discovery of new knowledge and its effective use in the marketplace; to aid the movement of new knowledge across industrial, disciplinary, and regional boundaries; and to contribute to the development of better means of transferring knowledge from its points of origin to other points of potential use.

As a part of this program an experimental study was carried out to evaluate the feasibility and value of an index to NASA Research and Technology Resumes. These are statements of project objectives and plans that have been prepared and used internally by NASA. The feasibility of indexing the technology likely to result from a project, using only the Research and Technology Resume as an information source, was evaluated. Those working in the NASA Technology Utilization Program could use such an index to identify NASA programs producing technology relevant to specifically defined nonaerospace needs. Presently available indexes, on the other hand, are used to identify NASA reports of relevant technology after it has been developed.

The experimental study itself consisted of three steps. First, statements of the technical objective and approach for several NASA projects were analyzed. These statements were contained in the documents discussed above, the NASA Research and Technology Resumes. As an example, the original statements of objective and approach for a cryogenic insulation research project are reproduced in Table 1. Such statements were prepared by the NASA project manager prior to initiation of the project. The significant point for technological forecasting is that only information available prior to the beginning of the NASA project was used in analyzing the technical objective and approach.

Second, descriptors of technology expected to result from each project were forecast. The forecasting procedure is discussed below. Essentially it consisted

Table 1. Research and Technology Resume Excerpt.

Title:	Cryogenic Insulation Research
Date:	June, 1964
Objective:	The objective of this program is to subject proposed concepts of high performance insulation to environments critical to thermal and structural performance of an operational spacecraft system. This continuation will provide for continual screening of recently developed insulations and for exploratory development into corrective measures on systems previously tested and rejected.
Approach:	The test specimen screening tanks and test procedures developed and used in Contract NAS8–11397 will be utilized for additional testing of untested system concepts. Promising systems which have failed or performed poorly during earlier tests will be improved by using knowledge gained in post-test analyses from these tests. Application of insulation systems to full scale tankage may then proceed with confidence.

of the following steps. A brief written description was prepared of the sequence of events expected to take place in the experimental project. Then using this written description as well as the original statement of technical objective and approach a list of key words and phrases was prepared. This list was intended to describe the new technology expected from the project. Finally, a set of descriptors for this technology was chosen from the fixed vocabulary used to index NASA documents. This vocabulary is contained in the *NASA Thesaurus*. Descriptors were chosen from the *Thesaurus* which corresponded to and amplified the key words and phrases.

Third, the descriptors chosen in the second part of the experiment were compared with the actual technology which resulted from each project. This was done by collecting all project reports published. These were studied and also the descriptors used by NASA to index these reports were studied. The NASA descriptors for the reports were compared with the descriptors of the projected technology selected in the second part of the experiment. The accuracy, completeness, and specificity of the projected technology descriptors were determined by this comparison. In considering the results of the experiment, it is important to keep in mind that one is comparing two sets of descriptors. The descriptors of projected technology were chosen by the writer using only information available prior to the initiation of the project; the descriptors of reports were chosen by NASA after the project was complete.

We were concerned with the accuracy of forecasting technology expected from a project. To measure this a search was made of project reports to determine whether or not technology was developed which corresponded to a given descriptor of the projected technology. It was found that ninety-five percent of the descriptors chosen corresponded to new technology actually resulting from the project.

A second question concerned completeness of forecasting. This was measured examining each technological concept reported and indexed by NASA. For each such concept it was determined whether or not corresponding projected technology descriptors had been chosen. It turned out that eighty-eight percent of the technological concepts reported and indexed by NASA were covered by the descriptors chosen in the forecast.

A third question concerned the specificity of forecasting. That is, could predicted technology be described more specifically after completion of the project. If the form of the forecast were a written detailed description of the technology, then, of course, this description could always be more specific after completion of the project. In the NASA experimental study, the form of the forecast was a list of technical terms intended to be descriptive of the projected technology. This list was chosen from the fixed vocabulary in the *NASA Thesaurus*. It was not known a priori whether or not a more specific list of descriptive terms could be chosen once the technical project was complete. In the experiment it was found, in fact, that fifty-three percent of the descriptors chosen in the forecast covered the corresponding technological development at least as specifically as it was indexed by NASA after project completion.

These results indicated that a methodology had been found which could forecast a set of terms descriptive of technology that would result in the future from a given research project. The experiment indicated that such forecasts were accurate and complete using only information available prior to initiation of the research project. The major type of prediction that is made using the methodology is pinpointing areas of technological advance. For example, in the NASA experiment it was predicted that new technology would be developed as regards the design of radioactive sources which produce beams of ionizing radiation. A report of such technology was first made two years after the information was available upon which this prediction was based. As another example, it was predicted that new technology would be developed as regards accommodation coefficients and aerodynamic drag in the interaction between gas molecules and a solid surface. A report of such technology was first made four years after the information was available upon which the prediction was based.

Given an entire collection of research plans, for example, a research program, forecasters can thus predict the areas of technological advance that will be brought about by that program. Information is available on many of the important government funded programs which would enable such a prediction using the above methodology. It is thus proposed that the technological impact of government funded programs could be forecast in this way. The methodology for achieving this is discussed in the next section.

METHODOLOGY

The starting point for forecasting the impact of government programs is a statement of technical objective and/or approach for each project in a particular area. Such statements are available in various forms from a number of government agencies. An example was seen in Table 1 of the Research and Technology Resume. NASA also has a set of RTOP (Research and Technology Objectives and Plans). These are prepared at the program level by the responsible technical manager. In Table 2 is shown an example of an RTOP summary. A collection of such technical summaries for each NASA Research and Technology Program for fiscal year 1970 may be purchased from the U.S. Department of Commerce, Clearinghouse for Federal Scientific and Technical Information, Springfield, Virginia 22151. Summaries of research funded by the Department of Health, Education and Welfare are available at the program level. An example is shown in Table 3. Technical objectives of future defense projects are published in the *Commerce Business Daily*. These are also available from specific agencies within the Department of Defense. An example is shown in Table 4. These documents contain the same type of information about future research projects as the NASA Research and Technology Resumes used in the experiment described in the previous section.

After collecting the information, it is then necessary to select a set of

Table 2. RTOP Summary.

RTOP No. 129–03–20

Title: High Temperature Materials

Organization: Lewis Research Center

Monitor: Ault, G. M.   Tel. 000–000–6387

Technical Summary

The objectives of this research are to develop new high temperature materials with superior properties for various aerospace applications and to extend processing technology so that advanced materials can be effectively exploited in such applications. A separate objective is to develop suitable low-cost materials for application to afterburner devices intended to reduce the level of automobile exhaust gas emissions. To achieve these objectives research is underway to extend the high temperature capability of a number of metallic alloy systems and several nonmetallic alloy systems. Alloying, dispersion strengthening, prealloyed powder technology and thermomechanical processing are major techniques under investigation to enhance the capability of various metallic systems. Specific joining techniques such as solid state welding are among the promising techniques being studied. Implicit in each research area are basic studies designed to contribute to the understanding of fundamental material behavior. This program couples with other in-house and contract work under RTOP's 129–03–21, 129–03–23, and 129–03–28.

descriptive terms which correspond to the technology that one is predicting will result from the project. In the NASA experiment a particular approach was found useful in selecting a set of descriptors. First, one would prepare a project description based on the original statement of technical objectives and/or approach. The description would discuss all events of the proposed R&D

Table 3. HEW Program Summary.

Title: Solid Wastes Research Grants

Date: August, 1969

Nature of Program Activity: This program provides grants for research in new and improved technological methods for solid waste collection, storage, utilization processing, salvage or final disposal.
Program Accomplishments: Grants under this program are currently being used to develop nine parameters for use in characterizing solid waste composition. Investigations are also being made into technology of a) collection and transportation, to possibly eliminate solid waste traffic in the street and storage in city buildings; b) incineration, solid waste reduction providing sterile, usable residue at reasonable cost; c) recycling, economical reuse of natural resources now being discarded as solid waste; and d) disposal, means of land fill most economical in terms of area of land and usability. Research contracts (under direct operations) are also being used to study waste problems in seven industrial classifications.

Table 4. Defense Project Objective.

Title: Ceramic Matrix Composites

Date: May 19, 1970

Objective: Development of ceramic matrix composites having low dielectric constants, temperature stability approaching 1500–1700°F. in air, mechanical and acoustical loading capability and integrity, and ease of fabrication in flat and curved shapes.

Sponsor: Air Force Materials Laboratory, Dayton, Ohio

project in chronological order. In preparing a chronological list of events, cause and effect relationships can be considered in arriving at the proper order and in predicting events which must take place but may not have been described in the original statement of project objectives. In listing any particular event one can ask whether the events already listed are sufficient to enable the event being listed to take place. It is not necessary that expert technical knowledge be applied to such a consideration, although whenever one possessed such knowledge he would utilize it. General logical relationships are used as a giude in extracting from available statements as much information as possible. For example, if the event being listed is the assembly of a piece of hardware, one can conclude that the parts being assembled must already be available. The origin of the parts should be deduced and discussed. If the event being listed is testing the corrosion resistance of a new material, one can conclude that the material would have already been prepared. This preparation may involve complete synthesis of the material or may be limited to modifying, such as coating, an existing material. Using the available information one would choose from among such alternatives and then include these as particular events in the project description. In Table 5 is given an example of a project description based on the original information in Table 1.

After preparing the list of predicted events or project description one would then select key words using both this list and the original resume. The key words would be chosen to cover all areas of technology expected to result

Table 5. Research and Technology Resume Analysis.

Title: Cryogenic Insulation Research

Project Description: In this program cryogenic insulation systems will be tested and experimental work will be done to improve those systems which fail the test.

The test methods used have not been previously applied to these systems. The test procedures are specially designed to evaluate the thermal and structural characteristics of these insulation systems in terms relevant to the actual environment in which they will be used. Knowledge gained from failure analysis will be used to experimentally modify systems.

Key Words	Descriptors
Cryogenic storage	Cryogenic fluid storage
Insulation materials	Insulation
Insulation systems	Insulated structures
Composite materials	Heat shielding
Heat transfer	Composite materials
Test methods	Heat transfer
Environmental testing	Performance tests
Thermal properties	Environmental tests
Structural properties	Thermodynamic properties
Failure analysis	Structural design
Cryogenic equipment	Failure
Spacecraft	Cryogenic equipment
	Spacecraft components

from the program. For this purpose the checklist shown in Table 6 has proven useful. Such a check list serves principally as a reminder of the factors about a project which should be considered. The particular check list in Table 6 has been applied many times to NASA projects and found to be complete in the sense that very few key words found suitable for describing a project were not covered by one of the questions in the check list. In choosing the key words equal emphasis should be given to all parts of the anticipated research program. That is, the experimental technique should be covered as well as the experimental results. In Table 5 are shown the key words selected for a typical project.

The key words are next used with the *NASA Thesaurus* to arrive at a final set of descriptors for the technology being forecast. Every descriptor is taken from the *Thesaurus*. The *NASA Thesaurus* also gives related terms after each listing in it. After locating the equivalent of a key word in the *Thesaurus*, these related terms are screened. Those considered relevant to the research project are included among the descriptors. In many cases a key word will thus result in the choice of more than one descriptor.

The use of the *Thesaurus* in this manner has two effects on the final list of descriptors. Since descriptors are chosen from a limited vocabulary there will

Table 6. Checklist for Descriptor Selection.

Background
    What new technology, if any, has led to this project?
    What is the aerospace need giving rise to this project?

Experimental methods
    What type of experimental procedures will be used?
    If analysis is to be done will novel mathematical techniques or computer programming be used?
    What special characteristics will be required of the experimental equipment?
    What procedures will be used to test or control the quality of products or processes developed in the project?

Novel materials
    What kind of novel materials, if any, will be involved in the project?
    What will be their composition or form?
    How will they be made?
    What will be their desired novel properties?
    How will they be applied?

Novel equipment
    What kind of novel equipment, if any, will be involved in the project?
    What will it do and how will it work?
    What novel materials or components will be used in this equipment?
    How will it be assembled?
    What will be its applications?

End result
    What will be the end result of the R&D project?
    If a new product or process is to be developed, what will it do?
    What problems must be solved to accomplish the end result?
    What will completion of the R&D project make possible?
    What are the anticipated applications of the work to be done?

be many variations in research results all covered by the same descriptive term. The use of that term as a predictor of the results will thus be accurate. The more limited the vocabulary, however, from which the descriptive terms are chosen, the less specifically can the anticipated technology be forecast. The *NASA Thesaurus,* however, contains a sufficient number of terms that it is used by NASA for indexing all of their reports. Thus, restricting the choice of descriptors to the *Thesaurus* would not restrict specificity any more than for the future indexing of the technology.

The second effect of using the *Thesaurus* is that the choice of descriptors can be broadened beyond the list of key words. The latter are limited to the words used in the statement of technical objective and to the words that the technology forecaster would associate immediately in his own mind with the anticipated technology. For the equivalent of each key word, the *Thesaurus* will contain a list of related terms. Some of these may not have been previously considered and yet still be appropriate descriptors for the anticipated technology.

For the examples in this paper the *NASA Thesaurus* was used in forecasting the impact of NASA research programs in certain specific areas. For general forecasting of the impact of all government programs it would be preferable to use the *Thesaurus of Engineering and Scientific Terms,* published by the Engineers Joint Council. It, however, is quite similar in organization and content to the *NASA Thesaurus.*

The final step in forecasting the impact of government funded programs would be the use of the descriptors to rank order areas of technology in which the greatest progress would be expected. Each area would be described by a set of closely related descriptors. Each of these descriptors would be chosen from a fixed vocabularly, as discussed above, for specific research projects that are planned. The technological area with the highest rank would be the one for which its corresponding descriptors were chosen the most often.

An example is provided in the next section where forty-one NASA programs in the area of materials science were analyzed as above. The technological areas of greatest impact have been predicted and rank ordered using this technique. Prior to the presentation of this example, the prediction for a single past project is compared to the actual developments that resulted in that project.

EXAMPLES

To illustrate the type of prediction that can be made from a single project consider the statement of objective and approach for such a project given in Table 1. This statement was prepared in June, 1964. The information given in Table 1 was analyzed as discussed in the previous section. In Table 5 is a description by the writer of the events that were expected to take place in the project. Also there is a list of key words and a list of descriptors for the technology expected to result from the project.

After the information in Table 1 was prepared by NASA a contract was awarded for carrying out the project. Eventually the results of the work were published [R. F. Crawford, et al., *J. Spacecraft,* IV (1967), 1598–1602]. The descriptors of the predicted technology in Table 5 have been compared with the reported technology in this article. Technology corresponding to each descriptor in Table 5 is discussed in the article. In Table 7 each descriptor is listed along with the type of technology developed corresponding to it.

This example also illustrates the effect of limiting the vocabulary from which the descriptors are chosen. When the descriptors in Table 5 were chosen by the writer, the details of the technology were not known. For exam-

Table 7. Comparison of Predictions with Actual Developments (cryogenic insulation research project).

Projected Technology Descriptors	Actual Technology Developed
Cryogenic fluid storage	The insulation developed was applied to the long-time space storage of liquid hydrogen.
Cryogenic equipment	A test apparatus for evaluating insulation at liquid helium temperatures was developed.
Spacecraft components	The insulation developed in the project was for specific application to cryogenic propellant tanks of advanced spacecraft.
Structural design	Different structural designs for multilayer thermal insulation were evaluated in the course of the project.
Performance tests	Apparatus was devised to carry out final performance tests on the insulation developed in the project.
Environmental tests	Methods were devised for carrying out large scale environmental testing of insulation during the course of the project. An entire section of the paper discussed these tests.
Failure	Previous work had shown that helium-purged, multilayer insulation systems failed to evacuate to the low pressure level required for efficient insulation performance. Part of the project involved determining the mechanism of this failure.
Insulation	The entire project involved the development of new insulation.
Insulated structures	The project involved primarily the development of insulated structures rather than new insulating materials.
Heat shielding	The effect of perforations in sheets of insulation on radiation transmittance was analyzed.
Heat transfer	The heat transfer rate was analyzed and measured for different insulated structures.
Thermodynamic properties	Properties of insulating materials, such as density and conductivity, were varied so as to reduce the overall heat transfer rate.
Composite materials	The developed insulation involved a multilayer structure rather than a single homogeneous material.

ple, it is stated in Table 1 that one objective of the project is to subject proposed insulation concepts to critical environments. This led to the use of "environmental tests" as a descriptor of the anticipated technology. At that point nothing was known about these tests. In the article published after the work was done there is an entire section devoted to the details of environmental testing. For subsequent information retrieval NASA chose descriptors for this same article from the same limited vocabulary. The only term chosen relevant to environmental testing was "environmental tests." Even though more specific information about the technology was then available, the identification of that technology in terms of descriptors was no more specific after the project than before.

This example illustrates that the methodology allows one to predict those technological areas on which a given set of programs will have the greatest impact. The specific nature of this impact cannot be predicted any more precisely than the limited vocabulary from which descriptors are chosen would allow. As a second example of applying the methodology, the impact of current NASA programs classified by NASA as Materials Research Programs has been predicted in this paper.

The starting point for applying the methodology was collecting a technical statement of objective and approach for each project. These statements are available in the Summary of the NASA Research and Technology Objective and Plan. There are forty-one materials research programs. For each a chronological description of the anticipated events in the research was prepared. A list of key words was then selected based on this description and the original statement of research objectives. Finally, corresponding to each key word, descriptors were chosen from the *NASA Thesaurus*. These descriptors are forecasts of the technological areas on which impact is expected from these materials research programs.

The total set of descriptors for all forty-one programs was next analyzed. Closely related descriptors were grouped into technological areas. There were 165 areas for which the corresponding descriptors were chosen only once. One would expect relatively low technological impact on these areas when examining the entire set of projects as a whole. There was one group of descriptors dealing with solid surfaces which appeared with greatest frequency. This corresponds to the technological area on which one would expect the greatest impact from the entire set of projects. In Table 8 the ten technological areas of greatest expected impact are listed. Specific descriptors for each area are also shown. Next to each descriptor in parentheses is the number of projects for which that descriptor was chosen. The total of all these numbers is called the "rank" for that technological area.

The number, rank, is considered significant only to order areas of technology according to the relative number of developments expected in these areas from a given set of programs. There may be, of course, a highly significant development in an area of very low rank. Also, the relative amounts of money

Table 8. Predicted Areas of Technological Development.

Technological Area	Rank	Specific Descriptors
Solid surfaces	42	Surface defects (6), surface distortion (2), surface energy (1), surface finishing (4), surface geometry (1), surface layers (1), surface properties (11), surface reactions (9), surface roughness (1), metal surfaces (3), gas-solid interfaces (3)
Alloys	23	Alloys (15), high strength alloys (3), refractory metal alloys (1), nickel alloys (2) chromium alloys (1), cobalt alloys (1)
Single crystals	22	Crystal defects (5), crystal growth (6), crystal structure (2), crystal surfaces (3), crystallization (1), crystallography (1), crystals (1), single crystals (3)
Polymers	19	Polymer chemistry (3), polymer physics (1), polymerization (5), polymers (7), plastics (2), elastomers (1)
High temperature	15	High temperature (10), high temperature gases (1), high temperature tests (4)
Environmental testing	13	Environmental tests (13)
High strength materials	13	High strength (1), high strength alloys (3), strength (9)
Metallurgy	12	Metallography (2), metallurgy (5), microstructure (5)
Mechanical properties	12	Mechanical properties (9), mechanical measurement (3)
Mechanical stress	11	Stress analysis (2), stress concentration (1), stress corrosion (4), stress cycles (1), stress functions (1), stress relaxation (1), stress-strain diagrams (1)

being spent on each program will vary. In the above example most of the programs range between four and ten professional man years. Relative emphasis within a particular program is not considered. These factors work against one measuring quantitatively and precisely the degree of technological advancement in a particular area. The results of the NASA experiment discussed in a previous section, however, imply that the forty-one materials research programs in the example will produce technology which is quite accurately described by the descriptors in Table 8. This same set of descriptors should also correspond to those areas of technology in which new developments will arise the most often. This latter conclusion follows from the fact that the descriptors were found to be complete in the NASA experiment. That is, if a project developed new technology in a particular area, there was a descriptor corresponding to this area. It is thus unlikely that a large number of new developments would be reported in an area with a low rank.

In conclusion one can state that a methodology has been structured based on experimental results in forecasting technology. This methodology provides one with a list of specific technical areas in which to expect advances from a given set of programs. The methodology is particularly applicable to predicting the impact of government funded programs.

Mapping–
A System Concept for
Displaying Alternatives

DONALD L. PYKE

The first industrial attempt to apply the Delphi technique was conducted by Harper North, Vice President of Research at TRW, and his assistant Donald Pyke. Their approach was described in the predecessor to this volume. They called their effort Probe I. Like all pioneering work it was not completely satisfying to its sponsors, and so Probe II was launched. That concept was described at an early stage in the *Harvard Business Review,* and in later stages in the unpublished papers of Donald Pyke. Pyke believes that Probe II now has led to something far more significant than simple technological predictions. He calls this concept "mapping."

Consider that a very broad set of forecasts about technological events provides a scatter pattern of events against time. These technical predictions can be considered in relationship to each other, and this further suggests a kind of R&D "PERT" chart of possible future developments. Pyke shows that this concept can be much refined by establishing a coordinate system, using markets and other environmental factors. One could then "map" a whole region of technological interest and societal impact. This idea, and its relationship to management needs, is the essence of his new concept.

A number of years ago my physics classmates and I were intrigued by the thought that one could insult less-informed colleagues with relative impunity by referring to them as "third derivatives of displacement with respect to time."

Most people are aware that the time-rate of change in displacement is called velocity and that the time-rate of change in velocity is called acceleration, but few have been exposed to the fact that the time-rate of change in acceleration (the third derivative of displacement with respect to time) is called the jerk.

I am reminded of this because I suspect that technological innovation is being introduced today at a rate which must be measured in jerks. This conclusion can be derived in a straightforward way.

First, it seems reasonable to assume that the number of technological innovations is somehow proportional to the number of practicing scientists and engineers. Derek J. DeSolla-Price[1] has noted that eighty to ninety percent of the scientists and engineers who ever lived are alive today. This would imply that the number of innovations per unit of time is increasing, that is, technological innovation has "velocity."

Second, it is clear that the computer has increased the productivity of the average scientist or engineer. Thus the rate of innovation per individual should be increasing; that is, technological innovation is accelerating.

Third, John McHale[2] has made a study which indicates that the rate at which typical technological innovations move from conception to practical use, is also increasing as shown in Exhibit 1. Thus one may infer that technological change is proceeding in "jerks."

This observation has important implications for both industry and government. No longer can we afford a "wait and see" attitude, which generally results in "management by response to crisis."

The industrial decision maker must not only anticipate the possible in-

1 *Little Science, Big Science* (New York: Columbia University Press, 1963).

2 *World Facts and Trends* (Binghamton, New York: Center for Integrative Studies, School of Advanced Technology, State University of New York, 1969).

Exhibit 1. Selective Illustrations of the Speed for Introducing Technical Development into Social Use.

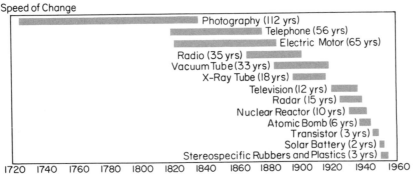

Speed of Change

Photography (112 yrs)
Telephone (56 yrs)
Electric Motor (65 yrs)
Radio (35 yrs)
Vacuum Tube (33 yrs)
X-Ray Tube (18 yrs)
Television (12 yrs)
Radar (15 yrs)
Nuclear Reactor (10 yrs)
Atomic Bomb (6 yrs)
Transistor (3 yrs)
Solar Battery (2 yrs)
Stereospecific Rubbers and Plastics (3 yrs)

1720 1740 1760 1780 1800 1820 1840 1860 1880 1900 1920 1940 1960

Source: Center for Integrative Studies, *World Facts and Trends* (Binghamton, New York: School of Advanced Technology, State University of New York, 1969).

novations which may obsolete his products or the way in which he produces them, but also he must be alert to the opportunities for new and more profitable utilization of the resources which he has programmed for the future.

The government policy maker must not only anticipate the side effects of potential innovations which may represent threats to other aspects of the environment, but also he should be alert to opportunities to encourage those potential innovations which have an overall positive effect on the future environment he envisions.

The requirements of both of these individuals are similar to those of one considering a trip. He must review the array of possible choices which are open to him and the nature of alternate routes to each, then compare the features of each option. For this kind of consideration he finds a map almost essential.

The purpose of this paper is to describe a concept for a comprehensive system which utilizes available forecasting techniques to generate a map of the technological future which will provide for the planner or assessor a tool similar to that available to the traveller. As envisioned this map would display technological alternatives and their environmental consequences in such a way as to enable the coupling of near term technical activities to long-range forecasts.

The analogy extends to the way in which both types of maps are prepared. Some of the major steps in each of the processes are:

1. Selection of the area to be mapped and the establishment of an appropriate coordinate system.
2. Detailed reconnaissance of the area.
3. Transfer of reconnaissance data to the coordinate system.
4. Topography to provide details of the "terrain" and the relationship between features of the map.

Let us examine each of these steps as applied in the preparation of maps of the technological future.

As noted earlier, in any type of cartography, the preparation of maps begins with establishing the boundaries of the area to be mapped and the layout of an appropriate coordinate system. For maps of the future, one axis of the coordinate system is time and the other is the category under investigation. An hierarchical system is envisioned in which the summary level consists of the environment subdivided into the eight aspects identified in Exhibit 2. Any meaningful technological forecast must include consideration of the interaction between technology and each of the other aspects of the environment.

For concentrated study, a particular aspect can be further subdivided to the desired level of detail. For technology, two levels appear adequate as shown in Exhibit 3. At the second level, technology may be subdivided into scientific and engineering capabilities. At the third level, science may be further sub-

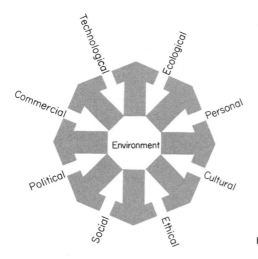

Exhibit 2. Many Elements Combine to Form the Environment.

divided into the specialties list used by the National Science Foundation in its National Register, and engineering may be further subdivided into the categories used by the Engineers Joint Council in its survey of the engineering profession. Applications, which constitute the interface between technology and other aspects of the environment, may be subdivided into categories compatible with those contained in the SIC (Standard Industrial Classification) codes and also those used in input-output analysis.

This level of detail is far beyond that required by most users. Nevertheless,

Exhibit 3. Hierarchy of Environmental Aspects.

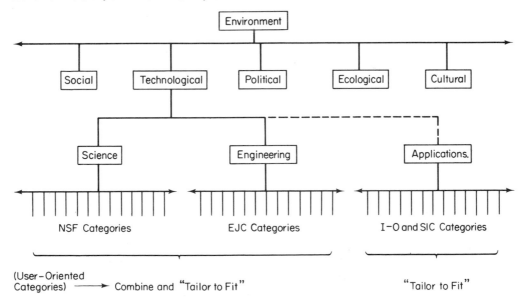

in order to permit comparison with national data banks, "tailored" lists of user-oriented categories should be able to accommodate any entry, though not uniformly to the level of detail provided in the "untailored" lists. For example, using the appropriate "keys," all industries, products, and services listed in the input-output and SIC codes may be related to TRW's application codes listed in Exhibit 4, and all specialties of science and engineering may be related to its research and engineering codes listed in Exhibit 5.

Exhibit 4. Probe II Categories: Applications (products or services/purchased or sold).

TRW Code	Title
31	Agricultural, forestry and fishery; PREAS*
32	Mining and quarrying, stone, clay, and glass; PREAS
33	Petroleum industry; PREAS
34	New construction, maintanance and repair; PREAS
35	Ordnance and accessories
36	Food, tobacco, drug, cleanser, and paint; PREAS
37	Textile, rubber and leather; PREAS
38	Household, furniture and fixtures, paper and related industries
39	Plastics and synthetic materials; PREAS
40	Chemical industry (excl. plastics and synthetics); PREAS
41	Metallic materials and primary parts
42	Industrial tools, machinery and equipment (including hardware)
43	Engines and turbines
44	Office and service industry, machinery and equipment
45	Radio, TV, and communications equipment and services
46	Computers and peripheral equipment and services
47	Electronic components and accessories
48	Electrical equipment, other
49	Motor vehicles, equipment, parts, and services
50	Aircraft, Equipment, parts, and services
51	Spacecraft systems and equipment
52	Transportation systems, equipment, parts, and services (other)
53	Professional/scientific controlling instruments and supplies
54	Miscellaneous products and services
55	Commerce
56	R&D and SETD
57	Government

*PREAS = products, related equipment, and services.

Exhibit 6 provides an example of a coordinate system in the form of a three level overlay. The ordinate, time, is common to all three levels. All aspects of the environment except technology appear as the abscissa of level one. The tailored lists of technologies and applications provide the abscissas for levels two and three.

The next step is reconnaissance, with the objective of establishing sufficient "triangulation points" in those subcategories which are of particular concern. This is accomplished using exploratory forecasting techniques to develop a comprehensive list of anticipations related to the subcategories under study. Ex-

Exhibit 5. Probe II Categories: Technologies
(scientific and engineering capabilities).

TRW Code	Title
01	Biological, agricultural, social, medical
02	Physics
	A — Acoustics
	H — Holography
	L — Lasers
	M — Magnetics
	N — Nuclear
	O — Optical
	S — Sensors
03	Chemistry and chemical processes
04	Atmospheric, earth and marine
05	Materials metallurgy, ceramics, coatings
06	Engineering science, mechanics, thermo, fluids
07	Electric power—generation, conversion and conditioning
08	Information processing, data handling, computer equipment
09	Instrumentation and control
10	Communication and navigation
11	Electronics, circuits, components
12	Structures and mechanical equipment
13	Manufacturing processing and automation
14	Materials handling and logistics
15	Environmental control, ecology, pollution
16	Defense and weaponry
17	Aerospace, rocket propulsion, spacecraft
18	Transportation
19	Systems engineering

trapolative and speculative techniques, such as Delphi and brainstorming, are particularly useful in this phase of mapping. Supplementary literature searches will help to make the list more complete.

"Triangulation points" are anticipations developed during reconnaissance which can be uniquely related to a specific category of the map such as: engineering developments and the prerequisite scientific discoveries, applications of one or the other in the form of products or services, and either precursors or consequences related to other aspects of the environment.

The third step in mapping is transfer of reconnaissance data to appropriate portions of the map in the form of:

1. Activities (represented by lines) which consume time and require the commitment of resources, such as scientific investigations or engineering developments.

2. Events (represented by circles), the occurrence of which can be recognized at a particular point in time.

3. Dependencies (represented by dotted lines), which indicate the dependencies between events and activities.

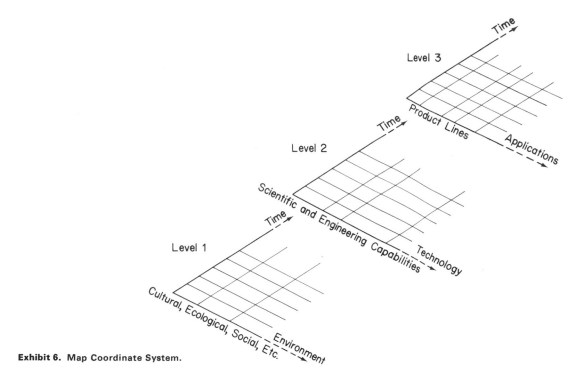

Exhibit 6. Map Coordinate System.

For example, Exhibit 7 shows an anticipation drawn from TRW's Probe II. The direct implications of this event as related to the symbolic coordinate system of Exhibit 8 are as follows:

1–2 *Research in dielectrics,* an activity in physics will lead to

2 *New materials with high energy storage capabilities discovered,* an event both in physics and materials technology.

Exhibit 7. TRW's Probe II Sample of Final Output.

Event Description	S E E A P P	Desirability D̄	Feasibility F̄	Probability P̄	Probability Dates						2000 and Beyond			
					1 9 6 9 0	1 9 7 7 0 1 2 3 4 5 6 7 8 9	1 9 8 8 0 1 2 3 4 5 6 7 8 9	1 9 8 8 0 1 2 3 4 5 6 7 8 9	1 9 9 9 0 1 2 3 4 5 6 7 8 9	1 9 9 9 0 1 2 3 4 5 6 7 8 9	. 1	. 5	. 9	
Breakthrough in dielectric research will yield materials of high energy storage capabilities. They will be in common use in electric cars.		+.43	−.29	.39	. .	. .	. .	. .	. .	. .	. .	. .	. .	

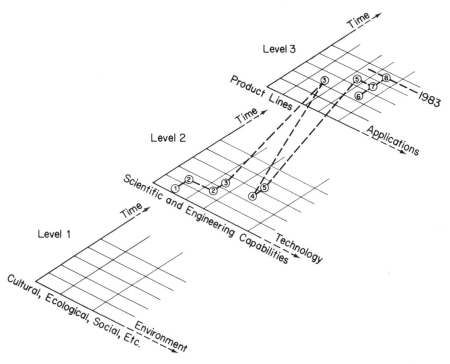

Exhibit 8. Transfer of Reconnaissance Data to Map Coordinate System.

2–3 *Development of processes for the production of materials with high energy storage capability,* an activity in materials technology will lead to

3 *Materials with high energy storage capabilities available,* an event in materials technology and in metallic materials and primary parts, an application.

4–5 *Incorporate material with high energy storage capability in storage batteries,* an activity in power conversion will lead to

5 *Batteries with improved high energy storage capability available,* an event in both power conversion technology and electrical equipment.

6 *Electric cars introduced,* an event in transportation systems.

7 *Electric cars utilize new batteries with high energy storage capability,* an event in transportation systems.

8 *New batteries with high energy storage capability in common use in electric cars,* an event in transportation systems.

Each anticipation generated during the reconnaissance phase is analyzed in a similar fashion and the results are entered in appropriate sections of the coordinate system. Preliminary estimates of timing are determined by the median date estimate associated with the key element. Thus in our example event Number 8 is established at 1983 and other activities and events are spaced throughout the interim period at intervals judged to be consistent with the time and/or effort required.

At the conclusion of this activity if reconnaissance was effective, entries should be sufficient for "topography" to proceed.

The fourth major step in mapping, topography, consists of several activities:

1. Each subcategory must be inspected carefully. It is likely that key development and/or events will have been overlooked during the reconnaissance phase; these must be added. Relevance tree techniques are applicable in this activity.

2. Each "application" must now be examined to insure that all scientific discoveries and/or engineering developments which must precede it appear in proper sequence in the technology level of the map and that interdependencies have been indicated. Phase one SOON Technique[3] can be used here.

3. Each scientific discovery and engineering development should then be studied to insure that all conceivable "applications" toward which they might contribute are added to the map and that appropriate dependencies are indicated. Phase two SOON Techniques can be useful during this process.

4. It is then time to examine the interface between technology and its applications and the rest of the environment. One should ask the following questions concerning each application or manufacturing process: What environmental factors will enhance or inhibit the probability that this anticipation will be realized? What are the possible side effects of each anticipation on other aspects of the environment? Responses and their relationship to these anticipations should be added to the map as additional events, activities and dependencies.

5. The final step in topography involves calibration. During the reconnaissance phase, additional data related to each anticipation may be obtained. For example, in TRW's Probe, we investigated: (a) both the desirability and feasibility of each event, (b) the probability of event occurrence, and (c) the probable time of occurrence (assuming that the event would occur). Since these estimates were made independently of one another, calibration is required to eliminate the resulting inconsistencies. Correlative techniques such as input/output analysis and cross impact matrix analysis should prove most effective during this phase of mapping.

The end result, if all has been done properly, will be a "map" (illustrated symbolically in the Exhibit 9) consisting of a computer-based data bank from which the assessor or decision maker can draw:

1. A surprise-free, time-phased array of the viable alternative options relevant to a particular area under consideration.

2. The sequence of scientific discoveries and engineering developments prerequisite to each alternative.

3. The environmental precursors which may influence the choice among alternatives.

4. The consequences of each choice in terms of its impact on the rest of the environment.

[3] Harper Q. North and Donald L. Pyke, "Probes' of the Technological Future," *Harvard Business Review* (May-June, 1969).

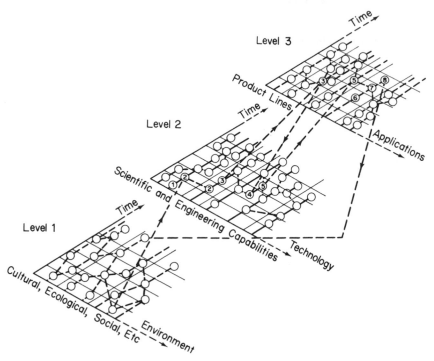

Exhibit 9. Symbolic Representation of Completed Map of Technological Future.

Displays will be organized so as to facilitate comparison of information related to each option.

It is at this point, as I see it, that the responsibility of forecasting in the planning or assessment process is fullfilled as indicated in the flow chart of Exhibit 10. Since we will all share the same future environment, one map of that environment should serve the purposes of both the planner and the assessor. The same holds true for planners in competitive companies. For example, opposing military commanders can, and probably do, use identical maps without revealing their strategies. So it is in business. We will all operate in the same future; however, the strategy adopted by each competitor will vary depending upon his interests and the resources available to him.

The dynamic aspect of mapping is also illustrated in Exhibit 10. The impact of man's decisions will affect the future environment. Therefore, the modification of his plans and policies and their implementation must be reflected by appropriate changes in the anticipated future. Mapping, therefore, must be viewed as an ongoing activity characterized by constant surveillance of both progress and prognosis followed by updating of the map as appropriate.

Less than 200 years ago our forefathers migrated to the West without maps to guide their way; today we use maps to plan a trip across town. I suspect the day is not far off when leaders of government, business education, and industry will use "maps" to aid their consideration of the future.

Exhibit 10. Role of Technological Forecasting in Planning and Assessment.

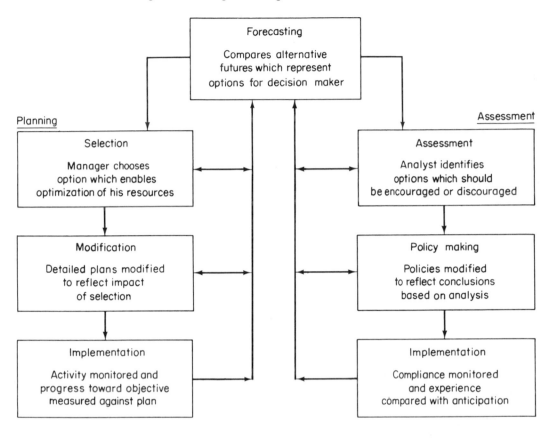

The Technological
Progress Function:
A New Technique
for Forecasting

ALAN R. FUSFELD

Much trend extrapolation forecasting is based on the assumption that technological progress varies with time. Fusfeld offers a radical departure. He suggests that some kinds of technological parameters vary with cumulative numbers of devices produced. Then, if production is not uniformly distributed over time, a different forecast will result. Fusfeld's "technological progress function" is analogous to the manufacturing progress function and the classical learning curves used for many years in airplane manufacturing.

Both theory and practice lend support to his hypothesis. Is it not logical that technological performance improves with "practice" as dozens of refinements and improvements are made over long production runs? And isn't it true that R&D budgets for further technical work often are based upon a percentage of sales income, which is proportional to units produced? At any rate, here is a new forecasting concept, with supporting data, that deserves further consideration.

BASIC ASSUMPTIONS

The core of technological trend forecasting has most often been based upon a plot of a technical parameter against time. This study concerns the implications for forecasting, when these technical parameters are plotted against cumulative production quantities.

For the examples covered in this study, the relationships between technical characteristics and cumulative production all have the mathematical form:

$$(1) \qquad\qquad T_i = a(i)^b$$

where T_i = value of parameter at the ith unit,
  i = cumulative production quantities,
 and a,b = constants.

This dependence of a technical value upon production quantities is defined here as the technological progress function. This paper relates the development of that function, the implications of the mathematical form just mentioned for trend forecasting, and information compiled from a variety of case studies.[1]

It is of interest to note that the form of the technological progress function is similar to that of the common industrial "learning curve," as well as the learning curves of psychologists. The industrial "learning curve" relates cost to production [1–3].

$$(2) \qquad\qquad r_i = r_1(i)^{-b}$$

where r_i = cost of the ith unit,
  i = cumulative production,
  r_i = cost of 1st unit,
  b = constant,
and the psychological learning curve relates the efficiency of performing a task to the number of repetitions [4–7].

$$(3) \qquad\qquad E_N = k(N)^b$$

where E_N = efficiency of performing Nth task,
  N = cumulative task number,
 and k, b = constants.

They are similar to the technological progress function, $T_i = a(i)^b$ because all three show the measure of some characteristic, for which improvement is desired, as an exponential function of the cumulative number of repetitions or production units.

The environment in which technology develops is clearly of additional interest and appears to affect the rate of learning through discrete changes in the learning constant, b. It must be specified here, that environment refers to the

[1] Studies completed included:
civil aircraft—speed
military aircraft—speed
turbojet engines—specific weight
turbojet engines—specific fuel consumption

automobiles—horsepower
electric lamps—lumens
computer programs—figure of merit
hovercraft—figure of merit

external economic factors which surround the production process. An example of this effect was the change in the government investment and market potential relating to the automotive industry in the late twenties [8, 9], which multiplied the rate of progress, *b*, by a factor of 6.2.

INTRODUCTION AND BACKGROUND INFORMATION

The development of the technological progress function, T_i, characteristic of technological improvement, evolved from a consideration of: (1) problems and background factors inherent in industrial progress relationships, (2) indications from psychology that general phenomena of "learning" were present, and (3) difficulties in existing techniques of forecasting.

Progress Functions

Briefly, industrial progress relationships are functions, such as production costs per unit, maintenance costs per unit, and manufacturing costs per unit, which can be written as a function of costs associated with unit number one, the cumulative unit number, and a learning constant. They are termed progress functions because a reduction in costs indicates a gain in efficiency. These functions, as previously mentioned, have been found to have the precise form [1, 2, 10, 11]:

$$(1) \qquad\qquad r_i = a\,(i)^{-b}$$

When plotted on log-log paper this gives a straight line whose slope, $-b$, is the rate of progress, that is, the rate at which efficiency is improving. It should be noted here that the technological progress function has the same form and expression, the difference being a positive rather than a negative slope.

Various studies of industrial progress functions have shown that there are a number of causes or factors which are common to most of the different types [1, 2, 11, 12, 13].

The more direct factors are: (1) engineering or design improvements, (2) servicing technician progress, (3) job familiarization by workmen, (4) job familiarization by shop personnel and engineering liaison, (5) development of a more efficient parts supply system, and (6) development of a more efficient method of manufacture.

The innate factors are: (1) the inherent susceptibility of an operation to improvement, and (2) the degree to which this can be exploited.

These functions can be called MEPF (microeconomic progress functions) because: (1) they are similar in mathematical form, and (2) each relates to decisions of the "firm." However, their applicability to development decisions is not as simple as it would seem initially since the functions contain two areas of problems. Both areas are relevant to the subject of the technological progress function because the first helped to prompt its development and the second helps to explain its behavior.

The first area involves certain MEPF characteristics which present problems of decision that have been unsolvable because of their link with technological innovation. The problems begin when one attempts to determine the component costs of the first unit of production and finds that one must arbitrarily set a limit upon the background development expenses. A mechanism which background and development costs would be linked to all of the generations of product might help to alleviate this difficulty of first unit costs; of course, this would first require a model for technological change.

Further and more fundamental complications become apparent when the development of new generations of equipment is considered, as opposed to the difficulties of measuring cost for any one generation. Technical changes and design modifications are part of the inherent forces that cause the MEPF to behave as it does; but there are no provisions for setting a limit on these changes, the engineering change procedure of firms being arbitrary in nature. That is to say that, after a certain number of modifications in production procedure or design, the product in reality will be a "new" product and will be so labeled by sales and management. This is often the case with developments resulting from both conscious and unconscious defensive research, which is designed to improve a product's competibility.

The locating of an "old-new" product line or determining where a major change begins are questions posed by users of the MEPF and seem to beg a progress function which cuts across individual generation lines. *The technological progress function is such a relationship and contains the same factors of engineering learning and motivation of the MEPF, which not only cause problems in its use but also provide a major stimulus for technological advancement.*

The other problem area relates to deviations from normal (linearity on log-log paper) MEPF behavior. The explanations of these differences are important, since they help to clarify the effects of external factors of the environment on technological progress. Some of these deviations are due to changes in the psychological motivation of the personnel involved, as would be the case where a product line is suddenly scheduled for discontinuance or where the success of a new development is announced to the employees [1, 2, 14]. Others would be due to changes in design or the introduction of new people to the job. Finally, if the product or system is being constructed on parallel assembly lines or shifts, the addition or subtraction of new lines, perhaps with changes in demand, will cause anomalies in the MEPF [14]. All of these deviations cause a change in the slope of the progress function. Thus it is implied that *changes in the rate of progress* (for example, the slope of the progress function) *are not random but are a function of critical parameters forming part of the external environment.*

Psychology

In addition to the information supplied by analysis of progress functions, further insight has been gained from psychological learning theory. Since there is considerable dispute over exact definitions of what learning is or is not, let me point out that, here, psychological learning refers to perceptible gains in performing given tasks. These tasks in academic studies have ranged from solving puzzles and going through mazes to simple studies of response time for given stimuli. It has been noted in a variety of projects concerning both animals and people that the efficiency of performing a given task increases with the cumulative number of repetitions [4, 6, 7, 14, 15, 16]. It should also be pointed out that none of the studies differentiates between the frequency of the repetitions as long as the frequency was within the maximum retention interval. The studies found that,

$$(2) \qquad\qquad E_N = K (N)^a$$

where E_N = efficiency in performing the Nth task,

$\qquad N$ = the cumulative task number,

and $\quad K, a$ = constants.

Such a relation was particularly apparent in a 1934 study with children by Melcher [7] and in a 1955 study with dogs by Bush and Mosteller [4]. A similar but different relationship regarding repetitious activity can be derived from work presented by Frank Logan in his book, *Incentive* [16]. Logan noticed of his subjects, that they. "...behave in such a way as to maximize reward while at the same time minimizing effort." *Such behavior is not only identical with the aims of man's economic endeavors but is also a causal factor of technological progress.* Since the progress functions discussed previously show a similar dependency on repetition of tasks, it would appear that the same type of learning discussed above is involved.

Forecasting

The third area that has contributed to the present development of the technological progress function is technological forecasting. The primary methods already available, such as Delphi, trend extrapolation, trend correlation, and growth analogy, do not allow for the easy or precise handling of environmental factors [17]. This provided further incentive to find a progress function that would be as competent as other techniques under constant forces and yet allow for environmental change.

TECHNIQUES USED IN ANALYSIS

The work involved in investigating and defining the technological progress function was completed through two separate of reasoning. Both paths were developed while taking note of: (1) an observation, through analysis of the MEPF, that the independent variable should be the cumulative production quantity: (2) an implication in learning theory that the form of the function should be similar to that of the MEPF and the general structure characteristic of improvement functions; (3) a need of forecasting techniques for a term through which environmental change could be introduced.

The paths chosen for development are common to most scientific endeavors. They were those of theoretical derivation from known relationships and empirical model building from real world data.

The theoretical side bases itself on the equivalence, for at least certain areas, of the rate of patent output with technological growth or progress. Through relationships derived or shown by Schmookler [18] and Villers [19], and rate of patent output was related to investment and expected profit functions. From this point, substitutions were made from investment profit functions, denoting technical advance. The result was a relationship that equated the level of technological improvement to a constant multiplied by an increasing quantity-dependent function raised to a positive exponent, for example,

$$(3) \qquad\qquad T_i = K\,(f(i)\,)^c$$

where $T_i =$ the level of technology at the ith unit of production,

$f(i) =$ a function which varies with the production quantity, i,

and $K, c =$ constants.

Although it seemed to substantiate earlier hypotheses, it could only be taken as a further indication that this might be the correct approach, since it was not precise enough in nature to stand by itself.

The results of the theoretical derivation cleared away doubts from the proposed directions of the empirical phase of the study. At this stage, it was clear that the work should attempt to correlate technological improvement with the cumulative quantity of production. In addition, thoughts of incorporating the results with other progress functions motivated the gathering of data designed for more extensive correlations.

Real data were then sought to provide information on technical parameters, production, and costs with background information to be provided where possible. One could then observe the behavior of technological parameters with respect to production quantities under the sets of conditions forming the background environment.

However, before describing the case studies and conclusions, some of the difficulties involved in this work should be pointed out. The most difficult problem was that of obtaining accurate data, particularly with regard to cost information. This was solved by combining data from several sources and where possible having the material validated by someone familiar with the field. Other problems arose concerning the choice of technical parameters. In this case decisions were made from background information and observation of changes in the parameter with respect to cost changes.

PRESENTATION OF DATA AND CONCLUSIONS

The empirical studies discussed here verified both the preliminary hypotheses and the theoretical derivation and were based upon analyses of data regarding the aircraft industry, the electric lamp industry, and computer programming.

Aircraft Data

The data from the aircraft industry concerned turbojet engine development over a period of nearly twenty years and were synthesized from two Air Force Institute masters theses [20, 21], supporting information supplied by the Pratt & Whitney Division of United Aircraft [22], and *Aviation Facts and Figures–1958* [23]. From these sources, production cost per unit, production by year, cumulative production, and the technological parameters were observed.

The results have shown, in Exhibit 1, that technological progress, as represented by specific weight (dry engine weight per pound of thrust) and specific fuel consumption (pounds of fuel per hour per pound of thrust) is log-linear to a logarithmic quantity axis, where quantity indicates the cumulative production of engines within the turbojet family. It also appears that an arithmetic time axis may be used in place of a logarithmic quantity axis on many occasions for the same accuracy, whenever production undergoes constant percentage increases with respect to time.[2]

2 The bracketed points at the end of the "specific weight" curve were obtained from current advertising information of Pratt & Whitney presented in *Aviation Week* during March, 1969.

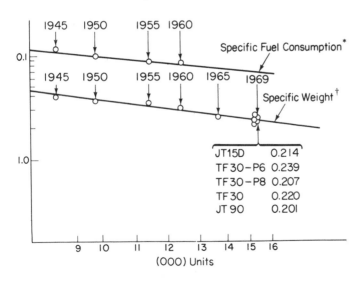

* In Lbs. of Fuel/Hr./Lb. of Thrust

† In Lb. Engine Wt./Lbs. of Thrust

Exhibit 1. Turbojet Engine Progress.

The electric lamp industry data were obtained by combining information from James Bright's book *Automation and Management* [24] and Arthur A. Bright's book, *The Electric Lamp Industry* [25] for sixty-watt lamps. These sources enabled the study to be concerned with technological progress as represented by the output of the lamp in lumens, production by year, and cumulative production. The many development changes made it unneccessary to examine each individual generation of lamp. The data examined covered the period from 1912 to 1940.

The results confirm the existence of a technological progress function behaving in accordance with the generalized MEPF principles. Exhibit 2, where this is illustrated, actually indicates the existence of two rates of progress.

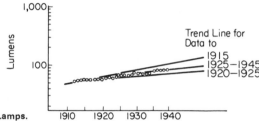

Exhibit 2. Lumen Output of Sixty-Watt Electric Lamps.

One extends from 1912 until 1920 and the second from 1920 until 1940. This would imply that the external environment affecting the lamp industry underwent a shift. Since the production which had been increasing at a steady 9.39 percent per year throughout the first period concurrently suddenly shifted downward to a rate of 2.74 percent growth at the same transition point (Exhibit 3), it may be concluded that one or more changes in the environment increased the value of the exponent b in the technological progress equation, while at the same time decreasing the rate of growth of production. It is interesting to note the comparison between the progress function for lamps as shown in Exhibit 2 with the more usual forecasting method shown in Exhibit 4. These data sub-

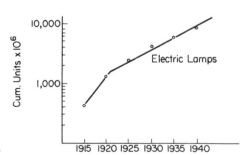

Exhibit 3. Production Versus Time for Electric Lamps.

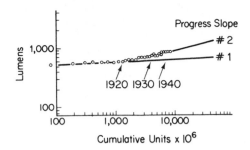

Exhibit 4. Lumen Output of Sixty-Watt Electric Lamps, Traditional Trend Forecast.

stantiate a possible claim that forecasting results with an arithmetic time axis are as good as they have been because of the substitution effect. That is, an arithmetic time axis may be substituted for a logarithmic quantity axis when production quantities increase in reasonably constant percentages with time, as in Exhibit 3.

Computer Programming Data

An additional example has been drawn from a specific experience in programming. Data for this example were made available through the help of John Cleckner, a student at The Johns Hopkins University. The example represents the rate of progress in the development of a single computer program developed solely by him under special arrangement with an area firm. A figure of merit was determined which would best represent his own aims while developing the program. This figure of merit equals the lines of output divided by the running time. The computer program dealt with an estimation of bakery goods to be sold according to the day of the week, store, and particular item.

In an analogy to the development of products, it was decided to regard the last best figure of merit (Exhibit 5) for a particular run as the figure of merit to be processed. This may or may not be the true figure of merit for that run, but it would be an accurate representation of the figure of merit of the best program available at any given point in its total development.

The results illustrate several areas of interest. The first is the applicability

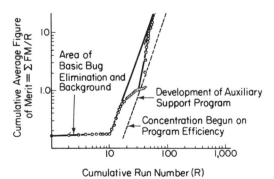

Exhibit 5. A Computer Program Development.

of the technological progress function to the area of programming. There is also the observance of different slopes corresponding to different development phases. Finally, since the work was done by one person, the analysis indicates the possible role of psychological "learning" theory in technological progress.

Additional Examples

Other cases studied demonstrated similar technological progress functions, that is, linearity on log-log paper when the technical progress parameter is plotted against production. Those cases not discussed here are civil aircraft speed, military aircraft speed, automobile horsepower, hovercraft figure of merit, and an office-machine-development figure of merit.

One could get the impression, from the data just discussed, that the trends observed were basically limited to items of hardware, but this was not found to be the case as evidenced by studies of the totally different area of agricultural efficiency. From information pertaining to the pounds of rice produced per acre in Japan over a period of twelve hundred years, three log-linear trends were noticeable [26, 27]. The two major lines were a "long-term" or ancient trend line and a modern trend line of much steeper slope. By calculating backwards, it was found that they intersected at a period of great external change—the "opening of Japan." The third line was a moderate decrease in slope, which prevailed during the period of World War II.

TECHNOLOGICAL PROGRESS FUNCTION

Thus, if we assume that initial studies are correct, technological progress functions exist and are of the same general nature as microeconomic progress functions. It is then further implied, that technological progress, as denoted by a positively increasing function is logarithmically linear with respect to a logarithmic quantity axis under constant external forces. Over periods of time during which external forces vary, the slope of the line, that is, rate of progress, may undergo discrete shifts. Such a function is denoted as,

$$(4) \qquad\qquad T_i = a \, (i)^b$$

where T_i = the value of the technological parameter,

i = the cumulative production quantity,

a = a constant associated with unit number one,

and b = the rate of progress, a variable, which is a function of the external environment.

Mathematics

A brief glance shows that where b is a constant:

$$(5) \qquad\qquad \log (T_i) = b \log (i) + \log (a)$$

and regarding the differential of (5)

$$(6) \qquad \frac{dT_i}{T_i} = b\,\frac{di}{i}.$$

Equation (6) indicates that, b being constant, the percentage change in a technical parameter is a linear function of the percentage change in cumulative production.

In addition, it should be noted that:

(7) if $\qquad i = (\text{constant})\, e^{Kt}$

(8) then $\qquad \log(i) = Kt + \log(\text{constant})$

(9) and $\qquad \dfrac{di}{i} = Kdt.$

Equation (9) represents a percentage change in production as a linear function of time, that is, what has been referred to as a "substitution effect."

Where $\qquad \dfrac{di}{i} = Kdt$

$$(10) \qquad \frac{dT_i}{T_i} = bKdt$$

which is the traditional trend forecasting relationship [28].

Rate of Progress

From analysis of background information involved in the work discussed above and current studies, the rate of progress, b, has been designated as a function of:

L, the effective size of the technical labor force;

α, an intelligence factor, average educational level;

β, a prelearning factor, average experience level;

I, the level of investment;

i, the rate of change of the level of investment;

δ, a maturation factor, durability of item compatability;

m, the anticipated rate of change of market demand, slope of sales curve;

and c, the communication or diffusion rate.

Currently, work is being done to define the exact nature of the function just described within the inherent constraints that b is greater than, or equal to, zero and never infinity and that if L, α, δ, I, or m go to zero, then b also goes to zero.

SUMMARY

This study has been concerned with the technological progress function. The study proposes that the technological progress function may be based on the cumulative production unit number on a logarithmic scale, as opposed to the arithmetic time series base used for most forecasts. In this form the technological progress function allows for environmental changes to act upon the rate of progress in a precise manner, that is, discrete changes in the slope of the function. However, it should be noted, that no less care must be taken when employing this technique in addition to or instead of existing methods, particularly with regard to physical constraints imposed upon the system.

FUTURE WORK

The technological progress function is a tool to be used with other techniques by the foresighted corporate and military development planner, who can no longer overlook the effect of technological change upon his proposals for future development. Although it is doubtful that a true "Newton's Law" for predicting technological advancement will ever be made, it is believed that the development of the technological progress function is a step toward more effective predictive methods.

Considerable work remains to be done in exploring the nature and anomalies of the technological progress function itself. There is, of course, the need for a precise definition of the rate of progress function. In addition, there are avenues of research concerning macroeconomic implications and ways in which the function might be of maximum value to technologically oriented firms.

REFERENCES

1. ASHER, H., *Cost-Quantity Relationships in the Airframe Industry,* Project Rand–291, July, 1956.
2. HIRSCHMANN, W. B., "Profit from the learning curve, *Harvard Business Review,*" Jan.–Feb., 1964.
3. KOTTLER, J. L., "The learning curve—a case history in its application," *The Journal of Industrial Engineers,* AIIE, July–Aug., 1964.
4. BUSH, R. R. and F. MOSTELLER, *Stochastic Models of Learning,* John Wiley and Sons, Inc., New York, 1955.
5. DEESE, J., *Psychology of Learning,* McGraw-Hill, New York, 1967.
6. HOVLAND, C. I., I. L. JAVIS, and H. H. KELLEY, *Communication and Persuasion,* Yale University Press, New Haven, Conn., 1953.
7. MELCHER, R. T., *Children's Motor Learning without Vision,* The Johns Hopkins University, Ph.D. Dissertation, 1934, p. 333.
8. OWEN, W., *Automotive Transportation,* The Brookings Institution, Washington, D.C., 1949.
9. STUART, F. (ed.), *Factors Affecting Determination of Market Shares in the American Automobile Industry.* Hofstra University Yearbook of Business, Series 2, Vol. 3, New York, October, 1965.
10. ABRAMOWITZ, J. G. and G. A. SHATTUCK, Jr., "The Learning Curve," IBM Report No. 31.101, 1966.
11. KNEIP J. G., The maintenance progress function, *The Journal of Industrial Engineers,* AIIE, Nov.–Dec., 1965.

12. CONWAY, R. A. and A. SCHULTZ, The manufacturing progress function. *The Journal of Industrial Engineers,* AIIE, Jan.–Feb., 1959.

13. SALVESON, M. E., Long range planning in technical industries, *The Journal of Industrial Engineers,* AIIE, Sept.–Oct., 1959.

14. RUSSELL, J. H., "Predicting Progress Function Deviations," IBM Technical Report TR 22.446, August 28, 1967.

15. HULL, C. L., C. I. HOVLAND, R. T. ROSS, M. HALL, D. T. PERKINS, and F. B. FITCH, *Mathematico–Deductive Theory of Rote Learning,* Yale University Press, New Haven, Conn., 1940.

16. LOGAN, F. A. *Incentive,* Yale University Press, New Haven, Conn., 1960.

17. CETRON, M. J., Forecasting technology, *Science and Technology,* Sept. 1967.

18. SCHMOOKLER, J., *Invention and Economic Growth,* Harvard University Press, Cambridge, Mass., 1966.

19. VILLERS, R., *Research and Development: Planning and Control,* Rautenstrauch and Villers, New York, 1964.

20. BURCKHARDT, R. E., Major, USAF, Cost Estimating Relationships for Turbojet Engines, Masters Thesis, GSM/SM/65–3, December, 1965.

21. GOULD, R. P., Major, USAF, Turbojet Engine Procurement Cost Estimating Relationships, Masters Thesis, GSM/SM/65–10, August, 1965.

22. TURKOWITZ, N., letter to author, November 5, 1968.

23. *Aviation Facts and Figures-1958,* American Aviation Publication, 1958.

24. BRIGHT, J. R., *Automation and Management,* Harvard University, Boston, Mass., 1958.

25. BRIGHT, A., *The Electric Lamp Industry,* McGraw-Hill, New York, 1949.

26. KAHN, H., and A. J. WIENER, *The Year 2000,* Macmillan Co., New York, 1967.

27. LOCKWOOD, W. W., *The Economic Development of Japan,* Princeton University Press, Princeton, New Jersey, 1954.

28. LENZ, R. C., Jr., *Technological Forecasting,* Air Force Systems Command, Wright Patterson Air Force Base, Ohio, AD 408 085, ASD*TDR*62–414, June, 1962.

BIBLIOGRAPHY

Books

ALMON, C. JR., *The American Economy of 1975,* Harper and Row, New York, 1966.

Automation and Technological Change, The American Assembly at Columbia University, Prentice-Hall, Englewood Cliffs, New Jersey, 1962.

AYRES, R., *Technological Forecasting and Long Range Planning,* McGraw-Hill, New York, 1969.

BRIGHT, J. R., *Research, Development and Technological Innovation,* Richard D. Irwin, Homewood, Ill., 1964.

BRIGHT, J. R. (ed.), *Technological Forecasting for Industry and Government,* Prentice-Hall, Englewood Cliffs, New Jersey, 1968.

BROWN, MURRAY, *On the Theory and Measurement of Technological Change,* Cambridge University Press, Cambridge, England, 1966.

HITCH, C. J. and ROLAND N. MOKEAN, *The Economics of Defense in the Nuclear Age,* Harvard University Press, Cambridge, Mass., 1967.

Human Relations in Industrial Research Management, Robert Teviot Livingston and Stanley H. Milberg (eds.), Columbia University Press, New York, 1957.

JANTSCH, E., *Technological Forecasting in Perspective,* OECD, Paris, 1967.

MANSFIELD, E., *The Economics of Technological Change,* W. W. Morton and Co., New York, 1968.

Mansfield, E., *Industrial Research and Technological Innovation*, W. W. Norton and Co., New York, 1968.

Quinn, J. B., *Yardsticks for Industrial Research*, Ronald Press, New York, 1959.

Schon, D. A., *Technology and Change*, Delacorte Press, New York, 1967.

Serven–Schreiber, J.-J., *The American Challenge*, Atheneum, New York, 1968.

Technological Innovation and Society, Dean Morse and Aaron W. Warner (eds.), Columbia University Press, New York, 1966.

Articles

Bright, J. R., Can we forecast technology?, *Industrial Research*, March 5, 1968.

Cetron, M. J., R. J. Happel, W. C. Hodgson, W. A. McKenney, and T. I. Monahan, A proposal for a navy technological forecast, Naval Material Command, Washington, D.C. Part I: Summary Report, AD 659199; Part II: Back Up Report, AD 659 200, May, 1, 1966.

Cetron, M. J. and A. L. Weiser, Technological change, technological forecasting, and planning R&D: A View From the R&D manager's desk. *The George Washington Law Review*. Vol: 36, No. 5, July, 1968.

Darracott, H. T., M. J. Cetron, *et. al.*, Report on technological forecasting, Joint Army Material Command/Naval Material Command/Air Force Systems Command. Washington, D.C., AD 664 165, June 30, 1967.

Duranton, R. A., Quelques remarques sur les courbes d'accoutumance, Internal IBM Report, May 13, 1965.

Mahanti, B., Progress report/manufacturing progress function report, Internal IBM Paper (German Office), April 9, 1964.

Technology and World Trade, U.S. Department of Commerce, NBS Misc. Pub. 284 (Symposium—11/16, 17/66).

The Dynamics of Automobile Demand, General Motors Corp., New York, 1939.

The growth force that can't be overlooked, *Business Week*, McGraw-Hill, New York, August 6, 1960.

ACKNOWLEDGMENTS

The author offers special thanks to John Cleckner, Class of 1970, The Johns Hopkins University, for his suggestions and help in preparing the data on computer programming. The author is also indebted to Lisa Geiser, Class of 1972, Goucher College, and Nancy Smith, Class of 1972, Goucher College, for their help in developing additional data. In addition. Norman Turkowitz of Pratt and Whitney must be noted for supplying the supporting information for the aircraft engine study for which the author is also very grateful.

Trend Extrapolation

JOSEPH P. MARTINO

Although it seems incongruous to describe any methodology as classical when referring to technological forecasting, one is tempted to apply this descriptor to trend extrapolation. Certainly trend extrapolation has been the most widely displayed and most published technique. However, a rigorous exposition of technology trend extrapolation has not yet been made.

Here one of the leading students of the field gives his views on the selection of trend parameters, the philosophy or rationale for using trend extrapolation, and interpretation of the forecast. He discusses special kinds of trends including those with limits and those correlated with other trends and introduces the notion of qualitative trends. His use of significant new examples of technological trends adds to the importance and relevance of this chapter. The curves in charts have been fitted by the least squares method.

INTRODUCTION

The usual behavior of the growth of functional capability, exhibited by a single technical approach, is the classical S curve. The growth of functional capability is slow at first, then rises rapidly as the initial difficulties are cleared away, and finally saturates as the practical upper limit of that approach is reached. As the growth of functional capability from this technical approach slows down, the technical approach is superseded by a new technical approach which, in turn, goes through the same slow start, rapid growth, and saturation, producing an S curve at a higher overall level.

It appears, then, that if the forecaster restricts himself to forecasting the growth of the current technical approach, he will be unable to forecast beyond its practical upper limit. His forecast, as well as the technical approach itself, will likely be rendered obsolete by a successor technical approach. Somehow, he must be able to forecast beyond the upper limit of the currently available technical approach.

But, it may be argued, how can he do that? How can he forecast beyond the limits of the current technical approach without knowing what the successor technical approach will be? To forecast beyond the limits of the current technical approach is equivalent to inventing the successor technical approach.

The whole notion of trend extrapolation is based on a rejection of this attitude. The central concept is one of continuity. If, in some area of technology, there has been a continuous progression of technical approaches, each one surpassing the limitations of the previous one, it is not unreasonable to expect the rate of innovation to continue. If the succession of technical approaches has brought about some regular rate of improvement, it is not unreasonable to expect this rate to continue. To argue the contrary, in fact, is to say that the present is a point of discontinuity; that despite the fact that there has been more or less regular innovation in the past, this innovation is coming to a halt at the present level of functional capability.

In making trend extrapolations, we assume that the present is a point of continuity and that the future behavior of the technology area will resemble past behavior. There are two cases, of course, where this approach is unreasonable. The first is the case where there is some known natural limit, such as achievement of 100 percent efficiency, which cannot be surpassed. Obviously, one should not extend a trend past some known natural limit. The other case is that in which it is known that certain of the conditions which produced the trend in the past are going to be changed by some outside agency. In this case, the present really is a point of discontinuity and projecting the trend will produce a misleading forecast.

EXAMPLES OF HISTORICAL TRENDS

We will next look at some historical examples to illustrate the growth of technology in some area through the introduction of a succession of technical approaches to achieving the same functional capability. Each of the examples covers a time period of several decades and involves a cumulative advance of several orders of magnitude.

Productivity of Transport Aircraft

Neither speed nor size alone are adequate measures of the functional capability of a transport aircraft. The product of the two, however, is such a measure since it permits the direct comparison of aircraft of differing sizes and speeds. When a number of aircraft have been designed for roughly the same military or commercial purposes, the productivity of successive aircraft, as measured by either ton-miles per hour or passenger-miles per hour can be used as a measure of technological advance.

Exhibit 1 shows the growth in ton-miles per hour of United States civil and military transport aircraft since 1935. The earliest point on the plot represents the DC-3, the latest point the 747F Jumbojet. Over a period of nearly thirty-five years there has been a remarkably steady growth in transport productivity. This is a period which has seen some radical changes in aircraft design. There has been a shift from piston engines to jets and a shift from manual operation to a virtual take-over by electronics. Many functions, such as communications and navigation, have been dramatically altered by electronics. Despite these innovations, however, there have been no radical shifts in the growth trend.

Exhibit 2 shows the growth in passenger-miles per hour of United States civil and military transport aircraft since 1926. As with the case of ton-mile productivity, there has been a steady growth during this forty year period. This growth has not been shifted, either faster or slower, by any of the technological advances cited above. Nor has it been affected by the advent of coach and economy class seating on commercial airlines (less space per passenger than in

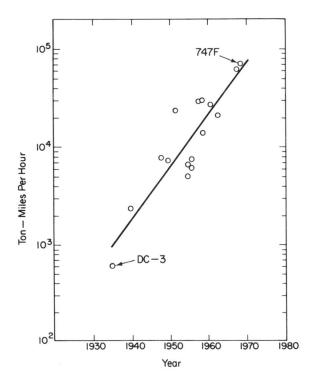

Exhibit 1. Productivity, in Ton-Miles per Hour, of Civil and Military Transport Aircraft.

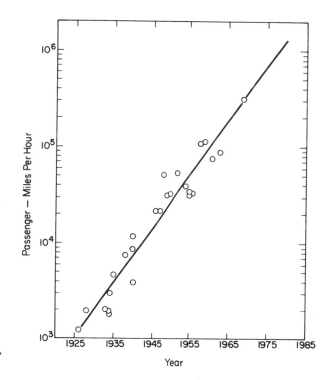

Exhibit 2. Productivity, Passenger-Miles per Hour, of Civil and Military Transport Aircraft.

the case of first class seating). One might even conclude that the economic changes in the airline industry (reduction in fares and a broadening of the clientele) were required to fill the larger aircraft which technology was making available.

One other point deserves mention. The productivity plots in the two figures are obtained by multiplying top speed by capacity. Since the late 1950s, the top speed of both civil and military transports has been limited to about 600 miles per hour. At moderately higher speeds, drag increases to the point that operation becomes uneconomical. To have gone to much higher speeds would have required a level of technology not yet fully proven in military aircraft and, hence, unlikely to be economical. Thus for economical reasons the speed of transport aircraft was limited to a level much lower than would have been possible from a technological standpoint. Nevertheless, productivity continued to increase at about the same rate as it had in the past when the possible top speed was constrained only by technology, not by economics. During the period when top speed was limited, aircraft size increased sufficiently to make up for the static top speed and thus continued a previous trend in productivity increase.

Illumination

Exhibit 3 shows the efficiency of light sources since 1850. Here we have an especially good example of a succession of different technical approaches to

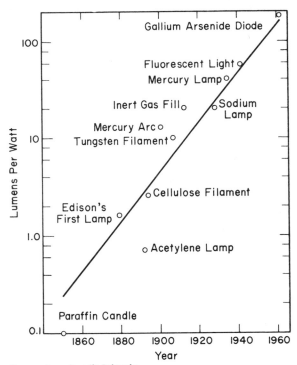

Source: Encyclopedia Brittanica.

Exhibit 3. Efficiency of Illumination Sources.

the same functional capability. Most of the technical approaches shown differed radically from their predecessors. Yet the rate of progress shows remarkable constancy over a period of a century. The plot shown was computed on the basis of all the sources except the gallium arsenide diode. The fact that the gallium arsenide diode falls so close to the projected trend is extremely encouraging. However, it should not be taken as too much encouragement. The gallium arsenide diode emits radiation in the infrared, so it is not really an illumination source. Nevertheless, it does convert electricity to what is essentially optical radiation, and its efficiency does fall near the projection of the preceding trend.

Fighter Aircraft Weight

Exhibit 4 shows the gross take off weight of United States single-place fighter aircraft since 1918. Here again we see a well-behaved trend which continued through a succession of radical changes in aircraft technology from the fabric covered spruce-wood frame of 1918 to the aluminum structure of the modern jet fighter. Note that the points for the F–4 and the F–111 were not included in the computation of the trend curve shown. They are two-place aircraft and, strictly speaking, do not belong with the rest. The reason for

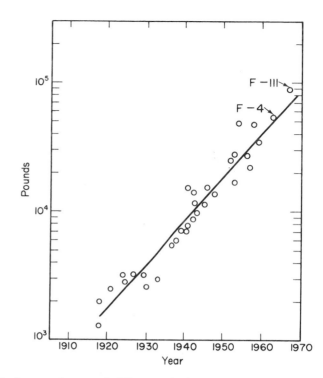

Exhibit 4. Gross Take-off Weight of United States Single-Place Fighter Aircraft. (The labeled points show the F-4 and the F-111, which are two-place aircraft.)

including them is to show that they do fall close to the trend. The conclusion one can draw from this is that the primary determinants of weight in a modern fighter aircraft are the electronics, the engines, and the structure needed to carry out the mission, not the crew size.

Speed of United States Combat Aircraft

Exhibit 5 shows the top speed of United States combat aircraft, both fighters and bombers, since 1909. The rate of progress of the speed trend seems to be quite steady, regardless of such innovations as the closed cockpit, the monoplane, the all metal airframe, and the jet engine. This last is of particular significance. The jet engine, introduced in 1944, did not bring about a jump in aircraft speed. On the contrary, the jet engine was introduced at the time it was needed to keep the previous trend going. Also of interest is that bombers and fighters are not distinguishable by their speed. Successive bombers usually had top speeds in excess of the speeds of the preceding fighter design. We are going to need the regression equation for this plot later, so it will be given here:

$$Y = -118.30568 + .06404T$$

where Y is the natural logarithm of speed in miles per hour and T is calendar time.

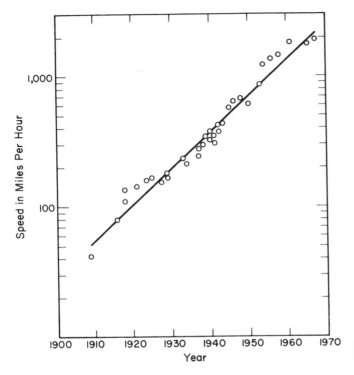

Exhibit 5. Top speed of United States Combat Aircraft.

Rockets

The next three examples illustrate progress in rocket engines. Exhibit 6 shows the maximum thrust available from United States built liquid propellant rocket engines since 1942. The growth of thrust level covered three orders of magnitude in about two decades, showing a fairly steady growth pattern during this period. Achievement of these thrust levels required major advances in the technology embodied in the engines themselves, such as methods for cooling; advances in accessories, such as pumps and turbines for auxiliary power; and theoretical advances in the level of understanding of combustion, so that combustion instabilities could be eliminated through better design.

Exhibit 7 shows the maximum thrust available from solid propellant rocket engines. While the large solid propellant rocket engines got a later start then did their liquid propellant competitors, they caught up quite rapidly. This advance involved not only improvements in fuels, such as high energy additives, but also involved such other features as liquid injection in the rocket in order to obtain thrust vector control since solid propellant engines cannot be mounted on a gimbal as can liquid propellant engines.

Exhibit 8 shows the maximum total impulse available from solid propellant engines, where total impulse is the product of thrust and burning time. Not only did the thrust level of solid rockets increase but the amount of fuel

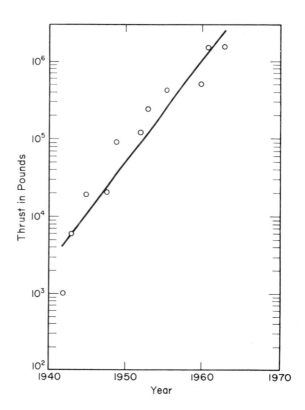

Exhibit 6. Maximum Thrust from United States Built Liquid Propellant Rocket Engines.

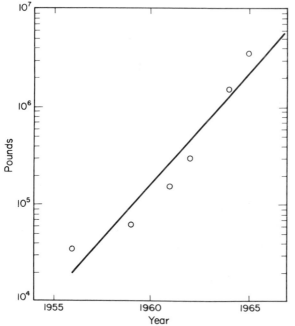

Exhibit 7. Maximum Thrust from United States Built Solid Propellant Rocket Engines.

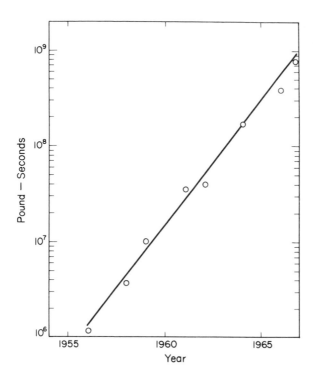

Exhibit 8. Maximum Total Impulse of United States Built Solid Propellant Rocket Engines.

in the engine increased also to give equal or greater burning time at the higher thrust levels. The increase in fuel content was based largely on the development of methods for casting large solid propellant grains so that they would be free of cracks and cavities. When the burning surface of the propellant grain reaches a crack or cavity, the available burning surface is suddenly increased, which leads to an explosive increase in pressure, temperature, and burning rate. At best this will cause unstable burning and can even lead to destruction of the engine. Thus advances in solid rocket impulse arose as much from advances in manufacturing technology as from technology embodied in the rocket engines themselves.

Installed Technological Horsepower

Exhibit 9 shows installed technological horsepower in the United States since 1840. This is defined to include all power produced by other than animal sources and includes both fixed and mobile power sources. The figures include electrical generators but not motors to avoid double counting the same kilowatt. Plotted on the same figure are United States population and horsepower per capita. The plot shows that not only has the total technological horsepower available to the nation grown rather steadily but also the average horsepower available to each individual.

One interesting sidelight of this exhibit is that this is one of the few technological trends which appears to be affected by economic conditions. The fifty-four-year Kondratieff wave of the business cycle shows up clearly in the plot for installed horsepower. It is evident that the rate of installation of technological horsepower was influenced by business conditions. However, on a percentage basis the influence of the business cycle was not very large since the peaks of the swings never deviate much from the trend line.

In each of these examples we have traced a parameter representing some functional capability through several successive technical approaches. The plots show that in each of the examples progress continued at what was essentially a steady pace, despite the introduction of radically different technical approaches. Each new technical approach allowed the technology to reach a higher level of functional capability. Nevertheless, the introduction of these approaches seemed to follow and extend previous trends rather than cause sharp breaks with the past. Thus even though the present technical approach used in some area of technology is approaching a limit, we can forecast beyond that limit by extending previous trends. We can have more confidence in our forecast if the technological area in question has a history of steady growth, a history which includes several instances of a radical change in technical approach.

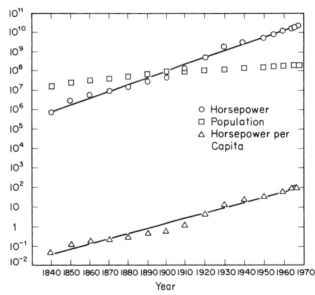

Exhibit 9. Installed Technological Horsepower in the United States, Population of the United States, and Installed Technological Horsepower per Capita.

Source: Mr. John Waring.

In each of the exhibits shown above, we plotted a trend line. This trend line was based on obtaining a regression fit between the natural logarithm of the level of functional capability and calendar time. A straight-line fit of this kind indicates that the technology in question grew at a constant percentage per unit time, or, expressed in an equivalent fashion, it grew exponentially.

Empirically, it turns out that many technologies do in fact grow exponentially. This idea seems quite acceptable today and usually evinces no particular surprise. We see exponential growth going on all around us and do not find it too hard to accept that technology should behave in a similar manner. However, the idea of the exponential growth of technology is not a very old one.

It appears that the earliest expression of the idea is due to Engels, one of the fathers of Marxism. Petrov and Potemkin, writing in the Russian journal *Novy Mir* (New World) for June, 1968, attribute to him the first formulation of the exponential law of growth in his controversy with the ideas of Malthus. Thus the idea of exponential growth of technology has a respectable intellectual history going back at least a century. However, there has been no equivalent understanding of why this should be so. Exponential growth was a brute fact; it simply was so, with or without reason. However, in the June, 1969 issue of *Technological Forecasting,* Robert C. Seamans, Jr. presents the first really well-developed explanation of exponential growth. It is not a complete explanation, and his theoretical model has a number of shortcomings. But it definitely appears to explain why technology will tend to grow exponentially in a competitive situation so long as there are no fundamental barriers inhibiting growth.

It should not be assumed, however, that all technological growth is exponential. Exhibit 10 shows an example of a technological trend which is linear rather than exponential. This figure plots the ratio of wingspan to fuselage length for bombers, fighters, and transport aircraft over four decades. For each of the three classes of aircraft, there is a fairly well-behaved linear

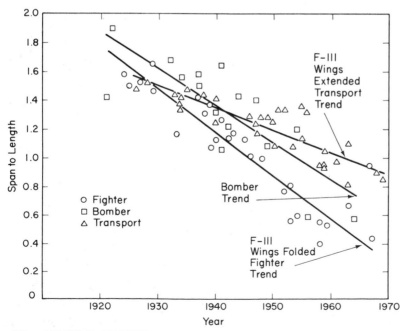

Exhibit 10. Ratio of Wing-Span to Length for United States Built Transport, Bomber, and Fighter Aircraft.

trend. It is apparent that bombers tend to lag fighters and that transports actually cut across both. This is a reflection of the differing requirements for streamlining in the three classes of aircraft. It is interesting to note that the F–111, with wings folded, falls near the fighter trend. The same aircraft with wings extended falls on the transport trend. We can conclude that the advent of swing-wing aircraft technology will render these curves meaningless in the future.

SELECTION OF PARAMETERS

Having satisfied ourselves that a number of technologies do exhibit regular trends in their growth despite changes in the technical approach used to achieve a given level of functional capability, we can now turn to the question of choosing parameters to be projected. Clearly, choice of an inappropriate parameter will lead to a misleading or erroneous forecast. To avoid this problem the parameter chosen must meet the following conditions.

First, it must be an *operational* parameter. That is, it must be one which can be measured in objectively meaningful quantitative terms. Parameters which involve a subjective judgment about the level of performance of some device do not meet this requirement. For each of the successive devices considered to be elements of the technology being forecast, it must be possible to measure the parameter, either in terms of the physical characteristics of the device itself or in terms of characteristics of its operation (for example, flow per unit time, and so on).

Second, the parameter must actually *represent the level of functional capability* of the technology being forecast. There are basically two ways in which a parameter can fail to do this. One is that it fails to tell the full story. In the example of transport aircraft given above, neither speed nor capacity alone would give a complete picture of the level achieved by transport aircraft technology. At many points in time there were small aircraft available which were faster than some larger aircraft of greater actual productivity. Thus speed alone would be a misleading parameter. Capacity alone would also be misleading. Some of the transport aircraft included in the analysis were military transports of considerable size (and therefore productivity) which were too slow to be commercially competitive. Thus both factors must be included in the parameter describing the state of the art. The second way in which a parameter can fail this requirement is that it be subject to a design trade-off. If, for instance, the designer has considerable freedom to trade off some of one parameter in order to get more of another, neither of the parameters by themselves will serve to represent the state of the art. A good example is the gain of electronic amplifiers. For any given level of the state of the art of amplifying devices, the gain of an amplifier can, in general, be increased by decreasing the bandwidth of the amplifier. Likewise, the bandwidth can be increased at the expense of gain. The electronics designer can make a trade-off between these two variables in terms of the requirements of his design problem. Neither alone is a representative measure of the level of functional capability of amplifying devices. However, the gain-bandwidth product is a good measure of the capability of an amplifying device, since over a wide range the trade-off between gain and bandwidth is linear. For other technologies where the trade-off is not linear, a more complicated function that the product may be needed. In any case, however, the parameter to be forecast must represent a composite of all the device parameters which can be involved in a trade-off.

Third, the parameter chosen for forecasting must be *capable of application to differing technical approaches*. In the example of the illumination trend given above, all the devices could be compared on a single measure of efficiency. The rate of consumption of paraffin or acetylene can be converted to an equivalent of watts electrical consumption and an efficiency determined for all devices in the entire class. If, however, the parameter chosen is one which is peculiar to a single technical approach, it will not be possible to compare devices based on differing technical approaches. For instance, it is not possible to define a meaningful measure of total impulse for a liquid propellant rocket engine. An engine can be connected to a fuel tank of any desired size without making any changes in the engine.

Fourth, the parameter chosen must be one for which there is *adequate historical data available*. While the theoretical considerations given above may seem restrictive, in actual practice this consideration may prove to be the most demanding. Especially if statistical confidence limits are to be applied to the forecast, there simply is no substitute for large quantities of data. If the forecaster is faced with a choice between two parameters, for one of which there is significantly more data available than for the other, he might be well advised to choose the former even if it means some sacrifice in terms of the theoretical considerations given above. This judgment must be made in each case.

Fifth, *the historical data points must be selected consistently*. They should all represent the same development status, such as first commercial introduction, first production article, first demonstration of laboratory feasibility, or some other explicit point in the life cycle of the technology. If some of the data points represent one stage of development and some another, the trend fitted to them can be badly distorted, especially if, say, the early points are concentrated at one stage in the development cycle and the late points at a much different stage. This advice, of course, has to be tempered with the reality of data availability. Nevertheless, to the extent that he can, the forecaster should select the data points in a consistent manner.

INTERPRETATION OF FORECAST

Having selected a parameter, gathered historical data, and fitted a trend to the data, the forecaster is now in a position to make the forecast. The forecast, of course, is simply the projection of the trend which he has fitted to the data. However, the fitting process is only a small part of making the forecast. Once the trend projection is available, it must be interpreted.

Initially, the forecaster must have the courage of his convictions. He has taken historical data, determined a regularity in the past pattern of innovation, and projected this into the future. Undoubtedly, the projection will show a forecast of level of functional capability which is well beyond the maximum feasible limit of the current technical approach. If this limit is set by some natural limit on the entire technology area, the forecaster should not even have projected a trend. Since he did project the trend, he must have determined that there was no natural upper limit. He must then interpret the forecast as meaning that some new technical approach will be invented which will surpass the limits of the current technical approach and permit reaching the level of functional capability given by the forecast. In particular, he must avoid the temptation to "revise" his forecast downward in view of the approaching upper limit to the current technical approach. If he does this, attempting in a misguided way to make his forecast more credible, he will have failed completely in his function, which is to warn his clients that something new is coming.

Next, the forecaster must recognize that the historical data points did not all lie on the trend he has fitted, hence he has no right to suppose they will all lie on the trend in the future. He must expect at least as much scatter in the future as he found in the past. He must therefore give his clients warning of this fact. One of the best ways of so doing is to put statistical confidence limits on the trend. This is particularly easy if regression methods have been used to fit the trend since techniques are readily available for deriving confidence limits. Deriving confidence limits in this manner is far better than drawing in some freehand limits on the plot, and every forecaster using trend extrapolation should familiarize himself with the appropriate statistical techniques.

Finally the forecaster must make every attempt to get behind the trend and identify the major factors which interacted to produce the trend. If it is clear that one or more of these factors are going to change radically, he is obliged to warn his clients of this fact, even though he cannot predict the precise impact. He ought, at the very least, to be able to give a qualitative estimate as to whether the change will increase or decrease the rate of growth and, therefore, whether his trend is likely to be an upper limit or a lower limit on the range of possible growth.

Having discussed the forecasting of trends, the choice of parameters to be forecast, and the interpretation of the forecast once obtained, we will now take up some special cases of trend extrapolation. These are a trend with a limit, the correlation of trends, and a qualitative (as opposed to a quantitative) trend.

TRENDS WITH LIMIT

In some cases it is clearly unreasonable to project a trend indefinitely over the time period of the forecast. The forecaster will be aware of some natural limit to growth which will cause the growth to slow down and level off. In this section we will take up the case of what to do with a trend which is approaching such a limit. The basic answer is to carry out a transformation on the variable being projected, so that for the transformed variable the limit is at infinity.

The simplest transformation to use in cases such as this is a logarithmic transformation. We will let y be the parameter we wish to project that is approaching a limit. We will let A represent a reference value of y which is to be used in the transformation. We will let L be the limiting value. Finally, we will let Y be the transformed parameter.

In the case that L is an upper limit, we used the transformation:

$$Y = ln \frac{A}{L - y}$$

In the case where L is a lower limit, we use the transformation:

$$Y = ln \frac{y - L}{A}$$

Although the natural logarithm is shown in the expressions for the transformation, the common logarithm (base 10) can be used also. The forecaster should use the one which appears the most natural or convenient under the circumstances.

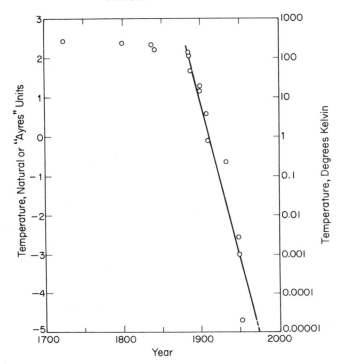

Exhibit 11. Lowest Temperature Achieved in the Laboratory by Artificial Means.

To illustrate the use of this transformation, we will consider the lowest temperatures achieved in the laboratory through artificial means. Exhibit 11 shows the data plotted against time. The first point is the temperature reached by Fahrenheit, using a freezing mixture of ice and salt. The data points through the 1830s represent improved freezing mixtures. The data points between 1850 and 1910 represent temperatures achieved through Joule-Thompson cooling, and the points since 1930 represent use of magnetic cooling. Clearly, the points resulting from freezing mixtures represent a different regime from those of the other methods. The transformation is shown in terms of a lower limit of $L = 0°$ Kelvin and $A = 1°$ Kelvin. The left-hand scale shows the transformed units, the right-hand scale is in degrees Kelvin. Despite the fact that the level of functional capability is approaching a limit, we have eliminated the difficulty by transforming the limit to minus infinity and have fitted a trend which appears to explain the data quite well. Extrapolating the regression line, it appears that a temperature of 10 microdegrees Kelvin or $-5°$ in natural units will be reached about 1975.

TREND CORRELATION

There are many cases where one technology appears to be a precursor to another. If, in fact, the precursor technology actually pioneers some advances which are later adopted by a successor technology, it would appear to be possible to forecast the successor on the basis of the achievements of the precursor and the observed lag time. The major difficulty in using this approach is that of defining a lag time. Usually there are few, if any, cases where devices of the precursor and follower technologies exhibit the same level of functional capability. Thus it is not possible to determine the lag in years between achievement of the same level. Likewise, there are few, if any, cases where devices of both precursor and follower technologies are introduced in the same year. Thus it is not possible to determine the year-by-year lag in capability. What is needed is some means of interpolating between the values of either precursor or follower so that the lag time between achievement of identical levels can be obtained.

It appears to be most logical to do the interpolation on the precursor. This can be done by fitting a trend curve to the data for the precursor technology. If the trend is linear we have an equation of the form

$$Y = A + BT$$

where T is the time of achievement of precursor capability. If we now substitute in this equation the level of functional capability achieved by a specific device of the follower technology, we can solve for T, the year in which the precursor "would have" had the same level if a device had been introduced with that specific level of capability. We can now subtract this year from the year in which the follower device with that level of capability was actually introduced. This gives us a lag for each follower device. We can now fit a trend to the lag data by regressing lag on the year of introduction of the follower device. Once this lag is obtained, we proceed as follows.

We want to obtain an equation for the follower device of the form:

$$Y = A + B(T - D)$$

where D is the lag for the follower. If the lag is not constant but is itself a function of time, we may have:

$$D = a + bT$$

Substituting this in the prior equation, we obtain:

$$Y = A - aB + B(1 - b)T$$

Here the A and B are from the trend line for the precursor, the a and b are from the expression for the lag of the follower, and T is the year the follower was introduced.

To illustrate how this works, consider Exhibit 12, which shows the lag between the top speed of combat aircraft and the top speed of transport aircraft. We had already obtained the trend line for combat aircraft as:

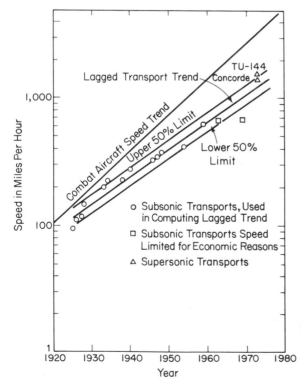

Exhibit 12. Correlated Lag between Combat Aircraft Speed and Maximum Transport Aircraft Speed.

$$Y = -118.30568 + .06404T$$

Taking the values for top speed and year of introduction of a series of pace-setting transport aircraft (for example, the fastest available at time of introduction), we obtain the lag equation:

$$D = -443.79507 + .23225T$$

Substituting this in the previous equation, we obtain the lagged transport speed trend equation:

$$Y = -89.88504 + .04917T$$

The exhibit shows the speed trend of combat aircraft and the lagged trend for those transport aircraft which were not subject to an economic upper limit on their speed. The exhibit also shows two subsonic transports, introduced in 1963 and 1969, which were designed for a lower speed than they would otherwise have achieved. As expected, they fall below the trend line. In addition, two supersonic transports are shown—The Anglo-French *Concorde* and the Soviet TU–144—both scheduled for commercial introduction in 1972. Both fall in the neighborhood of the fifty percent upper confidence limit shown for the transports.

We might conclude that the achievement dates for transport speed are therefore optimistic. However, if we look more closely at the exhibit, we see that the combat aircraft whose speeds correspond most closely to those of the SST's were several years ahead of the combat speed trend. Therefore it is not too surprising to find the SST also ahead of the lagged trend. Thus we can have somewhat more confidence in the forecast of transport aircraft speed, basing it on the demonstrated achievements of combat aircraft.

QUALITATIVE TRENDS

In the section on Choice of Parameters, it was stated that a parameter suitable for projection had to be one which could be stated in quantitative terms. Sometimes, however, this isn't possible. Not all trends can be described quantitatively. This does not mean they are any less real. It does mean, however, that it is harder to define them and that forecasts of them will of necessity be less precise.

Exhibit 13 illustrates a qualitative trend. This figure is extracted from an article by William E. Lamar which appeared in the July, 1969 issue of *Astronautics and Aeronautics,* a publication of the American Institute of Aeronautics and Astronautics.

Since the time of the Wright brothers, aircraft have been designed so that more and more of the total aircraft can be adjusted, varied, or otherwise moved in flight. The wing-warping capability of the original Wright Flyer was only the first step in a long sequence of such capabilities. Attempts to quantify this

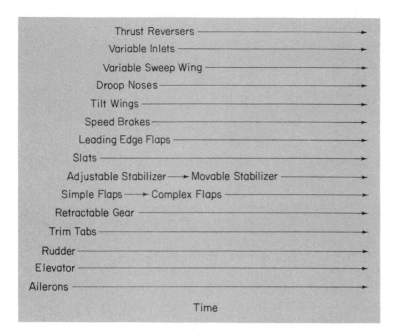

Thrust Reversers

Variable Inlets

Variable Sweep Wing

Droop Noses

Tilt Wings

Speed Brakes

Leading Edge Flaps

Slats

Adjustable Stabilizer → Movable Stabilizer

Simple Flaps → Complex Flaps

Retractable Gear

Trim Tabs

Rudder

Elevator

Ailerons

Time

Exhibit 13. Qualitative Trend of Aircraft Components Movable in Flight.

trend (for example, in terms of the percentage of empty weight which can be varied in flight) may not be completely successful since one movable portion may later be broken down into several movable portions, as illustrated in the figure by the step from simple flaps to complex flaps.

What can we do with a qualitative trend such as this? As with the other trends we have been considering, it is probably safe to extrapolate it. Aircraft will continue to be designed with more and more parts which can be moved or adjusted in flight. However, it does not appear to be possible to forecast just which parts will next be designed to be adjustable. In other areas where qualitative trends might be discernible, the same limitation is likely to apply. It will be possible to extrapolate the trend, in the sense that we can say the phenomenon represented by the specific elements in the trend will continue and increase; but it will probably not be possible to forecast the specific manner in which this increase will take place.

SUMMARY

Any given technical approach to achieving some functional capability will have an inherent upper limit. We have shown that in many cases it will be possible to forecast beyond this upper limit by trend extrapolation. However, the forecaster does not and need not invent the means by which the forecast will be fulfilled. Furthermore, since trend extrapolation is still largely an art rather than a science there is no rigorously logical method for selecting the parameters to be extrapolated.

It almost goes without saying that more research is needed in this field. The greatest need is for a more adequate data base from which many trends could be derived. With such a data base, it might be possible to identify and classify various technologies as to the types of trends they produce. We might be able to develop a better understanding of why certain types show exponential trends and why others exhibit other trends. In any case, we need additional research to get us beyond the "brute fact" stage of simply recognizing that a certain technology does in fact exhibit a certain kind of trend.

The Relevance Tree Method for
Planning Basic Research

THEODORE J. GORDON
M. J. RAFFENSPERGER

Planning basic research is an elusive and troublesome task, since the very nature of the activities involves a high degree of technical uncertainty as to both technical and economic success. Even at best, there will be an extremely distant payback horizon. The relevance tree method described here[1] suggests that the research planners begin first by considering the organization's goals. These then are viewed as leading to subordinate missions, and an integrated flow of activities, with each succeeding technical goal dependent upon the completion of one or more predecessors. By establishing a linkage between activities, one can consider and weigh the impact of completing each activity on the remainder of the network. In this way, different basic research activities may be evaluated in the context of eventual contribution to organizational goals.

The authors wish to thank Dr. H. L. Wolbers and S. Enzer for their help in preparing this paper.

[1] Any views expressed in this paper are those of the authors. They should not be interpreted as necessarily reflecting the attitudes of the McDonnell Douglas Corporation or the National Aeronautics and Space Administration. This paper was presented to the 2nd Annual Technology and Management Conference on Technological Forecasting, March 18-22, 1968, Washington, D.C.

Scientific tradition holds that it is essential to permit individual researchers almost complete freedom in their selection of research projects. As a result, structured research planning has not played a major role in determining the course of the sciences. By "structured research planning" we mean the methodical establishment of goals and the identification of research routines which apparently accord priority to highest value goals while minimizing the cost and time of their attainment. With increasing competition for limited fiscal and intellectual resources this kind of planning may become more necessary. Let us cite three specific examples: First, the individual researcher is often faced with the question of how best to spend his time in pursuing a particular line of investigation. He must allocate his personal resources of time and talent to work which appears to hold the promise of highest payoff. Second, on a larger scale, the competition for resources also raises the question of priority among alternative government-sponsored research projects. Individuals proposing equally alluring projects sometimes seek the same funds; government planners need some sort of objective criteria (hopefully based on a rational strategy) in order to judge the proposals. Furthermore, government planners face the problem of justifying their request for "big science" resources; a planning strategy which defines the use and expected output of the requested resources can help in the design and promotion of their concept. Finally, there is a growing sentiment that at least some research should be directed toward the solution of current and anticipated societal problems, to find means of circumventing or minimizing the impact of potentially dangerous problems of civilization. These needs, the choice of individual research, allocation of national resources among competing projects, and selecting research of societal importance, suggest that structured research planning might be valuable.

The need for initiating a cohesive policy for Federal scientific research was recognized by Congress in 1967 when the House Committee on Science and Astronautics undertook a critical review of the operations and functions of the National Science Foundation.[2] This investigation indicated that Congress was concerned about the directions into which science was carrying us and the mechanisms of its transport. The Committee made it clear that in the future Congress was likely to pursue questions related to the establishment "of a broad strategy for science in public affairs, . . . a framework for considering the effect of science and technology in regard to the national economy . . . priorities for choice among different possibilities for Federal basic research . . . (identification of issues) before they reach the crisis stage. . . ."[3] They concluded that focusing on "areas of appropriate research and education could be a major factor in maintaining the stability of a civilization which is today seriously threatened by the surfeit and concentration of people and their problems."[4]

2 *The National Science Foundation; Its Present and Future,* a report to the Committee on Science and Astronautics, U.S. House of Representatives, February 1, 1967.
3 *The National Science Foundation,* p. 99.
4 *The National Science Foundation,* xiv.

While the examples given in the preceding paragraphs are related specifically to Federal sponsorship of basic research, it should be made clear that the authors of this paper deal with the question of strategy in planning basic research not only on the level of massive government projects but on the basis of individual "little science" as well. Most of the systematic issues are similar: establishment of the goals of research, determination of the most effective ways of attaining those goals, and defining programs of resource allocations which appear to be most likely to promulgate the intended results.

The desirability of planning scientific research is challenged by those who hold that science, intrinsically, is not amenable to forecasting or planning. They argue that the uses made of the products of research are so obscure at the time research is performed that planning cannot possibly be effective. Using Kistiakowsky's example, who could have foreseen at the time of their research, "that the Curies would make a major contribution to the cure of cancer. Similarly, no man today is wise enough to know from what field will come a critical discovery that, directly or indirectly, will solve the problem of the control of insects."[5] Sometimes the research which seems to be the least promising proves to be the most fecund. The future, they observe with justification, is particularly obscure in the region of new applications of scientific research.

Some philosophers of science such as Thomas S. Kuhn have argued that science has progressed at its impressive rate precisely because it responds to forces other than logical planning.[6] Normal scientific activity, in Kuhn's sense, is directed toward the "determination of significant fact, matching of facts with theory, and articulation of theory." The effort is bounded by the paradigms, the common law, of the discipline. Normal scientific research refines the discipline's paradigms. When this refinement process uncovers anomalies, revisions to the paradigms or new constraints and laws may be required. This entails a period of crisis, a period of turmoil, between competing ideas. In Kuhn's view, the developmental processes of science are "characterized by an increasingly detailed and refined understanding of nature. But nothing...makes it a process of evolution *toward* any thing...if we can learn to substitute evolution-from-what-we-know for evolution-toward-what-we-wish-to-know a number of vexing problems may vanish in the process." In other words the processes of normal scientific research which probe existing paradigms and the crisis of paradigm revision represent a survival of the fittest ideas and an evolution toward an ultimate state of knowledge dictated *not by need* but by the mechanics of the scientific method itself.

Clearly science has produced tremendously impressive progress in almost all of the fields dealing with the physics of nature. Progress in the social sciences has been much less spectacular. Kuhn believes this disparity results

[5] G. B. Kistiakowsky, "On Federal Support of Basic Research," Basic Research and National Goals, a report to the Committee on Science and Astronautics by the National Academy of Sciences, March, 1965.
[6] *The Structure of Scientific Revolutions* (Chicago: The University of Chicago Press, 1962).

from the desirable isolation of the natural scientist and his ability to choose his research using criteria other than those established by society. The social scientist does not enjoy the isolation of the natural scientist. He must choose his problems and defend his solutions not only among his peers but in front of society as well. This concern with goals, the argument goes, has cost him progress.

These then are the contrasting views on the organization of scientific research: one holds that a goal structure is becoming more necessary in our modern society, the other, that a visible goal structure is inimical to scientific progress. It is apparent that many scientists responsibly recognize this dilemma. As only a single example, there has been a running dialog in the letters column of the journal *Science* recently, under the heading, "Does Science Neglect Society?" One letter read, "Alas, it is not enough to respond to critics of the Age of Technology with *scientia gratia scientiae,* for some of the critics pose difficult questions. As they observe the electronic affronts to human dignity, the threat of nuclear extinction, the fouling of the environment, and the citizen revolt against the vastness of our technically based institutions, these critics wonder whether science is an unalloyed blessing. . . . nor can one pass off such critics as a collection of chronic malcontents since they include the likes of Louis B. Sohn of Harvard, whose, report, *The United Nations and Human Rights,* was recently submitted to the U.N. by a committee of distinguished Americans. The report, dealing with the threat to individual freedom by applied science, says, 'There is cumulative danger in the merry march of technology and science without adequate consideration of the social effects.'[7]

We implied earlier that some of the methods of operations research could be used to explore strategies for research. Systems analysis for example is the process by which alternative solutions to a problem or alternative means of achieving a goal are compared on an objective basis and an optimum selected. The techniques grew out of the military sciences of the second world war; it was used to predict bombing effectiveness, compare weapon mixes and forecast the results of various military strategies among other applications. Since World War II the tools of systems analysis have grown tremendously in scope, power, and number; in its new societal applications it is becoming a plowshare of impressive promise. The process begins with a statement of *goals,* as clear cut and precise as possible. A *model* is constructed; this may be in the form of mathematical equations, a game in which the rules constraining the actions of the players simulate real life, or a computer program: Whatever its form the model serves the purpose of making explicit the assumed relationships between interacting elements of the system under study. The model serves to illustrate the effects of *alternative* courses of action or inaction, in terms of their *costs* and *benefits.* Costs, of course, include more than money; they include in concept the expenditure of all resources which might have otherwise been available for other actions. Benefits resulting from the alternative courses of action can be

[7] David W. Kean, *Science* (December 1, 1967), pp. 1134 and 1136.

weighed against one another; those which most closely approximate the goals with least cost would presumably be the most desirable. The process can be easily described but may be immensely difficult to implement, particularly if the system being tested is complex and many disciplines are involved in its description.

The McDonnell Douglas Corporation, under contract to the National Aeronautics and Space Administration, recently applied systems analysis methods to the determination of an optimum research strategy in a basic science area.[8] The study called for the definition and analysis of astronomical research programs to be conducted from a manned orbital observatory. In addition to the substantive results of the study which will be published elsewhere, the work showed that analysis methodologies could be applied in the area of basic science to evolve research programs which will most likely produce the most needed information first while minimizing the costs of its collection. The analysis process followed in the study also produced some interesting byproducts. The relevance tree chart used in the study became an analog of the discipline itself; this structure made visible the linkage between the theoretical and experimental facets of the science; it indexed contemporary models implicit and under test in the discipline; it exposed the areas of research receiving great attention and others of promise but as yet untapped; it served also as a pedagogic tool. Some preliminary work indicates that the approach of this study may have general applicability to other disciplines; for this reason the methodology followed will be described in some detail.

The purpose of the astronomy planning study was to derive a consistent set of astronomical research objectives, to translate these objectives into illustrative apparatus designs, to test these for engineering and economic feasibility, and finally to group these into an evolutionary plan for effectively accomplishing the research objectives. We knew from the outset that the capability of observing the universe from above the earth's atmosphere would bring new research capabilities of tremendous scope. Astronomical observations made from the surface of the earth are impeded by three types of phenomena; opacity, wave front distortion, and scattering. The terrestrial atmosphere permits visible light to reach the surface of the earth, but it is totally opaque to radiation of short wave length, for example, the ultraviolet, x-ray, and gamma ray bands of the electromagnetic spectrum. This radiation is absorbed by ozone, oxygen, and nitrogen in the atmosphere. In the long wave length region water and carbon dioxide absorb in broad bands leaving only scattered wave length windows of varying transparency. As shown in Exhibit 1, the earth's ionosphere also attenuates radio waves longer than thirty meters in a highly variable fashion and becomes totally reflective to extraterrestrial radia-

---

8 NASA Contract NAS8-21023, Orbital Astronomy Support Facility, performed for the Marshall Space Flight Center, Huntsville, Alabama. NASA personnel involved in the study included: Mr. Jean Oliver, Contracting Officer's Representative, and Dr. Charles Huebner, Office of Manned Space Flight.

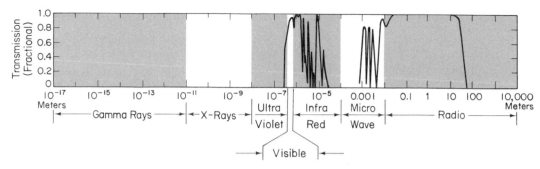

Exhibit 1. Vertical Transmission of Electromagnetic Radiation by the Earth's Atmosphere and Ionosphere.

tion of wave lengths longer than sixty meters. Furthermore, variations in atmospheric density create wavelength distortion. Astronomers refer to this as "bad seeing." There density changes cause the light passing through the atmosphere to be refracted and to indicate apparent motion of the observed body. These density changes also cause rapid brightness variations and size distortion of the image. The effect becomes worse as the dimensions of the optical instrument increase. The earth's atmosphere also scatters light from the sun, stars, and other optical sources because of reflections from suspended dust and other particles in the air. Finally, the atmosphere itself glows and contains the aurora. As a result the sky is not black even on the darkest moonless nights. All of these atmospheric effects would be absent from orbit; the promise for astronomical research above the earth's atmosphere is great indeed.

We were tempted at the outset to say simply that future orbiting astronomical observatories should capitalize on the new vantage point; that is, they should contain instruments which would permit observations in the regions of the spectrum opened by virtue of being above the atmosphere. Indeed, a number of our astronomer consultants took the position it was impossible to derive a strategy better than simply, "give us a platform stabilized to the highest possible accuracy, containing the largest possible optical and radio instruments, and let us scan the heavens in the new regions of the spectrum." This position is coincident with Kuhn's; the research program for such a facility would consist of the normal scientific enterprise of paradigm probing. While a case could be made for this kind of facility, real questions still remained such as priority among alternative astronomy research programs and the relative benefits of instruments of various capabilities. This latter question, for example, could be posed in real operational terms: It will be considerably more expensive to place a three-meter telescope in orbit than a one-meter telescope. Is the data return from the larger telescope worth the investment? The normal science approach, then, adds very little to our conception of what ought to be done.

Conversations between the Douglas study team and their consultants indicated that many astronomers had relatively narrow regions of interest within their discipline. For example, a solar astronomer might have intense current interest in solar corona phenomena but might be relatively unfamiliar with planetary or galactic astronomy. The programs of research which they suggested were, understandably, closely allied with their current interests and for the most part based on instrumentation which they thought it possible to produce. These discussions led us to construct our first "research model" shown in Exhibit 2. This structure is a relevance tree. This approach to integrated decision making was proposed by Churchman, Ackoff, and Arnoff as an aid to industrial decision making.[9] In its application to astronomy the discipline was broken down according to objects of interest to the study's consultants; these objects were in turn divided topographically. For each subdivision of the objects being reviewed, phenomena of interest were defined, and for these phenomena research programs were stated.

The lines connecting the boxes of the relevance tree chart indicate the hierarchy of the discipline. That is, the sum of the entries at each level should,

[9] C. W. Churchman, R. L. Ackoff, and E. L. Arnoff, *Introduction to Operations Research* (New York: John Wiley and Sons, 1957).

Exhibit 2. Object-oriented Approach.

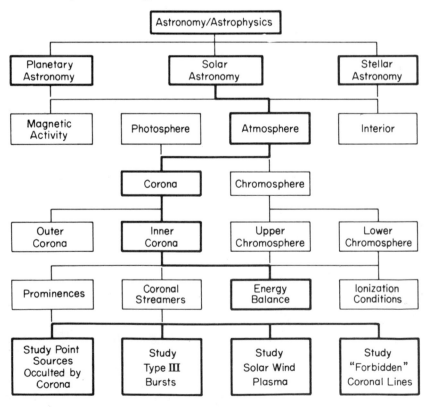

in theory, provide a complete set of information about the subject box above, to which they are connected. For example, in Exhibit 2 the corona consists only of outer and inner coronal elements; the study of solar astronomy presumably would be complete with complete knowledge of its magnetic activity, photosphere, atmosphere, or interior. Clearly, such a structure is based on contemporary models of the discipline and the personal concepts of its authors.

This first relevance tree approach, based on objects, was rejected because it did not establish criteria for assessing priority among the major divisions of the discipline. It was very sensitive to the personal convictions of the scientists who constructed it and is likely to change as the interests of astronomers change.

In inexact areas, expert testimony is admissible.[10] The study team felt that important information about research goals could be derived by collecting opinions from their astronomy consultants with regard to the contemporary "burning issues" of astronomy. If these issues could be stated with relatively high degree of consensus, perhaps research programs could be based on their solution. Through informal interviews with six of the consultants, eleven issues on which general agreement would be obtained were defined. These are listed in Exhibit 3. It was anticipated that a technique resembling the Delphi method

1. Evolution of solar system	6. Nature of x-ray sources
2. Evolution of galaxies	7. Extend observations to far UV
3. Origin of cosmic rays	8. Origin evolution of quasars
4. Origin of radio sources	9. Nature of interstellar medium
5. Nature of extragalactic objects	10. Establish cosmic distance scales

11. Origin of gamma ray sources

Exhibit 3. Contemporary Astronomy Problems.

for the collection of expert opinion could be used.[11] This approach, which has proven so valuable in other applications involving technological forecasting, was not completed. The questions of priorities among the critical problems and the optimum kinds of research programs required could have been posed to the consultants. However, as an industrial corporation engaged in a systems analysis, the study team was not prepared to mediate between scientists holding different points of view.

The third approach followed in an attempt to define a reasonable research plan was similar to the morphological approach of Zwicky.[12] This method has at its heart the mapping of the open options of the discipline to obtain

[10] For an expansion of this point of view see: O. Helmer and N. Rescher, *The Epistemology of the Inexact Sciences*. The RAND Corporation, February, 1960.

[11] N. Dalkey and O. Helmer, "An Experimental Application of the Delphi Method to the Use of Experts," *Management Science*, Vol. IX, (1963).

[12] Fritz Zwicky, "Morphology of Propulsive Power," *Monographs on Morphological Research No. 1*, Society for Morphological Research, Pasadena, California, 1962.

"a systematic perspective over all the possible solutions of the large-scale problem."

In a scientific discipline the full application of the morphological approach would involve the listing of all possible experimentation and a "filtering" to identify those tasks which can be accomplished with current state of the art or which should be scheduled in the future, in anticipation of certain advancements in technological capability. This is obviously a large, difficult job if it is to be done with a reasonable degree of completeness.

In the study of astronomy this technique was used by constructing several three-dimensional matrices whose dimensions consisted of: astronomical bodies, portions of the electromagnetic spectrum, and particular parameters of interest such as angular resolution or telescope aperture. Exhibit 4 is an example of one of these matrices. Presumably every cell of the matrix represents a potential series of measurements. Some of these cells could be eliminated because the observations implied were possible and more easily accomplished on the earth.

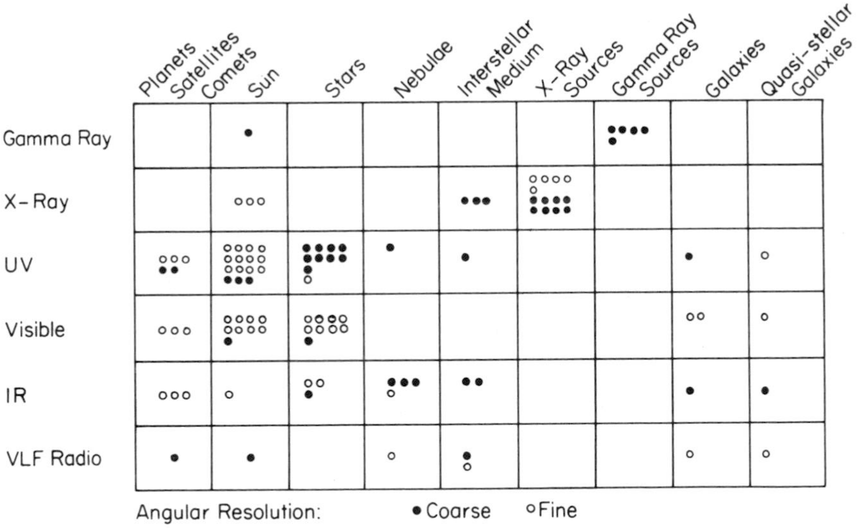

	Planets Satellites Comets Sun	Stars	Nebulae	Interstellar Medium	X-Ray Sources	Gamma Ray Sources	Galaxies	Quasi-stellar Galaxies
Gamma Ray	•					•••• / •		
X-Ray	ooo			•••	oooo o / •••• / •••			
UV	oo o / •• oooo oooo oooo ••• •••• •••• ••• / o	•	•				•	o
Visible	ooo oooo oooo • oooo oooo •						oo	o
IR	ooo o oo / •	••• / o	••				•	•
VLF Radio	• •		o	• / o			o	o

Angular Resolution: • Coarse o Fine

Exhibit 4. Parametric Matrix.

Other intersections represented questions which were as yet meaningless, for example, the measurement of gamma ray activity from planets. Other intersections represented technological challenges and required observations which were as yet not possible, for example, measurements with high angular resolution of the energy emanating from galaxies in the very low radio frequency spectrum. This morphological approach was of value in discriminating interesting opportunities in space. It led the study team to ask important questions and prompted dialog which defined some of the admissible terrain of the discipline.

But this approach and the previous methods failed to relate individual experiments to the real crises (in Kuhn's sense) of astronomy. In effect, these analysis methods, as applied, yielded only a systematic cataloging of potential experiments and observations with little cohesive structure to indicate logical relationships between experiments. This missing infrastructure represented the connection between the theories, the hypotheses, the models of the discipline, and the experimental programs which devolve from these concepts.

To express this concern in another way, while cataloging of data is a necessary preliminary step in the scientific search for knowledge, it is by itself probably not sufficient to give adequate direction to research. As an example, by 1860, Secchi had studied the spectra of over 4,000 stars, and by the start of the twentieth century, numerous catalogs and lists of star objects were in existence. It was not, however, until an attempt was made to organize these data by plotting spectral type against stellar magnitude that a basis was established for the organization of data in a form which could reveal intrinsic characteristics of stars. In the early 1900s, by asking a basic question regarding the relationship of stellar magnitudes (luminosities) and spectral types (temperature), Hertzsprung and Russell independently provided an organizational device by which theoretical implications of the data could be examined. The so-called H-R diagram shown in Exhibit 5 has had a tremendous impact on the organization of previously unrelated data as well as upon cosmological theory. As illustrated by the Hertzsprung-Russell analogy, categorization of data alone is not necessarily sufficient to permit the discovery of relationships that exist among phenomena. The truly interesting and significant work occurs when questions of relationships and similarities and differences are formulated.

Accordingly, the team studied a fourth approach which involved the interplay between the concepts implicitly under test in the discipline, the experiments which these suggested, and the inventory of possible experiments which result from "normal" scientific observations. The discipline was viewed as a relevance tree consisting of two major branches: theoretical and experi-

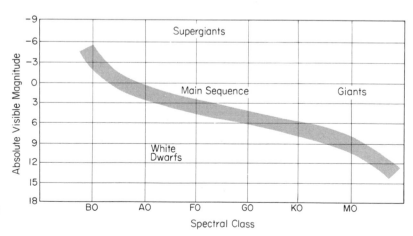

Exhibit 5. Hertzsprung-Russell Diagram.

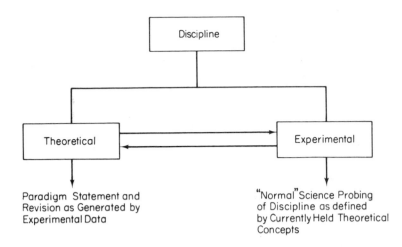

Paradigm Statement and
Revision as Generated by
Experimental Data

"Normal" Science Probing
of Discipline as defined
by Currently Held Theoretical
Concepts

Exhibit 6. **Theoretical-experimental Relevance Tree.**

mental. The theoretical line consisted of statements of the paradigms of the discipline and revision and refinement to these as new data are derived by experiments. The experimental branch consisted of a spectrum of potentially feasible experiments within the limits set by currently held views of the discipline. Clearly, there is an adaptive feedback between these elements in which new or unexpected experimental data causes theory revision and revised theories suggest new experimental domains. This is shown in Exhibit 6.

We applied this model to astronomy, labeling the theoretical model-building aspects: definition of the origin and future of the universe (evolution) and establishment of principles of change and order of the universe (laws). The experimental branch was called observation of the present state of the universe. This breakdown served as the top level of our final relevance tree format and was used in the form shown in Exhibit 7.

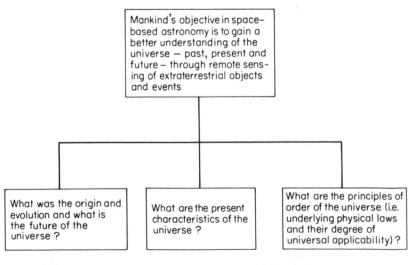

Exhibit 7. **User-oriented Objective Approach Astronomy/Astrophysics.**

These three categories represent three points of departure for the discipline. Taken together they fully define its present state of knowledge and are capable of being expanded to include new knowledge as it is collected. The division between these points of departure is coincident with the contemporary sub-disciplines of astronomy: cosmology and cosmogeny, observational astronomy, and astrophysics. This breakdown of evolution, state, and laws of order apparently has general application to relevance tree structuring of many other scientific disciplines.

To move from these top level questions to a statement of the requirements of research programs necessitated the development of logic processes which were peculiar to the theoretical and experimental domains. As indicated earlier, the state column (experimental) involves the generation of a relatively complete set of potentially feasible observation and experiment requirements within the constraints of currently held astronomical models. The experiments derived in the study assumed, for example:

1. The interchangeability of energy and matter.
2. The ability to ascribe to all bodies in the universe the properties of location, size, shape, mass, and energy.
3. The ability to obtain astronomical data with present technology or sometime in the future by means of sensing electromagnetic energy and distant fields.
4. A division of the universe which included the universe as a whole, integalactic matter, galaxies, interstellar matter, stellar bodies, planetary systems, satellites, and interplanetary matter.

An example of the particular logic used is shown in Exhibit 8. At the lowest level of this chart, some 3,000 potential measurement requirements were identified. These research objectives, as with Kuhn's "normal science," were generally devoted to increasing the precision of astronomical constants, comparison of the results of observations with those forecasted by the discipline's working paradigms, and for the refinement and articulation of the discipline's paradigms.[13] The measurement requirements which were formulated from this structuring typically included a statement of the:

1. Object of interest.
2. Nature of data desired.
3. Spectral range of interest.
4. Source characteristics (estimated size, distance magnitude).
5. Special descriptors (epoch, period, and so on).

The theoretical lines (the origin and evolution of the universe and the laws of change) required the development of a different sort of logic. Here the plan moved from articulation of operational theories, to statement of consequences of the theory, to development of critical tests of the theory. Exhibit 9 is a simplified illustration of the derivation used. If complete, this structure

[13] Kuhn, *The Structure of Scientific Revolutions*, pp. 25, 26, 27, and 33.

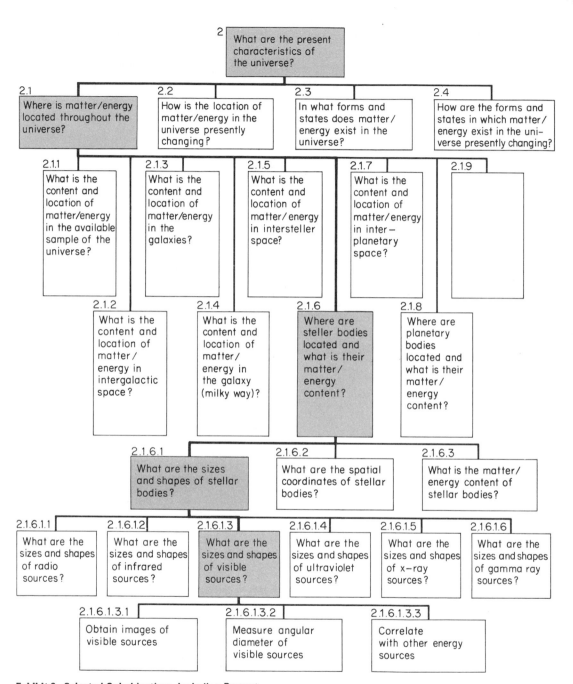

Exhibit 8. Selected Subobjectives, Including Present Characteristics of Universe Categories.

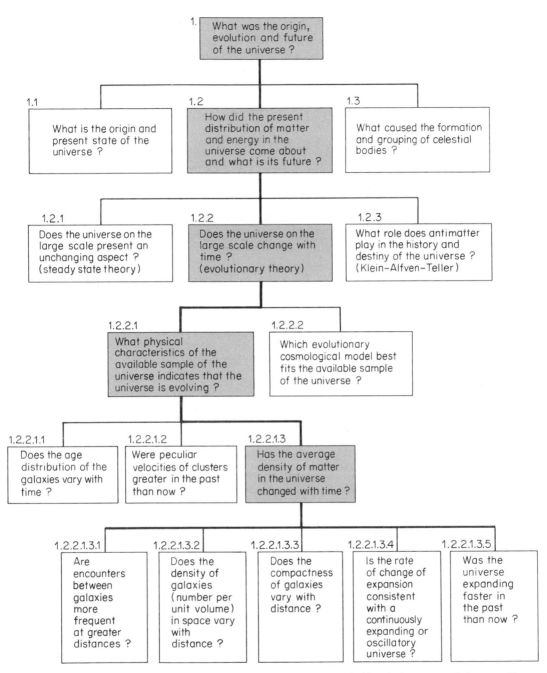

Exhibit 9. Selected Subobjectives in Astronomy/Relevance Chart, Including Evolution of Universe Categories.

would index all of the theories under test in the discipline and the crucial tests which would establish their superiority over competing concepts. In the NASA study, it was estimated that 1,000 crucial tests could have been defined.

Bunge has proposed the characteristics of a sound physical theory.[14] He asks that theories exhibit both *internal* and *external consistency,* that is, that they be contradiction-free and largely compatible with the body of previously observed data. The entries should have *unity of reference,* that is, they should refer to objects, not other concepts. *Conceptual connectedness* is the desirable property of a theory which refers to its logical relationship with the basic concepts from which it presumably draws its strength. Hopefully the theories and models entered in the two theoretical columns during the study satisfied these criteria.

Thus the relevance tree for astronomical research consisted of (1) two columns which stated the theoretical basis for the discipline and moved in their logic from theory to consequence to crucial test and, (2) an experimental column which provided an index of potential observational and experimental measurement requirements permitted by the generally accepted paradigms of the discipline.

With the relevance tree completed to this level the analysts were in a position to seek a strategy for determining priority among the alternate research programs suggested. They found that this structure, which made visible the models of the discipline and the open research, also could be used to a degree to assess the relative importance and urgency of observational and experimental program requirements, on the assumption that a gain in scientific knowledge resulting from an experimental or observational program is a valid measure of its intrinsic worth. Research which appeared to be both feasible (in the state column) and crucial (in the theoretical columns) was taken as more important than research which satisfied only one of these criteria. The greater the number of crucial issues a measurement requirement was likely to probe, the more urgent it was taken to be. Furthermore it was relatively easy to discern that class of measurement requirement in the state column which should be accorded priority because it would provide data in new regions of the electromagnetic spectrum, open in orbit but not on earth. Finally certain potential measurement requirements were found to be time or sequence dependent; that is, they had to be performed during certain celestial epics or prior to or following other experiments. In their interval of importance these measurement requirements had to be accorded priority. With this structured approach then, the assignment of priority could be based on feasibility, potential contribution to model building, epic dependency, and the uniqueness of the opportunity afforded by the orbital position of the observatory. Each of these criteria could be assigned a weight and the measurement requirements compared on the basis of accumulated value. Clearly other strategies could have been employed; for

[14] M. Bunge, *Foundations of Physics* (New York: Springer-Verlag New York, Inc., 1967).

example, a cost parameter could have been introduced if this were important to the planners, so that the measurements might be compared on the basis of value per unit cost. Or perhaps availability of competent and interested scientific sponsors might have been included in the determination of relative values. Dr. A. Wilson of the Douglas Corporate Laboratories has suggested that priority might be based on paradox: a measurement which has the maximum potential of blowing theory to bits should be valued more highly than one of lesser potential consequence. Regardless of the exact mechanism, it is clear that planning strategy can be introduced at this point and on the basis of objective analyses rather than subjective judgment only. Clearly no all encompassing prescription or formula can be specified for this analysis; judgment will be required to specify the priority-determining ground rules.

Systems analysis techniques can also be used to determine the best of the alternative means of accomplishing the measurement requirements. For example, suppose a highly valued measurement requirement was specified as: the determination of the apparent angular diameter of distant galaxies in the visible spectrum. How can this measurement best be accomplished? The analysts experimented with two methodologies. The first was a consensus system in which scientist consultants were asked to designate the required experimental constraints such as the necessary angular resolution, magnification, attitude control limits and the like to satisfy the measurement requirement. The other system was more rigorous and involved a systematic investigation of alternative measurement techniques. For example, Exhibit 10 shows the relationship between the angular diameter and distance for various sized bodies in space. The capability of the 200-inch Hale telescope is shown as the shaded area in this figure. Its angular resolution is limited by the intensity of the source. The unshaded region then is the acceptable region in which space measurements will produce data not obtainable on earth with current instruments. The line on the chart which depicts the relationship between angular

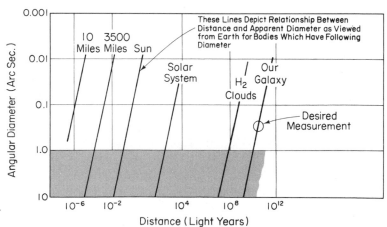

Exhibit 10. Size/Distance/Angular Diameter.

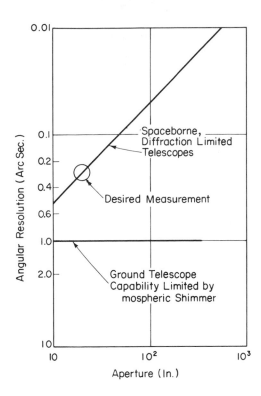

Exhibit 11. Angular Resolution Capability.

diameter and distance for bodies with diameters approximating the estimated size of galaxies represents the domain of the desired measurement requirement in this example. This information alone is not sufficient to designate the range of acceptable space instruments. Exhibit 11 shows that for space telescopes angular resolution is uniquely related to the diameter of a diffraction limited instrument. Thus to determine the apparent angular diameter of distant galaxies in the visible spectrum will require a telescope system capable of angular resolutions less than half an arc second. If 0.2 arc second can be achieved, it would be consistent to select a telescope diameter of fifteen inches. Through similar analysis focal length and sensor sensitivity can be selected.

These kinds of estimates can be made across the range of desired measurement requirements and the derived instruments viewed as a group. Exhibit 12, for instance, shows the percentage of measurement requirements satisfied as telescopes of increasing diameter are selected for the orbital program. With this kind of data at hand cost trades can be conducted to determine the expenditures required for achieving various percentages of the desired data.

In summary, from the study of astronomy research objectives, it was found that a properly constructed relevance tree could provide:

1. A method for communicating the way individual research contributes to the objectives of the discipline.

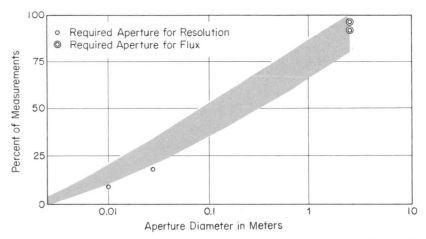

Exhibit 12. Stellar and Planetary Observations (23 ORDS).

2. A means of displaying the assumptions and interdependencies and the operational models and theories in use and under test in the discipline.

3. A basis for objectively assessing priorities with a minimum of bias.

4. A means of discovery of research enterprises likely to be important but not currently being pursued.

5. A pedagogical tool which draws together current views, directions, and interests of the discipline and its practitioners.

6. A framework for comparing alternative large scale programs involving a multiplicity of research projects and for testing their sensitivity to alternate experiment requirements.

This approach is to a degree an attempt to return to a natural philosophy in which the relevancy of a single action can be connected to the objectives of the discipline as a whole. It is a departure from the shotgun, one-experiment-after-another approach, where philosophy disappears into technique, and hopefully substitutes order, through a method which, in microcosm, is a simulation of the scientific method itself.

The model of the discipline which the final relevance tree represents is without doubt inaccurate and incomplete. It will change as new models are proposed and as new data are collected. This formulation leans heavily on science's perception of its own mission and is not necessarily representative of nature. As we know from studies of organizations and the history of science this perception is often fallacious. For these reasons we can expect the chart to change with time. It would probably be desirable to seek means to make the taxonomy adaptive, that is self-learning, but the study team has not been able to imagine a systematic way to accomplish this, short of periodic undating. If such a system could be designed, the self-learning would almost certainly improve the isomorphic similarity between the diagrammatic representation of nature and nature itself.

It was pointed out earlier that we have attempted this type of analysis in other disciplines. Oceanography, meteorology, and agricultural science have been examined in some detail: physical science has been sketched but less completely. While each of these exercises introduced peculiarities of its own, the methodological framework described for astronomy has been found to be useful. In the case of oceanography, for example, the central question was to acquire knowledge so man can realize optimum use of the oceans and its shores and to understand its history and predict its future. The "understand" portion of this question led to a structure which closely followed the astronomy reasoning; it included consideration of the evolution of the oceans, the measurement of the present state of the oceans, and the analysis of the laws by which the oceans change. The "use" part of the central objective introduced a new element: however, there was no analogous "use" of the universe considered in the study of astronomy. This portion of the objective led to the subsidiary questions: What are the current and potential uses of the oceans by man? What is the effect of the oceans on man? These in turn were subdivided and in the end produced a set of measurements needed to establish oceanography forecasting models, such as a theory of fish location by observing certain features of the oceans' surfaces such a temperature, current, salinity, phytoplankton, and so on. The value of these measurements might have been stated in terms of economic worth if fish position could indeed be predicted, then the anticipated return could be weighed against the cost of obtaining the measurements needed to establish the paradigm. The value criteria described earlier for astronomy could be applied also. A portion of the analysis of this discipline is shown in Exhibit 13.

Erich Jantsch, in surveying technological forecasting techniques in use today, reported a growing feeling among systems analysts at least that the organization of research should and can be studied.[15] The physicist Dennis Gabor has suggested that "a thorough scrutiny be made of all physical methods which have not been applied to cancer research." Bronowski has suggested that biological objectives of importance be analyzed to determine high priority research. Professor Burton Dean of Case Western Reserve, has developed a model for priority ranking of research and development projects.[16]

It seems to the authors that investigation of the problems of aging might be particularly amenable to systematic approaches. Included in the currently competing theories of aging are: calciphylaxis, the increasing sensitivity of the body to calcium: molecular cross-linkage within cellular proteins: and auto-immunization caused by chemical or mutation processes. These could be analyzed to determine consequences and critical tests. These in turn could be compared to possible state measurements in order to assess priority for

[15] *Technological Forecasting in Perspective* (France: Organization for Economic Co-Operation and Development, 1967).
[16] "Stochastic Networks in Research Planning," in *Research Program Effectiveness,* ed. M. C. Yovits (New York: Gordon Breach, 1966).

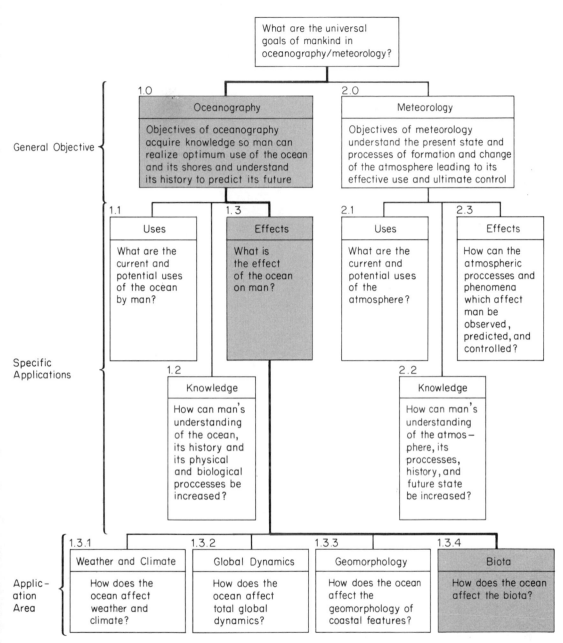

Exhibit 13. Systems Analysis Logic Flow.

research and perhaps government funding. It is not beyond possibility that new avenues of research will appear in this analysis which are not currently being pursued.

The authors believe that the planning of research may provide the scientist new insights into the definition of valuable work within his discipline. Rather than offering a choice between "planned research" and "free science," planning should be viewed as a decision-making tool which the scientist himself can use to his benefit. For the government planner who is charged with the responsibility of understanding, initiating, and sponsoring the most promising scientific work, these techniques hold the promise of communication which will bring him into closer synchronization with the communities he serves.

Honeywell's PATTERN:
Planning Assistance
Through Technical
Evaluation of
Relevance Numbers

MAURICE E. ESCH

Honeywell's PATTERN is perhaps the best documented industrial adaptation of the relevance tree approach to allocating resources to future technical activities. Essentially, it is a normative (goal-oriented) forecasting technique using relevance trees to establish the step-wise interconnection of current actions and future output. The author provides a description of the technique along with examples in both a military context and a medical one. Some eight years of experience have shown Honeywell that PATTERN is a useful aid in selecting long range R&D investment programs.

There are a great many techniques currently being used to forecast technological change, ranging from intuitive thinking (brain storming) to sophisticated systems analysis techniques using dynamic mathematical models. We found that we needed something in between these two extremes that would permit us to adequately structure our problem and yet be simple enough for our small planning staff to use on a continuing basis.

To meet this need, we developed a normative technique employing a relevance tree for several reasons. First, we are in the military technology business which lends itself well to mission oriented analysis. Second, this approach permitted us to structure the problem in such a way as to break out the many complex interrelated variables affecting system development into small manageable elements for decision making. Third, it helped to insure that we considered all types of alternatives and, at the same time, provided considerable insight into the various alternatives during the actual structuring process.

We should note at this point that two conditions must be present in order for a normative forecasting technique to produce meaningful results. First, the impact levels for technological transfer must be sufficiently closed either naturally or artifically by consensus. In our case, we closed the levels by consensus. Secondly, there must be a greater number of opportunities at each of the levels than can be exploited under a given budget or other constraint, making it necessary to optimize which implies selection.

At present, there are generally many more feasible technical alternatives available to solve any particular problem than can be supported economically by either Government or industry. As a result, new military, medical, and/or many other program development and procurement decisions are being weighed very carefully by the Government in terms of how well a particular systems approach supports the overall national objectives with heavy emphasis on cost effectiveness. In this situation, a large number of interrelated technical, political, military, economic, and social factors affect each decision made by Government officials. Faced with this environment, industry needs some auxiliary aids to help allocate R&D dollars in order to continue its growth.

Several years ago, Honeywell's Military and Space Sciences Department developed a forecasting technique called PATTERN (Planning Assistance Through Technical Evaluation of Relevance Numbers) which uses a requirements or need-oriented relevance tree to aid corporate decision makers to identify critical technology areas requiring upgrading to support the development of future system programs. Although PATTERN was originally developed and used most extensively for planning in the military area, the methodology is very flexible and can be applied to almost any area where decisions must be made under conditions of uncertainty. As an example, Honeywell has also used this technique to forecast future technology requirements in the biomedical field. Before getting into the medical field, I will review briefly how the methodology is used in the military area and compare its use here with its application to the problem of developing future biomedical technology needs.

There are a number of problems common to both areas which must be dealt with in producing any long range technology forecast. First, and probably the most difficult task, is to develop and maintain communications between the corporate objectives level and the technology needs/transfer at the lower levels which support these objectives. This communication problem exists between people at the various levels in the corporate structure and affects their understanding of the interrelationship of technology and the corporate objectives. This is essentially a relevance tree structuring problem involving both scope and definitions. The Government recognized the problem when it decided in 1965 to implement the Planning Programming Budgeting System (PPBS) in an attempt to improve communications among and within the various departments by stressing mission oriented programming and budgeting and attempting to bring to light the commonalities that exist both in technology and objectives across a broad spectrum of Government activities.

A second difficult problem is putting the forecast into the correct time frame. I will discuss how we attack this problem a bit later.

A third major problem is the selection of criteria by which the relative importance of various alternatives at a given level in the structure can be assessed. Certainly the criteria selected must be appropriate to the level at which the decision is being made; also criteria at the various levels should be mutually exclusive and above all they must be understood by the decision makers. I will discuss each of these problems in greater detail later.

It has been said that normative technology forecasting is an attempt to invent the future. Perhaps this is so. However, when we as a company with limited resources are considering making research and development investments in just a few of the many technical areas open to us, we certainly want first to establish the need for a particular technology in the future and the priority of that need. Assuming that rational men will attempt to fulfill these needs on a basis consistent with their priority and the available resources, one is next faced with the problem of which alternative solution best meets the need on a cost effective basis. This leads us to the problem of how to make a knowledgeable decision faced with "X" number of alternatives. Let's examine this probability (Exhibit 1). Evaluations of the capability of the human mind to interrelate

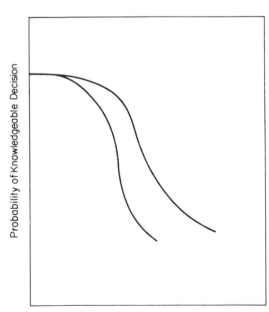

Exhibit 1. Probability of Knowledgeable Decisions as a Function of Number of Alternatives.

Number of Criteria X Alternatives

and properly correlate a large number of variables has shown that there is a very sharp knee in the curve in the region of about twenty alternatives. The curve shown in Exhibit 1 reflects the probability that the individual will make a knowledgeable decision versus the number of alternatives being considered. The knee of this curve falls in the region around twenty alternatives depending upon the latent ability of the individual. As the number of factors considered goes much beyond twenty, the probability of making a knowledgeable decision is greatly reduced. Therefore, it would seem that it is necessary to reduce the number of variables considered at any given time in a decision process to something under twenty, and preferably under ten, in order to assure that the mind can assimilate all the relevant factors associated with the decision to be made. Further, we feel that it is necessary to express the relative importance of the various items in quantative terms which are not only more exact than qualitative expressions, but which also can be used in a computer program.

The PATTERN methodology can be used to assess the relative importance of technology needs and select those technology areas where upgrading our capability will produce the greatest benefit toward meeting established goals or objectives.

PATTERN METHODOLOGY

PATTERN used in the military area can best be described as a decision aid consisting of three basic parts shown in Exhibit 2: first, a Relevance Tree which measures the relative importance to national objectives by upgrading a particular mission or technical area; second, Cross Support which measures the degree of technical growth that will result from solving a specific problem; and third, Status and Timing which measures industry's capability to solve identified technical problems. A scenario which projects the world environment (including military, political, and economical factors) that might be expected in the next decade and a state-of-the-art technology assessment are used as basic inputs in the evaluation process.

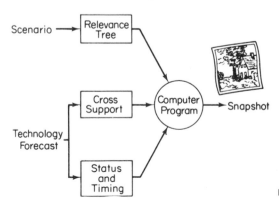

Exhibit 2. PATTERN Is a Decision Aid.

At each node of the tree, a team of experts, using matrices, decision criteria, and subjective probability assign quantitative values relating to the relative importance of upgrading that item in terms of its contribution to meeting the overall national objectives for the next decade. Cross support and status and timing factors are also assigned numerical values. These numerical inputs are fed into a computer program and the computer readout then gives a snap-shot of any particular area of interest in terms of its relative importance to the national military stature as compared to all other areas considered at the same level in the evaluation process.

First, let us examine the relevance tree in more detail. It is basically a structured decision network consisting of eight levels beginning with national

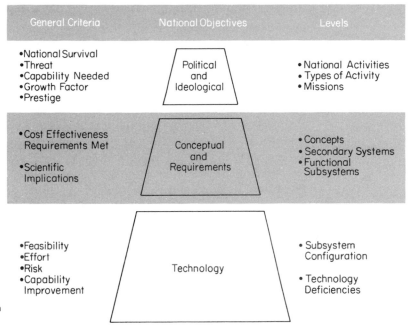

Exhibit 3. Basic Structure of a Relevance Tree.

security objectives and progressing through types of conflicts, forms of conflict, missions, system concepts, functional subsystems, subsystem configurations, and technology deficiencies in a tree-like organization. Initially, we divided the structure into three major sections shown in Exhibit 3. At the top supporting the National Objectives is the political and ideological area, where the President and his advisors make policy type decisions. The second section, the conceptual and requirements area, is the level where the Secretary of Defense, the Administrator of NASA and the individual Service Chiefs make decisions. The third section, the technology area, is where the individual service laboratories and the laboratories of industry make technical decisions. This basic three-part structure was then subdivided into the seven levels shown on the right of

Exhibit 3. In order to make value judgements in the horizontal plane at each level of the tree, it is necessary to apply specific definitions to each item on the tree as well as to develop a set of ground rules for the game situation. The ground rules take the form of criteria. Key words representing the various criteria are shown on the left of Exhibit 3. In developing the criteria it is necessary to insure that they are appropriate to the particular level at which the decision is to be made and that they are mutually exclusive to that level insofar as possible.

A more detailed breakdown of the top four levels of the tree to the mission level is shown in Exhibit 4. At the top we see the national objectives. Directly below that, the three types of activities are identified as those having to do with the challenges that we can expect to face in the future. On one side of the cold war is the spectrum of active hostilities and on the other is the scientific challenge in the field of exploration. In order to meet national security objectives (namely, to control aggression at any point in the world, to win the ensuing conflict if control fails, and to establish scientific preeminence in the field of exploration), the United States must have the technical capability to meet the challenges in all three areas. The various ways in which pressures may be brought to bear against the United States are shown at the third level of the relevance tree which are called "forms of activity." For example, in the military area we could be pressured into a general strategic global type conflict or various tactical theater situations of varying intensity; whereas, in exploration we could be subjected to pressures principally involving the scientific areas having to do with earth-related and space-related activities and so on.

Under each of these forms of conflict is a spectrum of missions which we must be capable of carrying out if we expect to counter the challenge in each specific area of competition. For example, under Limited War II, if we were able to meet all of the requirements of the eight missions shown in Exhibit 4, we should be able to engage in that form of conflict with a reasonable expectation of winning.

It is very important that each of the areas be carefully defined in order to determine the type of equipment and the technology upgrading required to achieve the specific objectives of the successive upward levels of the tree. The advantage of this type structure is that it divides a large many-faceted problem into smaller segments, allowing us to make comparisons among smaller numbers of items at any single level. At the same time we retain the capability of again looking at the total package as any entity once the individual value judgements have been made.

I mentioned earlier the necessity for each level of the tree to be "closed," that is, represent a complete set. We accomplish this by consensus, which is to say, we agree that the items as defined at each level represent a complete capability to accomplish the objectives at the next higher level. The basic principle of the tree structure, therefore, is to determine the area to be investigated (universal space) and then subdivide it (into sets and subsets) to

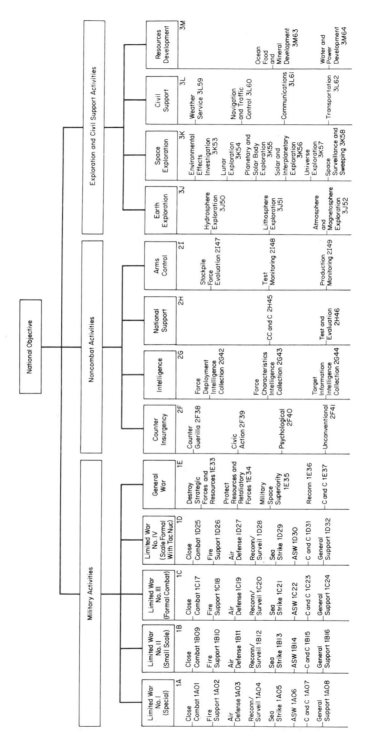

Exhibit 4. Relevance Tree Structure to the Mission Level.

a degree that the data can be evaluated in manageable pieces. We attempt to cull out overlapping regions, with appropriate definitions, so that they can be handled with straightforward thought processes. There is no mathematical constraint that the divisions be mutually exclusive; there are only data insertion and processing constraints. Furthermore, there is no mathematical constraint on the number of sets or their size. Neither is there a requirement that the relevance tree be composed of a specific number of levels since almost any type of structure can be used.

As a practical constraint, only those areas that are judged to need improvement are included on the relevance tree, with existing capability essentially factored out and included in the scenario or technology base. Hence, the tree is one of upgrading, with the relevance numbers used to measure the marginal utility of the need to improve.

The tree structure is therefore a way of organizing data such that all factors of interest are included, with as little redundancy as possible, through flexibility in the use of stipulative definitions of the scope required.

Exhibit 5 shows the lower four levels of the tree on a concept worksheet. The postulated concepts represent the fifth level of the tree. At the sixth level, each of the concepts was divided into seventeen functional subsystem areas together with associated requirements needed to meet the concept performance goals. There is no constraint on the number of subsystems used other than the total must result in a complete set as defined by consensus. To do this, it is necessary that each item be defined in considerable detail in order that every study participant understand precisely what is included in each particular area.

In most cases, there were several alternate ways by which the functional subsystem requirements could be met. These alternatives were called subsystem configurations and constituted the seventh level of the tree. Contained within the competing subsystem configurations were the technological deficiencies to be solved. These deficiencies represented the eighth level of the tree. Note on this chart that status and timing are entered at both the functional subsystem and at the technical deficiency level.

Just as in structuring the tree, the selection of criteria for the assignment of relevance numbers is made on the basis that they be as nearly mutually exclusive as possible at each level (practical constraint) as well as from level to level (logic constraint). The relevance number is simply the measure of the size of the subsets. It reflects the judgment of the individual as to the degree of improvement needed in the area of interest. All the decisions are made by experts using the same criteria and documented data, and drawing upon their experience. The relevance number is, therefore, no better than the data, criteria, and the judgment of the individual. Considerable effort needs to be made on developing a good scenario which is then read and discussed by the participating team members.

The requirement that the sum of the individual relevance numbers within any branch of the tree total one is necessary if the elements of each branch

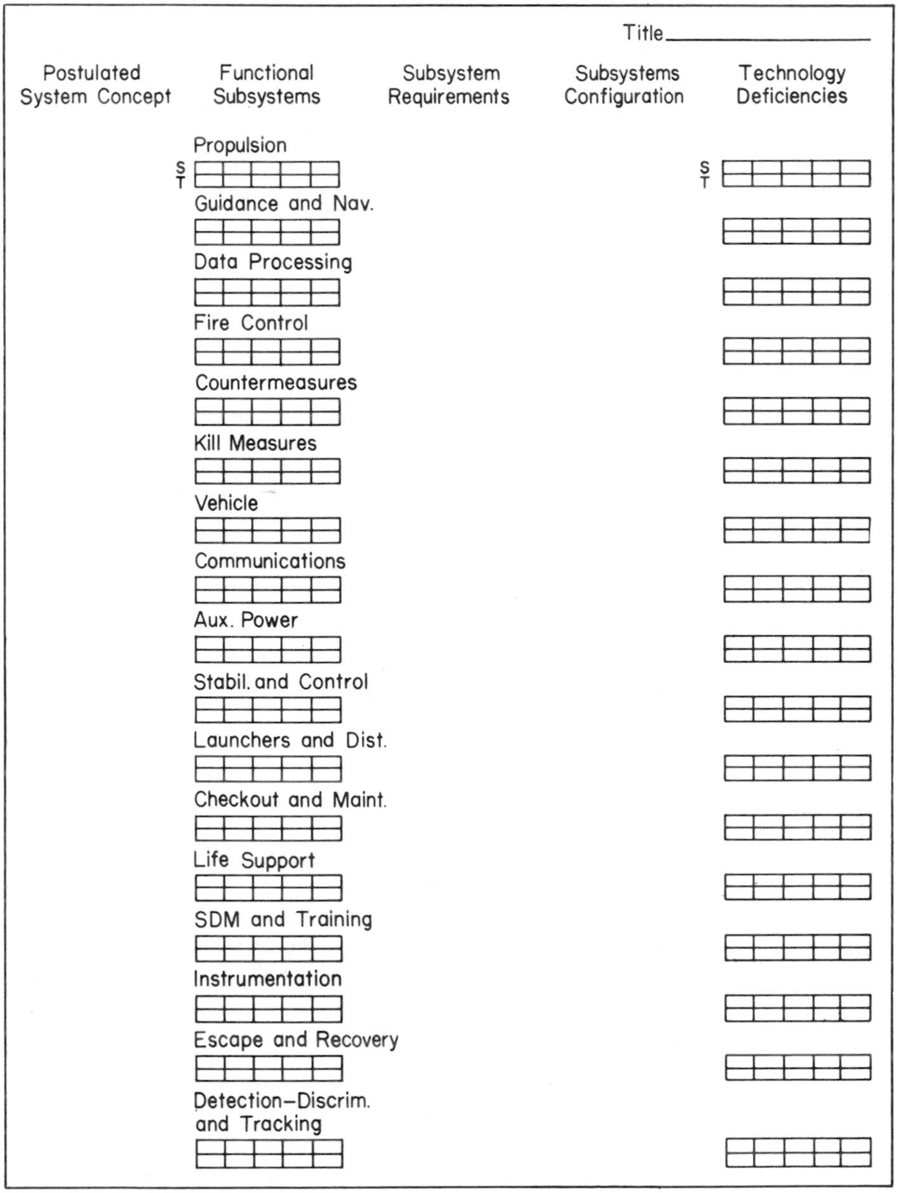

Postulated System Concept	Functional Subsystems	Subsystem Requirements	Subsystems Configuration	Technology Deficiencies

Propulsion

Guidance and Nav.

Data Processing

Fire Control

Countermeasures

Kill Measures

Vehicle

Communications

Aux. Power

Stabil. and Control

Launchers and Dist.

Checkout and Maint.

Life Support

SDM and Training

Instrumentation

Escape and Recovery

Detection—Discrim. and Tracking

Exhibit 5. Lower Four Levels of the Tree.

are to form a complete set. This is also a mathematical constraint required to meet the upgrading logic of the tree structure. If, in the judgment of the evaluation team, the number of elements at a given level do not cover a complete set of alternatives necessary in the decision process, adjustment is possible

Exhibit 6. Relevance Number Assignment Technique.

	Likelihood of Technical Challenge 0.5	Capability Needed to Counter 0.3	Contribution to International Posture 0.2	E (r_i)
Space exploration	.50	.45	.22	.43
Earth exploration	.30	.30	.55	.35
Civil support	.10	.15	.13	.12
Resources dev.	.10	.10	.10	.10

either by modification of the total weight of the identified elements or by the addition of supplemental (x) elements.

The procedure for assigning relevance numbers is described by Exhibit 6. This shows a typical relevance number assignment matrix at the second level of the tree under the Exploration and Civil Support Activities area. Here we are making a judgment regarding the technical upgrading required in the four areas shown at the left of Exhibit 6 using the three criteria shown at the top of the chart. The first task is to weigh the criteria. The sum of these weights in a horizontal direction should equal one. Once the relative importance of each criterion is assigned, we then make value judgments among the four items on the left of the chart using the first criterion only. Once this is accomplished, we make another value judgment using the second criterion. Finally, a third value judgment is made using the third criterion. In each case, the values assigned in the vertical direction must sum to one under each criterion used. Once the number assignment is completed, these data are inserted into the computer program where the weights given to the criteria are multiplied by the numbers assigned to each of the items, and the products are then summed across the horizontal axis for each item to give a final relevance number E(r_i) shown on the right of the chart. The same relevance number assignment technique is used at each successive node of the tree. In case of arithmetic errors in the assignment of the relevance numbers, the computer normalizes the figures in such a manner that the sum of the assigned numbers equals one and the computation continues.

A computer flow diagram for relevance computation is shown in Exhibit 7. After the structure and criteria data are inserted in the computer, the matrix for the assignment of relevance numbers is generated by the computer. The computer also calculates the averaged relevance value assigned by each member of the evaluation team and finally calculates the total relevance at each tree node and prints out the relative ranking for each item at the appropriate level of the tree.

Let us now put the entire process into perspective by taking the specific example, shown in Exhibit 8, down to the functional subsystem level and show the representative numbers that resulted.

The various tree levels are shown on the right of the chart, and the rele-

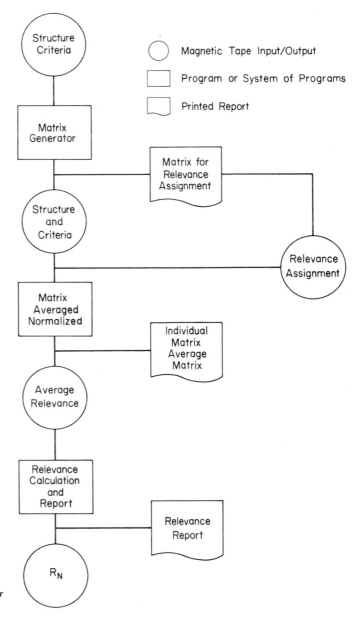

Exhibit 7. Computer Flow Diagram for Relevance Number Assignment.

vance numbers are shown on the left with the associated total rankings of the various situations and configurations. For example, at the mission level the relevance number was 0.01745 for the Fire Support Mission and it ranked thirteenth out of a total of sixty-four missions needing improvement to meet the national objectives. At the secondary system level we show the necessity for the development of another generation bomb. Note that when used

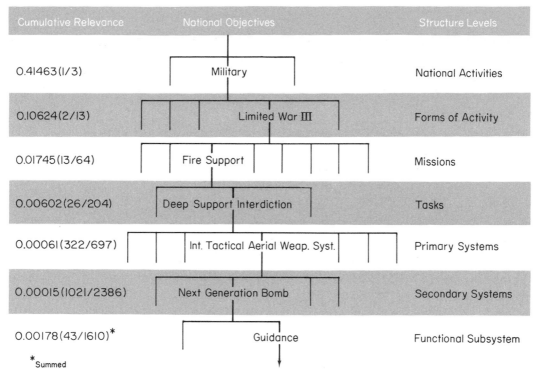

Cumulative Relevance	National Objectives	Structure Levels
0.41463(1/3)	Military	National Activities
0.10624(2/13)	Limited War III	Forms of Activity
0.01745(13/64)	Fire Support	Missions
0.00602(26/204)	Deep Support Interdiction	Tasks
0.00061(322/697)	Int. Tactical Aerial Weap. Syst.	Primary Systems
0.00015(1021/2386)	Next Generation Bomb	Secondary Systems
0.00178(43/1610)*	Guidance	Functional Subsystem

*Summed

Exhibit 8. Example Computation of Relevance Numbers.

only on this specific airplane, the bomb need ranked relatively low, about midway in the total secondary system grouping. However, the guidance system for this particular bomb had wide utility across so many different applications that it ranked very high (forty-third) out of a total of 1,610 functional subsystem categories.

Notice that the relevance numbers are smaller at the lower levels of the tree. However, if you sum across the tree, the relevance numbers for all the elements at each level would total one. These relevance numbers reflect a probability density function of priority of the particular node at the level on the tree.

A second major problem in forecasting technology needs is that of placing the forecast in the correct time frame. The second part of PATTERN has to do with the status and timing of technology needs.

Once the technology needs have been assessed without particular regard to the nation's ability to perform these functions, it is necessary to get an evaluation of the capability of industry to achieve technical solutions to the identified problems in some time frame. A basic input to this evaluation is an assessment of the current status of each of the technologies being considered as well as the likely rate of improvement in performance of each technology area over time. This information on the status of the various technologies and their projected rate of growth is used in determining the time and capability required to meet the identified technical needs of the nation. Status and timing data are inserted at the functional subsystem and technical deficiency levels in the relevance tree.

Exhibit 9 shows the development hierarchy used in assessing the status and timing of a given technology as it moves from research through exploratory development, advanced development, and product design to the available or off-the-shelf category. These categories are very similar to those used by the Government in defining their development phases. In assessing a given functional subsystem or deficiency, the team placed an "X" in the category where the technology under consideration was determined to be in its development cycle. In the example shown in Exhibit 9, the team felt that the technology being discussed was in the exploratory development phase. They were then asked to determine how long it would take a normally prosecuted program to move the technology from exploratory development to advanced development. This was considered to be three years. Two more years would be required in advanced development before it was ready to go into product design, and so on. So, the total time required for the technology to progress from exploratory development to inventory availability in this case was eight years. Thus, these data in the status and timing section basically indicate the capability of the country, time-oriented, to meet its military/exploration type technology needs.

Exhibit 9. Status and timing.

1. Available 2. Product design 3. Advanced development 4. Exploratory development 5. Research						
	S				X	
	T	1	2	2	3	

At this point we have discussed two of the three significant parts of PATTERN. The relevance tree and the status and timing problem. The third major part of PATTERN consists of the cross support information. Cross support represents the total gain in technical capability, across all subsystem areas, accumulated by working a specific subsystem problem area. Time was used as the common denominator and a zero to one scale was used in assigning the cross support values. Thus, the subsystem configuration cross support number represented the degree to which solving a specific technical problem on a particular subsystem configuration reduced the effort needed to solve similar or related subsystem technical problems within the same functional subsystem class for other concepts. Judgment factors considered by the technical specialists in the assignment of cross support numbers were the relative reductions in cost, time, and man-hours made possible because of the additional knowledge acquired. As stated the total gain was defined by a single common denominator time when the number was actually assigned.

A number of questions come to mind when considering the value of this forecasting technique such as: Isn't the PATTERN method merely a subjective judgment process and hence of questionable value? Isn't the selection of team personnel critical to the output? Can representative people be selected to make the judgments? Are the results repeatable? Aren't the results strongly influenced by tree structure?

Considerable effort has been spent on independent appraisals of the technique by analysts from within the company and from the consultant community. Drawing on experience generated during applications of PATTERN over the last few years, these evaluations have shown that:

1. The judgments, though subjective in nature, are representative of the type of managerial decisions required in resource allocation, program planning, and so on, where the manager must weigh all factors influencing a decision. Hence, the techniques used closely parallel and contribute to the managerial function of decision making.

2. When qualified personnel (generally the type which line managers usually seek for advice and opinions) are used in the judgment process, the results achieve a high degree of value to managers, are quite specific, and provide data very useful to the ultimate decision.

3. Repeated tests of the data have shown that with similar information and discussion, the results of these judgments are highly repeatable when averaged for a group of at least five experts in the field, and that they accurately reflect the consensus of the consulted community.

4. Comparative conclusions from entirely different tree structures have demonstrated that the results are not sensitive to tree structure, provided each person understands the specific definitions of each tree decision node.

5. These studies show that the tree can be used to divide and identify specific objectives useful to planning, without fear of deteriorating the value of the judgment process.

Several evaluations of the consistency and credibility of information contained in the assignment process have been made using various techniques. For example, results from voting matrices were randomly distributed in smaller groups, and correlation coefficients and levels of significance were calculated. In general, the significance numbers range between 0.01 and 0.1 where significance is equal to the probability that a correlation coefficient equal to or greater than the one observed could only have occurred by chance. The correlation coefficients normally run between 0.85 and 0.99. To further evaluate the consistency of the data base, relevance number assignments were made several months apart with some new members replacing members from the original team. Calculation of the results in these cases, while using the same tree structure, criteria, and scenario, showed that values in the later iterations were extremely close to those calculated several months earlier. The differences in expected value were only one to two percent.

In other studies, trial runs were taken with different tree structures and criteria, and with different groups of people. The results in all cases were generally equivalent. Basically, we found that groups of five or more people tend to have a predictable average understanding or common misunderstanding (we have devised nonparametric computer tests to flag the cases of misunderstanding) of the basic data and voting criteria. There is good agreement between groups of this size in the application of the criteria to the assignment process itself. The results, of course, depend heavily upon the validity of the scenario, the selection of valid criteria, and the experience of the team making the value judgments.

PATTERN METHODOLOGY APPLIED TO BIOMEDICINE

As stated earlier, PATTERN methodology can be applied to any problem area where decisions must be made under conditions of uncertainty. Some time ago, a study was performed to identify future needs in biomedicine. This study was given the acronym MEDICINE (MEDical Instrumentation and Control Identified and Numerically Evaluated), and used the tree structuring and relevance number assignment techniques developed earlier in project PATTERN. The study was conducted by a group of medical doctors, company personnel and external consultants.

The team defined the principal national objective in the biomedicine study as the need to maximize human life span with optimal health and activities in all environments. The generalized tree structure shown in Exhibit 10 illustrates the thought processes used in going from the national objective down to the technologies and tools required to perform the various functions. For example, at the task level of the tree, *diagnosis* would be one of the primary activities necessary to achieve the objective. The approach level contains the procedures to be applied to scientific tasks, such as surgery under *treatment*. At the system level are those bodily activities carried out by several organs operating together under integrated control (that is, reflexive control of locomotion, circulation of blood, and so on). The subsystem, or organ level, contains those units either natural or artificial, which perform a specialized function. The factor level lists those parameters which need to be measured, analyzed, and controlled to evaluate or cope with the normal or abnormal operation of the organ. The tool level lists the devices in the form of hardware or software used to acquire or process data or to treat, cure, or prevent disease.

After the tree network was structured, the evaluation team doctors used selected criteria at each level to assign appropriate relevance numbers. For example, one criterion used at the task level measured the need for upgrading of capabilities to meet the biomedical objectives based on the influence of locale and types of mass threats on population dynamics.

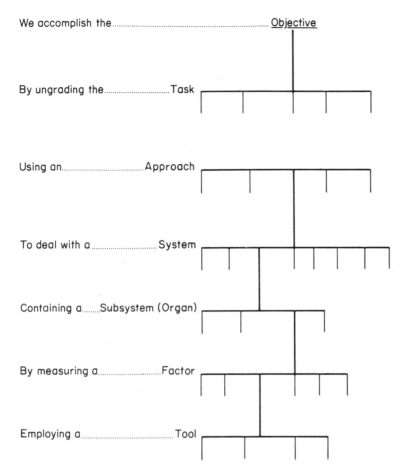

We accomplish the..Objective

By ungrading the...........................Task

Using an...................................Approach

To deal with a...........................System

Containing a........Subsystem (Organ)

By measuring a..........................Factor

Employing a....................................Tool

Exhibit 10. Relevance Tree for Biomedicine.

MEDICINE VS. PATTERN

Obviously there were some differences between the two studies; however, the basic problem was the same. How to make better investment decisions under conditions of uncertainty. Probably one of the best ways to compare the two studies is to analyze the differences in data characteristics. The basic data available to the PATTERN and MEDICINE studies offer an example of sharp contrast. In PATTERN, an evaluation of a series of man-originated systems concepts and hardware was made. In MEDICINE, the problem was not to evaluate competing systems but rather to evaluate maintenance and repair methods as applied to the human system. In the case of PATTERN, the system concepts and hardware were reasonably well understood, in that they were man-originated, and we had a fairly complete development of cause and effect in systems performance. In the MEDICINE study, we dealt with the human system on which substantially less hard data is available. Most of that which was available would have to be categorized as associative in nature, where the relationship of cause and effect is less well understood.

In essence PATTERN was an evaluation of hardware concepts which could be designed and built by men to meet requirements established by man's thought processes in response to world political and economic pressures. It included the system design, factors such as reliability, repairability, cost, and the degree to which a basic national need was met. MEDICINE, on the other hand, dealt with an evaluation of the medical instrumentation and control systems, subsystems, and associated hardware that contribute to maximizing the human life span with optimal health and activity. The objectives of MEDICINE could be met primarily by upgrading the use, maintenance, and repair of the human body. Since we do not, as yet, have the option of design or significant modification of the human body, the only avenue open to us for upgrading this case appears to be in the development of a more complete understanding of the human mind and body and the remedial actions required to correct possible malfunctions.

After the relevance part of the study was completed and the computer outputs reviewed, the doctors were again asked to give their personal opinions of the greatest medical needs in the one to seven, seven to fifteen, and fifteen to twenty year periods. It is significant that their individual judgments at this point followed the relevance tree quite closely. It is also interesting that, at the start of the study, there was no concensus among the doctors on any such need listings. This seems to strengthen the validity of the group judgments represented by the study outputs.

In summary it can be said that the problem structuring techniques and the quantitative evaluation methods developed and employed in PATTERN are of significant value in aiding the decision making process in that they:

1. Provide a basic framework within which the many complex interrelated variables which affect today's decisions can be structured and evaluated on a quantitative basis.
2. Furnish the results of these evaluations to the decision maker in a form usable to him.
3. Provide a relatively simple method of assessing the impact of changing conditions by periodically and systematically updating the inputs which in turn modify and update the outputs.

Perspective Trees:
A Method for Creatively
Using Forecasts

WILLIAM L. SWAGER

Intuitively or explicitly, many people suggest that a forecast of a single technological parameter, system, or concept is of dubious validity. Too many technical alternatives are possible, and a host of interactions with other developments—political, social, and economic—can alter the materialization of such a forecast.

Swager and his colleagues at Battelle have been groping for a way to anticipate at least some of these future developments. He proposes a kind of relevance tree that attempts to help one anticipate relevant factors in three domains:

1. The external environment, including technology in general.
2. The technology under consideration and available to the institution in question.
3. Utilities and functions as seen by the user.

A seven-step process is described, which leads to a broad morphological analysis. The goal of this analytical procedure is not a listing of technical options. Instead the perspective tree provides a way of thinking about technical alternatives against potential changes in the environment and the user's concerns. It becomes a way of structuring miniscenarios about numerous aspects of the future.

Over the past twenty years, my associates and I at Battelle have been attempting to help our sponsors take into consideration technological change in corporate, marketing, and R&D plans. In the past seven years we have had about thirty major programs involving technological forecasting and planning. During the course of these programs, procedures now called perspective trees were evolved.[1] The evolution of these concepts was strictly pragmatic. What worked was considered good. What failed to give insight and guidance was discarded no matter how theoretically appealing. Within the past year, the procedures and experience on these programs have been reviewed systematically, and this has resulted in a better defined step-by-step procedure, which is described in this chapter.

Our concept deals with a particular type of relevance tree which Erich Jantsch classed as a horizontal relevance tree[2] and which we now choose to call a "perspective tree." It is distinct in its purpose and use from the vertically structured relevance tree such as objectives trees, decision trees, alternative trees, and resource-allocation trees, and from the quantitative nature of relevance-tree concepts such as PATTERN (Honeywell).[3] It is, rather, a qualitative approach to thinking about the future for purposes of planning and policy development that seeks to discover and anticipate situations rather than to devise means for reacting to them. The concepts and methods described here are useful in identifying threats and opportunities or policy options, as suggested by Swager,[4] which are subject to further ordering and evaluation.

A perspective tree, as indicated schematically in Exhibit 1, is a diagram which graphically links relevant factors through three domains of concern. The environment domain is external to the company and is the one which includes changes and events—economic, social, political, and technological (except the technologies directly applicable to the business or technology being studied). Technology is the domain of technical changes available or potentially available to the company or its competitors for accomplishing technical advances in the business or technology under study. Between the two domains of environment and technology is a domain of utilities and functions. In one sense, the three domains in Exhibit 1 are simplified versions of the eight technology-transfer levels suggested by Erich Jantsch.[5] Jantsch suggests that there are four impact levels:

---

1 Important contributors to the concepts and practice of perspective trees included P. R. Beck, E. Bortis, D. Buckelew, F. H. Buttner, J. E. Burch, E. S. Cheaney, B. Goobich, H. R. Hamilton, H. S. Kleiman, E. S. Lipinsky, W. J. Sheppard, and R. Toelle.

2 Jantsch, Erich, *Technological Forecasting in Perspective,* Organization for Economic Cooperation and Development, Paris, France, 1967, p. 231.

3 Esch, Maurice E., "Planning Assistance Through Technical Evaluation of Relevance Numbers", *Technological Forecasting,* edited by R. V. Arnfield, University Press, Edinburgh (1969), pp. 197–210, European Conference on Technological Forecasting, June, 1968.

4 Swager, William L., "Technological Forecasting in R&D", *Chemical Engineering Progress,* LXV, No. 12, December, 1969, pp. 39–46.

5 Jantsch, *Technological Forecasting in Perspective,* p. 24.

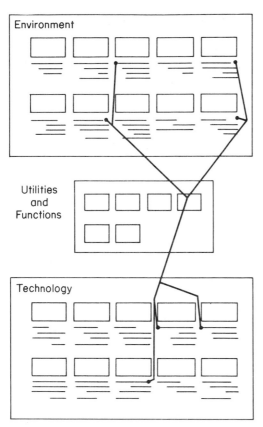

Environment

Social, Economic, Political
and General Technological
Events and Changes

Utilities
and
Functions

Technology

Technological Events and
Postulated Changes Directly
Contributing to Utilities and
Functions Defined by Scope

Exhibit 1. General Structure of Array and Perspective Tree.

He also defines four development levels:

Jantsch goes on to discuss conceptually a three-dimensional technology-transfer space as one of the means of indicating relationships and directions encountered in technological forecasting. We simply have not found it useful to think about technological change in either of these more complicated ways. Rather, we have chosen to think of all four of Jantsch's impact levels as one domain and his four development levels as another. The domain of utilities and functions was not included in Jantsch's technology-transfer levels. In Battelle's experience it is essential to define these utilities and functions explicitly in order to discover potential avenues of technology transfer. In the course of our work it has also been necessary to define more precisely what we mean by the terms: "utilities and functions." A utility is an attribute of a system deemed of value by the user of the system. A function is an attribute of a physical system subject to change or improvement through one or more technologies.

Jantsch indicated the primary direction of technological change as initiating at the development levels and having impacts at the higher levels. Of course this is true when technological change is the primary motivating force of technology transfer. However, one can view the transfer space of Jantsch or the simplified domains used by Battelle as multidimensional space, with each change being a vector force. Each such vector acts selectively on other vectors within its domain and on vectors in other domains. In such case, there is no unidirectional causative relationship implied by Jantsch's conceptualization. Rather, in some cases there are sets of environmental change which tend to have an overriding influence on a set of potential changes in technology. In other cases, there are sets of technological changes which tend to have an overriding influence on related sets of changes in the environment. The "seeds" of change can be discovered in either domain. The search for consonant changes in the other domain is aided by the logic provided by the utilities and functions. Cetron on many occasions has facetiously referred to this situation as a need looking for a solution or a solution looking for a need. The array provides a convenient visual display of thoughts to aid in that two-way search.

A set of perceived changes in the environment linked with a set of possible changes in technology through one or more of the functions and/or utilities depicts a tree. Each tree represents a set of related changes—the impact of which could threaten present products and services or could provide a foundation for building an area of new business. In other words, each tree represents a partially defined threat or opportunity to be considered further. The analysis and ordering of these business threats and opportunities or policy options are essential inputs to a planning process.

In the very earliest stages of forecasting and planning, the most critical stages, numerous authors of technological forecasting have pointed up the need to identify what should be forecast. It is this problem to which the perspective-tree analysis is addressed. What forces of change will affect the business? What possibilities for new businesses will be opened? What forecasts are needed to define better what these possibilities might be? Note that this approach does not initially depend on formal data. It relies first on intuitive judgment. It does so not because intuition is preferred but because adequate data are seldom available. Further, we are searching not only for the most probable future but also for all reasonable possibilities. To develop detailed data and forecasts on all possible factors in complex situations is plainly impractical even for the largest of business firms. In long-range forecasting, the point of diminishing returns, both in terms of costs and reliability, is passed too quickly. Some selection from among the virtually infinite combinations must be made if subsequent forecasting and planning are to be meaningful.

Furthermore, in this search for potential threats and opportunities, the nature and purpose of forecasts is different from that inferred from most of the literature of technological forecasting. One would infer that progress in technological forecasting is achieved by developing methods for making "hard predictions" of future technological events and trends and that the judgment of their usefulness will be made ultimately relative to the fulfillment of the predictions. In later steps involving the setting of priorities on policy options and in selecting program alternatives, the contingent forecasts required here can be "hardened." For the purposes of monitoring and surveillance, however, progress is not judged by the sophistication of methods for predicting a particular (or even several) pacing parameters further into the future. Rather, it is judged by insight and clarity by which today's options are identified. Unless the combination of forecasts clarify perceived options and help discover new ones, the forecasts are mental exercises that are not tied to business realities. The process of developing an array and noting relationships among preliminary forecasts of change by perspective trees provides an approach by which such surveillance and option definition can be accomplished in a reasonable and logical manner.

Specific applications of perspective trees to a variety of fields have been

Table 1. Steps in Development and Use of Perspective Trees.

Steps	Brief Description and/or Comments
1. Define initial scope	An identifiable portion of a business or technology of concern is the framework within which R&D plans are being developed.
2. Develop initial lists of relevant factors	Factors are listed in three classes or domains—environment, utilities and functions, and technology.
3. Make an array using an initial categorization scheme	Categories conceptually simple and germane to the business and/or technology are created.
4. Fill gaps and identify new factors	This step uses the categorization as a creative device to search for new and more meaningful factors.
5. Assemble initial forecasts and purge array	Each factor listed is implicitly a forecast. Make them explicit and, if possible, quantitative.
6. Search for relationships and identify perspective trees	This is a search for sets of changes in the environment that can be related to sets of changes in technology through the logic of the utilities and functions.
7. Translate to specific threats and opportunities	Each perspective tree represents an area of potential change. As such, each represents a threat or opportunity, depending on the present position of the company.

reported, including coatings in packaging,[6] petroleum technology,[7] steel technology,[8] paint technology,[9] and the technology of food additives.[10] These previous papers emphasize the general results derived and the usefulness of the planning method. In those references no descriptions were given of the procedures for conducting such programs or how perspective trees are developed and used.

Looking back now on those pragmatic efforts, we see a structuring of seven steps as indicated in Table 1. Each of these steps will be discussed with examples illustrating the principles and practice involved.

STEP 1. DEFINE INITIAL SCOPE

It is essential to identify a specific bounded area of business or technology on which to focus attention. This step of defining initial scope may appear to be an obvious one hardly worthy of mention. However, it is a step, often overlooked, with unrecognized effects on the forecasting and planning effort. The breadth of vision and thought patterns established by the initial definition are critical. In some ways this initial definition of scope emphasizes the question, What business are you in? as phrased many years ago by Peter Drucker.[11] The answer to that question can be given in terms of the description of the physical product, or it can be given in terms of the characteristics of that product, or it can be given in terms of what the product will do for customers. The scope of inquiry suggested by answers to these questions differs substantially.

Although the importance of defining the initial scope is being emphasized, there is no reason for making it into a major study. Give consideration to possible definitions covering the several levels that appear to be realistic. Select the level that appears appropriate and define precisely the scope selected and then be on with the study. If concentrated thought is given to the conceptualization of the scope at various levels, a few days of effort at most would be required. The significance of defining an initial scope may not be apparent from the above generalized discussion. Two examples reinforce the critical nature of this often overlooked step.

[6] F. H. Buttner and E. S. Cheaney, "An Integrated Model of Technological Change," in *Technological Forecasting for Industry and Government: Methods and Applications*, ed. James R. Bright (Englewood Cliffs, N.J.: Prentice-Hall, Inc., 1968).

[7] William L. Swager, "Technological Forecasting for Practical Planning," *Petroleum Management* (July, 1966).

[8] William L. Swager, "Materials," *Science Journal*, (October, 1967).

[9] Edward S. Lipinsky, "Application of Relevance-Tree Techniques in Technological Forecasting in Paint Technology," *Journal of Paint Technology*, XLI, No. 533 (June, 1969), 17A-22A.

[10] William J. Sheppard, "Relevance Analysis in Research Planning," *Technological Forecasting*, I, No, 4, (Spring, 1970), 371–9.

[11] Drucker, Peter, *The Practice of Management*, (New York: Harper and Brothers, 1954).

For the first example, a company that had been for many years a manufacturer of writing instruments wanted to insure its future leadership position. It asked Battelle to consider a program of technological forecasting and planning in the field of writing instruments. During consideration of this request, the significance and importance of carefully defining the initial scope became clear to us. One way to define the initial scope would be to develop forecasts bearing on the future of hand-manipulated devices that transfer opaque fluids to paper substrates. Traditionally, "writing instruments" have been so conceived—from the quill pen of middle ages to current instruments such as pencils, fountain pens, ball-point pens, and flow pens. Is an electronic light pen or light wand used on a cathode-ray-tube display a writing instrument by that definition? The "fluid" is changed and so is the substrate. What other fluids might be visualized? What other substrates might be useful? It was obvious that the definition of writing instruments initially was too narrowly defined. Rather than the technologies associated with the transfer of an opaque fluid onto paper, the definition was revised to "flexible symbolic generating systems and instruments for business and social communications." With this broadened definition, the scope of vision of the investigation encompasses systems and instruments for flexibly generating symbols that go beyond the use of pencil or pen on paper. It allows consideration of future alternative methods of transferring information from the mind of one person to the mind of another. Yet with the restrictive connotations of the term "flexible," computer printout devices such as line printers, plotters, computer output microfilm, and other machine-driven output devices could be excluded from detailed study and forecasts. If the definition had been broadened further to communications, the subsequent effort would have lost its focus. The effort would have been diluted. The creative visualization of forces of change affecting flexible symbol-generating instruments and systems was deemed proper to encompass the threats and opportunities facing a conventional writing instrument manufacturer.

Another striking example evolved during the technological forecasting course sponsored by the Industrial Management Center, Inc., in May, 1968. Professor Bright asked me to develop in a classroom situation a perspective-tree array and analysis on the subject of electric automobiles. As a part of that exercise, the initial scope was defined "to develop technological forecast bearing on the future of electric automobiles." At that time the steps in the procedure for developing and using perspective trees had not been formalized to the extent presented in this paper. I, too, overlooked the importance of Step 1.

The class was asked to list the events and changes in the environment affecting the future of electric automobiles. Later, it was asked to list the utilities and functions associated with electric automobiles, and finally it was asked to list the technical changes that might affect their future. At some point in the ordering of the lists, one of the students suggested that forecasting only those changes associated with electric automobiles provided too narrow a focus. He pointed out that the electric automobile is one alternative system for fulfilling business and personal needs for transportation primarily in urban areas. He argued that pursuing the narrow focus of electric automobiles we would not even ask for environmental or technical events and changes related to urban transportation systems that could be considered competitive with the electric automobile. The initial definition of scope was changed to "forecasts relative to urban transportation systems for business and personal purposes." This change brought about a polarization of the students. Some thought this move avoided the critical issue of forecasting the time at which some fraction of automobiles would be electric. Others took the stand that such forecasts could be meaningful only if at the same time forecasts were made bearing on the future of other viable concepts for urban transportation systems.

Battelle's experience supports the latter stand. Forecasts must be believed in order to be useful. An isolated forecast of the future of electric automobiles does not have credibility. Forecasts of progress of one technology are meaningless unless placed in perspective to comparable forecasts of progress in other technologies competing to fulfill the same utilities and functions.

STEP 2. DEVELOP INITIAL LISTS OF RELEVANT FACTORS

The purpose of this step is to assemble, quickly, intuitive understanding and insight concerning all of the possible forces—social, economic, political, and technical—that may have a bearing on the future of a business or technologies defined in the initial scope. We have found that small groups of people can stimulate one another's thinking in sessions similar to those popular several years ago called brain storming. In this case, however, attention is focused on the forces of change rather than on potential solutions to a particular problem. It is desirable to have a broad representation of experience in each of the groups so that a variety of points of view are reflected. A typical group may include a business manager, an engineer, a physicist, a sociologist, an economist, a marketing specialist, and a physical chemist.

The leader first calls the group's attention to the initial definition of the scope of the business and/or technology of concern. Then he asks members of the group to suggest events or forces of change in the environment that could conceivably have a bearing on the business or technology under study. Concentrating first on change in the environment—social, economic, political, and general technological—the suggested events or forces are recorded, usually on a large pad on a standing easel so that all in the group can see what is being recorded. The leader usually needs to provide no further direction. After an hour the group may have produced five to ten tear sheets from the easel. Then the leader focuses the group's attention on the functions and utilities associated with the business or technologies defined in the initial scope. Some of the functions and utilities may be tangible performance characteristics that are readily measured, such as cost, product uniformity, yield, quality, and reliability. Others may be much less tangible, but nonetheless real, attributes considered of value by the user, such as convenience, well-being, self-esteem, or prestige. And finally, the group's attention is turned to the domain of technology. The leader asks the group to identify engineering concepts that are in use in other technologies that might be transferred and applied to the one under study. He asks then for major trends or changes in science that anyone can suggest that could conceivably have an influence on the functions and utilities as defined previously. And finally, he asks the group to stretch their imaginations to list theory and recently identified phenomena that also could be used in one way or another to improve or change the functions and utilities previously identified.

You will note that the sequence of questioning by the leader follows to some degree Jantsch's technology-transfer levels. Any strict adherence to them, however, tends to retard the "stream of consciousness" of the group. For example, suggestion of a new phenomenon by one person may trigger thoughts by another on the applicability of existing engineering concepts.

The random nature of events and changes identified during these sessions is indicated in Exhibits 2, 3, and 4. Exhibit 2 indicates some of the factors listed in the three domains in a study of flexible symbol-generating instruments and systems. Exhibit 3 indicates the trends and changes visualized in a study of fragrances in cosmetics. Exhibit 4 presents some of the events and forces of change identified in a study of hot-melt extruded fibers.

As can be seen in Exhibits 2, 3, and 4, the "stream of consciousness" generated by one of these sessions in turn generates a random list of changes that have been visualized as having potential bearing on the future of the business

Exhibit 2. Lists of Potentially Relevant Factors Influencing the Future of Flexible Symbol Generating Systems.

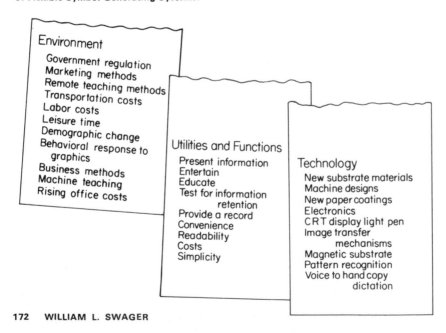

Environment
Government regulation
Marketing methods
Remote teaching methods
Transportation costs
Labor costs
Leisure time
Demographic change
Behavioral response to
    graphics
Business methods
Machine teaching
Rising office costs

Utilities and Functions
Present information
Entertain
Educate
Test for information
    retention
Provide a record
Convenience
Readability
Costs
Simplicity

Technology
New substrate materials
Machine designs
New paper coatings
Electronics
CRT display light pen
Image transfer
    mechanisms
Magnetic substrate
Pattern recognition
Voice to hand copy
    dictation

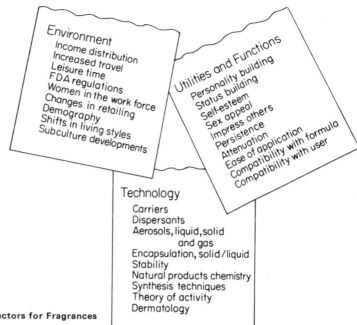

Exhibit 3. Lists of Potentially Relevant Factors for Fragrances in Cosmetics.

and/or technology defined by the initial scope. The point of diminishing returns has been reached after the third group has been queried. The first group may list sixty-five percent of the factors initially conceived; a second group, another twenty to twenty-five percent; and a third, ten to fifteen percent.

Exhibit 4. Lists of Potentially Relevant Factors Influencing the Future of Fiber System.

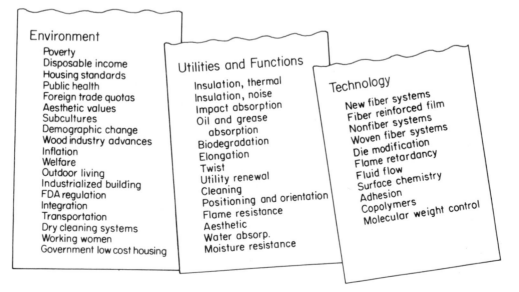

It may be well to stop here to contemplate what has been produced in these lists. Various words have been used to refer to the items in the list—factors, events, and changes. The items listed are primarily qualitative in nature. In the domains of environment and technology, each of the words or phrases is a crude qualitative forecast. It may be a generalization that can be later supported with data in the form of a prediction; it may be an erroneous assumption of a future change that cannot be supported, or it may be a contingent forecast: if the causative factors do not change, the trend will result in x in ten years. The main point here is that each item in the lists is a forecast of sorts. A priori, no one can prove that a refinement of any one of the crude forecasts would be relevant and justify the time and expense of making it. Perhaps even more important, it should be recognized that the crude forecasts as represented by the words are not a consistent set nor a complete set. The next four steps, although enumerated in sequence, are generally accomplished by a thought process better described as a cycling and recycling of the four in order to develop judgments of relevance and significance. In other words, the next four steps constitute an effort to answer the question: What forecasts need to be refined?

STEP 3. MAKE ARRAY USING INITIAL CATEGORIZATION

Lists of initial factors in the environment domain include broad aggregate terms such as government regulation, business methods, and leisure time. They also include more specific micro and disaggregate terms such as women in the work force, dry-cleaning systems, FDA regulations, and rising office costs. Even though the factors in the lists have been suggested by knowledgeable people as having some bearing on the future of the business or technology under consideration and are, therefore, crude qualitative forecasts, the precise relationship of such forecasts to the future is not apparent.

As a means of searching for more precise terms for which relevance can be postulated, a systematic categorization of the initial list is undertaken. The broad aggregate terms usually suggest broad categories which may include some of the disaggregate terms in the lists. Other terms suggest disaggregate or more micro terms that might be applicable. For example, referring to Exhibit 4, one natural way, although it may not be the best for the purpose at hand, would be to classify the "forecasts" in the environment domain into social, economic, political, and general technology categories as follows:

Social	Economic
Housing standards	Poverty
Aesthetic values	Disposable income
Subcultures	Inflation
Outdoor living	Transportation
Integration	Demographic change
	Working women

Political	General Technology
Public health	Industrialized building
Foreign trade quotas	Dry cleaning systems
Welfare	Wood industry advances
FDA regulation	Air conditioning
Government low-cost housing	

A preferred categorization approach to the same initial list would start as follows:

Demographic change	Government low-cost housing
Subcultures	Housing standards
Aesthetic values	Industrialized building
Disposable income	
Women in the work force	
Outdoor living	
Integration	

Such an approach to categorization begins to develop a thought pattern regarding the influence of shifting-value systems of subcultures and their influence on future fiber systems. It also raises some questions. If government low-cost-housing programs might have an influence, what about private low-cost housing? This question is recognition that a term in the list is a member of a set of terms for a category that had not been suggested in the initial listing —public and private housing. Each term in the initial list is examined in order to identify a broader category in which it is only a member, and, similarly each term is examined in order to subdivide it into a number of terms for which the original term is descriptive of a category. For example, the term "outdoor living" is but one aspect of changing "patterns of activities in the home," and "housing standards" is encompassing a set of more detailed standards more directly influencing fiber systems.

In recent projects suggestions have been made by sponsors to start with an a priori set of broad categories for the environment, such as social, economic, political, and general technology, which were presented initially. Thus far we have resisted any attempts to standardize this step of categorization because we have recognized that there is an opportunity for a creative structuring and categorization that is not conventional. It is often this unconventional categorization that provides the new insight needed to discover previously unrecognized relationships. Although this approach cannot insure such discoveries it can set the stage for the possibility of zigzag thinking as discussed by deBono[12] or creative synthesis, relating the apparently unrelated, as noted by Editor Luberoff in the same issue of *Chemtech*.

[12] Edward deBono, "The Virtues of ZigZag Thinking," *Chemtech* (January, 1971), pp. 10–14.

Similarly, a categorization scheme for the domain of technology is developed. Here, again, it is natural to use an initial categorization similar to the technology-transfer levels suggested by Jantsch. In such case the technology domain indicated in Exhibit 4 might have been categorized under the following headings:

Systems
    Fiber systems
    Composites
    Nonfiber systems
Engineering
Materials
Applied research
Scientific phenomena and disciplines

Often it is convenient to do so. For example, if such a categorization scheme were used, it would have been almost self-evident that the list of "scientific phenomena and disciplines" from Exhibit 4, which initially included only fluid flow, surface chemistry, and adhesion, should be expanded to include crystallization, cohesion, and nucleation. Here again the conventional categorization does not of itself stimulate identification of relationships. A new, more revealing, categorization might be built around possible advances in several kinds of systems, for example, spun-bonded systems. In such case, a categorization might include:

Spun-bonded systems
    Temperature control
    Spinning techniques
    Materials
        Polyesters
            Molecular weight control
            Bonding
            Crystallization
            Adhesion
    Flame retardancy

Similarly, the factors in the utilities and functions domain are placed in what intuitively appear to be useful categories. This domain will be covered in more detail in the discussion of the next step.

This step of conceiving an initial categorization in the making of an array cannot be separated from the next, that of filling gaps and identifying new factors. We have chosen to list it as a separate step because it provides a framework and a logic pattern for Step 4.

STEP 4. FILL GAPS AND IDENTIFY NEW FACTORS

In a study of coatings completed in 1965, an initial categorization of terms in the environment domain included: transportation, distribution and marketing concepts, shifts in technology of the processing of products, government regulations, automatic dispensing, packaging machinery, changes in forms of products, shifts in demands for products, and per capita disposable income. At that stage we recognized that the categories of factors listed gave only hints of the true factors that were relevant to the future of coatings in packaging. That suggested the possibility of a morphological exercise as a means of continuing the search.

Morphological analysis was introduced by Fritz Zwicky, a pioneer in jet-engine design.[13] He used the term to describe a standard engineering method for identifying all possible devices to achieve a specific functional capability. Zwicky dealt only with technical dimensions. He formulated a technique for identifying, indexing, and organizing the parameters affecting the design of a physical device.

Some would consider morphological analysis limited to the ordering of technical factors. Hence, the thought process I want to describe now is probably not properly defined as a morphological analysis. Whether one chooses to call it semantics factoring, a morphological exercise, or common sense, the purpose is to search for more relevant and meaningful factors by continuing to consider disaggregated terms related to the aggregate ones and aggregated terms encompassing the disaggregated ones. The exercise involves going from one level of abstraction or aggregation to others in order to identify significant and relevant change.

In the search for relevant change in coatings in packaging, it became apparent that any transportation changes would be subsidiary to distribution and marketing changes, transportation being only one of several factors influencing distribution and marketing. To extend the example, the various ways were explored to describe the morphology of "distribution and marketing" as specifically related to coatings in packaging.

Data were collected and preliminary forecasts were made on disaggregated terms that were felt to be important, such as transportation methods, warehousing, and other factors. Each of the more micro forecasts were examined to judge whether or not the forecast condition could have a significant influence on the future of packaging and/or coatings of packaging materials. In the cyclical process, all preliminary forecasts judged as not significant or not relevant were set aside. Those judged significant and relevant are given in Exhibit 5. From the obvious condensation accomplished in arriving at Exhibit 5, the purpose of the morphological analysis is not to expand the array but rather to reduce it to include only significant and relevant changes that can be visualized. Judgments of significance and relevance and irrelevance are made in relation to the utilities and functions which, in the case of coatings in packaging were:

[13] Fritz Zwicky, *Morphology of Propulsive Power,* Society for Morphological Research, Pasadena, 1962.

Functions of packages	Functions of coatings
Protection	Grease barrier
Containment	Water barrier
Distribution	Water-vapor barrier
Facilitate product use	Gas barrier
Merchandise	Odor barrier
Advertise	Ultraviolet barrier
Brand continuity	Chemical resistance
	Printability
	Decoration
	Smoothness, brightness, and gloss
	Abrasion and scuff resistance
	Antiskid
	Lubricity
	Release
	Sealability

Exhibit 5. Preliminary Forecasts in an Array Related to Distribution and Marketing.

| Transportation Methods | Distribution and Marketing | | | |
	Warehousing and Materials Handling	Retailing	Industrial and Institutional Merchandising	Other Factors
△Increase use of air cargo	△Automated warehousing △Bulk handling systems	△Self service	△Unit packages	△Advertising △Branding △Vending

Throughout the morphological exercise, each new breakdown or aggregation was searched for potential changes that would increase or decrease the future weighting, importance, or requirements for these functions and utilities. Otherwise the morphological analysis would have expanded the "distribution and marketing category" unmanageably by including tables of contents of books on transportation, warehousing and materials handling, retailing and merchandising, etc.

A morphological exercise in the technology domain is similarly accomplished. There is a tendency for most engineers to use a textbook approach, i.e., list from a textbook the morphology of a particular technology or discipline, or seek some standardized structuring of all science without the critical pruning based on consideration of relevance to the utilities and functions.

Utilities and functions, too, may be subject to a morphological analysis. Buckelew[14] suggested a way to view the utilities and functions of communications systems by a morphology of what communications systems do. A portion of that morphology is given in Table 2. Referring now to Table 2, conventional radio is described by a combination of the following: sound, educate/in-

14 Donald Buckelew, Battelle's Columbus Laboratories, private communications.

form and entertain, speech, electronic, directed, one to multiple, simplex, and audio. Similarly, conventional television can be so identified. It is obvious that other combinations of these functions and utilities represent systems not now in being. One of the roles of developing a morphology of the functions and utilities is to stimulate a search in the other two domains, first in the technology domain to identify potential step functions and order of magnitude changes in technologies that would make possible major advances in the utilities and functions, then in the environment domain for major changes that will affect the future weighting or importance of the utilities and functions. Experience has shown that the "cross-over" logic between the environment domain and the technology domain provided by the functions and utilities domain is important. A new way of classifying and subdividing the functions and utilities has stimulated a search mechanism for identifying a set of potential changes in the environment related to a potential set of changes in the technical domain.

Table 2. Tentative Expanded Structure

Functional Characteristics	Elements
Transmission sense employed	Sight, sound, touch, smell, and taste
Purpose	Persuade, educate/inform, direct/control, and entertain
Transmission mode	Speech and symbology
Transmission method	Manual, mechanical, electronic, optical, and telepathic
Nature	Random and directed
Dispersion	One-to-one, one-to-multiple, multiple-to-one, and multiple-to-multiple chain
Direction	Simplex and duplex
Reception data form	Audio, symbolic, replica, sense experience* and concept

*Not otherwise included.

STEP 5. ASSEMBLE INITIAL FORECASTS AND PURGE ARRAY

The point has been made that the words in the array are crude qualitative forecasts. During the morphological analysis, readily available data are assembled quickly in order to make as quantitative as possible the initial forecasts. With initial forecasts available, it is possible to purge the array, eliminating the irrelevant. The criteria used for this purge generally are: (1) the forecast change will become significant beyond the time horizon of concern; (2) forecasts are not relevant because they are included implicitly in other grosser forecasts, and (3) even major change of a factor will not have significant impact on any of the functions or utilities.

A few examples will illustrate these points. In a recent study of equipment and process control in the paper industry, one of the environmental factors included in the initial lists was synthetic fibers. The Japanese had recently reported the preparation of a sheet of paperlike qualities made from synthetic fibers. Initial forecasts relative to this factor included costs of synthetic fibers relative to costs of wood fibers. Assembly of readily available information on the demand for feed stocks for lead-free gasoline, sulfur-free fuels, and polymer requirements for other purposes indicated that the cost of raw material for synthetic fibers, petroleum feed stocks, would remain at a relatively high level or at least would not drop in price by a factor sufficient to make synthetic fibers competitive with wood fibers. Similarly, readily available information on the supply of wood was collected which indicated that supplies of wood are still growing, and the technology of wood harvesting is keeping real costs of wood fibers at about a constant level. The initial crude forecast, implicit in the term "synthetic fibers," was that the development of synthetic fibers would have an impact on paper-making equipment, processing, and control. This crude forecast was rejected on the basis of initial forecasts using readily available data. It was placed in a list of rejected factors so that any new information regarding synthetic fibers would be considered in light of the judgments and forecasts made.

Another forecast implicit in the initial list of factors in the environment was "electronic communications versus hard copy." Preliminary data were assembled regarding facsimile systems and the growth of such systems. It was concluded that any shifts toward electronic communication will affect, first, periodicals; magazines and newspapers. It was further estimated, however, that significant impact on periodicals will occur after 1980, the time horizon of the study. This factor, therefore, was considered to be irrelevant and placed in a list for further review. Another factor in the initial list was plastics in packaging. Since this is a crude forecast related to demand for paper and a forecast of a general nature had already been made relating demand for paper to the growth of GNP and since no major breakthrough in the use of plastics in packaging could be identified that would significantly modify the gross demand for paper, this factor also was considered to be irrelevant and placed in a separate list for further review.

During the course of this step, factors are either included in the array or placed in a separate list of factors that are considered to be irrelevant. This provides a basis for subsequent review and modification on the basis of any new information.

At this point in the analysis, it is usually a good time to reconsider the initial scope. If, during the course of developing the morphology of utilities and functions and the making of initial forecasts, it appears desirable to either broaden or further restrict the initial scope, such a step can be taken. The array is modified physically to reflect that revised scope.

First note the obvious relationships. Experience is a great teacher. Men with insight can intuitively sketch out quite a number of relationships or perspective trees relating a set of changes in the environment to a set of changes in technology through the logic of a set of functions and utilities.

After the obvious ones have been noted, a search pattern can be established. Kleiman[15] has suggested that each of the changes in the environment domain can be considered as a deterministic forecast, a conditional forecast, or a possibility. Similarly, each of the changes in the technology domain can be considered a deterministic forecast, a conditional forecast, or a possibility. The search pattern is conducted in six steps, starting first with the deterministic forecasts in the environment. The question is asked: What other forecasts in the environment are related and reinforcing or countervailing? Next, the questions are asked: What deterministic forecasts in technology are related? Through which utilities and functions? Next: What conditional forecasts in the technology domain are related? Through which utilities and functions? And finally: What possibilities of technical change are related? Through which utilities and functions? Wherever relationships are found or postulated, they are noted as trees or partial trees.

Next, similar questions are asked, starting with the deterministic technological forecasts. What other forecasts in technology are related and reinforcing or countervailing? What deterministic environmental forecasts are related? What conditional forecasts in the environment are related? What possible changes in the environment are related? Similar series of questions can be formulated, starting with conditional forecasts in the environment, conditional forecasts in technology, possibilities in the environment, and possibilities in technology. Examples of postulated relevance or perspective trees are given in Exhibits 6 and 7.

The partial array and perspective tree in Exhibit 6 (perspective tree for fragrances) are only indicative of the line of reasoning used. Both would be expanded to provide more detail concerning the likely changes in both the environment and in technology. The simplified tree notes the social forces unleashed by the Women's Liberation Movement and the rise of women in diverse occupations in the working force. These are reinforcing trends that may give rise to a rapidly increasing number of women in new kinds of job classifications. These social and demographic factors are related to utilities and functions that fragrances can provide in cosmetics. In particular, more persistent fragrances may increase as women undertake certain specific new occupations. Within

15 H. S. Kleiman, Battelle's Columbus Laboratories, private communication.

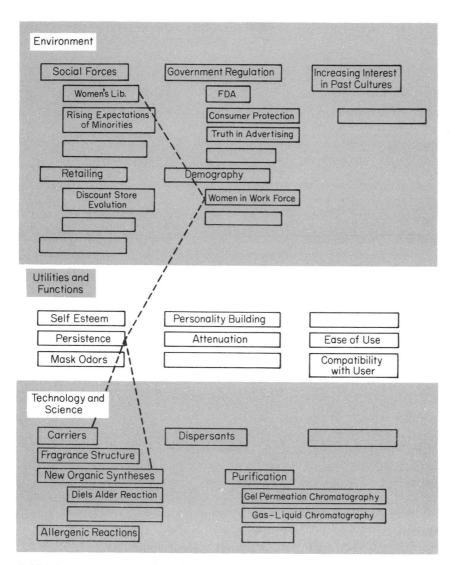

Exhibit 6. Partial Array and Example of Perspective Tree on Future Role of Fragrances in Cosmetics.

science and technology, there are a large number of alternative means for providing a greater degree of persistence without relying on a gradual decrease of an overwhelming initial aroma. Some of these alternatives are indicated in the technical domain.

Referring now to Exhibit 7 (perspective tree for equipment in paper industry), most business economists forecast that the cost of capital in the next ten-year period will be generally higher than it has been in previous time periods. Return on capital in the industry decreased from 1950 to 1963 and has remained lower than normal. A reasonable forecast is that a trend upward will be achieved. Of the many functions of process equipment, control systems, and associated instruments, two are uniquely related to the above environmental factors: increased capacity and increased yield. If, through technological change the capacity of existing equipment can be increased by reducing cycle time or yield can be increased by better process control substantially improved capital efficiency in the paper industry can be achieved. New pulping processes, continuous digesters, and direct digital control hold promise for increasing capacity and/or yield. Here again the array and the tree are only partially developed to indicate the logic of the perspective tree.

The above search for relevance emphasizes relationships among domains. In the normal course of the selection process accomplished in our educational system and in the compartmentalization in our institutions, most specialists are comfortable and able to perceive relevance within a domain. An economist,

Exhibit 7. Example of Perspective Tree in Partial Array of Equipment in Paper Industry.

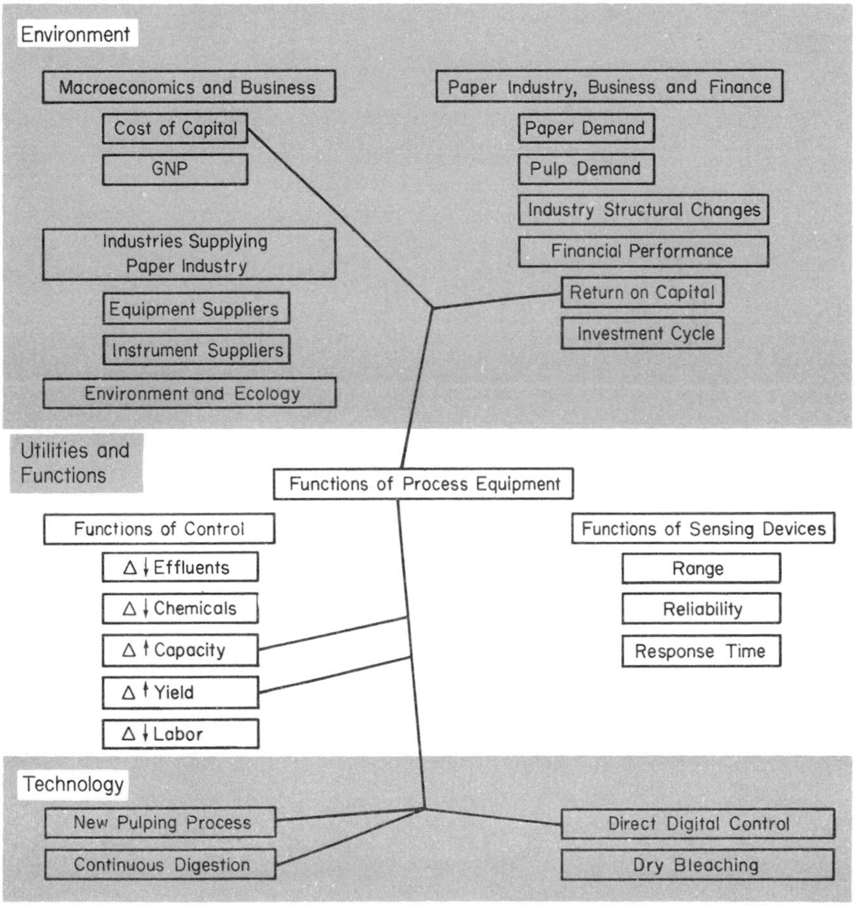

sociologist, or marketing man, for example, sees quickly the relationship among demographic change, income redistribution, and shifting demands for certain kinds of products, and similarly a chemist or chemical engineer readily perceives relationships among Nylon 6, polyesters, molecular weight control, and methods of fiber formulation. The array and search pattern described above at least emphasizes relationships among domains. Relatively few technically trained people are comfortable or able to conceptualize and creatively relate forecasts in the three domains. Recent discussions with managers of technical planning and innovation teams indicate agreement that better understanding is needed to train and select people for this kind of analysis. That, perhaps, could be a subject for a whole new paper.

STEP 7. TRANSLATE TO SPECIFIC THREATS AND OPPORTUNITIES

The array and perspective trees are never completed in any absolute sense. There is always the possibility that a factor has been overlooked or a new tree can be discovered with more effort. One can only say the analysis reaches a point of usefulness.

Each of the perspective trees identified might better be called "an area of potential change." In a way, the tree could be described as a shorthand miniscenario needing further interpretation from a particular institution's point of view. Each area of change is either a threat or an opportunity, depending upon a company's position. An example of a perspective tree drawn through an array involving coatings in packaging also involved food processing and packaging. The implications of the changes identified in the array for coatings in packaging is given in Exhibit 8. Environmental changes, such as rising labor costs, the pressure of potential savings from economies of scale, recent innovations in food transportation and warehousing, and recent trends in acquisitions and mergers of companies in snackfood businesses have positive relationships to the possible future centralization of production of snack foods such as potato chips. Yet today there is no discernible trend toward the centralization of production facilities. One of the main barriers to the centralization of production facilities of such snackfood items is the shelf life of present packages. Protection with longer shelf life required for a large food transportation and warehousing system is the "utility" of packaging that must change if the structure of the industry is to change. As indicated in Exhibit 8, water-vapor-barrier, gas-barrier, and strength functions are positively related to the protection utility. It is recognized that these functions can be achieved through metallic and other rigid systems at a cost. Opportunities for improving rigid systems could be diagrammed in a similar fashion. We have chosen, however, to diagram in the tree only the potential technical advances related to flexible systems.

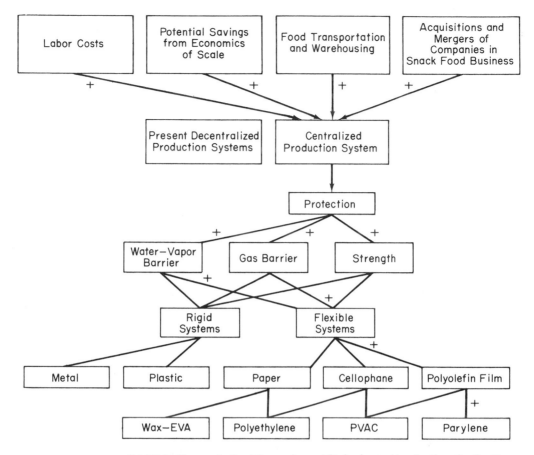

Exhibit 8. Changes in Food Processing and Packaging and Implications for Coatings.

The systems of flexible packaging include the substrate materials such as paper, cellophane, and polyolefin films. To achieve substantial improvements in water-vapor-barrier properties, gas-barrier properties, and strength, such substrate materials must be coated. In the past these coatings have been wax-EVA (ethylene vinyl acetate), polyethylene, and polyvinylidine chloride. A new polymeric material, Parylene, has been visualized as a potential coating material to achieve superior barrier properties. At present Parylene is several hundred dollars a pound and cannot be considered a commercially usable material, yet a specific forecast of advances in the performance and cost of Parylene or any of the competing materials that would allow a major improvement in the barrier properties of flexible packaging systems would result in an opportunity for centralized production of snack food. Such a tree provides the logic for noting the pacing parameters of water-vapor-barrier properties, gas-barrier properties, and strength. Forecasts of these may be made on the basis of historical data, or they may be made on the basis of expert judgment. All of these forecasts are obviously related to the forecasts of the factors in the environment such as labor costs, economies of scale, food transportation and warehousing, and so on. Forecasts of major improvement in barrier properties of flexible packaging systems could be a major threat to the continued operation and viability of a small manufacturer of snack-food items. Similarly, it could be a major opportunity for a well-financed small manufacturer of snack-food items to expand by acquisition. Similarly, the tree can be viewed as a threat to conventional flexible materials, producers, and coaters.

Another example is indicated in Exhibit 9 which covers changes in packaging and distribution of petroleum products and their implications for coatings. In revealing the dynamics of competition among materials for the packaging of motor oil, it has been observed that several environmental factors result in the overwhelming use of unit containers. These factors include "advertising, branding, and vending" as well as "government and industry pressures for consumer protection," both of which have positive influences on the use of unit-container systems. Another environmental factor, "social and political pressures for control of solid waste," has a negative influence on the use of unit containers. For

Exhibit 9. Changes in Packaging and Distribution of Petroleum Products and Implications for Coatings.

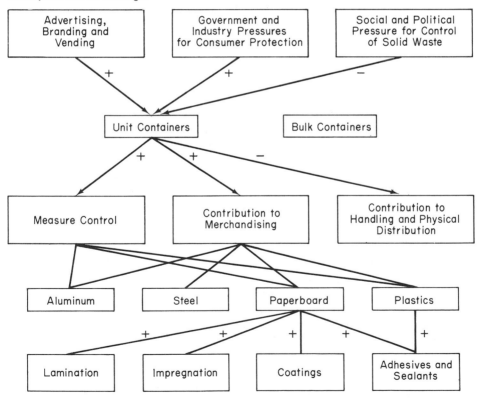

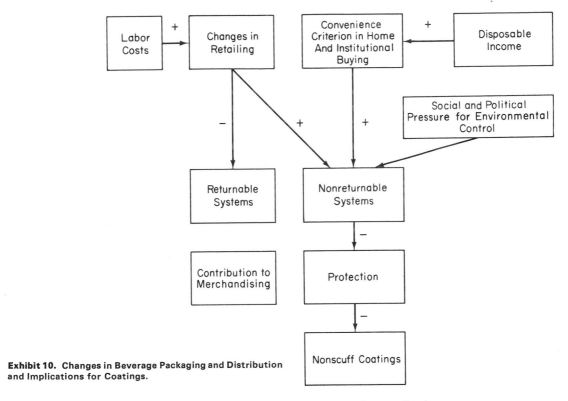

Exhibit 10. Changes in Beverage Packaging and Distribution and Implications for Coatings.

unit containers, the utilities and functions are measure control, contribution to merchandising, and contribution to handling and physical distribution. Note that some have positive influences and others have negative influences. The utilities of measure control and merchandising have been fulfilled historically first by steel, then aluminum, and then paperboard, and aluminum foil containers. Efforts to improve the usefulness of paperboard and aluminum foil containers have been directed through laminations, impregnations, coatings, and adhesives and sealants. Thus far, the social and political pressures for control of solid wastes have not been sufficient to make major changes in the unit container systems. Distributors of soft drinks have solved the problems of branding and adversiting in their introduction of vending machines (using bulk syrup, ice, and water), which points to the possibility of similar dispensing units with positive control of quality and quantity from bulk containers.

This is obviously a major threat to manufacturers of unit containers for automotive-engine oil. It may be, at the same time, a major opportunity for vending-machine manufacturers and other manufacturers of pumping and dispensing equipment to develop systems for bulk transportation and unit dispensing at the point of use.

Still another example is given in Exhibit 10. The tree describes the forces of change affecting beverage packaging and distribution and their implications on coatings. Rising labor costs, demands for space in retailing establishments, and changes in retailing policy have a negative influence on returnable beverage-packaging systems and have a positive influence on nonreturnable systems. As disposible income has increased, the convenience criterion has become more important in home and institutional buying, which in turn has had a positive influence on nonreturnable systems. In addition, the social and political pressures for environmental control have had a negative influence on nonreturnable systems. Nonreturnable beverage-packaging systems require only a one-way corrugated box, compared with a solid fiber box required for a returnable system. The amount of protection in the way of nonscuff coatings on a returnable container was eliminated in the nonreturnable systems. Here again, this is an area of change with the salient aspects of both environmental and technical changes noted. Whether these changes are a threat or an opportunity depends upon whether you are a manufacturer of nonreturnable containers, a box manufacturer, or a supplier of coating materials. It has been our experience that the identification and elucidation of these strategic options is a major advance. Experienced managers can usually rank order the threats and opportunities so identified if they are clearly communicated. Any problems in ordering of the strategic options and establishing priorities may call for more detailed forecasts, but in such cases there is a ready answer to the question, "What should be forecast?"

FORECASTING METHODS IN A PERSPECTIVE-TREE FRAMEWORK

A seminar on technological forecastıng was held at the Battelle Seattle Research Center in November, 1969, to examine critically some of the methods of technological forecasting and planning. Participants in the seminar were encouraged to criticize available methods, to look for relationships among various approaches, and to point directions for development of new methods. Three of the six conclusions reached during the seminar were: (1) the methods used for technological forecasting when explicitly tied to planning tend to converge; (2) multisteps are involved in forecasting and planning, and part of the previously emphasized differences between methods stemmed from the sequence of steps; and (3) as forecasting attempts consciously to link social, economic, and political potentialities to technical possibilities, explicit modeling of structural relationships is a necessity. In this seminar, the structural relationships took the form of identification of relevance both within and among domains. The domains were defined as environment, applications, and technology. The presentations by Donald Pyke and William Sheppard, among others, led to those conclusions. Pyke[16] explained the sequence that he was

16 North, Harper Q., and Donald L. Pyke, " 'Probes' of the Technological Future," *Harvard Business Review,* May-June 1969, pp. 68–76.

working toward in using Probe II. In an attempt to develop a normative structure, Pyke has grouped each of the events identified in Probe II in one of the following domains: technology, applications, or environment. He is using contextual mapping with morphological analysis within these domains in order to identify gaps and to develop a better estimate of interdependencies among future forecast events. Following the mapping exercise, he visualizes the need for identifying and assessing the interrelationships among domains, which involves notions of relevance. Once these steps are complete, he sees the possibility of relating the technological forecasts and relationships to an input-output model developed by TRW, Inc., and subsequently to the company's marketing research efforts. Sheppard[17] presented the iterative process used in searching for relevance among exploratory forecasts in an environmental and technology array, the detailed steps for which have been outlined above. The discussion focused on the similarity of contextual mapping among the three domains of technology, environment, and applications as suggested by Pyke and the perspective-tree approach denoting relationships between environment domain and the technology domain through the logic of utilities and functions. In the ensuing discussion it became clear that the major differences between the two approaches derived from the sequence of forecasting and planning steps. In one case the forecasts are generated and then related. In the other case, crude forecasts are generated and then related, and finally detailed forecasts are made selectively.

It is clear that the perspective-tree procedure as outlined here is not a method of technological forecasting, but rather a method for using forecasts in a planning process. A Delphi exercise, for example, could be used to generate the preliminary forecast events in the environment domain and the technology domain. Similarly, all methods of trend extrapolation and models such as projection by analogy, substitution, and simulation may be used in the technology domain. In general, however, the formal methods for forecasting require substantial effort in data collection, analysis, and interpretation. When such methods are used in connection with the perspective-tree procedure, they are used selectively for refining and adding credence to forecasts of specific events and trends. The justification for the time and cost required for making the forecasts can be judged in light of the decision-making deadlines and the adjudged relative importance of the threat or opportunity being evaluated. Furthermore, the perspective-tree analysis provides detailed consideration of the utilities and functions which, in turn, provide a basis for selecting the pacing parameters to be used in such forecasts.

This paper has focused on only the first of four major steps in rationalizing business plans with R&D plans. Experience and practice in the other major steps will be covered in several papers soon to be released.

[17] *Technological Forecasting*, I, 371–79.

SUMMARY

One of the roles of technological forecasting is the identification and evaluation of potential change to which a business must respond. Steps outlined here in the perspective-tree procedure provide one way to develop a surveillance procedure that gives reasonable assurance that the threats and opportunities facing a company will be at least identified. With such identification, a management team has the opportunity to establish its business strategies and concommitant R&D strategies in a rational way.

The perspective-tree procedure is presented as one way to accomplish this surveillance step. It is not the only way; I am sure it is not the best way; it is merely a way that Battelle has found of value.

A Scenario Generating
Methodology

CLARK C. ABT
RICHARD N. FOSTER
ROBERT H. REA

Although the writings of Kahn and Weiner, particularly in *The Year 2000,* first popularized the concept of scenarios, little methodology has been offered on how to develop them. These authors offer us an explicit procedure, which they have applied to industrial and government problems.

The relevance of scenarios lies in the following: Plans and forecasts are based explicitly or implicitly on assumptions about the future. Hindsight shows that our assumptions about that "true" future have been, usually, far from correct or complete. Therefore, the scenario concept also could be described as contingency planning, which has long been practiced by governments. A scenario means a *possible* future not, however, the most *probable* future. By studying a number of scenarios, hopefully one can identify those future events that would be truly significant and which would call for a major shift in plans. For example, a United States oil company should consider scenarios that include different assumptions about our oil import quota policy; an automotive firm should consider scenarios that include different degrees of technical progress and social acceptance of various types of automotive power.

Variables such as environmental inter-user relations, with their political and economic implications, are included by the authors to provide scenarios that deal with possible future economic, technological, political, and social aspects of the world.

Fourteen different potential "world types" are generated by this computerized methodology. Alternative scenarios develop by allowing the variables in each world type to evolve randomly. The explicit descriptions of the results in some instances make one realize how potent our vague and unimaginative assumptions about the future can be to a proposed course of action.

INTRODUCTION

Implicit in every forecast and every decision about the future is a scenario or set of scenarios. The implicit train of logic which generates a scenario and its permutations is very much like generation of scenarios by "wise old men." From the early tribal witch doctors to the President's Council of Economic Advisors, each society has men who, using their knowledge and intuition predict a future environment beginning with a set of conditions in the present. If the wise old men could really predict the future with accuracy, then forecasting would become a relatively simple task. Unfortunately, wise old men have the same problem that the businessman planning a trip has—their derivation of a future environment grows out of knowledge and logic based on their understanding of past and present conditions. Events that do not apparently follow from the present are missed in the process. In this case the scenario is an imaginative extrapolation of present assumptions rather than a picture of a number of possible environments.

Given the historical fallibility of wise old men's forecasts, industry and governments have turned to more objective techniques to create vignettes of the future world. Among the better known techniques for generating "more objective" scenarios are: consensus techniques, iteration through synopsis, and cross-impact matrices. In the consensus technique, groups of experts in various fields are asked what major events they might anticipate in a specified future period. The experts interact through a Delphi process. They remain anonymous to each other and answer questionnaires which seek reasons for minority positions. These positions are then fed back to the group members who are then asked to evaluate their estimates. One could summarize the method as a consensus of "wise old men." The disadvantages are that the biases introduced by the present are not completely netted out; only certain types of items are included and mutual interactions and dependencies are obscure. In most cases the data is also nonquantitative, thus limiting its usefulness for indicating such things as market size.

Ronald Brech in his book *Britain 1984*[1] used iteration through synopses to develop his vignette of Britain in 1984. This method is designed to increase interdisciplinary consistency in the scenario. It consists of developing independent scenarios for each discipline and then modifying the descriptions through

---

[1] Ronald Brech, *Britain 1984: Unilever's Forecast—An Experiment in The Economic History of the Future,* (London: Darton, Longman and Todd, 1963).

an iterative process which makes the scenarios compatible with one another. Problems arise when disciplines, such as sociology, must remain compartmentalized because of the difficulty of estimating their effects on other fields.

A third, and potentially powerful method of scenario generation is the use of cross-impact matrices. Gordon and Helmer describe the technique as a:

> ...method of analysis which permits the orderly investigation of the effects of potential interactions among items in a forecasted set of occurrences. It requires a methodical questioning about the potential impact of one item, should it occur, on the others of the set in terms of mode of linkage, strength of linkage, and time when the effect of the first on the second might be expected.... Having collected the judgments or data linking all possible combinations of items in terms of mode, strength and time, it is possible to perform an analysis which revises the initial estimates of probability of the items in the set.[2]

The principal problem is the use of subjective probabilities and subjective measures of effect which occurs in a compound iterative process. Constant multiplication of subjective numbers creates a scenario whose probabilities are highly subject to error. A second problem is the built-in bias generated by starting with fixed relationships which assume cause-effect relationships to be identifiable in the future on the basis of today's thinking.

The advantages and disadvantages of each of these three techniques, consensus, synopsis, and cross-impact matrices, are summarized in Table 1.

Table 1. Advantages and Disadvantages of Current Scenario Generating Techniques.

Method	Advantages	Disadvantages
Consensus technique	Decreases the bias in "wise old men" forecasts	Only certain items are considered; interactions are obscure; nonquantitative
Interaction through synopsis	Increased interdisciplinary consistency	Lack of definable relationships between certain disciplines
Cross-impact matrices	Internal consistency	Subjective assignment of values, biases result

AN ALTERNATIVE SCENARIO GENERATING METHODOLOGY

Each of the three methods described has inherent advantages and disadvantages. All three share three common faults: the generation of a single picture of the future (albeit one can alter variables to test their effect) instead

[2] T. J. Gordon and O. Helmer, "Generation of Internally Consistent Scenarios Through the Study of Cross-Impact," (presented at the Industrial Management Center, January 27–30, 1969).

of a range of possibilities; bias in future interrelationships stemming from the use of today's perceptions; and lack of emphasis on variable identification. This chapter will discuss an alternative scenario generation methodology which Abt Associates developed and the application of this method to current work on an industrial scenario generator. The methodology was designed to avoid the common faults shared by the other methods.

Environmental Control Study

In years past, environmental systems in the United States were designed both according to their improvements on existing components and to intuitive projections of future environmental requirements. Little attention had been given to discontinuities or reversals in current political trends or even exponential growth in technology. The common observation "engineers always overestimate what they can do in five years and underestimate what they can do in twenty" illustrates the fallibility of technical planning. Here the fault is only in misjudgment. In environmental control, however, misinterpretation may lie at the root of error, since rarely is undistorted information available on national behavior and attitudes. When it comes to so vital an issue as the natural environment, careful forecasting is essential.

In light of the multiplicity of factors and levels of uncertainty that pervade the environmental planning perspective, educated guessing by even the most expert economists and political scientists is not a reliable projection technique. Wittingly or not, the prejudices of their authors enter into forecasts of the future and areas of little emotional interest are often dealt with in incomplete detail. The rapid diffusion of environmentally "interesting" locations will compel planners to regard the political-economic environment as an organic unity rather than as a limited number of points, and no location or aspect should be rejected out of hand from future environmental forecasts.

In response to the need for improving upon the scenario forecasts now available to environmental planners, *A Scenario Generation Methodology* was prepared by Abt Associates Inc. Its objective was to provide a common projection technique for the analysts who would be engaged in exploring the long-range possibilities of environmental control systems.

The procedures developed were designed to replace selective bias in environmental forecasting with self-consistent and uniform coverage. In particular, these techniques include:

1. the selection of only those variables in the environment that bear directly on the design of environmental control systems and the structuring of environmental control organizations.
2. the generation of scenarios of sufficient detail for planners to identify the threats to the environment and derive the missions, tasks, and systems needed to counter them.
3. the circumscription of a manageable set of scenarios ranging across the entire planning period involving all levels of environmental threat and every major pattern of environmental interaction.

4. the development of a computer procedure for generating unbiased scenario forecasts.

Before detailed and organized scenarios could be developed, the environment had to be decomposed into the political relationships between its dominant actors and the physical characteristics relevant to the constitution of control activities. Typical industries, communities and control agencies are selected for consideration in the scenarios by virtue of their likelihood to enter an environmental conflict within the next two decades. For each of these environment users, solution potential and economic and technological capability are projected across the planning period (1972 to 1985) and integrated into forecasts of alternative environmental control postures. From these projections, the envelope of possible futures was reduced to the set of environmental worlds most plausibly derived from the world of today. Given these future-world guidelines, the next step was to select variables from the environment.

Since a scenario focuses on only the variables which are selected by the designers, it was vital to develop a method which would produce the most relevant variables in terms of the planned use of the scenarios. The dual objectives of the scenarios require relating the environment directly to government and industry control establishments. Since an environmental control activity is designed to neutralize or reduce pollution, the criterion used for translating polluter behavior into terms commensurate with control responses was the perception of environmental threat. As threats vary in magnitude and direction, the incentives and laws structured to deter or control these threats vary correspondingly. There is always an observable coincidence between changes in the environment, changes in the perception of threat, and changes in controls. The variables in the scenario, therefore, were selected for their individual contributions to the definition of environmental threat. The method used to determine contributions, and thus the variables, was based on content analysis of the forecasts of "wise old men." Rather than focusing on the forecasts themselves, this method identifies the variables which were part of the logic process of the forecasters and indicates the existence, although not the strength, of relationships and interdependencies.

Three of the variables chosen for inclusion in the scenario were the competition parameters: physical interactions, economic interactions, and political interactions. Physical interactions represent the physical behavior patterns between polluters and polluted. They account for the fundamental competition for the use of the environment: the struggles for least-cost environmental use by industry and government, versus highest quality environment by other governments and communities. Since threats are extensions of policy conflicts, they need not necessarily originate in environmental arenas. Environmental competition may be the most apparent menace to environmental quality, but crises often arise over economic and technological issues as well. These too can escalate to environmental conflict and mismanagement. To insure that the spread of policy conflicts and potential environmental threats was accounted for in the scenarios, economic interactions and political interactions were added to physical interactions to complete the set of competition parameters. All four dimensions along which instruments of national policy could conflict—environmental, political, economic, and technological—were thus incorporated into the scenario format.

Organizations and individual users compete in the environment in accordance with their own interests: the assurances of environmental, economic, and political survival (needs) among environment users indicate the extent of competition and the attitudes toward competition, but they do not identify substantive user motivations. What the scenario structure needed was parameters for the value each user placed on its environment and on its commitments to other users. Under the title National Policy, environmental protection and inter-user policy were added to the variable list. Environmental protection identifies a nation's willingness to commit its resources to the deterrence, control and alleviation of environmental degradation on any level. This variable takes into consideration each user's predilection for balanced environmental use, economic or cultural requirements, and political unity among his co-users. Inter-user policy is a measure of a user's commitments to other users, which can only be pursued if user economic and political survival is reasonably assured. This variable consists of a user's aspirations, such as development of quality environments and economic resources, projected onto the national stage. Together, environmental protection and inter-user policy represent the user's interests in power over the environment and inter-user involvement.

User policy provides the incentive for competitive user action within the scenario. As long as there is a difference between what a user feels he needs and what he actually has, the user will attempt to eliminate the disparity. This means that he will have to compete with the users he perceives to be responsible for the inequality. If competition devolves into conflict, it can do so on several levels of intensity, from friendly communications and suggestions through public opinion mobilization to direct litigation and legal conflict, and criminal actions. Levels of conflict, then, were needed in the scenario to identify the height to which antagonism over the environment in a given world had risen and the likelihood that conflict would escalate to criminal acts.

These five variables—physical interactions, economic interactions, political interactions, national and user policies, and levels of conflict—constitute the substance of the scenario description. They indicate why environment users feel the way they do about one another, how they behave toward one another, what policy conflicts and threats exist between them, and what they are likely to do about them in terms of their goals. Integrating the variables for the community of users yielded a strategic environment in which the effectiveness of any environmental protection policy could be assessed.

Unlike most scenarios, the environments are both quantitatively and qualitatively indexed. This means that the value of each variable in the scenario can be indicated by number as well as by description. The advantage of numerical, over prose, descriptors is that large quantities of scenario data can be scanned with a minimum of effort, and overall relationships can be quickly derived. With quantitative indexing, there is no need to trudge through cumbersome, and often irrelevant, scenario text. Numerical expressions facilitate the classification of scenarios and, as is shown later, provide a shortcut to the derivation of threats to the environment.

Quantifying a qualitative variable, like inter-user interactions, involved formulating an index scale corresponding to degrees of difference, real or theoretical, of the variable quality. Mutual relations between user pairs were used to develop the index scale in order to simplify its numerical descriptions. The scale identifies only the important nodes along what is presumed to be a continuous spectrum. Though the nodes are simplifications of the subtle variations in inter-user relations, the advantage of index scaling is in locating user pairs along a finite set of easily distinguishable relationships. The index scales for the three "competitiveness" variables—environmental control of physical interactions, economic interactions, and political interactions—appear in Table 2.

The fourth variable, national policy, is divided into environmental protection and interuser policy, each with its own numerical designation. Environmental protection is a user's commitment to the prevention and deterrence of environmental threats, the avoidance of other user violations, and the minimization of pollution damage to itself. It is defined here as the percentage of annual expenses a user allocates to pollution control. Environmental control policy is an indicator of a user's social interaction orientation measured along an index scale ranging from laissez-faire through forceful expansion of controls to reduction of opposed user political influence and litigation. The control policy scale of values is presented in Table 3.

The final variable, level of conflict, is subdivided into six categories: Pollution, location and scope, duration, user commitment, environmental conflict intensity, and conflict issue. Pollution control activity type designates the highest level of environmental protection operations a given conflict has reached and ranges from friendly suggestions to litigation. "One" represents the broad band of environmental consultation; "two" designates educational and opinion-mobilizing operations; "three" denotes limited administrative actions; "four" designates general litigation. The location and scope variables identify the environmental area and type fought over. This variable is limited to the areas included in the projection, assuming all have retained their user-

Table 2. Index Scale Descriptors for "Competitiveness" Parameters.

Political interactions
1 = Coordinated environmental actions with other users.
2 = Cooperative policy arrangements with other users.
3 = Neutral policy toward competing environment user.
4 = Policy conflicts with opposed environment user.
5 = Strong hostility toward opposed environment user.
6 = Become total environmental adversary of.

Economic interactions
1 = Bilateral agreement with other users.
2 = Common environmental market agreement with other users.
3 = Free use arrangement with other users.
4 = Favored user relationship with other users.
5 = Restrictive use with other users.
6 = No economic relations with other users.
7 = Legal restraints imposed on or by other users.

Physical interactions
1 = Received technical support from other users.
2 = Controls sharing agreement with other users.
3 = Mutual restraint agreement with other users.
4 = R & D sharing with other users.
5 = Test sharing with other users.
6 = Production sharing with other users.
7 = Location controls with other users.
8 = Operations controls with other users.
9 = Unabated pollution in pure competition for environmental resources.

related identity. Duration describes the time from initiation to scenario date. The range extends from just-initiated to extended recurrent for environmental conflicts that have either operated for a long period of time or habitually recur. User commitment identifies the principal adversaries, the conflict resources per user, the most sophisticated techniques employed by each user involved, and user commitment as measured by the highest level of conflict intensity in which a user is willing to engage. Principal user adversaries and number of conflict resources are straightforward text and numerical entries, but technique employment ranges from primitive to advanced. "One" equals primitive techniques; "two" designates conventional techniques and equipment; "three" denotes moderately advanced; "four" symbolizes the employment of the most advanced. Conflict intensity is measured by the ratio of daily pollution

Table 3. Environmental Control Policy Index Scale.

1 = Laissez-faire.
2 = Co-existence with opposed users with mild suggestions.
3 = Containment and avoidance of adversary users by physical and economic relocations.
4 = Cooperative expansion of controls.
5 = Unilateral, forceful expansion of controls, by legal action if needed.
6 = Forceful expansion and active reduction of opposed user political influence.

Table 4. Levels of Conflict.

1. *Pollution Control Activity* designates the highest level of environmental operations a given conflict has reached.
   1 = Consultation
   2 = Education and opinion mobilization
   3 = Limited administrative actions
   4 = Litigation
2. *Location and scope* identifies the land area in environmental conflict.
3. *Duration* describes how long a given conflict has been salient.
4. *User Commitment* identifies:

   a. Principal environmental adversaries
   b. Conflict resources per user
   c. Conflict resources employment
   User's commitment is also measured by the orientation of the methods employed:
   Education/persuasion
   Public opinion mobilization
   Administrative measures
   Litigation/violation

5. *Conflict intensity* is indicated by the amount of resources expended divided by the total resources committed and by the dollar value of daily environmental damage.
6. *Conflict issue* defines the principal arena of dispute.

   a. Political — User violations
                — Autonomy of user decision-making
                — Jurisdictional disputes
                — Ideological—equity issues—who should pay?
                — Internal organizational conflicts
                — Settlement of inter-user disputes
                — Ownership issues concerning environmental values
                — Regional power struggles over environmental control
                — Information sharing/denial and confidentiality
                — Unification of user groups

   b. Social — Social costs and benefits
             — Distribution of costs and benefits
             — Esthetic and cultural issues

   c. Economic — Effects on relative industry growth and competition
               — Impact on marginal enterprises
               — Impact on rate of technological change
               — Impact on employment
               — Impact on per capita productivity

costs to total worth of abatement resources committed, and by the dollar value of daily environmental damage. The last variable, conflict issue, defines the motivations of the user and reflects on his respective environmental policies. Issues may be user-internal, local, or regional in scope and political (for example, a power struggle), economic (for example, a user sharing agreement), or social (for example, a quality of life dispute) in nature. Usually, combinations of technical, political, economic, and social factors contribute to an environmental conflict. The levels of conflict categories appear in Table 4.

Range Limitations on Scenario Variables

A larger or more detailed selection of scenario variables would only have produced diminishing returns in the planning applicability of the scenarios. As the level of scenario abstraction decreases from the mere identification of user alliance patterns to the detailed description of environmental control equipment components, an absolute maximum in planning perspective is reached at the interface between the determination of environmental threats and the derivation of control mission/tasks. At either extreme, complete abstraction or complete detail, the planner falls victim to a paradox: in the first case, "he can't see the trees for the forest" and in the second, "he can't see the forest for the trees" (Exhibit 1). Too much detail in the scenario limits its utility to the specific systems being considered without any attention being given to the wealth of alternative system applications implicit in any environment; not enough detail, and mission tasks cannot be identified. A scenario is optimally useful when it simply lays the groundwork for problem perception.

Exhibit 1. Range of Planning Applications Versus Level of Scenario Abstraction.

Base Scenarios

Generation of a scenario context began with the selection of one of the fourteen world types shown in Table 5. To each of the four major environmental power blocs values for the five scenario variables were assigned in accordance with the designated user alliance structure. Then other user groups were included in the scenario by simply assigning variable values to them in relation to one another and to the four power centers. Since the time range for the world type was specified, political shifts between user pairs were not allowed to exceed a specified maximum. Economic and technical capability estimates were then specified for each user and the scenario was ready to be organized for written presentation.

In a fully-conceived scenario, values exist for all sets of user pairs in inter-user relations, economic interactions, and central policies. To get an

Table 5. Fourteen Environmental World Types.

World type I: One world of environmental cooperation
 1985—Cooperation of middle income and poor consumers/producers/industry/government

World type II: Three and one
 1977—IIA : Middle income and poor consumers *vs* industry *vs* government
 1981—IIB : Poor and middle income *vs* industry *vs* government
  IIC : Middle income and industry *vs* poor consumers *vs* government
 1976—IID : Poor consumers and industry *vs* middle income *vs* government

World type III: Two and two
 1978—IIIA : Middle income and poor consumers *vs* industry *vs* government
 1973—IIIB : Middle income and industry *vs* poor consumers *vs* government
 1980—IIIC : Middle income and government *vs* poor consumers *vs* industry

World type IV: Two and one and one
 1979—IVA : Middle income and poor consumers *vs* industry *vs* government
 1974—IVB : Middle income and industry *vs* poor consumers *vs* government
 1972—IVC : Middle income and government *vs* poor consumers *vs* industry
 1982—IVD : Poor consumers and industry *vs* middle income *vs* government
 1983—IVE : Poor consumers and government *vs* middle income *vs* industry
  IVF : Industry and government *vs* middle income *vs* poor consumers

World type V: Multipolarity
 1975—Conflict among all four groups

World type VI: Wild card (appended)
 1984—Examples :
  Technological breakthroughs
  Bio-environmental disasters
  Economic crises
  Urban socio-political crises

overall picture of the situation described by the values, two-dimensional matrices of user-pair relations should be employed rather than individual bilateral listings. This matrix notation permits greater facility in reviewing data for internal consistency than itemized lists.

For levels of conflict and user policy, however, the information contained in the variables is not conducive to matrix application. For these variable values, individual lists for each conflict, in the first case, and each political unit, in the second, are advisable.

In Tables 6 and 7 sample outputs (World type IIA) for the three matrix variables and for the two list variables are given. These outputs are intended as models for scenario display.

The scenario may be converted into prose format for presentation by translating the variable values back into text according to their index scale settings. An historical introduction to the scenario (Exhibit 2), tracing the devolution of today's world to that of the scenario date, is recommended as a lead-in to the context itself.

In order to keep scenario outputs consistent with one another, a format was developed for the textual translation. Following the historical introduction

Table 6. Sample Matrix Output (World type IIA).

Inter-User Relations

	Middle Income Consumer B	Poor Consumer A	Poor Consumer B	Government Level A	Middle Income Consumer C	Industry A	Industry B	Government Level B
Middle Income Consumer A	1	1	4	6	5	5	5	5
Middle Income Consumer B	—	2	3	5	5	5	5	5
Poor Consumer A		—	5	5	5	5	5	5
Poor Consumer B			—	5	5	3	3	3
Government Level A				—	2	3	3	3
Middle Income Consumer C					—	4	3	4
Industry A						—	2	2
Industry B							—	3
Government Level B								—

Economic Interactions

	Middle Income Consumer B	Poor Consumer A	Poor Consumer B	Government Level A	Middle Income Consumer C	Industry A	Industry B	Government Level B
Middle Income Consumer A	2	5	5	6	6	6	6	6
Middle Income Consumer B			—	6	6	5	5	5
Poor Consumer A								

Interaction matrix (upper table):

	Middle Income Consumer A	Middle Income Consumer B	Poor Consumer A	Poor Consumer B	Government Level A	Middle Income Consumer C	Industry A	Industry B
Middle Income Consumer B	2							
Poor Consumer A								
Poor Consumer B								
Government Level A	—							
Middle Income Consumer C	2	—						
Industry A	5	5	—					
Industry B	5	5	5					
Government Level B	5	5	3	5	—			

Control Policy

	Middle Income Consumer A	Middle Income Consumer B	Poor Consumer A	Poor Consumer B	Government Level A	Middle Income Consumer C	Industry A	Industry B
Middle Income Consumer B	2							
Poor Consumer A								
Poor Consumer B	9	—						
Government Level A	9	9	—					
Middle Income Consumer C	9	9	1	—				
Industry A	9	9	9	9	—			
Industry B	9	9	9	9	2	—		
Government Level B	9	9	9	9	2	2	—	

Table 7. Sample List Output.

Levels of Conflict

1. Limited political conflict (ad. campaign)
2. Poor consumer B vs Industry A
3. Just initiated
4. Industry A vs. Government Level B
   2 = administrative actions

	Costs
Industry A	5
Government Level B	5

5. Conflict intensity   5
6. Economic (income and employment impacts)
   Social (have vs. have-nots)

User Policy

	Abatement & Control Spending (Fraction Resources)	Control Policy
Middle Income Consumer A	1/6	1
Middle Income Consumer B	1/24	1
Poor Consumer A	1/24	1
Poor Consumer B	1/100	3
Government Level A	1/8	6
Middle Income Consumer C	1/12	5
Industry A	1/10	5
Industry B	1/10	4
Government Level B	1/10	4

(which illustrates one way of getting from "now" to "then"), the scenario text should include in order:

1. inter-user relations and user policies
2. economic inter-relations
3. control policies
4. levels of conflict

A sample text, corresponding exactly to the sample outputs in Tables 6 and 7, is provided in Exhibit 3.

For each of the fourteen world types, a typical base scenario, with references to user technological capabilities, was generated. These scenarios are only sample situations, and a host of alternative scenarios, equally valid, can be produced for each world type using the procedures developed in this section.

The Computer Generator

For each of the fourteen sample worlds, a set of decision-critical situations was derived for a panel to use as models in investigating environmental threat

Exhibit 2. Sample Introduction to World Type IIA.

Middle-income consumer groups' defense of suburban water quality dragged on until 1979 when Nixon's successor responded to the public's antipathy toward intervention and withdrew government attention (Attorney General's Office and EPA) from industry intervention. For the next seven years, middle-income consumers gradually seceded from inter-user politics and concentrated on environmental defenses. Investment in polluting industries was virtually eliminated and government interests turned toward increased efforts to equalize the distribution of environmental quality. Industries B and poor consumers joined the middle-income consumers in an economic alliance and the three user groups became totally preoccupied with local interests and mutual environmental security. Both the middle-income and poor consumers resigned from the Environmental Defense Association which devolved into the leftist Federation of Non-Polluters, a primary economic target of technologically-minded users.

Growing political consciousness in industry in the latter 1970's coupled with consumers' retreat from inter-user affairs has led to complete disaffiliation between the two groups. In 1981, Industries of Type A formed a trade organization which, a year later, developed into a political alliance to deter consumer threats to Industry self-regulation.

All three of these users have set their sights on the untapped environmental resources of the poor consumer areas to the total neglect of inner cities.

Increased commerce between Industries A and B in the late 1970's resulted in the establishment of a Common Environmental Market and a subsequent mutual environmental support alliance between the two users. The middle-income consumers' withdrawal from suburban water quality maintenance paved the way for the poor consumers' assimilation of the consumer cause, which counterbalanced their government lobbying failures in the inner cities. With this support and enhanced self-confidence, the poor consumers are now launching an aggressive campaign against "devouring pollutionists."

Relations between the major consumer blocs are anything but amicable. The middle-income consumers have become the outcasts of a nation dividing more and more along environmental consumption lines. Developing regions dominated by poor consumers are vigorously hostile to the middle-income consumer group although somewhat less unfriendly to the Industry group. Industries A, fearful of consumer intentions, have all expressed interest in increasing their respective spheres of environmental influence but are currently limiting their overtures to the Government B level. Here, they have been in direct conflict with consumer groups, and relations between the two have grown tense. The proliferation of polluting transport vehicles among all of these users and among several smaller users has made the threat of violation and litigation a very real possibility.

potential. These crises were inferred as logical developments of the policy conflicts existing in each of the sample contexts. The situations chosen are representative of the changes that could require critical environmental control decisions on the part of the consumer groups; they are no more exhaustive within the sample worlds than the sample worlds are within the world types.

To facilitate the task of the scenario panel, an analytic computer program was designed to generate analogous forecasts by transforming the scenario variables into a series of operating instructions. The computer output goes one step beyond the manual procedure by identifying the antagonists and likely crises either indicated or tacitly implied by the environment. The computer scenarios save the panel the task of deriving these crises by hand

Exhibit 3. Sample Scenario Text (World type IIA).

---

1 January 1976

Highly sensitive to a public opinion that advocates an environmentally permissive policy, consumer groups concerned themselves in recent years primarily with local environmental affairs and have gradually withdrawn political influence from the national arena. Stepped-up environmental defense efforts on the part of consumer groups have attracted a strong technical economic alliance with Industries B who have followed the consumer model in national dis-involvement. Between these three users technical interaction has been so pervasive that today there exists a near identity of user interests.

The introversion of the consumer groups environmental associations has alienated the B Industries who have joined in a political mutual support alliance with the poor consumers. Because these users are interested in expanding their political influence in rural areas independently, inter-user relations are not nearly as homogenous as in the consumer associations. Industry X is ambivalent toward Industry Y on whom it depends for technical support, but with whose political ambitions it often conflicts. Poor consumers have expressed a more positive commitment to Industry X than to Industry Y, but seriously suspects the sincerity of X's commitment to the cooperative agreement.

and perhaps overlooking some of the more subtle or involved policy conflicts. Theoretically, the computer generator projects the spectrum of potential decision-critical situations for each scenario and, moreover, is not constrained to the fourteen world types for its environmental output.

In preparation for the design of the computer program, several of the base scenarios were subjected to content analysis for their crises-sensitive variables. The variables derived from the analysis were then arranged in related groups and assigned ranges and units of measurement for alpha-numeric computer classification. Since the variables in each group were interdependent, constraints had to be imposed to assure consistent results. For example, it would be inconceivable for two users to be litigating against one another and, at the same time, be considered maximum cooperators. Coupling the constraints with the variable families precluded contradictory output values.

So there would not be subjective bias in the assignment of values to the scenario variables, the computer program was designed to randomly select values within the imputed ranges. The detailed contexts that were generated were checked for consistency and plausibility and, in the event that one was found to be infeasible, the rationale for its exclusion became an additional variable to the original program. A second computer run eliminated this scenario possibility and others that might have been generated because of the same oversight. The entire iteration procedure was then executed for new scenarios and the critical variable lists compared with the original. When all modifications had been made in the final variable set, the resulting computer program was an effective scenario generator. Exhibit 4 is a flow diagram of the iteration procedure.

This computer generator is a logical outgrowth of the manual procedure.

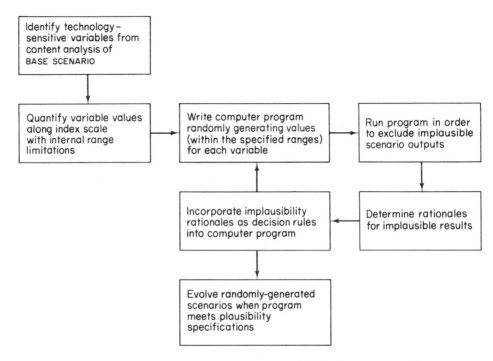

Exhibit 4. Computer Scenario Generation.

However, instead of beginning with a variable analysis of a Base Scenario, the program could easily have been initiated with any systematic identification of crisis-relevant variables. Its purpose in either case is to reduce the amount of work required to evolve self-consistent and varied scenario forecasts and decision-critical situations without any loss of reliability, range, or depth. The computer generator ultimately allows the planner to explore a substantially larger number of possible futures than could any manually-generated mechanism.

Range Limitations on Scenario Generation

Whether scenarios and their intrinsic crises are generated by hand or by computer, there is a finite limit on the number that will usefully contribute to the design of new control systems. At the very least, the number of scenarios considered should cover environments ranging from maximum to minimum control systems applications. Beyond this point, diminishing returns to scale will come into play since, as far as control is concerned, the state of the art imposes a constraint on the number of designable environmental control systems.

The generation of additional scenarios, therefore, would contribute negligibly to the definition of executable mission/tasks and the subsequent design of environmental control systems. In Exhibit 5, the slope of the curve relating system types to number of scenarios decreases as the latter variable increases, indicating the diminishing contribution of larger and larger numbers of scenarios. Scenario generation should be suspended at the point where the slope begins to decrease.

Exhibit 5. New System Types versus Number of Scenarios.

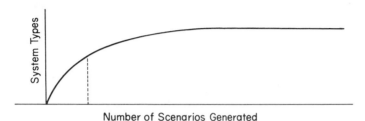

Number of Scenarios Generated

Parametric Relationships

All the variables in the scenarios are parametrically related. Economic interactions, control postures, and levels of conflict are derived from the inter-user relations. Technological capabilities and control systems budgets can be traced directly to available resources projections. Inter-user relations and user policies both originate in political forecasts. As a result of this structural interdependence among the variables, internal consistency is of prime importance in assigning variable values to the users. A random imputation might yield two mutually technologically supportive users competing in the conflict of two adversary users with a bilateral tax or purchase agreement. To preclude contradictory outputs, or ambivalent relationships, all the variables should be aligned and checked for compatibility. This cannot be emphasized too strongly if the scenarios are displayed in matrix format, where inconsistencies are not immediately visible.

Because the scenarios are organic descriptions of environments, that is, all their variables interrelate, a change in any one parameter is going to have correlative effects on all the others. Adjustments in user policy will be reflected in inter-user relations and perhaps even in levels of conflict. The converse is equally true. Dramatic examples of this overall structural interdependence are the effects of a coordinated environmental use agreement on control systems or the impact of protracted environmental defense system lead times on user policy.

In these situations, the scenario is a dynamic tool for exploring the possibilities of circumventing time constraints in favor of technological objectives or conceiving novel control systems to either stabilize deterrence of violation or promote compliance. By convincing a rational violator that the government has the ultimate deterrent—for example, adequate legislation for unacceptable cost imposition—he might be more inclined to come to the bargaining table.

In any case, identifiable parametric relationships impart to the scenario a wider range of planning applications than it would otherwise have had.

Summary

The base scenarios are synoptic descriptions of potential environmental control strategies. They consist of five quantifiable variables—inter-user relations, economic interactions, control technology, user policy, and levels of conflict—which represent the fundamental behavior patterns in the national control systems that contribute to the perception of environmental threat. Since the perception of environmental threats is the motivation for the structuring of controls, the scenarios provide excellent settings for the evaluation of effectiveness of formal government and informal user controls.

As models for the Scenario Panel, fourteen base scenarios were generated, corresponding to the fourteen world types described in Table 5. Though these models were only samples of potential settings, they ranged across the entire spectrum of inter-user environments. These fourteen scenarios thus constitute a broad set of environmental futures for planning applications.

The present scenario generating methodology has three advantages over the previous techniques for scenario generation described earlier. First, scenario variables are derived in an explicit manner designed to ensure completeness while reducing bias. This is done by using content analysis on "expert" forecasts to isolate the variables with which they are concerned. In effect this method says while the forecasts of wise old men do not consider all possible events and interactions, they *do* consider the critical variables which are subject to change. This method is independent of the overall methodology. There is no reason why other techniques, such as the Delphi method, could not use it as a first step.

A second advantage is that, by having a range of possible relationships between each variable, assumptions need only be made about the extent of the range rather than about item by item indicators such as in the cross-impact method. Use of a random number generator to determine paired values within the possible range thus reduces the bias due to interpretations of relationships based on present knowledge projected into the future. Iterative programming picks out "impossible" combinations and reprograms the generator to avoid unusable scenarios.

The third, and perhaps most critical advantage, is the concept of an envelope, or range, of possible scenarios, rather than one scenario. By grouping scenarios into classes, strategic action within any one class can be determined. Actions required across all classes may then be compared to determine the direction current plans must take if they are to meet a majority of, or the most likely, alternatives. This is, in effect, a way of saying that while one scenario forecast may be superior to that of an individual, no method has yet been developed that is that assuredly accurate. Therefore, knowing the range of possible environments is more informative than having a picture of just one. Even if one examines the effects of variable changes within the one scenario, the changes are based on current ideas of different ways the variable might behave in the future, rather than all possible variables combinations.

INDUSTRIAL APPLICATION

In our current work, applying the scenario generating methodology to industrial situations, we first examined a detailed model (Exhibit 6) of the process developed in the environmental control study to identify parallels for an industrialized model. Exhibit 7 presents the model adapted to the generation of industrial scenarios.

The environmental control study had seven key characteristics. It focused on the future world *resources, economy,* and *technology* to identify *events* which were *dangers* to continued *environmental qualities and capacities* and which required *investment* today if one was to be ready for tomorrow. Taking these characteristics, the following industrial schema was developed:

	Environmental Control Study	Industrial Study
Environment:	User group resources Economic profile	User group resources Economic market profile
	Technology profile	Technology profile
Events:	Shifts in balance of control power due to political and technological capabilities	Shifts in market share due to production capabilities
Goal:	Identify dangers to environment in order to plan current investment	Identify opportunities and dangers to continued growth in order to plan current investment

It is clear that the focus was on current investment: in the Environmental Control Model, investment in environmental research and development; in the industrial mode, product R & D.

The next step was to define the variables to be used in the model. Since this was an experimental situation we did not use the technique of content analysis of "learned" forecasts. We used a simpler method where people from a number of disciplines developed the variable list in an informal discussion session.

The first set of variables identified were the actors. The actors all had common characteristics which described their position as a user. In this model there appeared to be no one set of parallel actors. The final list of actors developed without this constraint was:

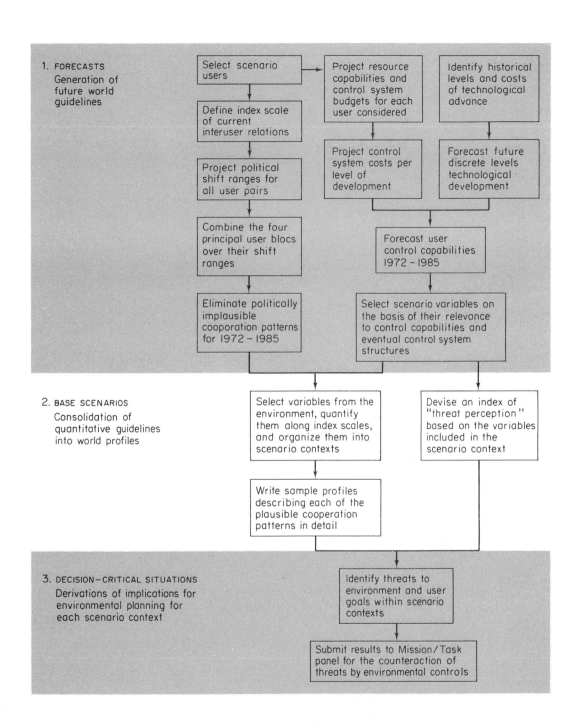

1. FORECASTS
Generation of
future world
guidelines

| Select scenario users |
| Define index scale of current interuser relations |
| Project political shift ranges for all user pairs |
| Combine the four principal user blocs over their shift ranges |
| Eliminate politically implausible cooporation patterns for 1972 – 1985 |

| Project resource capabilities and control system budgets for each user considered |
| Project control system costs per level of development |
| Forecast user control capabilities 1972 – 1985 |
| Select scenario variables on the basis of their relevance to control capabilities and eventual control system structures |

| Identify historical levels and costs of technological advance |
| Forecast future discrete levels technological development |

2. BASE SCENARIOS
Consolidation of
quantitative guidelines
into world profiles

| Select variables from the environment, quantify them along index scales, and organize them into scenario contexts |
| Write sample profiles describing each of the plausible cooperation patterns in detail |

| Devise an index of "threat perception" based on the variables included in the scenario context |

3. DECISION–CRITICAL SITUATIONS
Derivations of implications for
environmental planning for
each scenario context

| Identify threats to environment and user goals within scenario contexts |
| Submit results to Mission / Task panel for the counteraction of threats by environmental controls |

Exhibit 6. Manual Scenario Generation—Environmental Control.

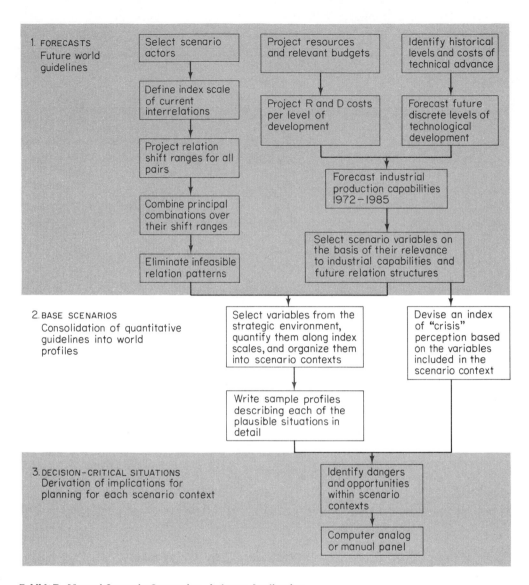

Exhibit 7. Manual Scenario Generation—Industry Application.

Consumers (one actor)
Government (one actor)
Labor in the specific company
Competition (number of potential competitors)
Company X (the company we're interested in)

To avoid cross-product and multiproduct market situations, company X was defined as (1) either a single product company or (2) a profit center based on a single product.

The environment base for the future-world profile was very similar to the environmental control model. Economic resources and the state of technology variables were parallel. A market profile based on demographic projections was added. Within this environment the actors could interact in a positive manner, which either lowered costs, raised prices, or increased market share, or in a negative manner which lowered prices, raised costs, or decreased market share.

Unlike the environmental control model, in the industrial model the company seeks *both* to protect itself from danger situations *and* to exploit favorable situations. Opportunities and dangers are identified in Exhibit 8.

Exhibit 8. Dangers and Opportunities in the Industrial Environment.

Dangers	Opportunities
1. Competitive companies gain market share	1. Competitive companies lose market share
2. Higher costs due to a. Labor b. Materials c. Market shrinkage	2. Lower costs due to a. Labor b. Materials c. Market shrinkage
3. Obsolescence of X's product	3. Technological breakthrough
4. Government action	4. Total market growth
5. Total market shrinkage	5. Government legislation

Generation of a scenario context began with the selection of one of 255 (2^8-1) market types (parallel to the world types). Values were assigned to each of the eight actors for their positive or negative interactions. Since the range chosen was 1980, interactions could occur only under environmental constraints projected for 1980. Production and technological capacities estimates were then specified for each corporation and the scenario was ready to be organized for written presentation. A sample text is provided in Table 8. The scenario is only a sample situation, and a host of alternative scenarios, equally valid, can be produced for each market type using the scenario generating methodology.

CONCLUDING REMARKS

While no forecast of the future environment can properly predict all relevant variables, the use of scenarios based on methods designed to avoid personal bias appears to be an excellent method for approaching an "ideal" forecast. The scenario generating methodology discussed in this chapter has the following advantages over other current techniques:

Table 8. Sample Scenario Text.

1 January 1980

Highly sensitive to a public opinion that advocates isolation, the Government has concerned itself in recent years primarily with domestic affairs. Consumer pressure on the Government has forced large-scale consumer-protection legislation raising the costs of production and packaging in your industry. Continued inflation resulted in wage-price restrictions in 1975 which are highly inflexible. Public opinion against teachers and municipal employees' strikes resulted in anti-strike legislation which has prevented any strikes in your industry. Raw materials prices have increased because of decreased international trading, a function of the prevailing desire for isolationism. Technological advances resulting from increased Government encouragement of private R & D have given your company an increased market share of 14%. The overall product market has expanded by 11%. However it is highly probable that both company A and C will come out with a new line, which will make your product obsolete, at the end of the year.

1. Scenario variables are derived in an explicit manner using content analysis of expert forecasts, thus reducing the biases of the scenario designer.

2. Use of a range of possible relationships between variables limits biases in assumptions concerning relationships to classes of possible ranges. A random number generation rather than assumptions determines the paired relationships.

3. By having an envelope, or range of possible scenarios, one can examine more than one possible future. In addition the range can be divided into classes of strategies formulated for each class and current plans designed to meet the set of probable futures rather than a single event.

Our experience with the environmental control model has shown that the scenario generator is both easy to work with and instructional in formulating the logic for variable interactions. In current industrial work, the adaptability and general applicability of the method are evident.

Analysis of
Industrial Behavior and
Its Use in Forecasting

W. H. CLIVE SIMMONDS

In 1968 Simmonds presented his first paper to our Industrial Management Center technology forecasting course as observations about the remarkably regular "step-functions" in the bulk chemical industry. In successive years he gradually refined his concepts to argue that step-functions reflected industrial behavior. The "steps" revealed how and when the particular industry tended to think, plan, and act. He then concluded that such patterns probably appear in other types of industries. If so, they might form an excellent basis for predicting just what technology steps the industry might take in the future.

This chapter presents such an analysis for the performance-maximizing industries, and shows how the resulting model can be used to study possible future situations and clarify intercountry competition and management decision making.

INTRODUCTION

The purpose of this chapter is to show how forecasts of company and industry futures can be made if current and past industry experience is analysed in terms relevant for forecasting. In its current state technological forecasting hardly fulfills this condition. It has been described as "a science without a discipline."[1] It has the appearance of a science, but lacks the central core of experiment, data, and theory characteristic of a physical science.

This central core of organized knowledge is what is missing. It can be supplied through:

1. A taxonomy, or classification, of industry relevant to technological forecasting.
2. An analysis of the three interrelated elements of industrial behavior, the technological, the economic, and the human and social.

This chapter describes the derivation and organization of such information for one class of industry, the performance maximizing (or, as they are often termed, the science-based) industries.

An Industrial Taxonomy

Ayres has proposed a classification of industry relevant to forecasting, in which he classifies industry as performance-maximizing, sales-maximizing, or cost-minimizing.[2]

Performance-maximizing industries compete on the basis of highest performance per unit price. They include the aerospace, computer, defence, electronic, nuclear, petrochemical, scientific instrument, and tanker industries and are typically science- and technology-based.

The performance-maximizing industries contrast with the sales-maximizing industries which compete for the consumers' dollar through intensive advertising. Examples of these industries include the auto, cosmetic, detergent, food and beverage, and apparel industries. Cost-minimizing industries compete to produce standardized products at lower prices. They include utilities, communication, transportation, metals, glass, cement, concrete, and petroleum products. *The thesis is therefore advanced that it is the common pattern of behavior of industries grouped in this fashion which is relevant for technological forecasting.*[3] This thesis will now be substantiated for the case of the performance-maximizing industries.

BEHAVIOR PATTERN OF PERFORMANCE-MAXIMIZING INDUSTRIES

The main driving forces in determining the behavior pattern of the performance-maximizing industries have been, and continue to be, technological

[1] T. J. Gordon, President, The Futures Group, Glastonbury, Conn.

[2] R. U. Ayres, *Technological Forecasting and Long-Range Planning* (New York: McGraw-Hill, 1969).

[3] Ayres' taxonomy is incomplete. It omits the capital goods industry (construction, machinery, tools), which competes on the basis of the lowest installed cost per desired activity or function; the resource-based industries, which compete on the basis of the

innovation, scale, and market demand. Continuing technological innovation has lowered the cost of any technical capability and usually increased its range as well; for example, semiconductors and solid-state devices. Lower cost has permitted lower prices, which in turn maintain market growth rates of ten percent per annum or substantially higher. Larger markets permit larger plants as in petrochemicals, larger reactors, rockets, and oil tankers, or larger or new capability as in computers and scientific instruments. Increasing size usually permits economies of scale and still lower unit costs. Thus the combination of lower costs due to advancing technology and lower costs due to increasing scale has provided a powerful driving force towards continually falling prices[4] and continuing market growth.

A specific and characteristic pattern of expansion by these twin factors of innovation and scale has thus been imposed on these industries, a pattern which has exerted a major influence on their economic behavior.

At low rates of growth of markets, production units can often be "debottlenecked" or "stretched" to keep pace with sales. However, at the high rates of market growth characteristic of the performance-maximizing industries, this is no longer possible. New plants must be sized larger than their initial markets require and can rarely be "stretched" beyond one or two years market growth.[5]

However, by the time that a new round of expansion becomes necessary, new technology enables *larger* size plants to be built. In the competitive market which exists in the United States, one company sooner or later has elected to take the risk of building one of these, and its competitors have usually followed suit. A stepwise pattern of expansion is thus set up,[6] the chief characteristic of which is that, at any time, there is a largest plant or unit size in existence which dominates the competitive situation by virtue of its potentially lowest unit costs.[7] Coexisting with these plants is a spectrum of smaller and usually older plants or units.

Examples of this stepwise expansion in the size of the *largest* plant or

---

lowest cost of discovery-plus acquisition-plus recovery for the desired resources; the personal service industries, which compete on the basis of greatest satisfaction per unit cost (that is subjectively); and the defence industries, from which governments require the maximum military capability by a given date at bearable cost.

4 In constant dollars.

5 Taking the aircraft industry as an example, the Boeing 707 could only be "stretched" 15 percent in capacity. The DC–8 was stretched 50 percent to the Super DC–8 (Series 61, 62, 63). The Boeing Company was therefore forced to consider a new, larger design, which became the 747. Not to be caught twice, Boeing has already designed stretched and "superstretched" versions of the 747, while McDonnell-Douglas explored the Airbus market against Lockheed. The average rate of growth of the international air passenger market during this period was 13.5 percent a year.

6 See Martino, J. P., "A Note on Seaman's Competitive Model of Technological Growth," *Tech. Fore. & Soc. Change, II,* (1970), 107–108, for an analysis of the situation among the leading aircraft manufacturers.

7 It should be noted that it is not the *actual* costs of the largest plant or unit which are relevant. Management decisions have to be based on the *expected* unit costs. The results have varied in practice from highly rewarding to horrific.

operating unit are shown in Exhibit 1 for passenger aircraft, a typical petro-chemical such as ethylene, computers, and space launching.

This continued increase in size appears to be related to the growth of the market. Table 1 shows this relationship for ethylene, and Table 2 that for ammonia.

Plots of other petrochemicals, aircraft, and oil tankers show similar relationships between increase in size of plants or aircraft or ships and increase in demand for their services.

These results suggest that, in the period from the end of World War II

Exhibit 1. Example of Stepwise Expansion of Largest Units in the Performance-Maximizing Industries.

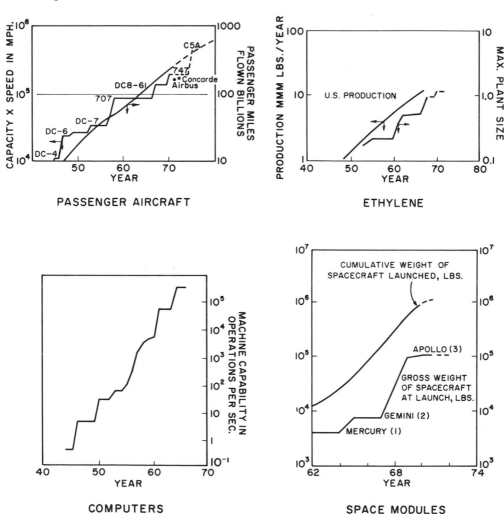

PASSENGER AIRCRAFT

ETHYLENE

COMPUTERS

SPACE MODULES

Table 1. Ethylene Situation in the United States.

Billions of Lbs./Yr.	1970	1963	1955	1945
A. Largest, single-train plant, nominal capacity	1.2	0.55	0.22	0.1
B. Production in year plant first came on stream	15.7	7.5	3.05	0.7
Ratio of B : A	13.1 : 1	13.6 : 1	13.9 : 1	7 : 1
No. of manufacturers	25	20	13	

Table 2. Ammonia situation in the United States.

Thousands of Short Tons/Yr.	1968	1965	1963	1956	1953
A. Largest, single-train plant, nominal capacity	510	340	210	105	52.5
B. Production in year plant first came on stream	12,120	10,660	6,690	3,378	2,270
Ratio of B : A	24 : 1	31 : 1	32 : 1	32 : 1	44 : 1
No. of manufacturers	73	65	54	31	16

to date in the United States, there has been a general tendency in the performance-maximizing industries for the size of the largest plant or operating unit to follow the growth in market demand.

Further examples of this stepwise pattern of expansion in size are given in Exhibit 2 in index form (year 0 = 100) to facilitate comparison. (For complete data refer to the appendix to this chapter.)

The highest rate of growth in unit size or capability is shown by computers. The plot of useful weight put into space is similar, but the growth will diminish or cease as funding of NASA is cut back. Petrochemicals such as ethylene, ammonia and benzene, the manufacture of elemental phosphorus, tankers for crude oil and liquified natural gas, civilian jet engines, and civil aircraft, which supply markets growing in the ten to thirty percent a year range, all show similar patterns. A somewhat different operation, such as the prepainting or precoating of metal coil, follows the same pattern.

The area in the graph below these examples represents a different situation. With average growth rates below seven to eight percent a year, it becomes easier to size plants closer to markets. In many cases the cost of freight or storage or both becomes increasingly important, as with heavy chemicals such as sulphuric acid and choralkali and in petroleum refining. Markets for such products are regional or local rather than national or international. The cost advantages of process innovation and scale diminish as markets shrink geo-

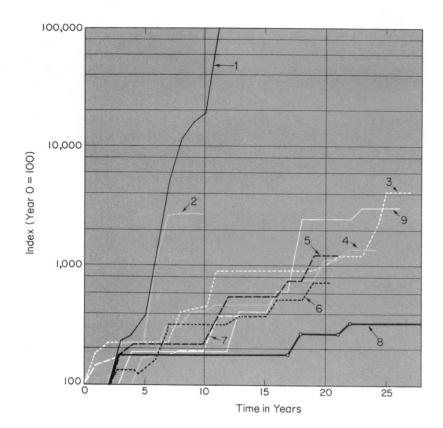

1. Computers: M Operations/sec.
   1950 = 100

2. Space Modules M lbs. wt
   1962 = 100

3. Coil Coating, in. ft./min. Capacity
   1946 = 100

4. Oil Tankers, M dwt. tons
   1950 = 100

5. Ethylene: MM lbs./yr Capacity
   1943 = 100

6. Passenger Aircraft, Speed x Capacity
   1950 = 100

7. Civilian Jet Engines: M lbs. Thrust
   1958 = 100

8. Phosphorus furnaces: Max. Power, kW
   1934 = 100

9. Steam–electric Turbo-generators: kW
   1904 = 100

Exhibit 2. Index Plots of Largest Plant/Unit versus Time in the Performance-Maximizing Industries.

graphically, and raw materials, distribution and storage costs, and factors such as seasonality rise in importance.

Summarizing then: *In performance maximizing industries in the United States, other things being equal, the size of the largest plant, operating unit, or unit capability will tend to increase with growth of the market in a stepwise pattern.*

The qualifications to this hypothesis ("the other things") are important. The number of companies competing in the United States in any one area is usually large enough to eliminate effects of location, company size and type, special company circumstances, and other similar factors which affect the situation in countries such as Britain, Canada, Australia, and Mexico. The majority of companies are United States owned, that is, are not subsidiaries. Price collusion is illegal and competition real. There is no overall state or federal control of planning, in contrast to the situation in Japan, Latin America, France, India, and the Communist countries.

The significance of the phrase—other things being equal—is examined further under Application of Performance-maximizing Industry Model to Future Possible Situations.

The twin stimuli of technological innovation and market growth have thus been shown to generate a similar pattern of behavior in a variety of different industries, which all have the same characteristic of being performance-maximizing.[8] Given this, it should be possible to conceptualize a general model and to use this model to understand both the existing behavior of these industries and their future possible behavior.

TECHNOECONOMIC-HUMAN MODEL OF PERFORMANCE-MAXIMIZING INDUSTRIES

A model of the technical, economic, and psychological characteristics of the performance-maximizing industries is given schematically in Exhibit 3. Curve A shows growth in demand.[9] Curve B shows the size of the largest unit in existence and how this has increased in steps or jumps with time. Curve C shows the industry operating ratio and the oscillation between under- and overcapacity due to overexpansion at each expansion step. Curve D shows the abrupt price declines set off by overexpansion in capacity, which are followed by a levelling off of prices during periods of undercapacity or supply-demand balance. Curve E shows changes in management psychology as a company passes through an expansion cycle. Curve F indicates the business cycle which influences and is influenced by the stepwise expansion pattern in the performance-maximizing industries.

The increase in size from one generation of plants to the next (150 to 250 percent every five to seven years in the petrochemical industry) greatly exceeds the rate of growth of the market (7 to 25 percent a year). If one company opts for the larger plant size, its competitors normally follow suit

8 This statement refers to the major characteristics of these industries. It should not be inferred that *every* element in these industries shows such behavior, nor that sales-maximizing or cost-minimizing industries do not show similar behavior in *some* of their operations.

9 In many cases the only or chief source of information is production data which must then be corrected for exports, imports, and so on to obtain demand.

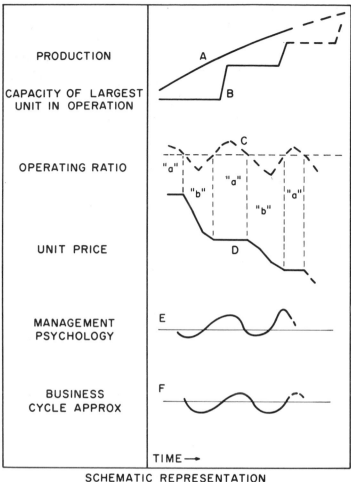

	PRODUCTION
	CAPACITY OF LARGEST UNIT IN OPERATION
	OPERATING RATIO
	UNIT PRICE
	MANAGEMENT PSYCHOLOGY
	BUSINESS CYCLE APPROX

TIME →

SCHEMATIC REPRESENTATION

Exhibit 3. Relationships in Performance-Maximizing Industries Versus Time.

under United States conditions. It then takes only so many of the next generation of plants to produce an overcapacity situation.[10]

[10] The change in operating ratio in the case of a petrochemical growing at fifteen percent a year would be as follows, where the largest plant in existence is assumed to be 100 million pounds per year capacity in year one, and the next generation plant size is 175 million pounds per year capacity in year four.

capacity, million pounds per year

Year	Market	Old	New	Total	Operating Ratio	Prices
1	756	1,000	—	1,000	0.76	
2	870	1,000	—	1,000	0.87	Firming
3	1,000	1,000	—	1,000	1.00	Firm
4	1,150	1,000	2 × 175	1,350	0.97	Firm
5	1,325	1,000	4 × 175	2,050	0.87	Cascading
6	1,520	1,000	—	2,050	0.74	Cascading
7	1,750	1,000	—	2,050	0.85	Firming
8	2,010	1,000	—	2,050	0.99	Firm

Abrupt price declines (the "b's" in Exhibit 3) are set in motion by attempts to sell off the excess output of the new larger plants. If excessive numbers of new plants are announced, as in the recent case of ammonia, serious price instability can develop. Order is restored through the shutting down or placing in mothballs of older plants and by expansion of the market, stimulated by the price declines. Start-up troubles with new plants may also assist. Such price declines can be prevented in only two ways: by not constructing excess capacity[11] or by accepting price regulation, as in air and other transportation.

Prices of products of these industries do not usually rise during periods of undercapacity or supply-demand balance owing to cross competition, substitution possibilities, and imports, that is, there are usually no price upswings to offset price downswings in these industries.

Loss of profit occurs because selling prices drop faster than predicted by smooth price decline formulae or projections.[12] They drop when the new generation of plants is being built or have just come on stream, that is, just at the time when a rapid recovery of investment is expected but does not occur.[13] This phenomenon has given rise to the phrase "profitless prosperity."

It is also apparent, and experience has confirmed, that management psychology goes through a cycle during an expansion step. Management becomes "bullish" as supply and demand come into balance, and new capacity is needed to maintain the company's position. It turns "bearish" if things do not go well during the expansion owing to strikes, start-up troubles, or too many new plants. If serious loss of profit occurs, a crisis situation may develop, leading to firings, cutbacks, and reorganization.[14]

Companies and the people in them have made substantial emotional as well as financial investments in successful lines of business. The stepwise pattern of expansion affects these emotions as well as dollars, and their interaction must be understood.

The business cycle and the stepwise expansion pattern will also clearly relate. Expansions are usually delayed during business downturns, and this produces an acceleration of expansion during the ensuing upturns. The capital goods industry waits for such opportunities to sell the next generation of larger plants, units, and capabilities. This leads to an unfortunate "bunching" of demand for capital and capital goods. A current example is the introduction

---

11 Since demand is finite, it is obvious that the decision to expand is not solely a company decision. It is also an industry decision.

12 See, for example, W. D. Nordhaus, *Invention, Growth and Welfare: A Theoretical Treatment of Technological Change* (Cambridge, Mass.: M.I.T. Press, 1969).

13 The original reason for undertaking this analysis when the author was with Canadian Industries, Ltd., was precisely this loss of profit and the reasons for it.

14 Readers familiar with Antony Jay's "Management and Machiavelli" (1967), pages 192 ff., will appreciate the following. When new capacity is needed, the hierarchy put on their headdresses, the tomtoms begin to beat, and the rhythm of pounding feet can be felt through HQ. "We big ———makers! Big share of market! Much competence! How...!" The war dances have begun in order to curry favour with the gods of profit and growth, and the way is laid for human sacrifices later if need be! Such pictures carry an element of truth as events have shown.

of the Boeing 747. Overloading of equipment-producing facilities may occur. The net effect is therefore to reinforce the business cycle, unfortunately in the wrong direction.

The conceptual model presented here does not represent any single industry or company in this group explicitly.[15,16] Its value lies in providing a rational framework within which the performance of a company, the behaviour of its competitors, and the achievement of its industry can be studied. The model has the additional merit of relating the experience of different industries in the performance-maximizing group and permitting useful cross comparisons. And it is this wider basis of correlation which establishes its value for the study of future possible as well as current situations. The future possible aspects will now be explored.

APPLICATION OF PERFORMANCE-MAXIMIZING INDUSTRY MODEL TO FUTURE POSSIBLE SITUATIONS

How can this model be used to investigate future possible situations, and what does "other things being equal" mean?

Two approaches are possible: The assumption that other things will continue equal, and the assumption that other things will be "unequal." Both should be used as a cross check on each other.

"Other Things Will Continue Equal"

This is the assumption underlying a conventional trend extrapolation, which is easily carried out with this model as shown in Exhibit 4. Assuming a continuing relation between largest plant size and demand, the next step is marked off as shown. This would indicate, in the case of ethylene, a demand

[15] See Max Weber's "ideal-types" in sociology, which he used to assist in the analysis of complex social situations.

[16] See R. Aron, "Max Weber," in *Main Currents in Sociological Thought,* (New York: Basic Books, 1967) II, iii, 201–4.

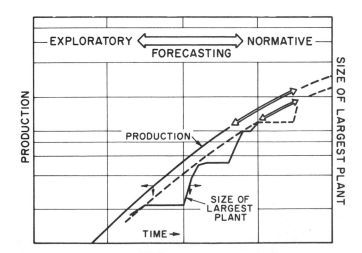

Exhibit 4. Exploratory/Normative Forecasting of Future Plant Size.

for 24 to 26 billion pounds in 1975–76, and the next generation ethylene plant size of around 1.8 billion pounds per year capacity.

Normative forecasting techniques can then be used to make such a forecast self-fulfilling,[17] by establishing the necessary steps to achieve the appropriate design, unit costs, and marketing developments. In commercial practice this becomes an iterative process.[18] Its merit lies in establishing R&D and other requirements soon enough for useful action to be taken. It challenges company engineers to obtain the required information at lowest cost, for example, from or through others where possible. Innovation and demand become directly coupled, as shown by Wade Blackman in the case of civilian jet engine development.[19]

The largest ethylene unit which could be ordered at the time of writing is 1.6 billion pounds per year capacity. However, the technology of building larger hydrocarbon cracking furnaces is advancing[20]; the chemical uses of centrifugal compressors are being extended; and the throughput of ethylene distillation columns is being significantly increased.[21] It is important therefore not to *"miss-estimate"* future advances in technology through preoccupation with current difficulties.[22]

A trend extrapolation of this type sets up a bench mark, to use the economists' term, where we will be if everything continues in the same way. Unfortunately the world is Orwellian. Some things are more equal than others, and the alternative assumption must now be considered.

"Other Things Will be Unequal!"

Can future changes be anticipated? Without answering this question explicitly, the existence of a conceptual model of industrial behaviour permits systematic examination of each step, as follows:

Theoretical Limits. Their existence and significance have been documented.[23]

Lagging Technology. An example of this is limitation on reactor size in the manufacture of ethylene oxide in the fifties (Exhibit 5).

[17] J. P. Martino, *Technological Forecasting for Decision Making* (New York: American Elsevier, 1972) Chap. xxi.

[18] E. B. Roberts, "Exploratory and Normative Technological Forecasting: A Critical Appraisal," (ed.) M. Cetron, *Technological Forecasting,* (New York: Gordon and Breach, 1969).

[19] A. Wade Blackman, "Normex Forecasting of Jet Engine Characteristics," Chapter 23 of this book.

[20] M. Spielman, "Steam Cracking at Esso: Yesterday, Today and Tomorrow," *Chem. Eng. Prog. LXV* (November, 1969), 43.

[21] Union Carbide has announced a forty percent increase in throughput of ethylene distillation columns.

[22] See, for example, J. A. Lofthouse, "Large Chemical Plants and their Problems," *The Chemical Engineer,* CE153-157, (September, 1969).

[23] R. U. Ayres, *Technological Forecasting and Long-Range Planning,* Chapter, VI

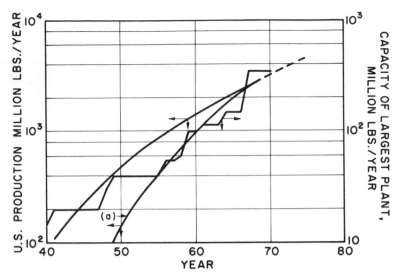

(a) PRODUCTION BY DIRECT OXIDATION PROCESS

Exhibit 5. Ethylene Oxide.

Leading Technology. New designs, new processes, and new equipment can facilitate an above average jump in plant or unit sizes. Exhibit 6 shows the jump in vinyl chloride monomer plant size when the oxychlorination process was introduced; Exhibit 7 the jump in oil tanker size when the switch was made from gasoline to crude oil shipments from the Middle East in the midfifties; Exhibit 8 the expansion to 600 short tons per day minimum size

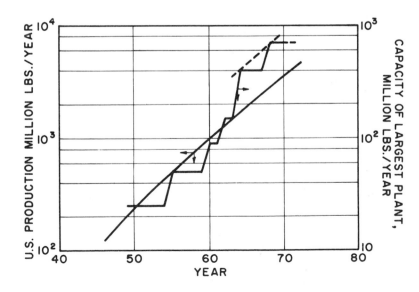

Exhibit 6. Vinyl Chloride Monomer Showing Jump in Largest Plant Size.

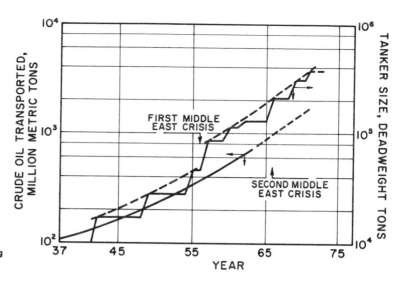

Exhibit 7. Oil Tankers Showing Jump in Largest Tanker Size.

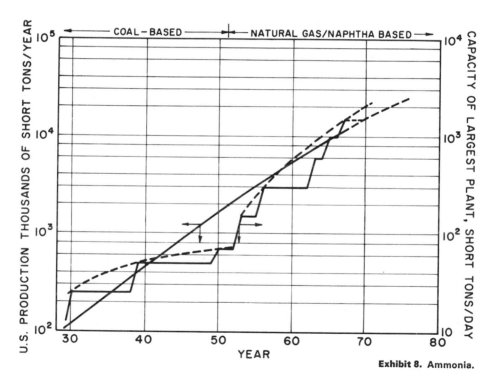

Exhibit 8. Ammonia.

ammonia plants when reciprocating compressors were replaced by centrifugals in the sixties.

Practical Limits. Water depth will limit the increase in size of oil tankers of current design, shown in Exhibit 7, to one million or so tons deadweight. Their design could, however, change significantly to avoid this limitation.

Fabrication and transportation techniques set limits to equipment sizes, diameters, and vessel thicknesses.

Technological Obsolescence. Two methods of detecting technological obsolescence exist. An obsolescing technology ceases to yield further reductions in unit cost with increase in scale, and is therefore ripe for new technology which can do this. Examples of such new technology include the introduction of the fluid bed, the centrifugal compressor, the transistor, and the jet engine.

Conventional technology can be extrapolated to a reductio ad absurdum. The radial aircraft engine had nowhere to go in the fifties except to be replaced; much current chemical technology appears inadequate when judged by potential requirements in the 1990s. The steel industry affords interesting examples of technological obsolescence and of technical changes made to offset them.

Technology Transfer. The answers to one industry's problems may already exist in another industry. Hence James Bright's emphasis on the need to *monitor* the environment for new technology.[24]

Technology Nontransfer. Technology transfer is *not* automatic. Recent troubles with new ammonia plants revealed that the petrochemical industry had apparently not followed electric utility practice of monitoring *equipment* as well as process performance; compressor bearings were of different length/diameter ratio than those used by utility and jet engine manufacturers[25];

[24] J. R. Bright, "Forecasting by Monitoring Signals of Technological Change," Chapter 14 of this book.

[25] Lord Hinton of Bankside, address to the 1967 Congress of Canadian Engineers, Montreal, May, 1967.

Exhibit 9. Low-density Polyethylene.

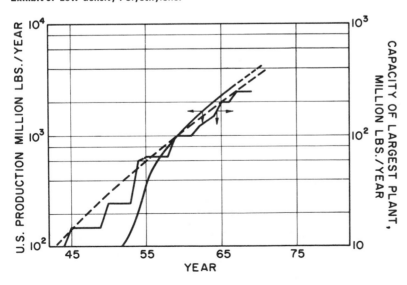

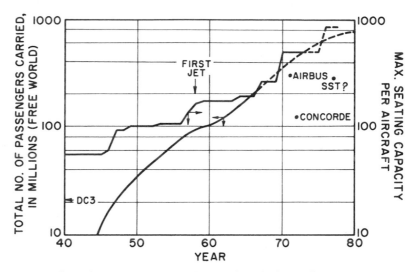

Exhibit 10. Air Transportation, Aircraft Capacity Versus Passengers Carried.

and the value of the aircraft industry's practice of analysing failure had not been appreciated.

Lagging Markets. An imbalance between technology and markets may occur as shown in Exhibit 9 for polyethylene and Exhibit 10 for air transportation. In both cases, technological capability in the earlier years appears to have been greater than that demanded by the market. As the market grew they came into better balance.

Government Intervention. The government-owned butadiene and styrene plants built during World War II were substantially larger than necessary to satisfy postwar demands. No increase in the size of these units occurred until civilian demand caught up.

Nontechnological Factors. The recent surge in public interest in the environment is a warning that nontechnological factors must also be considered for their technological and cost implications.

INTERCOUNTRY COMPETITION

The second major use for the model of the performance-maximizing industries is the competitive situation between companies in different countries and of the policies of their governments.[26]

If the largest plant in the United States follows the growth of the American market, what happens elsewhere? Can other countries with smaller markets build plants of comparable size and, if not, how can they compete?

[26] See W.H.C. Simmonds, "The Canada–U.S. Scale Problem," *Chemistry in Canada* (October, 1969), 39–41, for additional details.

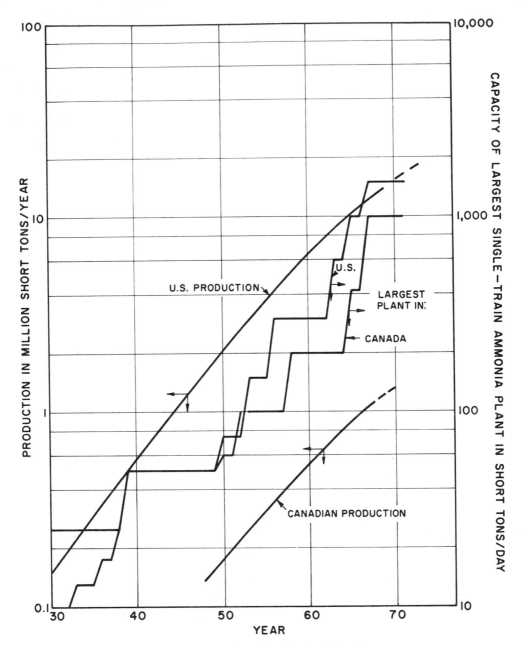

Exhibit 11. Relation Between Largest Plant Size and Production in Canada and the United States—Ammonia.

The situation between the United States and Canada is shown in Exhibit 11 for ammonia and Exhibit 12 for ethylene.[27] Ammonia production in Canada in 1969 was approximately one tenth that in the United States that is,

[27] Note that the scale for largest plant size on the right ordinate of each graph is one-tenth the scale for production on the left ordinate. The change in scales is made to bring the curves on the graph closer together to facilitate their comparison.

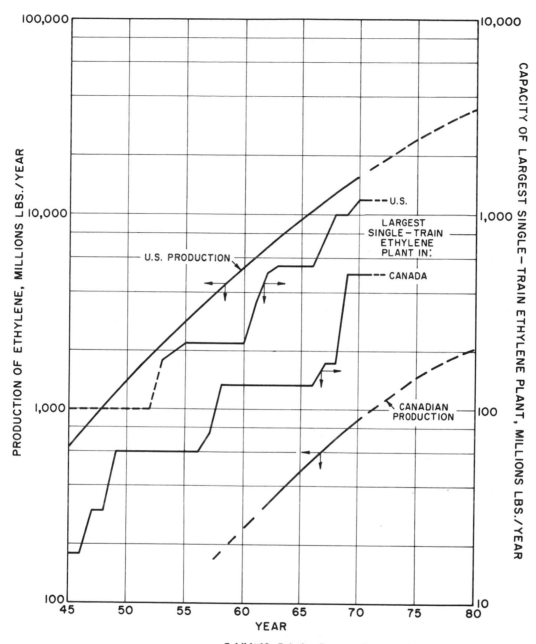

Exhibit 12. Relation Between Largest Plant Size and Production in Canada and the United States—Ethylene.

in the ratio of the populations of the two countries. This high ratio exists because fertilizers made from ammonia can be exported duty free to the United States from Canada.

The ratio of ethylene production in the two countries is more characteristic of the petrochemical situation, 1 : 20. Ethylene derivatives cannot be exported to the United States from Canada owing to high United States tariffs.

Exhibit 11 reveals that the largest ammonia plants in the United States have been 1.5 times larger than the largest ammonia plant in Canada; or, alternatively, there has been a two- to three-year lag in the construction of an ammonia plant in Canada of the same size as the largest in the United States.[28]

Similarly, Exhibit 12 shows that the largest United States ethylene plant has mostly been 2.5 times the size of the largest Canadian ethylene plant; or, in terms of time, the lag in building an ethylene plant in Canada equal in size to the largest in the United States has been six to seven years.

A continual and real cost penalty therefore exists as a result of the difference in size of the markets, and therefore according to this model, of the plant sizes in the two countries. Since the costs of capital and taxes are higher in the smaller country, the handicaps of scale, money cost, and taxation must be offset by differences in wages, tariffs, freight, and exchange.[29]

The *risk* of building a large plant is greater in the smaller country, as can be seen in Table 3 from the number of plants required in the two countries.

Table 3. Numbers of Plants Required in Canada and the United States.

	U.S.	Canada Potential	Canada Actual
Ammonia plants (1968)			
Capacity, short tons/day	1,500	1,500	1,000
No. required	24	2.2	3.3
Ethylene plants (1970)			
Capacity, millions lbs./yr.	1,200	1,200	500
No. required	13	1	2

Ammonia and ethylene are intermediates which must be converted into further products (derivatives) for sale. This is normally carried out on adjacent sites, giving rise to an integrated petrochemical complex in which several companies usually participate.

The large number of plants required to satisfy the United States market has resulted in the development of a petrochemical region along the Gulf Coast of Texas and Louisiana by means of pipelines. Similar developments are occurring in Western Europe, Britain, and Japan. They are not feasible in smaller countries such as Australia, Canada, Mexico, and India, where only one or two integrated sites are required to satisfy demand. These problems

[28] Except during World War II when standard 50 ST/D plants were built in both countries by agreement.

[29] Excluding indirect control of trade through quotas, border taxes, and so on.

are of concern to governments in these countries and the model proposed here assists in their examination.[30]

This situation varies from industry to industry. Thus Canada competes internationally in civil aircraft of up to twenty-four seats capacity, but has purchased its larger aircraft mainly from the United States and Britain. Smaller countries do not produce computer hardware but can compete in computer utilization. *The most critical point for the country with the smaller market is the step from one generation of plants or units of one size to the next larger size. Appropriate government policies at this stage are a "must."*

PROFITABILITY CALCULATIONS AND MANAGEMENT DECISIONS

The third area for which the performance-maximizing industries model has implications is the human aspects of profitability calculations in decision making.

Fifteen-year price forecasts are required for discounted cash flow calculations.[31] It will be shown that the accuracy of such forecasts is inherently lower than expected, and that the real functions of such calculations are definition of objectives and resolution of higher managerial relationships.

Publishable price data are difficult to obtain. Table 4 provides published trade estimates of the average price of a variety of grades of low- and medium-density polyethylene resin in current United States cents per pound. They are plotted in Exhibit 13 with the estimated operating ratio in the industry.

Table 4. Price History of Low/Medium Density Polyethylene in the United States.

Year	1943	44	46	47	48	50	52	54	56
U.S. ¢/lb	100	70	52	47	43	48	47	41	39
Operating ratio, %				60	60	75	87	68	90

Year	1957	58	60	61	62	63	64	67	68
U.S. ¢/lb	37	32	28	23	21	18	17	15	12
Operating ratio, %	98	108	75	75	83	86	93	75	84

[30] W.H.C. Simmonds, "The Economies of Scale through Integration," paper presented to the 20th Annual Conference of the Canadian Society for Chemical Engineering, Sarnia, October 21, 1970.

[31] See W. W. Twaddle and J. B. Malloy, "Evaluating and Sizing New Chemical Plants in a Dynamic Economy," A.A.C.E. 10th National Meeting, Philadelphia, Pa., June, 1966.

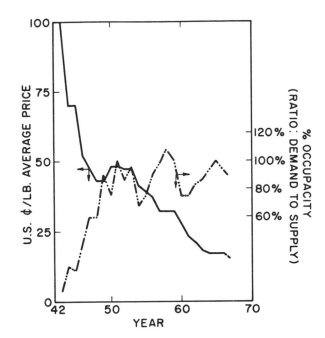

Exhibit 13. Relation Between Average Selling Price of Low- and Medium-Density Polyethylene and Overcapacity.

The price drops coincide with overcapacity as shown in Exhibit 13. No price rises occur except when polythylene was on allocation during the Korean war.

The Boston Consulting Group has shown that, in performance-maximizing industries, costs usually decrease by fifteen to thirty percent each time experience (for example, cumulative production) has doubled.[32] In the absence of published cost data, *prices* can be plotted as in Exhibit 14. Knowledge of the *cost curve* can be used to forecast future costs, but these do not necessarily provide future selling prices. As Exhibit 13 shows, price drops are a function of industry overcapacity, that is, of the action of competitors. They cannot therefore be forecast by any simple extrapolation, since there is no rational way of deciding when existing or new company managements will overexpand (as in the past) or limit expansion (as at present). The effects of such decisions on the magnitude and timing of price changes introduce an inherent uncertainty into price forecasts in this group of industries.[33]

Since companies blithely accept long-range price forecasts, it suggests that profitability calculations are fulfilling other purposes. One of these is clearly a better definition of the intended expansion or investment and its

---

[32] The Boston Consulting Group, "Perspectives on Experience," 1968.

[33] This is analogous to the Heisenberg uncertainty principle in physical science. It asserts that no amount of time or money can improve price forecasts beyond the limit set by the uncertainty in human decision making.

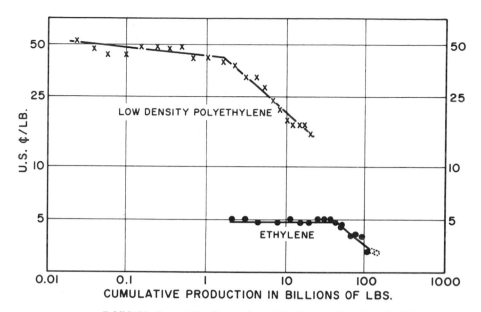

Exhibit 14. Cumulative Production of Ethylene and Low-Density Polyethylene Versus Average Selling Price.

objectives. The other is more interesting, since it throws further light on the interplay of the human and technoeconomic factors in business.

Two groups of people are involved in major company decisions, top management and divisional management. This human situation "parallels" the technoeconomic situation. Top management must determine whether its divisional managers will put themselves and their people on the line to ensure the success of their proposal. Equally, divisional management must find out whether top management will commit itself fully to the new venture and not "welsh" on its subordinates if the going gets rough, whatever the reasons. The medium used is the profitability calculation, the results of which swing backwards and forwards as the two sides probe each other's positions and intentions. The biggest variable is usually the price forecast. And when both sets of managers see eye to eye (that is, the human situation has been resolved), it is interesting to watch how the final price forecast happens to give just that discounted cash flow return which both parties are willing to accept. *Quelle bonne surprise!*

These comments indicate where some companies may be having trouble with technological forecasting. They have apparently attempted to treat technological forecasting as another expertise, similar to operations research, systems analysis, which you get an expert to do. The present analysis does not support this approach. Any consideration of the future involves human as well as technical and economic factors *within* the company as well as without. Forecasting is likely, therefore, only to be successful to the extent that it is integrated into the thinking and planning of senior management in a company.

In this respect, it is significantly different from *la technique*[34] and a great deal more consideration needs to be given to the question, "What is forecasting?"[35]

This chapter suggests that an essential first step in any examination of future possibilities is the classification and analysis of industrial experience in terms relevant to "forecasting." A classification of industries in ten groups appears realistic. The analysis of the behaviour of one such group, the performance-maximizing industries, has proved rewarding in disclosing the reasons that profit expectations were not met. It has also helped to explain why countries with smaller markets face considerable difficulties and limitations in competing in these industries. It has also thrown light on the interaction of the human and the technoeconomic in industry.

APPENDIX: DATA FOR EXHIBIT 2

Size of the Largest Plant or Operating Unit vs Time.*

	Civilian Jet Engines		Useful Weight Put Into Space		Computing Systems		Passenger Aircraft		Oil Tankers	
	Thrust M Lbs.	Index	M Lbs.	Index	M. Op.^{n}/Sec.	Index	Max. Cruise Speed × Capy.	Index	M Dwt.	Index
1950					0.30	100	2.8×10^4	100	28	100
1951					0.30	100	2.8×10^4	100	28	100
1952					0.30	100	3.0×10^4	107	28	100
1953					0.67	223	3.65×10^4	131	28	100
1954					0.77	257	3.65×10^4	131	28	100
1955					1.1	367	3.65×10^4	131	46	164
1956					3.8	1270	3.65×10^4	131	46	164
1957					15.6	5200	4.8×10^4	172	86	307
1958	9.5	100			34	11,300	8.9×10^4	318	86	307
1959	13.5	142			46	15,300	9.4×10^4	336	115	410
1960	14.6	154			56	18,700	9.4×10^4	336	115	410
1961	16.4	173	Mercury		631	210,000	9.4×10^4	336	130	465
1962	16.4	173	4.2	100	631	210,000	9.4×10^4	336	130	465
1963	16.6	175	4.3	102.5	631	210,000	9.4×10^4	336	130	465
1964	16.6	175	—	—	4091	1,360,000	10.4×10^4	370	130	465
			Gemini							
1965	17.9	188	8.0	190	4091	1,360,000	10.4×10^4	370	130	465
1966	17.9	188	8.3	198	4307	1,440,000	10.4×10^4	370	210	750
1967	17.9	188	Apollo				14.2×10^4	506	210	750
1968	35.4	372	45.2	1,080			14.2×10^4	506	210	750
1969	35.4	372	108	2,570			14.2×10^4	506	312	1,115
1970	35.4	372	110	2,620			20.0×10^4	710	312	1,115
1971	35.4	372	110	2,620			(Airbus 17×10^4	606)	380	1,360

* Represents the *expected* capacity, thrust, power, etc., of the largest unit in the United States at the time of ordering. *Actual* performance may, or may not, have borne out management's expectations.

[34] Jacques Ellul, *The Technological Society*, trans. J. Wilkinson, (New York: Vintage, 1964).
[35] W.H.C. Simmonds, "How to Make Sure you Have a Future," *Innovation, XII* (1970), 36–43.

	Ethylene Plant Capacity		Coil Coating Machine Capacity		Phosphorus Furnaces, Max. Power		Largest Steam Turbine	
	MM Lbs./Yr.	Index	In. ft./Min.	Index	M KW	Index	M KW	Index
1904							5	100
06							12.5	250
12							20	400
15							45	900
24							50	1,000
25							60	1,200
28							94	1,880
29							208	4,160
34					7.5	100		
37					24	320		
43	100	100						
46			600	100				
50			1,120	187				
52					36	480		
53	180	180					217	4,350
54	220	220						
56					45	600	260	5,200
57							275	5,500
58							350	7,000
59			2,400	400				
60							500	10,000
61	380	380						
62	550	550	3,600	600				
63							690	13,800
64			14,400	2,400				
65		′					1,000	20,000
67	750	750			60	800		
68					70	933		
69	1,200	1,200	18,000	3,000				

Forecasting by Monitoring Signals of Technological Change

JAMES R. BRIGHT

Coming technology casts shadows ahead and becomes visible to society long before it has significant impact. Therefore, we should be able to detect signals of coming technological change and monitor this progress in a systematic way. Philosophically, this is forecasting from events in being. Actually, monitoring is loosely done by many firms. The author offers a monitoring journal, and a system of analyzing signals from five environments is described. A specific example, the missile versus the bomber, is developed in part to demonstrate the method.

Since new technology emerges from idea into physical reality and then is applied and adopted at some gradual rate, rather than instantaneously, it follows that it may be possible and should be useful to monitor this progress so as to react wisely at some future time. Specifically, it should be possible to *monitor* the environment to detect the beginnings, the progress, and the likely consequences of new technological advances. Monitoring, by this definition, is based upon assessing events in being that impinge on technology. It includes four activities:

A partial presentation of this concept appeared in "Evaluating Signals of Technological Change," *Harvard Business Review* (January-February 1970), pp. 62–70.

1. Searching the environment for signals that may be forerunners of significant technological change.
2. Identifying possible alternative consequences if these signals are not spurious and if the trends that they suggest continue.
3. Choosing those parameters, policies, events, and decisions that should be followed in order to verify the true speed and direction of technology and the effects of employing that technology.
4. Presenting the data from the foregoing steps in a timely and appropriate manner for management's use in decisions about the organization's reaction.

Monitoring includes much more than simply "scanning." It also includes much more than collection of traditional library reference lists. The essence is evaluation and continuous review. Monitoring, philosophically speaking, includes the acceptance of uncertainty as to meaning and rate of change of developments. The forecaster "runs scared"; he tries to stay ahead of developments and their implications. His contribution lies in providing early warning and in remaining open-minded about possible significance. He presents his warning early and his conclusion relatively late in the forecasting process.

Individuals in high technology firms (electronics, drugs, chemicals, and so on) may point to the fact that their firms and/or trade associations have superb technical libraries, with the responsibilities of collecting information on pertinent topics. They can quickly muster a list of references and a critical topic if so requested. They even may automatically report papers and articles on selected topics. "Monitoring," therefore, "is nothing new." This position is not correct, as can be seen by contrasting the goals of monitoring with the role of the library.

1. Libraries collect and can collate information. They do not usually *evaluate* it. Evaluation is the goal, not collection per se.
2. Libraries collect conventional publications by specified labels. They cannot be expected to report things that do not crop up under usual sources and labels. For instance, would they present the contents of the writings and speeches of, say, a new secretary of HEW or sense new goals and efforts that might affect a particular business? Would they analyze the national budget for the technical significance in the distribution of national R&D funds? The list of signals in Table 1 would not be provided through a request to a librarian. No library can encompass all the sources of signals, for many of them are not reduced to public print.
3. Libraries cannot be expected to identify or deduce the second, third, and further order of effects of news events on technology and vice versa. For instance, would the library retrieval system link the appointment of retired Gen. Maxwell Taylor to Chairman of the Joint Chiefs of Staff with his book, *The Uncertain Trumpet?*[1] And would it link the General's plea for "nap of the earth" capability for the Army to a possibility of potential invest-

[1] Maxwell G. Taylor, *The Uncertain Trumpet,* (New York: Harper & Row Publishers, Inc., 1960).

ment in helicopter technology? This kind of analysis is not the product of mechanistic cataloging and retrieval, but of imaginative and perceptive analysis of many things and their interactions.

Monitoring includes search, consideration of alternative possibilities and their effects, selection of critical parameters for observation, and a conclusion based on synthesis of progress and implications. The feasibility of monitoring rests upon the fact that technology emerges from the mind of man into widespread economic reality, with resulting societal impacts, over a long period of time. Therefore, this process by which technological innovation emerges can be used as an analytical framework. There are always some identifiable points, events, relationships, and other types of "signals" that mark the passage of the new technology through this process. If one can detect these signals, it should be possible to follow the progress of that technology relative to time, cost, performance, obstacles, possible impacts, and other considerations. Then the manager would be given two more inputs to his decisions: awareness of the new technology and its progress and some thoughtful speculation and specific guidance for assessing timing and impact. Exhibit 3, p. 7 in Chapter 1 of this book, identifies eight stages in this process of technological innovation according to one analytical plan. Other schematic structures of greater refinement than this one are equally feasible for the purpose of monitoring.

Searching the technological environment alone is totally inadequate. Technological innovation often is influenced as to direction, timing, speed of emergence, rate of diffusion, and mode of application by events in four other environments. The traditional analytical procedures used by business and government (and currently taught in our best schools of business administration and engineering) are to methodologically examine and evaluate only factors in the *technical* and *economic* environments. Innovation studies show that such "two-dimension" analyses often have been pathetically inadequate. Many times the key events, changing values, and relationships that determined the ultimate significance of the technology and its timing lay in either the *social* or the *political* environments. In recent years the *ecological* environment is becoming a stronger source of pressures for change. Furthermore, interactions of events between these five environments often proved to be the significant force. Exhibit 1, p. 4, in Chapter 1 is a schematic representation of these environments. It follows that monitoring requires us to look for "signals" in all five of these environments, and we must become more skilled in examining and interpreting relationships affecting this very complex process.

It should be apparent that events in any of these environments may interact with the progress and economic significance of the new technology. For example, the closing of the Suez Canal affected the economics (hence, size) of supertankers. The discovery of natural gas in Europe and the North Sea seriously altered the desirability of marine shipment of liquified natural gas and refinery equivalents from Canada and Algeria to Europe. In the first case he factor was *political,* in the second, *ecological.* Similar relationships appear n other examples throughout this chapter, and in the examples given by Richard Davis in Chapter 35 of this book.

Exhibit 1, p. 4, should convey that these interacting forces may be multidirectional, of changing strength in each environment, and of changing significance over time. The "feedback" or "feed ahead" time lag to and within any environment may be of a very different order for one environment as compared to another. The time lag and force will vary with respect to different technological innovations, and at different times in history according to the social, political, and economic climates.

As an example of such an interaction, consider how the social environment now is pressing technology for solutions that will reduce automobile engine air pollution. However, the economic environment traditionally does not measure social costs. Therefore, air pollution (an ecological event) did not, until the mid 1960s, put much pressure on the engine-fuel technologist for technological changes to reduce air pollution. The automobile manufacturer traditionally has responded to demands for economy and power, both economic and social factors. Executives in the automobile and petroleum industries have conscientiously studied the technical and economic facts of the ICE (internal combustion engine) and its competitors. They sought guidance from this traditional analysis, and took comfort in the fact that no other vehicle power plant could provide the performance and economy demanded of the American family car. Furthermore, planned engineering refinements of the ICE will eliminate perhaps seventy to eighty percent of the emissions formerly generated.

Unfortunately, this conventional analysis neglects at least two potentially significant signals: (1) the Congressional *Hearings* on the Electric Car (March, 1967) indicate that there could be a very abrupt and powerful economic support for other automobile power sources not based upon economic logic. This support could arise out of the *political environment* in the form of mandatory government purchases of electric and steam cars (as proposed by Secretary Udall and Congressman Ottinger), plus the possibility of restrictive legislation on conventional ICE powered cars. (2) The trend toward multicar family ownership makes it possible to have special-purpose vehicles, the commuting or "around-town" car, the recreational vehicle, and so on. How are these second and third vehicles used? In a brief assignment one of my students found that the second and third cars in two of the three local families he studied each went less than a total of nine miles per day and in traffic that prohibited speeds of more than twenty-five miles an hour. Such power demand is well within the capability of today's electric vehicle power plant, provided that the user cares to accept the costs and other implications. At the same time, another student found that at least four couples (out of about 120) living in a Texas retirement-vacation resort had sold their second cars and were using their electric golf carts for local transportation. He then found that this practice was followed by fifty percent of the residents in one California retirement community.

This example provides another important consideration: conventional economic measurement sometimes impedes the consideration of technical alternatives, since it measures only one type of cost in one environment. Conventional business wisdom thus may badly mislead us on the forces at work.

The conclusion should be obvious, but it still deserves emphasis: There are signals in the political and social environments of factors that *may* eventually support vehicle power plant systems other than the traditional automobile engine. This possibility is emerging despite the ICE's improved pollution characteristics and in the face of its far superior economics and overall performance.

Any technological forecasting procedure, therefore, requires exploring these questions:

1. What changes and events in each of these environments are creating new needs or opportunities for technical change? What type and degree of support is being thrown behind these pressures? What changes in timing and pressure are possible and likely?
2. What is the impact of the new technology, on *each* environment, in terms of (a) requirements to produce the technology and (b) the effects that will follow its widespread use?

SEARCHING ENVIRONMENTS FOR SIGNALS

Within each environment there are different types of parameters and events that are indicators of potential change. Here are some phenomena that provide signals.

Technological Environment

Time series of technical parameters and figures of merit are very suggestive Projection of these trends often suggests possibilities and significances, as well a a rough idea of timing.[2] For example, the extension of the number of electrical circuits per unit of space or area in solid state electronic systems points to a coming capability for compression of system size to at least $1/100$ of the size of the 1960 circuitry and perhaps much more. Meanwhile, circuit cost also is declining exponentially. The product design implications for devices employing electronics, radio, radar, TV, the computer, and military equipment are clear and certain. These devices will become much smaller, cheaper, and more

---

[2] This is not advocacy of blind and never-ending extrapolation. The forecaster must consider whether technical, economic, political, and social factors will limit the trend in the near future. If he can foresee no limiting factors, the trend of the past possibly the guide to the near future conditions.

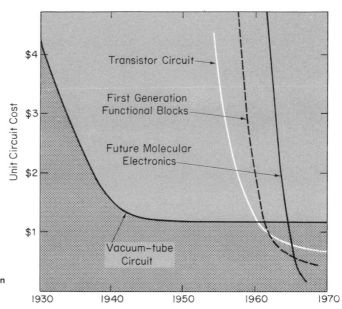

Exhibit 1. A Forecast of Cost Reduction in Electronic Circuitry.

Graph labels: Transistor Circuit; First Generation Functional Blocks; Future Molecular Electronics; Vacuum–tube Circuit. Unit Circuit Cost axis: $4, $3, $2, $1. Year axis: 1930, 1940, 1950, 1960, 1970.

reliable. What are the operational results? They certainly include portability, low cost, and, hence, widespread use. These devices will be available for use in many new situations. The curves in Exhibit 1, a forecast of the mid 1960s by the same U.S. Air Force Avionics Laboratory that underwrote the development and first application of integrated circuits, are suggestive. If these curves are anywhere near correct, they tell the electronics components and equipment designer and manufacturer that:

1. Almost incredible cost and size reductions lie just ahead.
2. Most electronic products are likely to be rapidly and continually obsoleted over the next five to ten years.
3. The wisdom of Solomon or the gambling instincts of a wildcatter will be needed, at times, to set production and marketing policy and tactics in this environment of rapid product change.

Reflections of interest and commitments to areas, problems, or phenomena also provide signals. The general direction of effort of the technological community may be revealed in many ways. A firm may announce that they are establishing a laboratory to develop holography. A company president may give a speech stating that his firm sees opportunity in home entertainment centers. A professional society (such as the American Society of Mechanical Engineers) may announce a conference on underwater technology. When 500 engineers enroll instead of the anticipated 100, this unusual response is our signal. Does it reflect merely wishful thinking, "band wagon" effect, or major progress and coming effort? We must monitor this technical activity to determine its true significance.

Demonstrations, announcements of progress, patent awards, and trade paper and professional society reports of physical materialization of new technology deserve scrutiny. Sincere but optimistic overstatements may be involved. For instance, one can recall many optimistic statements about such developments as the Curtiss-Wright Aerocar, the Wankel engine, RCA's Ultrafax transmission-reproduction facsimile demonstration (simultaneous with Haloid's 1948 demonstration of xerography), and the Allis Chalmers fuel-cell powered tractor. All were impressive demonstrations in the last ten or twenty years, yet none are economic successes in 1971. This is not to disparage these demonstrations. Some things fail for good technical or economic reasons, others are displaced by a superior technology. Still others, such as the Wankel engine (apparently) are merely delayed much longer than anyone had imagined. The point is that such demonstrations may signal that an important new technology is envisioned or has materialized. The Wright brothers' first flight was 110 feet, a joke in terms of performance. Yet its real significance was that man had proved that he could make a machine that would carry him through the air. From hindsight we know that the distance was irrelevant, the *nature of the accomplishment* was everything. Further technical progress could easily have been monitored.

Carlson's first patent on xerography was reported in the *New York Times* in 1940 and spotted by a young IBM marketing man. According to Carlson, during the next year and a half this young man tried to interest IBM in Carlson's invention. He failed, and so his firm failed to capitalize on the potential in the new technology that public disclosure had placed in their hands. The signal was there. Opportunity was not only knocking, it was literally begging for entry. Only the assessment of Carlson's first crude demonstration was inadequate. Similarly, Eastman Kodak's own periodic technical report *Patent Abstracts,* duly reported an abstract of Carlson's first patent in 1940 Although Kodak's librarians picked up the signal, no one monitored its progress; or if so, they were unable to persuade management to continuously observe and evaluate the new technology. A great opportunity was at hand and could easily have been Kodak's.

Xerography also provides dramatic examples of demonstrations that were signals of great impending technical success. In October 1948, the little Haloid Company held a major press demonstration in New York City. They showed with laboratory scale equipment, how xerography could be used to copy letters and other office data. They demonstrated a camera to take and develop pictures in two minutes, and they showed a model printing press running 1,200 feet per minute. All of these and other potential applications were widely discussed in the press, ranging from *Business Week* and the *New York Times* to various trade journals. Industry had its signal. History shows their response. Aggressive monitoring and awareness of the phenomena of technological innovation discussed in Chapter 1 could have made the fortunes of many firms.

It is most important to note that two of these xerography applications have not yet materialized after twenty years, while one has exploded! In other words, *an appraisal of potentials made at that time (1948) could easily have been wrong.* Progress should have been *monitored* to detect the ultimate true course.

Performance data of technological improvement also must be scrutinized. These data are often mentioned in trade journals, professional society papers, technical reports of government agencies, and in advertising literature. When related to earlier achievements in a time series they show (possibly!) the state of the art, and so they may suggest rates of technical progress. One of the great benefits in such reports is the identification of problem areas and limitations. These help to define what critical technical-economic factors must be monitored, and where the bottlenecks lie.

Usage and applications data, of course, tell the value, volume, and adoption of the device. By using consumption and application data one can build estimates of technological needs and opportunities. For instance, in the early 1950s it was commonly stated that the United States might possibly use as many as a few dozen computers. But adoption has been unbelievably great, and about 70,000 computers were in use in 1969, with over 50,000 reported to be on production schedules for the year. This signal (number of adoptions) should lead the manager and analyst to explore questions about the future material requirements, testing facilities, software, personnel, and supplies needed, as well as the impact of usage on related fields such as accounting, tax procedures, and financial analysis.

The *nature* of usage can sometimes forewarn one as to what other technology must be created. For instance, computer process control is growing rapidly. This means that instruments to sense and provide computer input for dozens of different attributes of production processes such as pressure, tension, color, viscosity, acidity, temperature, and so on must be developed. Today the instrument makers have strong signals of impending opportunity.

As another example, consider the growing use of plastics to replace metal. If plastics are to compete with cold-rolled steel sheet, vacuum-molding equipment must be developed to produce shapes of appropriate size and form for end product applications. Have the plastic press manufacturers yet identified this signal and considered their alternative courses of action? Has anyone yet done exploratory design of molding equipment large enough to handle parts of automobiles? This is not to argue that this necessarily should be done now. The point is that signals of coming conflict between cold-rolled steel and plastics are obvious. Concerned industrialists should be monitoring further progress to decide when and what to do. Table 1 below points to a few possible signals, available to even a nontechnical casual observer.

Table 1. Possible Signals of Coming Competition Between Cold-rolled Steel and Plastics, 1967–1970.

1967	Refrigerator cabinets and TV cabinets of vacuum molded plastics commercially available.
1967	Marbon Division of Borg Warner advertises a commercially available camper body for a pickup truck. It "won't rust, fade, chip, and so on," claims to be better than steel and is made in a few minutes by vacuum molding.
1967	Centaur Engineering Division of Marbon displays a "sports sedan" in full-scale vacuum molded of plastics in three minutes; it consists of only seventeen parts. Finished vehicle with conventional power plant will weigh around 2,600 pounds.
1967–68	Chemical engineering and petroleum journals point out that proposed petrochemical plants of far larger size promise enormous gains in capacity and cost reduction of plastic materials.
1968	German firm demonstrates an "all-plastic" body car. Pictures and story in *Business Week*.
1969	At annual meeting of the Society of Automotive Engineers a paper "Plastics vs. Steel" is presented. Similar papers appear in British engineering circles.
1969–70	The number and weight of plastic components of automobiles grow steadily. One 1970 Pontiac offers a plastic bumper.
1968–70	Vacuum-molded plastic bodies for all terrain vehicles and snow mobiles advertised widely.
August 1969	A plastic-bodied golf cart described in *New York Times* and *Business Week*.
July 1970	UAM labor negotiations with General Motors plus articles on reaction of automobile labor to assembly line work indicate social and economic pressure to reduce work force on assembly line tasks.
Mid-1970	Pressed Steel Fisher, Ltd. combines with Imperial Chemical Industries and Austin Motors to produce the Austin Mini automobile with a thermoformed plastic body. Construction of 500 car/week factory begun in Chile. Rumors of six more such factories planned for other countries.

Note: These "signals" are not the result of intensive search but simply of casual business reading. Critical monitoring would turn up many more and probably far more significant signals.

Economic Environment

There are, of course, many powerful signals emerging from purely *economic trends* and *events*. Since thousands of economists labor in this field, I will not presume to survey the possibilities. In general, economic signals can be found in time series of costs, capacities, demand, available resources, rates of production, consumption, and similar data.

Social Environment

Population trends are a prime source of signals relevant to the ten- to thirty-year horizon. Hindsight reviews of past technological forecasts suggest that population figures are one of the most neglected and unappreciated social signals affecting the economic significance of technical changes. For instance, the Paley Report on America's needs for materials in 1975, "Resources for Freedom," was published at President Truman's direction in 1952. It was largely based on 1950 statistics. Numerous forecasts of consumption and demand for various materials and, therefore, production facilities are under stated in this report because of what proved to be erroneous population forecasts. Business and government need to give much more consideration to the use of the very best demographic data in assessing the potential significance and direction of new technology.

Measures of *activities* such as leisure time usage, education, and occupational interests; measures of *social conditions,* such as the amount of disease, poverty, crime, and air pollution; and measures of *attitudes* or *values* such as preferences, interests, political opinions, and aspirations (for example, changing attitudes on overpopulation, hence new impetus to birth control technique) may provide useful parameters of social change.

We must also recognize the formation of special interest groups and the advent of books, speeches, and personalities that capture public attention, such as Ralph Nader's work on auto safety or Rachel Carson's *Silent Spring.* All these types of trends and events *may* presage the coming of technological change.

Social attitudes are one of the biggest stumbling blocks for all of us. It is absolutely vital that the manager or analyst divorce his own value system from his assessment of these phenomena. Although one may believe that Nader may grossly misrepresent the facts of auto safety or that Rachel Carson may exaggerate the effect of insecticides and although antiflouridation groups may be scientifically in error, their influences on technological progress stem from *attitudes that they create and not solely from facts.* The industrialist, secure in his technical-economic logic, may easily overlook the force of opinion when it differs from his perception of properly weighed truth.

Political Environment

Action follows resources, and today it is the political environment in which many technological directions are chosen, resources are committed, and usage is determined. It follows that governmental actions that support technological development, underwrite development, finance usage, or prohibit or limit applications must be watched very closely.

Earlier signals may be available because formal government actions almost invariably are preceded by discussion, committee reviews and recommendations, and by reports of debates about alternatives. If we can spot these signals of *considerations* and *possible choices,* we may be able to follow the evolution of a new technology. The convening of a strong committee to examine a national problem may well herald the beginnings of new technological approaches. I am indebted to Ronald Smelt, vice-president of Lockheed, for pointing this out so clearly in an actual industrial situation. He writes:

> . . . (my comment) was intended to dispel the popular illusion that a technological breakthrough is the sudden accidental stumbling over a new concept or principle in the laboratory, which immediately becomes translated into an improvement or new product for industry or defense. On the

contrary, the technological forward steps which are the ingredients of most breakthroughs in the aero-space industry can be traced quite clearly to research and development efforts which usually have preceded the breakthrough by several years. What seems to happen at the critical time is that a number of these technological improvements fall into place and are recognized, in concert, as providing the solution for a national or industrial need...this critical point at which the innovations fall into place can be identified in our industry (aero-space) by the formation of a national committee, a government task force, or other similar managerial device which brings the various experts of the technical areas and the exponents of the national need into the limelight together. The result of this multiple mating process is that the capability of technology to satisfy the need, at least approximately, is recognized at a sufficiently high level to stimulate effort and funding specifically aimed at exploiting the situation.

I suppose the best example of this program is the ICBM. Here the technologies for the compact thermonuclear device, the long-range liquid rocket, and the development of hypersonic heat transfer data, together with the materials capability to withstand the heating, had all been proceeding independently prior to 1953. The recognition that these had proceeded far enough to offer the prospect of fulfilling the need for an ICBM was signified by the formation of a few national committees, of which the most important was undoubtedly the special President's committee with Charles Lindbergh as its chairman. The report of this committee started the flow of effort and funds into the Air Force ICBM project, and to many people the year 1953 is identified as the "breakthrough" year for the ICBM.

Another more current example is occurring in the growth of the surface effect ship. Work on support by an air bubble has been going on for many years, all over the world; in a few years it grew into a specific program in England. In the United States the breakthrough occurred about three years ago when the Maritime administration and the Navy together formed a special committee which recognized that the state of technology would now permit the development of a new class of vehicles, satisfying special needs for high-speed water transportation.

The basic point which I am making is that the growth of science and technology is a relatively continuous process, with most technological advances casting shadows well ahead. (This is probably why the breakthrough seems to come in several countries at roughly the same time, even without conscious communication between the workers.) A definite stimulus is provided, in the majority of examples, by the formation of a government committee—probably the most effective method of bringing the technology face to face with the need.[3]

Even these public modes of decision making, government committees, have their individual instigators. There may be a very strong signal in the

[3] From a letter to the author.

appointment of a given person to a department, cabinet, or bureau from which he can influence the activity of a major agency. Identifying the interests, opinions, aims, and even obsessions of such individuals may help us to determine what technology is likely to be explored and supported. An agency such as the White House Office of National Goals or a Committee on Environmental Quality may be a new source of signals in the political sphere.

Ecological Environment

One might argue as to whether ecology provides its own signals or social perception of ecological conditions is the true signal. Certainly the latter often leads to social and political action that triggers technological change. Presumably, signals lie in:

1. *Quality* of air, water, food, noise level, and the appearance of our surroundings.
2. *Processes* that disturb the appearance and condition of existing natural resources, including wildlife.
3. Processes that generate wastes of substantial magnitude or impact.

This review of environmental signals is not complete for any given firm or industry, of course, but it does describe the types of things that one must monitor.

ANALYZING SIGNALS

Assuming a potential signal is identified, the analyst should go through these steps:

1. SIGNAL | What are the *facts*, meaning hard data, about this presumed signal?
2. A? B? C? | What are the major different alternative implications (A,B,C) that might be foreshadowed by this signal? (Note that those may be contradictory or supplementary)
3. A F_{A1}, F_{A2} B F_{B1}, F_{B2} C F_{C1} F_{C2} | What factors need to be followed to determine the progress of the possible alternative A; of B; of C?

SOME POSTULATES ABOUT MONITORING SIGNALS OF TECHNOLOGICAL CHANGE

The systematic collection and analysis of all signals are habitually done only by the intelligence agencies of military organizations. In industry many individuals, departments, and firms have at one time or another attempted careful monitoring jobs.[4] Obviously, a continual, comprehensive study of an area could be an extremely expensive job, requiring many man-hours. We cannot prescribe the amount of effort but only propose the need to adopt this procedure at some times and to some degree. Certain postulates about monitoring have been found useful:

1. The output of monitoring is additional information, not a decision.
2. The emergence of a technical capability is not equivalent to economic success. It merely tells that the technology is available, not that it will be used.
3. Technology rarely achieves major economic impact until it is adopted in significant volume. The problems of production and the time needed for diffusion of radical new technology are such that major economic impact will take at least five and maybe twenty years to occur. It would seem that there usually is ample warning of *economic* effect.
4. At times the role of an individual is decisive. An able and determined man in a certain position may completely dominate the direction or the timing to technology.
5. A given signal usually has a number of possible implications. Therefore, one must follow all implications until it becomes clear which are the correct ones. Do advances in television programming (color, live world-wide coverage, and better shows) mean that people will tend to stay home, or will they be inspired to travel more and see the things that TV showed them first hand?
6. Since we lack the wisdom to be certain in selection of relevant items, we must follow many things with care. This is impossible some say. Not so. It is difficult, and the results will be imperfect. But it is not impossible to gain some guidance. After all, when Sputnik I was launched it was rather late to decide that rockets were here to stay. It is not impossible for automobile manufacturers to monitor progress in several types of power systems that might compete with the internal combustion engine.
7. There are false and misleading signals, and important signals are sometimes obscured by "noise" in the environment. Systematic and thoughtful search of the environment is needed to make sure that real signals are not missed. We must dissect events which frequently obscure things by combining several signals. Furthermore, there are completely contradictory implications that can be drawn from a given signal or from concurrent signals. For example, what will emerge from trends of miniaturization of computer circuitry, drastic circuit cost reduction, and the growing capacity of super computers? Do these mean time-sharing or millions of special-purpose computers or both?
8. One tends to be too limited in his assumptions of implications by considering only the state of technology at the moment of review. We should

---

[4] The earliest systematic industrial effort described in print was launched by Herbert Sorrows of Texas Instruments. It appears in *Research Management,* March, 1967. However, the program focused on technology and did not seem to include strong recognition of signals in nontechnical environments. The program described by Richard Davis of Whirlpool is the most comprehensive and effective industrial effort yet to come to our attention. (See Chapter 35 of this book.)

mentally extend the nature and amount of technical progress observed and then consider the implications in order to get a better idea of potential impact.

9. Many technologies are offered to solve some problems, and most never materialize economically. There are more contenders than winners. Therefore, one must monitor many developments, realizing that only time will tell which is the truly significant technology.

10. It is useful to try to define "thresholds" or measureable points at which a conclusion will follow or major impact will occur.

EXAMPLE OF MONITORING

To demonstrate the concept of monitoring and to clarify the analytical approach needed, let us first look at a hindsight example. As a dramatic and quickly grasped case consider the coming of the missile and its impact on the "plane makers," the airplane manufacturing industry.[5] The interested reader can put himself in the role of a student, then grapple with these questions:

1. Granted all the wisdom of hindsight, what indicators or significant signals (technological, social, political, and economic) could have been suggestive warnings to the "plane makers" of possible serious future developments that might affect the profitability and technical validity of their product line, the airplane? Consider the 1944–1960 period. Exhibit 2 provides a partial list of signals.

2. Assume that, as a member of a major airplane manufacturer's top management, you had identified each of your signals at the time they occurred. (a) Explain exactly how you would continue to monitor this development. (b) At what point, amount, time, or other measure will this indicator call for action by your firm?

3. At what point was the total situation clear enough for an aerospace firm management to take action steps? (Obviously, this would vary in each firm and with each manager's risk-taking propensities.)

4. Who in your corporation is supposed to do this kind of thinking? How should it be fed back to the firm for review and action?

As each signal is tentatively identified, it can be entered in the manner shown on the monitoring journal form in Exhibit 3. In the second column, one should list the descriptive *facts*—technical, economic, and operational—to the extent known. Judgments about prospects must be omitted in this step.

Next, one should think about the major implications and the significant

---

[5] This example is not aimed at criticizing these earlier industry managers, and I certainly am not imputing superior judgment on my part. Rather, this hindsight example was chosen to help my students learn how this monitoring concept might possibly be applied. We need all the wisdom and guidance that we can get from history. It was first taught in my Harvard Business School course in 1961; and the above questions were used in the textbook J. R. Bright, *Research Development and Technological Innovation*, Homewood, Ill., Richard D. Irwin, Inc., 1964.

Exhibit 2.   Some Types of Signals Available to "Plane Makers" from 1944 to 1960. (not in chronological order).

Growth of military rocketry in late World War II; the German V-1 and V-2.

Jet engine.

A-Bomb, H-Bomb.

Seizure and recruitment of German rocket scientists by U.S. Army and Russia.

Growing interest in space as evidenced in technical and popular literature, and professional societies.

Coming of computer. Time series data on computation of missile trajectories (provided here to illustrate their nature):

Year	Computation Time	Computation Cost	Error Rate	Set Up Time	Set Up Cost
1949 (hand)	6 months	$5,000.00	$1/10^3$	None	None
1949 (computer)	3 hours	$ 100.00	$1/10^5$	2 weeks	$500
1956	1 minute	$ 5.00	$1/10^9$	2 days	$300
1961	1/2 second	$ 0.25	$1/10^{12}$	1 day	$200

Progress in inertial guidance systems, radar, and long-distance guidance systems.

Coming of transistor and solid state physics.

Cold war situation: Greece, Berlin airlift crisis, and so on.

Korean War.

"Limited war" concepts discussed in much political and military literature.

Effectiveness of SAC in deterring all war and political aggression.

Postwar commercial plane market.

Trends in construction and economic content of elements of military plane (example is provided):

Time	Percent of Fighter Plane Costs Going to Electronic Components
1946	12
Korean War	46
1960	70–80

Air Force and Department of Defense budget distributions for R&D and various types of hardware.

Subcontracting trends of plane makers themselves.

Early United States missile contracts: Navajo, Regulus, Titan, Atlas, and so on.

Technical and dollar content and different manufacturing needs of missiles vis-a-vis airplane.

Power struggle between U.S. Army and U.S. Air Force for control of long-range missiles.

Attitudes of Air Force and Department of Defense personnel on missiles as expressed in speeches and conversation.

Ramo-Woolridge chosen over Convair as systems manager for Titan missile contract. Sputnik.

International attitudes on nuclear warfare.

Note: This list is by no means complete.

Exhibit 3. Sample of Part of Monitoring Journal of Technical Progress in Missile Development from Hypothetical Aircraft Manufacturer's Point of View.

Date	Event and Technical-Economic Data	Possible Significance	Things to Consider
1944	V‑1 used. 400 mph; 150‑mile range. Can be deflected or shot down by interceptors and AA guns. Pilotless. Poor accuracy, small payload (2,200 lbs.). Different power plant (ram jet). Simple launch facilities. Cannot be recalled or redirected. Weather no limitation on use.	New method for delivering a warhead. Expendable and cheap; thousands might be used. Does not use conventional plane manufacturing skills or technology. Low-skill operating work force. No present threat to bomber. Probably very low cost.	Accuracy, payload, range, and speed are poor. If each capability were improved, is it an alternative to the bomber? Counterweapon available?
1945	V‑2 used. 3,600 mph; 200‑mile range. Pilotless. Rocket motor. Cannot be stopped by planes or guns. Poor accuracy, small payload. No warning. No sound. Germany launches about 1,300.	Conventional bomber defenses unable to stop this weapon. Uses chemical motor for propulsion; this technology not in airplane manufacturer's skills. V‑1 speed limitation surmounted.	Launching facilities? Material? If accuracy, range, and payload improve, this is real threat. What are U.S. Army, Navy, Air Corps attitudes re future adoption?
1945	A‑bomb. Explosive force is thousands of times greater than anything previously known. Can be delivered only by manned bomber because of weight and size. Expensive, scarce raw materials. Terrible effects and aftereffects.	Only a few bombers needed to deliver a given explosive force. Too heavy for use in a missile, but radius of damage makes need for accuracy less significant. Horror weapon, probably will have limited use. But what if explosive force were diminished or the device miniaturized?	What happens if size of bomb reduced to missile capacity? Availability of fission materials? Cost? Clean-up of aftereffects? Reduction in power? National policy? International reaction to atomic warfare?
War ends, 1945	Russians and U.S. Army race to grab German rocket scientists. U.S. Army ships back 200 German missiles and establishes White Sands Missile Range.	Military badly wants this technical knowledge. Army Air Corps apparently not interested. Army Ordnance deeply interested. Potential enemy deeply interested.	Air Corps still favors our products. But might the missile, in hands of Army artillery, reduce Air Corps mission, hence plane needs? Advances in missile accuracy, range, and payload?
1946	Regulus and Navajo missile contracts let by DOD. Range 5,000 miles. Nuclear warhead.	Real threat to bomber and SAC mission. Navy to obtain strategic defense role?	Technical success. Future funding. Response of Air Force to prototype performance.
1946	Electronic computer developed.	Computation speeded exponentially. Hence missiles can be guided and redirected to different targets more rapidly.	Effect on missile and guidance systems. Computation speed. Effect on target assignments.

Exhibit 3. cont.

Date	Event and Technical-Economic Data	Possible Significance	Things to Consider
1947–1948	Greek and Berlin crises, crystallization of communist policy.	Russians intend to advance communism. Arms race to continue. Technical progress of all weapons systems will have priority.	Limitations in missile (and aircraft) performance will be gradually removed. Planes needed for what missions?
1948	Convair proposes ICBM.	Would turn into a major program affecting traditional bomber missions. Plane makers getting into missile business in big way?	Air Force response? What skills needed to make missiles? What mix of planes and missiles will be used?
1948	Development of transistor is announced.	Miniaturization of control. Reliability improvement. Both lead to missile guidance improvement.	What technical effect on missiles, planes, computers, and control systems?
1950	Korean War. Nuclear weapons not used. Missile-carrying planes, ground-to-air missiles appear.	Confirms that airplanes still are essential. Bombers remain necessary for tactical support.	Trends in plane and missile usage. Role of each? Further technical progress.
1950	Electronic content of plane now up to 46% of cost.	Plane manufacturers' need for electronic skills and for facilities probable, or else dependence on subcontracting.	R&D budget of Air Force. Cost of missiles and bombers? Skills and facilities needed.

Source: Courtesy *Harvard Business Review*.

possibilities. These should be listed in the third column. They are speculative, of course, and in many cases one will see contradictory possibilities. All possibilities should be listed.

Finally, one should establish those factors, threshold performance levels, figures of merit, effects, decisions, and so on, that are pertinent to each possibility. These lead to identifying critical parameters and activities that one must follow. Over time it will become clear that some of the signals were wrongly interpreted and that others were missed. Some of the "possible significance" ideas will prove to be erroneous and even ridiculous. But gradually the picture will begin to define itself, and one will see more clearly the need to act. One can sense this by following Exhibit 3, and by applying his own consideration to signals listed on Exhibit 2 and others which he will recall.[6]

The ultimate impact of the missile on the airplane manufacturers was dramatically conveyed in *Aviation Week,* March 28, 1955, by the report of the first major address of the new president of the Institute of the Aeronautical Sciences. This was Robert Gross, president of Lockheed Aircraft, who said:

> Of the 22 major weapons systems now under development, only nine are controlled by aircraft companies.
>
> Shall we bend with the storm and see our industry splintered or face the wind and build more of the weapons systems ourselves?

He cited the growth of the guided missile field and technical advances such as new types of power plants and automatic flight as reasons for the situation now facing the industry.

> We do not necessarily have within the geographic confines of the earlier airframe companies the people with all the skills now needed.

That was the year that the Air Force first gave a major missile development system contract to an unknown little newcomer, the Ramo-Woolridge Corporation. *Fortune's* 1960 article, "The Plane Makers Under Stress," suggests that even 1955 was too late for action. In 1957, Sputnik blazed into space and international headlines. By 1960, the "plane makers," the six great "primes," all lost their profitability when the Department of Defense formally recognized that the missile had replaced the manned bomber as the primary

---

[6] The reader should appreciate that these signals were selected solely from an academic and outsider's viewpoint. Anyone in the industry would have much more intimate and profound insight to relevant signals. It is also conceivable that an industry participant would also be so obsessed with present values and busy with current problems that he would fail to spot or refuse to consider signals that had disturbing conclusions for his firm. For instance, Louis B. Mayer of MGM fame is known to have fired people for advocating television in his presence in the mid-1940s.

deterrent weapon. A tremendous industrial upheaval and technological shake-out took place from 1957 through the early 1960s. Yet, the first signals had been dramatically evident to all the world eleven years earlier.

Admittedly, we have not answered the hard, troublesome, and fundamental question of just what the leader of the organization should do about a new technology and when and how he should do it. The answer will vary with the firm, its resources, commitments, alternative opportunities, and risk-taking propensities and even the personal philosophy of its leaders. Monitoring makes no pretence at providing that answer. What it does is to provide a systematic approach to an "early warning system." It helps to identify areas for study, and to establish bench mark numbers and events that will be crucial to organization decisions.

Forecasting Through Dynamic Modeling

A. WADE BLACKMAN, JR.

One potential source of inaccuracy in a forecast may be its static nature. If a forecast is made at a given point in time and is not revised as varying circumstances influence the elements of the forecast, then there is little chance the forecast and reality will coincide. Another source of inaccuracy may rest in the distinction between what is desired and what is feasible. Together these two difficulties suggest the need for dynamic forecasting methodology and a resolution of normative (needs-oriented) and exploratory (capability-oriented) forecasting.

Blackman shows how the techniques of industrial (systems) dynamics were applied to simulate a representative industrial research laboratory. The intent is to forecast the effect on future laboratory operations by matching an exploratory forecast of the laboratory's output to an exogenous goal schedule set by normative forecasts of future requirements.

INTRODUCTION

Technological forecasting work to date has been focused on two distinct classes of methodologies: those having to do with exploratory or capability-oriented forecasts and those having to do with normative or needs-oriented forecasts. These methodologies have been critically appraised by Roberts [1] where the contrasts between the rather crude methodologies of exploratory forecasts and the relatively sophisticated methodologies of normative forecasts have been pointed out and where the desirability of the use of dynamic modeling techniques which would combine exploratory and normative forecasts in a dynamic feedback system has been recognized. The need for the development of dynamic modeling techniques has also been pointed out by Jantsch [2] who has stated that "the full potential of technological forecasting is realized only where exploratory and normative components are joined in an iterative or ultimately in a feedback cycle."

The application of dynamic modeling to the problems of R&D project selection and budget allocation was first accomplished by Roberts [3]. Other past work related to the application of dynamic modeling to technological forecasting is summarized by Roberts [1].

Although the work published to date has provided much insight into the dynamics of the R&D process, there has been no application of dynamic modeling using "real" data to examine managerial policies which are required to close the loop between normative and exploratory technological forecasts related to the operation of a representative R&D laboratory. Such a model would be extremely useful for planning laboratory operations and determining the desirability of undertaking proposed future R&D programs. For example, suppose normative forecasts indicate the desirability of maintaining an annual growth rate in R&D output at a specified amount over the next decade. To meet these goals the management of the R&D laboratory is then faced with the problems of planning resource allocations for facilities acquisition, the acquisition of personnel, planning proposal and sales efforts to obtain external support, developing new organizational structures, and so on. The processes involved in the operation of an R&D laboratory are highly complex and involve many feedback loops and nonlinearities. It has been shown by Forrester [4, 5] that intuition is unreliable when applied to high–order, multiloop, nonlinear feedback systems. Therefore, the only available alternative for planning future operations which will ensure smooth growth is to develop a dynamic model which sufficiently replicates laboratory operations and then use this model to forecast the future policies which will be required to achieve a normative-oriented goal schedule. The full implications of the normative forecast of future R&D needs can only be evaluated after the dynamic model indicates what is actually required (in terms of costs, manpower, disruption of the existing organizational structure, and so on) to produce a future technological capability schedule (that is, an exploratory forecast) which responds to the normative forecast. Once these implications are fully set forth, it may then be desirable to modify the normative forecasts or to achieve the normative requirements by alternative means more attractive in a benefit-cost sense, for example, the acquisition and/or licensing route for supplying a future technological need might be preferred to internal development.

The objective of the work described herein was to construct a dynamic model of a representative R&D laboratory and apply this model to determine operating policies required to produce a future technological capability in response to a normative forecast of future technological requirements.

To achieve the objectives of this study, an approach was necessary that allowed evaluation of the effect of a normative goal schedule on a number of different policies and strategies evaluated over a period of a number of years. The required approach would realistically evaluate the complex interactions between various segments of the R&D laboratory, allow for realistic time variations, and yet not be too complex or costly to preclude solution. A mathematical modeling approach based on the techniques of industrial dynamics[1] developed by Professor Jay W. Forrester and his associates at M.I.T. [6] appeared to possess the characteristics required for this investigation.

System dynamics is based on servomechanism theory and other techniques of system analysis and is predicated on the ability of high-speed digital computers to solve large numbers (hundreds) of equations in short periods of time. The equations are mathematical descriptions of the operation of the system being simulated and are in the form of expressions for levels of various types which change at rates controlled by decision functions. The level equations represent accumulations within the system of such variables as dollars, personnel, facilities, and so on, and the rate equations govern the change of the levels with time. The decision functions represent either implicit or explicit policies established for the system operation.

Mathematical simulation of a system can only represent a real system to the extent that the equations describing the operation of the components of the system accurately describe the operations of the real system components. It is usually impossible to include equations for all of the myriad components of a real system because the simulation rapidly becomes too complex. It is, therefore, necessary to obtain an abstraction of the real system based on judgment and assumptions regarding which components of the real system are those which control overall system operation.

The model construction involved the iterative procedure illustrated in Exhibit 1. A model was initially constructed based on a perceived mechanism of system operation. This model was then tested against the historical performance of the system. If disagreement existed, sensitivity analyses were conducted to indicate the controlling variables in the model and revisions were made in the model formulation until acceptable agreement was obtained between past system performance and model predictions. The model was then

[1] As pointed out by Forrester [4], system dynamics is a more appropriate name for the industrial dynamics methodology, because it applies to complex systems in a wide variety of fields. This nomenclature will be adopted in the discussions which follow.

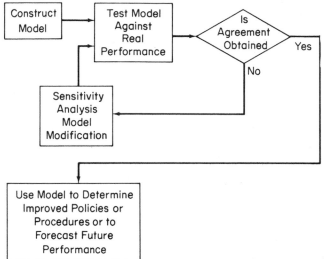

Exhibit 1. Dynamic Model-Building Procedure.

used to determine improved policies and to forecast future system performance with existing or revised operating policies.

The data and assumptions upon which this study is based were obtained from the records of a selected industrial R&D laboratory and from interviews with its management. The data presented herein, however, do not represent the actual operations of the laboratory but have been arbitrarily scaled to represent operation of a hypothetical research laboratory employing approximately 700 people and operating on an annual budget of approximately $12 million. Most of the data obtained from the "real" laboratory were from one operating section representing approximately one-fourth of the entire laboratory's research engineering effort.

The dynamic relationships between normative forecasts of desired future R&D requirements, the operating policies of the laboratory required to meet these requirements, and exploratory forecasts of the laboratory's future output with a set of operating policies are illustrated in Exhibit 2. In this illustration, the normative forecast is considered to have been established exogenously. With a given set of operating policies, exploratory forecasts are then made to determine the laboratory's future output capabilities. If the output capabilities do not agree with the normative requirements, revisions must be effected in the laboratory's operations in order to achieve the normative goal schedule. The major focus of the study described herein was to determine the operating policy revisions (as a function of time) which were required to respond to an exogenous normative output schedule.

The annual expenditure rate of the laboratory was adopted as a measure of the laboratory's output on the basis of historical relationships which were found to exist between this variable and the laboratory's output of papers and

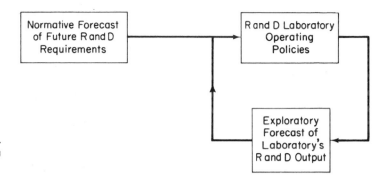

presentations as well as the number of patents produced. These relationships
are shown in Exhibits 3 and 4, respectively. The exogenous normative forecast
was expressed in terms of a desired annual growth in expenditure rate which
corresponded to an annual growth rate in the output of papers and presentations and patents.

ANALYSIS

The primary feedback loops assumed to affect the growth of the laboratory
are shown in Exhibit 5. The backlog of unallocated funds of the laboratory
(that is, the funds available for expenditure at the beginning of a time period)
is assumed to control the number of engineers and scientists employed which
in turn controls the overall employment level. The number of engineers and
scientists employed have an influence (Loop 1) on the efforts of manage-

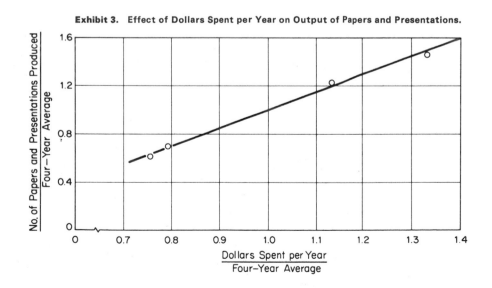

Exhibit 3. Effect of Dollars Spent per Year on Output of Papers and Presentations.

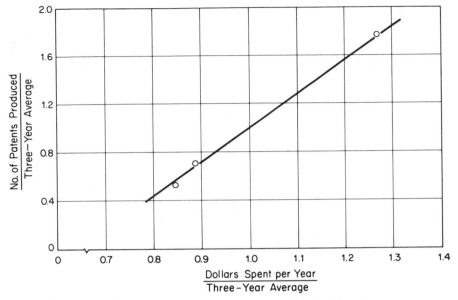

Exhibit 4. Effect of Dollars Spent per Year on the Number of Patents Produced.

ment to attract new business[2] (in the form of government contracts)—the larger the number of engineers and scientists employed, the greater will be management's efforts to attract new business to keep them productively employed. Also, the number of engineers and scientists employed will affect the rate of expenditure of funds which affects the output of the laboratory which will in turn affect new business (Loop 2). Similarly, the greater the number of engineers and scientists, the greater must be the inventory of facilities for their use. A larger facilities inventory (Loop 3) would be expected to have a favorable effect on attracting new business. The rate of expenditure will also affect the level of the backlog of unallocated funds of the laboratory which will in turn affect the level of the work force which can be supported (Loop 4). It can be seen that Loops 1 through 3 are positive feedback loops because an increase in the variables in the loop causes the other variables to increase in time; that is, increasing the backlog in Loop 1 generates new business which further increases the backlog. Conversely, Loop 4 can be seen to be a negative feedback loop; that is, an increase in the backlog increases the rate of expenditure which tends to decrease the backlog.

Because the extent of new business was a key variable in the study, it was necessary to investigate those factors that influenced this variable. It was hypothesized that the management effort to attract new business is related to the

2 In the sense used herein "new business" refers to the dollar volume of new business obtained. It is assumed that relationships between the expenditure of funds and laboratory output which have prevailed historically (see Exhibits 3 and 4) will continue to prevail in the future.

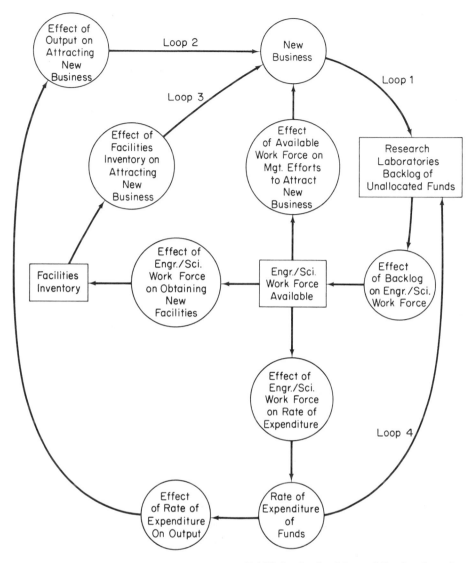

Exhibit 5. Feedback Loops Affecting Growth.

level of the engineer/scientist work force. Therefore, new business should be related to the professional work force available. It was also expected that the extent of new business attracted should depend on the general reputation of the laboratory for doing good work. This reputation was hypothesized to be related to the laboratory's output which in turn should depend on the rate of expenditure of funds. The facilities inventory was also expected to attract new business, because the greater the extent of facilities, the greater the probability of the laboratory having the special equipment required for a requested research job.

Hence, it was hypothesized that the key variables affecting new business are: (1) the engineer/scientist work force, (2) the rate of expenditure of funds, and (3) the facilities inventory. The dollars per year received from new government contracts was used as a measure of the extent of new business. To reduce fluctuations with time, values for the dollars received per year from government contracts were smoothed utilizing a two-year moving average. A line was fitted through the smoothed points by the method of least squares. Multiple regression analysis was then employed to derive an equation which related the least square, smoothed values of the extent of new business to the three key variables discussed previously. The equation obtained was

$$(1) \qquad y = 0.635x_1 + 0.363x_2 - 0.010x_3 - 0.189$$

where

$$y = \frac{\text{dollars/year received from new contracts}}{\text{four-year average of dollars/year received}}$$

$$x_1 = \frac{\text{dollars spent/year}}{\text{four-year average of dollars spent/year}}$$

$$x_2 = \frac{\text{number of engineers and scientists}}{\text{four-year average of engineers and scientists}}$$

$$x_3 = \frac{\text{net fixed assets}}{\text{four-year average of net fixed assets}}$$

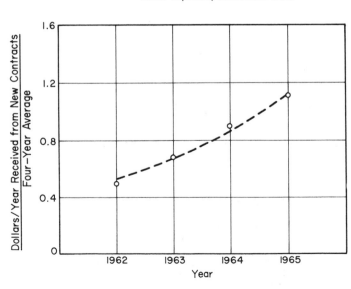

— — Calculated from Equation 1
○ Least Squares, Smoothed Data

Exhibit 6. Agreement between Smoothed Values and Values Calculated from Equation 1.

A multiple correlation coefficient of 0.988 was determined for the equation, and the agreement between the data and Equation 1 is shown in Exhibit 6. The agreement between the computed values and the smoothed data is excellent (the standard error of estimate was determined to be 3.16 percent) which gives justification to the assumption that the three key variables utilized are the controlling ones.

An examination of the relationship between output and rate of expenditure of funds gives further justification. It was believed initially that output should be related to the number of engineers and scientists on the work force. To test this belief, it was decided to use the number of papers and presentations made and the number of patents produced as measures of output. When these variables were plotted versus the number of engineers and scientists in the work force, relatively poor correlations (variations of greater than forty percent from a straight line through the data) were obtained. However, when these variables were plotted versus the expenditure rate, the obviously good correlations (variations of less than five percent from a straight line through the data) of Exhibits 3 and 4 resulted.

In summary, the key variables affecting the extent of new business appear to be (1) the rate of expenditure of funds, (2) the engineer/scientist work force, and (3) the facilities inventory. The output of the laboratories appears to be related to the rate of expenditure of funds to a much greater degree than to the engineer/scientist work force. The facilities inventory has a very small effect within the range of the data.

Further analysis of the assumptions made will be discussed in relation to the feedback loops shown in Exhibit 5.

Loop 1. Exhibit 7 presents a correlation (maximum variation approximately

Exhibit 7. Effect of Laboratory Budget on Engineer/Scientist Work Force.

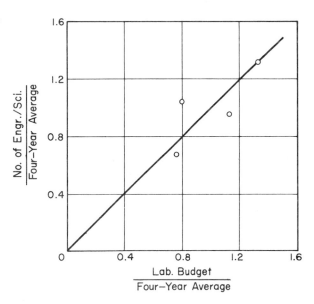

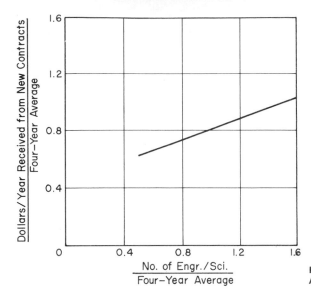

Exhibit 8. Effect of Professional Work Force on Attracting New Business Calculated from Equation 1.

25 percent from a straight line through the data) between the laboratory budget and the number of engineers and scientists in the work force. The curve appears to justify the assumption made in Loop 1 that the level of the professional work force depends on the backlog of unallocated funds.

It has been assumed that the management effort to attract new business is related to the level of the professional work force. The effect of the extent of new business with variations in the professional work force was calculated from Equation 1 with x_1 and x_3 held equal to one and the results are presented in Exhibit 8.

Loop 2. The effect of the number of engineers and scientists in the work force on the rate of expenditure is shown in Exhibit 9 (maximum variation

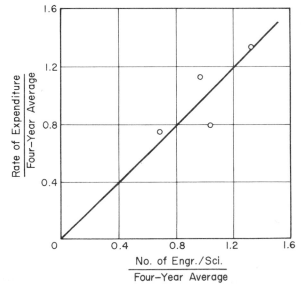

Exhibit 9. Effect of the Number of Engineers/Scientists on Dollars Spent per Year.

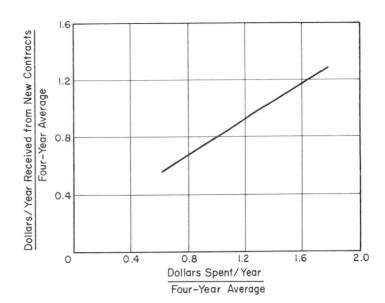

Exhibit 10. Effect of Rate of Expenditure on Attracting New Business Calculated from Equation 1.

approximately twenty-five percent) and the relationship between the expenditure rate and the laboratories output has been discussed previously. The effect of the rate of expenditures (and hence output) on attracting new business was calculated from Equation 1 with x_2 and x_3 held equal to one and is presented in Exhibit 10.

Loop 3. The effect of the number of engineers and scientists in the work force on the facilities inventory is shown in Exhibit 11 (maximum variation

Exhibit 11. Effect of Number of Engineers/Scientists on the Facilities Inventory.

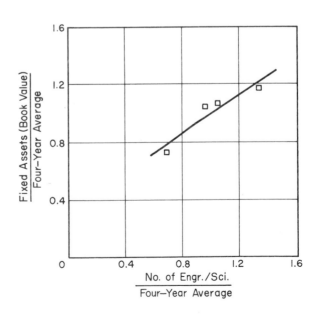

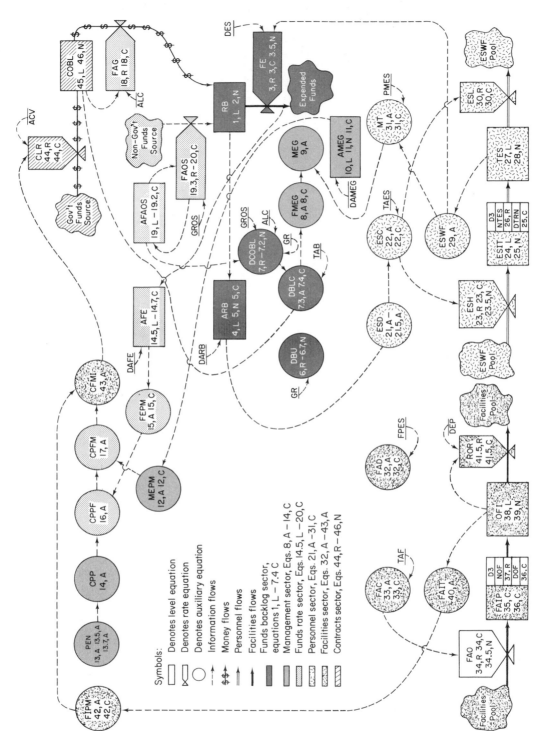

Exhibit 12. Flow Diagram of Model.

approximately eight percent from a straight line through the data). On the basis of this correlation the facilities desired by the engineers and scientists were assumed to vary linearly with the number of engineers and scientists in the work force.

The facilities inventory can be seen from Equation 1 to have a negligible effect on new business in the range for which data were available. However, at lower values of the facilities inventory some effect would be expected, because facilities limitations might then become a constraint on work output.

Loop 4. Because the level of the backlog of unallocated funds is a function of the rate of expenditure of funds, this last assumption is simply justified by definition.

The detailed system dynamics model shown in Exhibit 12 was developed from the assumed key feedback patterns. It contains about fifty equations which are developed and discussed in detail by Blackman [7]. The level of aggregation considered in the model is such that resource allocations are on the basis of total budgets or capital outlays and are not carried down to the level of detailed project funding.

The model shown in Exhibit 12 contains six sectors:

1. The *funds backlog sector,* Equations 1,L through 7.4,C, describe how the funds backlog responds to the organization's activities and desired growth rate.[3]
2. *The management sector,* Equations 8,A through 14,C, represent the effects of management's efforts on attracting new government business and describes the flow of contractual support from the government.
3. The *funds-rate sector,* Equations 14.5,L through 20,C, describe the exogenous flow of funds from nongovernment sources, the flow of funds from government sources, and the effects of the rate of expenditure on the inflow of government funds.
4. The *personnel sector,* Equations 21,A through 31,C, represent the hiring and firing decisions and their influence on the professional work force.
5. The *facilities sector,* Equations 32,A through 43,A, represent the inventory of facilities, its change to fluctuations in the professional work force, and its influence on the receipt of government contractual support.
6. The *contracts sector,* Equations 44,R through 46,N, describe the rate of receipt of government contractual support and the contractual support backlog.

RESULTS AND DISCUSSION

The system dynamics model as initially formulated was run on an IBM 7094 computer utilizing the Dynamo compiler simulator program [8].

Exhibit 13 presents the results of these calculations in the form of plots of the data obtained from the Dynamo output. The several variables that are

---

[3] Equation numbers refer to those given in Blackman [7].

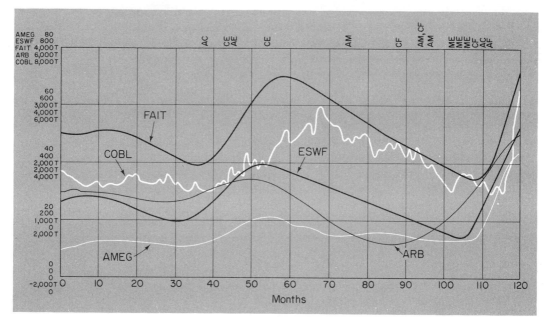

Exhibit 13. Ten-Year Simulation of Laboratory Operations,
Growth Rate Is 5% per Year.

plotted against time are: the average backlog of unallocated funds, ARB; the contract backlog, COBL; the average managerial effort toward attracting additional government business, AMEG; the engineer/scientist work force, ESWF; and the total facilities inventory, FAIT.[4] These plots represent the operation of the laboratories for a period of ten years for a normative growth rate of five percent per year. Conditions are assumed to be initially in a steady-state and operation of the model begins at time equal to zero. During the first months of operation the backlog of unallocated funds and the professional work force begins to expand slightly, but the contract backlog decreases because of a paucity of new contracts. This reduced contract backlog causes the unallocated funds backlog to begin to decrease. At about the same time, the facilities inventory begins to increase as a result of the earlier increase in professional work force. The reduced funds backlog causes a greater management effort toward attracting new government business which works with the increase in facilities to cause an increase in the contract backlog which causes the funds backlog to go up again around the twentieth month. This overall pattern repeats itself and oscillations of increasing amplitude (almost fifty percent increase in the funds backlog in the first cycle) characterize the growth

[4] The relationship between these variables and the symbols used in the plot is indicated on the ordinate of Exhibits 13 and 14.

pattern over the ten-year period. The period of the oscillations in the funds backlog increases with time. The first cycle has a period of approximately forty months and the second, approximately seventy-two months. The oscillations result from the negative feedback loop (Loop 4, Exhibit 5) and the increasing amplitude results from the positive feedback loops (Loops 1 through 3, Exhibit 5). Such oscillations although highly undesirable are nevertheless characteristic of the growth patterns of many organizations as discussed by Packer [9].

It is instructive to observe that oscillating growth behavior can be produced by a set of operating policies which appear to be logically consistent and which on the basis of intuition would have been predicted to produce stable growth. Stability can only be determined a posteriori from the simulation results, which makes obvious the usefulness of simulation techniques as a management tool and which emphasizes the unreliability of intuition (as discussed by Forrester [4]) when applied to the multiloop, high-order, nonlinear feedback systems which are generally characteristic of most management and social systems.

The initial formulation of the model failed to produce acceptable agreement when tested against actual historical performance of the "real" laboratory. As indicated in Exhibit 1, it was then necessary to conduct sensitivity analyses, identify the controlling variables in the model, and modify the model to obtain acceptable agreement between the model predictions and historical data. The results of these sensitivity calculations and the model modifications are discussed in detail by Blackman [7]. The salient features of the revisions required to obtain satisfactory agreement between model predictions and historical data are summarized below.

It was decided that a hiring policy for engineers and scientists which depended in large measure on the desired funds backlog and to a lesser extent on an average of the actual funds backlog would be desirable. Such a policy would insulate the hiring decisions from the short-range fluctuations in the funds backlog and make them depend to a greater extent on the longer range considerations of the normative growth rate. This would be consistent with the efforts to attract additional government business which also depend on the normative growth rate. In other words, a hiring policy based to some degree on the desired future growth would tend to compensate for the lag times which are inevitable in hiring competent professional personnel.

The assumption made earlier that the laboratory expenditure rate depended only on the level of the professional work force was reexamined. It was decided that it would be more realistic to assume that the expenditure rate depends both on the level of the professional work force and on the overall management policy concerning expenditures. For example, if in a given period few contracts were received and the addition of government funds to the unallocated funds backlog was at a reduced rate, management might exercise caution and expend funds at a reduced rate to maintain the desired funds backlog. This could be accomplished by reducing the support personnel used on the project, revising scheduling, deferring material and computational expenditures, or tighter cost control. Of course, the variation which could be achieved would have proscribed limits; for example, a lower limit would be imposed by the minimum costs of the existing personnel.

An additional revision was a change in the policy assumed concerning management efforts to attract additional government business. Instead of basing the efforts to attract new business solely on the difference between the desired contract backlog (that is, that required for a given growth rate) and the actual contract backlog, it appeared desirable to also have these efforts depend on the difference between the desired and actual funds backlog levels.

The results of the sensitivity runs were incorporated into a final set of revised policies which appeared to be the best compromise between the variables to produce the most stable growth patterns. The final set of equations is discussed in detail in Blackman [7]. The results of simulation runs for these policies for a normative growth rate of ten percent per year are presented in Exhibit 14. Acceptable agreement was obtained between the normative (desired) R&D output of the laboratory and the exploratory forecast of the laboratory's output capabilities (as indicated by the average research budget, ARB).

The agreement between the simulated results for a growth rate of twenty

Exhibit 14. Ten-Year Simulation of Laboratory Operations, Growth Rate is 10% per Year.

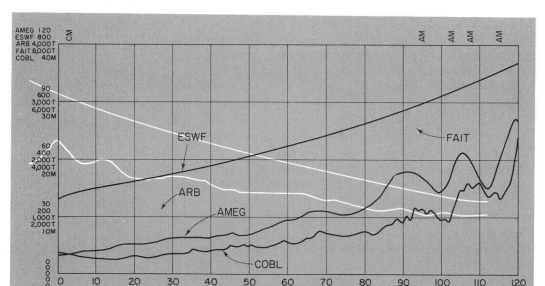

percent per year and the actual data obtained from the operation of the "real" laboratory used as a basis for this investigation is presented in Exhibits 15 through 17. The first four years of the simulated results are compared with historical data for: the funds expenditure rate, FE; the engineer/scientist

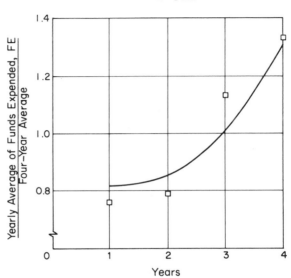

Exhibit 15. Comparison of Predicted and Actual Funds Expenditure Rates, Standard Deviation of ±7.4%.

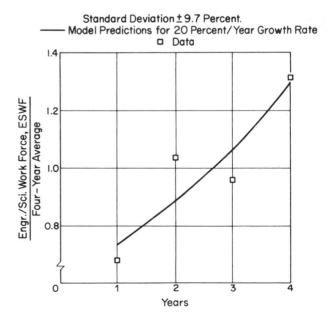

Exhibit 16. Comparison of Predicted and Actual Professional Work Force, Standard Deviation ±9.7%.

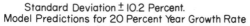

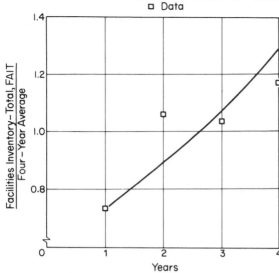

Standard Deviation ± 10.2 Percent.
Model Predictions for 20 Percent Year Growth Rate
□ Data

Exhibit 17. Comparison of Predicted and Actual Facilities Inventories, Standard Deviation ±10.2%.

work force, ESWF; and the facilities inventory, FAIT. The results agreed within standard deviations of ± 7.4 percent for FE; ± 9.7 percent for ESWF; and ± 10.2 percent for FAIT. This agreement is thought to be very good considering the scatter which exists characteristically in data of this type and provides ex post facto support for the assumptions upon which the model was based.

Several conclusions can be derived from these results which have useful implications for the management of R&D laboratories. If a high normative rate of growth is desired or required from competitive pressures and this rate of growth cannot be attained internally from the supporting organization or from external commercial sources, then the rate of government support is the most important single factor in achieving the normative (desired) growth rate. Management must seek this support aggressively. It is necessary to increase the effort devoted to attracting government business as the laboratory grows in size. The effort exerted in attracting new business can be expected to exhibit short-term fluctuations which will be subject to random variations in the incidence of government requirements for additional research. It will, therefore, be necessary to have a technical sales force which is flexible or can be easily expanded to respond to these varying requirements.

It is desirable to base personnel policies and facilities acquisition policies on the long-term desired rate of growth rather than short-term backlogs in funds. A consequence of this result is the necessity of assuming the risk of not having sufficient future support for the personnel hired. This, however, must be faced and planned for or smooth growth patterns cannot be achieved. It is also necessary to assume the risk involved for the commitment of funds for facilities as the professional work force increases. It is essential to have these funds available without too much delay because the effectiveness of the newly hired personnel will be limited if sufficient facilities are not available, their output will decrease, and it will be more difficult for the laboratories to attain the level of new government business necessary for their support.

Therefore, the most essential ingredient for growth would appear to be a management policy fully committed to long-term growth. Concomitant with such a growth commitment must be the willingness to assume the risk involved in the commitment of funds for the personnel and facilities required to achieve the desired long-term growth. Although such a policy may involve risk, such risks may be the price which must be paid to achieve stable long-term growth.

The model outputs which are considered most useful in indicating the effect on laboratory operations of closing the loop between normative and exploratory forecasts are summarized below:

1. The balance between government and internal support required to achieve a growth objective can be forecast for future time periods.

2. Personnel and facilities acquisition strategies can be investigated and requirements forecast.

3. Future capital requirements can be predicted.

4. Management information system parameters and decision criteria required for smooth growth patterns can be identified.

5. The long-term cost of various normative forecasts of R&D requirements can be determined to allow benefit-cost comparisons of internally produced R&D with R&D obtained through possible licensing and/or acquisition and merger opportunities.

REFERENCES

1. ROBERTS, E. B.. "Exploratory and Normative Technological Forecasting: A Critical Appraisal," *Technological Forecasting,* Vol. 1, 1969, p. 113.

2. JANTSCH, E.. *Technological Forecasting in Perspective.* Paris: Organization for Economic Co-operation and Development, 1967, pp. 15–17.

3. ROBERTS, E. B.. *The Dynamics of Research and Development.* New York: Harper and Row, Publishers, 1964.

4. FORRESTER, J. W.. "Counterintuitive Behavior of Social Systems," *Technology Review,* Vol. 73, No. 3, January, 1971, pp. 52–68.

5. ———. *Principles of Systems.* Cambridge, Mass.: Wright-Allen Press, 1968, Chapter 3.

6. ———. *Industrial Dynamics.* Cambridge, Mass.: The M.I.T. Press, 1961.

7. BLACKMAN, A. W.. "The Growth Dynamics of an Industrial Research Laboratory in the Aerospace Industry." Unpublished Thesis, M.I.T. Sloan School of Management, 1966.

8. PUGH, A. L., III. *Dynamo User's Manual,* 2nd ed. Cambridge, Mass.: The M.I.T. Press, 1963.

9. PACKER, D. W.. *Resource Acquisition in Corporate Growth.* Cambridge, Mass.: The M.I.T. Press, 1964.

Study of
Alternative Futures:
A Scenario Writing Method

LUCIEN GERARDIN

Goal-oriented or normative forecasting is much discussed by futurists. The scenario
is an even more popular and equally undeveloped methodology in this class of
forecasting. Here a leading French student of industrial forecasting gives his views
on a practical way to develop scenarios of possible alternative futures, the *futuribles*
of Bertrand de Jouvenel. The procedure calls for finding pathways for systems evolution at a sequence of time intervals and identifying those thresholds of tension where
social forces will alter or inhibit present trends.

Then it will become clearer to decision makers what paths and sequences of
actions should be pursued to achieve the desired goals. Gerardin's concepts are
intended for the planner dealing with the fifteen- to twenty-year time horizon.

FOUR ATTITUDES REGARDING THE FUTURE

For centuries the idea that the future could be different from the past did
not touch even the most intelligent minds. In an agricultural society subjected
to the rhythm of seasons, the future was quite naturally an indefinite repetition
of the past, therefore a *totally passive attitude* regarding the future. Industry

with its ever increasing power on the natural environment has forced man to revise his ideas. The future no longer resembles the past, and more and more the dynamic change is the normal and not the static perennial. To write this statement is one thing; to really assimilate it is something quite different. Therefore there is nothing astonishing that today's attitude concerning the future is more often an *opportunist attitude*. Carried off by the turbulent floods of life, man tries to avoid the dangers and to profit from the place in which he finds himself. He tries to organize his activities in order to extract the good (or more exactly not the bad) of the daily opportunities.

This opportunist attitude is completely outdated today where everything reacts to everything else. By wishing to profit by a local and temporary opportunity, one almost always will compromise on the future at a longer term. And if the immediate results of the opportunist attitude can be interesting for an individual, a small business, or a small collectivity, the long-run cost is disastrous for bigger collectivities. There is no need to give examples because they are now too numerous around us.

In order to see further ahead, we must try to forecast the future. Quite naturaly future extends present status according to past trends, as Exhibit 1 shows.

This third attitude regarding the future is an *adaptative attitude* since we try to *optimize the today actions in order that their consequences are better adapted to the foreseeable future*. Exhibit 2 schematizes this adaptive philosophy:

Exhibit 1. The Probable Future Extends the Past.

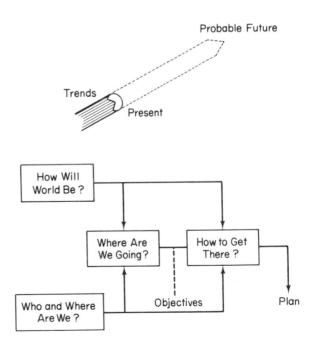

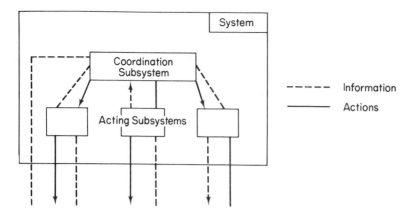

System

Coordination Subsystem

Acting Subsystems

- - - - - Information
———— Actions

Exhibit 2. The Adaptative Attitude Regarding the Future and the Long-Range Planning Philosophy.

We extrapolate the past trends to estimate what the probable future will be in which the consequences of the actions decided upon today will develop. The decision maker studies this extrapolated environment and tries to determine where he wishes to go, and he will set the objectives which we will try to reach according to *a plan, which is a set of decisions and actions logically ordered in time.* As Exhibit 2 indicates, the planning is more important than the plan: *planning is a dynamic process and not a static statement;* planning is a new decision-making philosophy which implies participation. Planning cannot be superimposed on an old decision-making structure without a profound reaction on that structure. Otherwise it would be only a costly ineffective tool. French National Planning in its Fifth Plan is an example of such a participative process. Planning requires harmonization of views to attain the common objectives aimed for. The common general interest must balance but not overcome the various divergent particular interests.

Adaptive attitude is better than opportunism, but adaptation is also outdated because it is based on illogism. What is a forecast? *A forecast is an estimation of the probable state of something at a given future time under the hypothesis that everything will continue as in the past.* But forecasts are made to act upon decisions. These decisions modify the environment, therefore the main hypothesis, according to which everything will continue as in the past, is no longer valid. Therefore, this is the illogism of adaptative attitude.

This illogism is fortunatly sometimes more apparent than real. It may happen that the actions decided upon following the forecasts are on so small a scale that the consequences have no noticeable effect on the general evolution of the environment. But it is quite evident that this is no longer valid when decisions are made at multinational industrial corporations level or at governmental level because the actions at this level have an important effect on the future. In that case the classical viewpoint according to which the value of a forecast is measured by the exactness with which the predicted future will be realized, is entirely false. Rather, *the value of a forecast is measured by the degree of influence it will have on the decisions taken today.* This new viewpoint is that of the *creative attitude* regarding the future. There are no future facts such as "past" and "present" facts but only a set of potentialities which outlines a set of possible futures (*futuribles,* to use the word coined by *Bertrand de Jouvenel*). We no longer try to forecast future facts because these facts do not exist. We try to answer the true question: *what are the alternative possible futures?* This creative attitude is that of the *futures creative* planning, to use the word coined by H. Ozbekhan [1]. In French we use the word *prospective* to qualify this creative attitude. If it is possible to find accurate translations in other European languages (*prognosis* in German, *prognosirovanie* in Russian), it is not so easy to find an equivalent in English. *Technological forecasting* has a restrictive meaning. Some people use *futures research,* others *futuristics* or *futurology* (a word coined by O. K. Fletcheim).

This lack of English translation is not surprising. Futures creative planning implies planning, meaning some general consensus toward national goals. This is not compatible with a true free-market economy.

The difference between the adaptative attitude of long-range planning and the creative attitude of the futures creative planning is very essential. This difference is between adapting (with delay) to the external change or causing (in advance) the wanted change. Long-range planning is at the strategic level, futures creative planning at the policy level [1].

FUTURES CREATIVE PLANNING PHILOSOPHY

The futures creative planning is linked to the necessity of voluntary actions to shape our future. In some way, voluntary action is equivalent to a force. In the mechanical world, force produces acceleration, therefore, modification of movement. The same thing happens in society: voluntary actions change the course of natural evolution. We must therefore substitute the Exhibit 3 diagram for that of Exhibit 1.

A certain system being given whose development we wish to organize has several alternative possible futures, several *futuribles* (A, B, C, and so on) of which one, A in Exhibit 3, is the probable future through trend forecasting. Long-range planning proposes to follow the most probable development; *futures-creative planning has the goal of fixing for each futurible a reference situated in the future and no longer on the past projected into the future.* The general philosophy of that creative attitude is summarized in Exhibit 4 which replaces the Exhibit 2 diagram. To the rather naive question, *how will the world be?* the futures-creative planning substitutes the true question, *How are the alternative possible worlds going to be?*

But be careful! The future remains malleable to human will. It is but

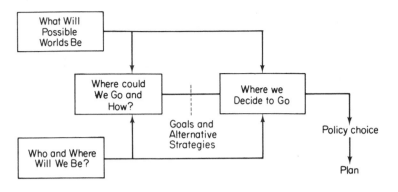

Exhibit 3. A Given Possible Future is Reached by Voluntary Actions on the Trends.

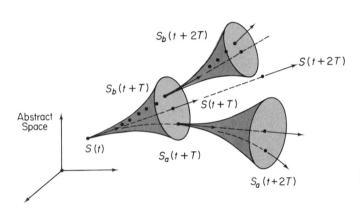

Exhibit 4. Construction of the Future on Futures-creative Planning.

prestructured by past trends. Futures creative planning has to be based on accurate forecasts. We must carefully study the strong trends that bring about the future, and strong trends are not obligated to be the most visible. This study of forecasts is quite different from the study made by long-range planners. To summarize the difference in two sentences: *Long-range planners examine the past, say 1960, in order to forecast today what 1975 will be.* Against this, *futures creative planners bring themselves to the year 2000 in order to decide today what they want 1985 to be.*

The choice for 1975 and 1985 is not arbitrary. The time horizon of forecasting is limited if accuracy is needed. On the other side, the time horizon of futures-creative planning is necessarily a long one. To voluntarily influence the movement, a sufficient margin for action has to be provided. Past trends influence very strongly the next five years, therefore futures-creative planners must look more distantly and take enough advance time so that the voluntary actions become noticeable. This is because the results of acceleration are proportional to time multiplied by itself and not simply proportional to time.

The actions for voluntarily constructing the future must respect certain constraints in order to remain realistic. As long as an aircraft flies at a constant speed in quiet sky, the passengers feel nothing. When the aircraft goes into a turn, the centrifugal acceleration becomes noticeable. If the pilot tries to turn too abruptly, he will have an accident and even break up the aircraft if the acceleration becomes too strong. The same applies to the voluntary action for building a wanted future. *It is necessary to remain within certain limits to avoid giving too great a shock which might unstabilize or even destroy society instead of constructing it.*

COMMON ERRORS TO BE AVOIDED

Having explained what the futures-creative planning philosophy is, let us now say that which it is not. Futures creative planning must not be confused with a more or less distant forecast. A fortiori, *futures creative planning is not to be confused with technological forecasting.* Certainly, our society is becoming more and more technologically oriented, but the technological component is only one aspect of society. Forecasting is almost uniquely technological only in fields such as space, advanced weapons systems, and so on.

A second error is made too often: *the search for innovation for innovation's sake.* Too often people confuse the desire for something new with the future itself. It is not because something is new that it is future-oriented. Man has too much of a tendency to be fascinated by innovation. We must resist the mirage of miraculous ideas.

Also *futures-creative planning has nothing to do with predictions.* Certainly the accurate knowledge of future environment would be very valuable. Then decision makers could operate without error. This dream is only nonsense. As we said before, there are no future facts but only trends and potentialities that outline a schematic future which becomes more and more nebulous as we move further away from the present.

Futures-creative planning is not futurology or vague descriptions of the future made without involving responsibilities and decisions. Future studies are done because decisions have to be made today whose consequences will happen tomorrow or even after tomorrow. Futures creative planners activities do not interfere with a decision maker's responsibility. Moreover the responsibility for policy choice is increased by the supplementary information that creative planners put at the disposal of those decision makers.

Finally, *futures-creative planning is not utopia.* It is not a tendency to dream the future by projecting the desires for reform upon it. Having defined a reference in the future, we must return to the present in order to verify that a pathway actually exists which permits going from the present to the future aimed for, and this pathway must take into account the constraints of the real world in which we live.

A system is a set of elements (or subsystems) organized in such a way that the system as a whole will reach some common objective. Here the general goal aimed for is a given *futurible*. The time evolution of a given system is either natural (according to past trends) or voluntary (acting on these trends). In order to determine the adequate voluntary actions it is first necessary to know the structure of the system under study. Structure studies are relatively easy in the case of a purely technological system (for example, a weapons system, an air-traffic control system, a communications network, an industrial process, and so on). In this case, the organization of subsystems is quite simple and very often is linear. There are not many conflicts between subsystems. The global judgment about the system is done in cost-effectiveness or cost-benefit terms.

The situation is much more complicated in the case of a complex social system, for example, educational system, health system, urban system, communication by mass media, and so on. Because people with divergent interests are involved, conflicts are also present. *In a social system, we always find internal conflicts between the subsystems because of diverging interests within the realization of the overall goal of the system.* These diverging interests are essential because they are the power for change. In cybernetic terms, these are the error signals which provide feedback within the social systems. This effect happens in every society.

The degree of liberalism in a social system cannot be endless, or the society will end up in the greatest incoherency. It is necessary to assume at least some internal coherence of the system. The conflicts between subsystems must not go beyond some threshold, otherwise the system will no longer work. Natural feedbacks reestablish more or less abruptly the balance if the conflicts become too strong. Feedback actions preferably should be of a voluntary and well-known nature. Existing social systems are very often hierarchies, as schematized in Exhibit 5, with at least two levels: subsystems of action upon the environment and an upper level of coordination. The study of multilevel hierarchical systems and their problems has hardly begun. We quote here only the important contribution of M. Mesarovic and his team at Case Western Research University [2]. One of the simplest examples of a hierarchized social system is the industrial corporation with its two levels: staff working operationally on the environment and line ensuring the functional overall coherence of these operations [3].

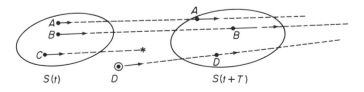

Exhibit 5. Multilevel Hierarchical System.

Another important difference between technological and social systems is the way of judging them. We always try to speak in terms of cost effectiveness. We can compute the cost, even though very often the true cost will not be only the direct cost. The internal conflicts inherent to all social systems involve *indirect costs* that are difficult to quantify in economic terms. The dead and injured on the highways is an excellent example of the indirect cost to be charged to automotive transport system. Social effectiveness is much more difficult to assess because there is no unique effectiveness parameter. Social effectiveness requires a set of values instead of a unique and simple measurement. We call these values *social indicators,* a subject on which more and more people begin to work seriously almost everywhere [4, 5].

It is relatively easy to study a technological system by modeling it on a computer, for example. It is much more difficult to model a hierarchical social system. However, it is necessary to try to do this if we want futures research to acquire some degree of objectivity. Many interesting attempts to model social systems are now made. Some are very ambitious. The organization known under the name of Club of Rome wishes to model world society. That model has recently been described [6]. In order to computerize the subject, conflicts have been discarded. How could such a theoretical model represent the real life? A more interesting exercise is conducted by the Japanese team of the Club of Rome which tries to survey social events that may have an effect on system behavior or which may become critical and to determine the variables dominating these events [7]. A matrix is used to structure the parameters according to their degree of correlation. This idea of using some kind of interaction matrix seems very interesting. A similar proposal was made in the same sense two years ago by O. Sulc [8]. Confusion with the cross-impact matrix [9] must be avoided. The cross-impact matrix is mainly a tool to make a computer run to produce inaccurate figures from inaccurate data.

SCENARIOS OF ALTERNATIVE FUTURES

Futures creative planning is above all an attitude to deal with the future, but an attitude is not enough to creatively study and construct the future. We also require appropriate methods and techniques. Scenario writing is an answer to these latter requirements. We must avoid both: a too rigid model which removes all the reality of the social system being studied and a purely verbal description which removes all the effectiveness of the study made.

An example of semiqualitative, semiquantitative alternative futures study is given by the first (and last) report of *The National Goals Research Staff* (U.S.A.) [10]. *Herman Kahn* was the first to introduce scenario writing in order to study the future of complex systems [11]. But *Herman Kahn* is much more a futurologist, and his scenarios are more intuitively than rationally constructed. The method of scenarios has been greatly perfected in France by DATAR

(La Delegation a l'Amenagement du Territoire et a l'Action Regionale) which currently uses this tool for creative thinking in preparing policy decisions at the government level [12, 13]. *The Ministry of Industrial Development* (France) also proposes using this method [14]. The use of this technique can be generalized to study the problems of sociotechnological systems of all kinds.

What is the basic philosophy for scenario writing? An adequate description of a system being studied permits, as Exhibit 6 shows, situating in an abstract space of n dimensions (n being the number of parameters which describe the system) the point $S(t)$ representing the status of S at time t. The values of the trends permit stating the probable direction of S evolution. If no voluntary action is applied, the system is going to evolve naturally by following that probable direction. But it is possible to make the system S evolve in a preferential direction by exercising an appropriate set of voluntary actions. These actions, equivalent to forces, modify the trajectory of $S(t)$ within the abstract space where the figurative point representing the system evolves. These actions are limited by external contraints which, at the limit, are linked to the stability of the society in which we live. The actions are also limited by internal constraints. The system must keep a certain coherence in order to preserve its individuality as a given system.

At a time posterior to t, the figurative point of S can be located within a cone which opens more and more as we move away from the initial time. At time $t + T$, the system S under study can thus be represented as any point of the section of the cone by the value $t + T$ of time parameter. All the points $S_i(t + T)$ of this surface represent the possible alternative futures of the system S, and so on, as shown in Exhibit 6. Note that forecasting considers only the most probable future $S(t + T)$, $S(t + 2T)$ ... ; while futures-creative planning tries to take into account all the possible futures (for example, the set of all the points of the sections of the development cones by the value $(t + nT)$ of the time parameter).

For practical reasons, it is not possible to consider all the possible futures, but only a discrete set, S_a, S_{bc}, S, and so on. Exhibits 3 and 6 correspond,

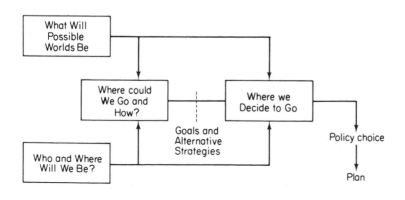

Exhibit 6. Evolution of a System Within Possible Abstract Space.

Exhibit 6 showing how to build Exhibit 3 in an objective way. The points S_a, S_b, S_c, ..., S_t will generally be what we call *contrasted scenarios,* that is futures where certain values of parameters will have been voluntarily exaggerated in order to better analyze what the importance of these parameters is in the evolution of the system under study.

If it is easy to theoretically represent the evolution of a complex system by the trajectory of a point $S(t)$ in an abstract space, the corresponding practical work is less simple. The goal aimed for is to determine sequences of actions in order to cause the system to evolve toward a reference point set in advance. We can only broadly study the movement of a complex system because only the individual evolution of each of the parameters that characterize the structure of the system S can be estimated. The overall evolution is the result of the combination of these individual evolutions. When we perform this kind of work we observe, as Exhibit 7 displays, that some parameters (such as A and B in the exhibit) evolve at different speeds. Internal tensions therefore appear that are going to break up the internal coherence of the system or which are going to disturb the coherence of the liaisons of the system with its environment. The study of these tensions is essential since these tensions are the motive of the actions (natural or voluntary) which are going to act to modify the systems evolution.

How can these tensions be evaluated? With the aid of what we call *economic and social indicators.* The definition of these indicators is never easy because they sum up into a single number a whole group of different parameters. These indicators must extend far beyond the economic aspect of things. The *homo economicus* is far from being the whole of mankind, and to consider only this aspect might be a tragic error.

HOW TO PUT METHOD IN PRACTICE

The scenario's writing method has to be put in practice with the aid of the two following alternative steps:

Exhibit 7. Development of the Tensions in the Evolution of a Social System.

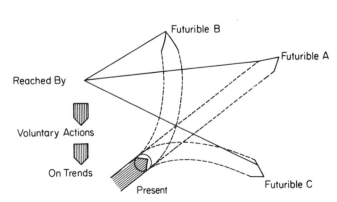

Reached By

Voluntary Actions

On Trends

Present

Futurible B

Futurible A

Futurible C

1. Description of system structure at a given time and verification of the internal coherence at that time.
2. Time evolution from one description to another under the natural effect of the trends and the voluntary effect of policy decisions.

These decisions are directed toward the realization of a given goal aimed for in the distant future. This goal is specific for the given *futurible* under study. We want to evaluate the more or less realistic character of this goal, so we must verify that a pathway really exists between that goal and present status and trends. Therefore the study is alternately made from the future to the present and from the present toward the future. The logical diagram of Exhibit 8 shows how to do this.

The description at the time t of the system being studied $S(t)$ permits evaluation of the values of economical and social indicators that are to be used for that particular system under study. To use a *profile* of such indicators seems very useful. We should also assign threshold values that we might call *the thresholds of dissatisfaction*. If indicators values cross these thresholds, a natural process is triggered for a return to better conditions. But it is preferable not to wait for these natural reactions which are generally brutal. We can anticipate them by determining the direction in which the indicators are evolving. This is done by comparing profiles at the instants $(t + T)$ and (t).

If we then see there is a chance of passing the threshold at a date close to the future, we can then decide whether or not to act in order to voluntarily stabilize the system evolution. Indicators evolution is also used to perform comparisons and to decide on other types of actions that will aim at the evolution and not at the stabilization of the system. The various actions have to be made coherent before being applied to the system $S(t)$ in order to give a new description that we may call $S(t_+)$ which is going to evolve to $S(t + T)$. Coherence studies are not straightforward. Automatization of these seems very difficult if not impossible. Group discussion is more appropriate and the best procedure is to build such a group around the various social groups that are involved in the system being studied [15][1]. The most interesting result of the scenario method is not so much the descriptions of possible alternative evolutions of S as *the better understanding of the system itself*. The need for defining indicators and dissatisfaction thresholds obliges us to think of system goals and of the degree to which these goals must be fulfilled.

A practical problem remains. The evolution process of a social system is continuous, but it is impossible in practice to consider this parameter t as being continuous. The study has to be made with discrete values. How to choose the time interval T between successive descriptions is a question. We have to define the time constant of the system, for example, the time needed for a deep

---

[1] The study of alternative possible futures with the expert consultation is done by H. Bianchi, Laboratoire de Construction Prévisionnelle, Centre de Recherches Science et Vie (2, rue de la Bauma, Paris, France).

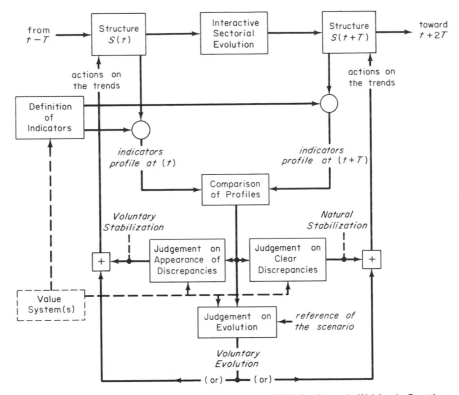

Exhibit 8. Scenario Writing in Practice.

change in system parameter values. More and more people claim today that everything is accelerating and therefore time constants are decreasing rapidly. The objective study of the past developments, either technological or sociological, shows that this is not completely true. *An average of twenty years is needed to go from a new idea to the widespread use of results coming from that idea.*[2] Certainly there is a spread around the figure of twenty from ten to forty years [16]. The figure of twenty years has a natural explanation: *The development of an innovation involves a change that cannot be fully realized except by a change in men.* Twenty years is the time span for one generation of men to be replaced with another.

The feeling of acceleration is linked up with the ever increasing number of evolving things. That ever increasing quantity creates the illusion of a decrease in individual time constants. Beware of this common error. In a given specific domain, careful studies show that the speed of evolution has not accelerated during the past decades, and that is important to define the time constant of the evolution of the existing world. Moreover human actions

[2] The 20 years empiric rule has been widely demonstrated by professor James Bright (Texas University).

never have instantaneous results. There is a certain delay, especially in liberal societies. This is an other reason in favor of the twenty-years rule [17]. By analogy with sampled data a value of 20/3 or seven years seems adequate for the difference T between successive steps.

Finally, what time horizon should be sought for scenarios? The same empiric rule suggests that to look fifty years ahead is almost unthinkable. Two generations of men are too much. The reasonable horizon limit for writing the scenario of a possible future is from fifteen to twenty years (three to five temporal steps).

REFERENCES

1. HASAN OZBEKHAN, "Toward a General Theory of Planning." *Perspectives of Planning*. Proceedings of the OCDE Working Symposium on Long-Range Forecasting and Planning, Bellagio, 27 Oct. to 2 Nov. 1968. (Paris: OCDE, 1969), pp. 47, 155.

2. M. D. MESAROVIC, D. MACKO, and Y. TAKAHARA, *Theory of Hierarchical, Multilevel, Sytems.* (New York: Academic Press, 1970), 294 pp.

3. A. DANZIN, "La Biologie de l'Entreprise." Conference given at the Superior School of Electricity, (Paris: 5 March 1971), 12 pp.

4. *Toward a Social Indicator* (Wash., D.C.: U.S. Department of Health, Education and Welfare, 1969).

5. I. B. BESTUZHEV-LADA, *OKNO V BUDUSHCHEE: SOVREMENNYE PROBLEMY SOT SIAL 'NO GO PROGNOZIROVANIA (A window toward the future, the present state of social forecasting).* (Moscow: 1970).

6. JAY W. FORRESTER, *World Dynamics.* (Cambridge U.S.A.: Wright Allen Press, 1971).

7. Y. KAYA, H. ISHITANI, T. TANAKA, Y. SUZUKI, Y. KODAMA, and M. NAOI, "An Approach to General Social Systems Simulation." Preprints of the AICA Symposium: *Simulation of Complex Systems,* (Tokyo: September, 1971), pp. I.2/1–I.2/6.

8. OTO SULC, "Interactions Between Technological and Social Changes: a Forecasting Model." *Futures,* (September, 1969) pp. 402, 407.

9. R. ROCHBERG, Th.J. GORDON, and O. HELMER, "The Use of Cross-Impact Matrices for Forecasting and Planning." Institute for the Future, Report R/10, (April, 1970), 63 pp.

10. *Toward Balanced Growth: Quantity with Quality.* Report of the National Goals Research Staff (Washington, D.C.: U.S. Government Printing Office, 4 July 1970), 228 pp.

11. H. KAHN, *De l'Escalade, Métaphores et Scenarios.* (Paris: Calman-Levy, 1966).

12. *Scenarios d'Aménagement du Territoire.* Travaux et Recherches de Prospective Collection—n° 12 (Paris: La Documentation Française, January, 1971), 120 pp.

13. J. C. BLUET and JOSEE ZEMOR, "Prospective Géographique: Méthode et Directions de Recherche." *Revue METRA, N°1* (1970) pp. 111—127.

14. J. DELEAGE, "Réflexions sur l'Utilisation de la Methode des Scénarios dans la Prospective Industrielle." *Cahiers of Centre de Recherches Science et Vie* (2, rue de la Baume, Paris, France), N°1 (Décembre, 1970), pp. 72, 79.

15. R. EKSL and P. ZEMOR, "Méthode de Consultation d'Experts et de Simulation du Futur en Temps Accéléré. *Analyse et Prévision,* (1969), pp. 551, 588.

16. J. C. FISHER and R. H. PRY, "A Simple Substitution Model of Technological Change." General Electric Report 70-C-215, (June, 1970), 8 pp.

17. JOHN PLATT, "How Men Can Shape Their Future." *Futures* (March, 1971), pp. 32—47.

TECHNOLOGICAL FORECASTING APPLIED TO CURRENT PROBLEMS

PART THREE

How are industry and government using technological forecasting today? In this section nine authors describe specific studies that have been directed across a wide range of concerns:

Field of Work	Function	Forecast Objective	Method Used	Subject of Forecast
Machine tools	Research planning	Providing guides to equipment & process design	Trend extrapolation	Accuracy of tools, automation
Instruments	Strategic planning (defensive)	Identify impact of coming technology	Trend extrapolation, monitoring, structural analysis	Electronic circuitry
Memory devices	Strategic planning (offensive)	Selecting optimum technology market target for firm	Trend extrapolation	Electronic components
Oil industry	Standards planning	Select optimum test procedure	Structural analysis (goal oriented and Delphi)	Testing apparatus

Field of Work	Function	Forecast Objective	Method Used	Subject of Forecast
Military supply	Management	Determine effect of future technology on Navy logistics management	Structural analysis	Supplies and procedures
Computer industry	Research	Analyze nature of technological innovation	Trend analysis	Computing time and cost
Jet engine industry	Market research	Relate engine performance to market share	Trend analysis and normative studies	Jet engine performance parameters
Research laboratories	Research planning	Improve project selection	Statistical analysis	Time and cost of research projects
Communication systems	Alternative futures	Identify fields for possible study	Morphological	Communication systems

Thoughtful reading of these applications demonstrates that technological forecasting can be a great aid to both sensing the general shape and impact of coming technology as well as to setting specific technical goals.

Further specific applications are included in chapters in other sections.

Lucien Gerardin closes this section with a chapter describing a method for studying alternative futures. The procedure involves finding pathways for systems evolution at a sequence of time intervals, each path leading toward a desired goal generally in the next fifteen to twenty years.

Technological Forecasting as an Aid to Equipment and Process Design

M. EUGENE MERCHANT

The expanding role of technological forecasting in industry is only recently being documented. Industrial experience with some of the methods is often too brief to provide credibility or confidence. Merchant's chapter, however, describes forecasts published in 1959–60 and shows contemporary results relative to those ten-year-old technological forecasts. The integration of forecasts into R&D decision making is described for areas such as tool materials, grinding processes, and metal removal.

It is Merchant's thesis that these trend extrapolations have provided very useful guides to equipment design goals.

INTRODUCTION

Nothing is as important in planning modern research and development programs as knowing what the technological future will be. We are interested in knowing two things about that future—what new technological events or capabilities it is likely to contain and how soon these events or capabilities are likely to materialize. However, important as such knowledge is to research and development planning, there is as yet no exact scientific way to obtain it.

Nevertheless, as a result of the increasingly urgent need for such knowledge in recent years, some qualitative and semiquantitative techniques to help us obtain it are beginning to be developed. Although these are still very crude, they are already able to give us a little more assurance about what lies ahead than our former method of pure intuition. These techniques are being gathered together and developed in a new branch of technology known as technological forecasting. In this paper we will examine the experience of the writer's company in using technological forecasting in the planning of research and development on manufacturing equipment and processes, as an aid to the design of greatly improved manufacturing equipment and processes. However, before we do this, let us briefly indicate our philosophy of the need for such forecasting and summarize our way of looking at the techniques presently available for doing it.

NEED

Certainly, in the past, the need for this activity did not appear to be very great. By taking a good look at the current state of the art in one's particular field of technology and the general direction in which that was moving, one could intuitively get a good feel for what lay some distance ahead in that field. Today all that is changing due to the fact that two new elements are exerting a major influence on technology, namely, the breaking down of barriers between different fields of technology and the increasingly rapid advance of technology. Why is this happening?

Today, we are at the beginning of a revolution; we are in the initial stages of the ascendancy of a major controlling factor in human and technological affairs, namely *creative innovation* (and at the moment, at least, creative innovation aimed primarily at satisfying basic human needs). We see evidence of this all around us in various forms, some even bizarre or frightening, such as the hippie movement and the student revolt, and some benign and grand, such as the beginnings of space travel. Although this factor has already begun to assume major control of human and technological affairs, it seems likely that it will not displace the control exercised by tested thought (in the way in which tested thought previously displaced authoritarianism) but will instead complement and couple with tested thought so that the two, in combination, become the prime factor controlling human and technological affairs.

The rise of tested thought generated and has produced a steady acceleration of technological advance. However, by itself it tended to keep different fields of technology quite distinct (even building walls between them for the sake of neatness and ease in the process of analysis and experimentation). Further, it tended to keep technological advance from becoming extremely rapid because of the repeated cycles of analysis and time consuming experiment (testing of the analysis) until a new technological event or capability was reached. On the other hand, creative innovation tends to break down the barriers between different fields of technology, using whatever it needs from each and combining it into the overall system. Further, it tends to greatly accelerate technological advance, again through the systems approach, by observing the rapidly obtained results of overall system simulation (with the computer, for example) before finally creating from them the innovated whole. Thus it is this new motive force, creative innovation, with the systems approach as its tool which is bringing about the mingling of technologies and the acceleration of technological advance noted above.

Under such circumstances the purely intuitive method breaks down almost completely as a means for defining what lies some distance ahead in technology. Without some better means of forecasting the technological future, the planning and thus the staffing, funding, or conduct, of meaningful research and development becomes an almost hopeless activity. The need for new and better techniques of technological forecasting thus becomes self-evident.

TECHNOLOGICAL FORECASTING IN MANUFACTURING EQUIPMENT AND PROCESS R&D

Let us now turn to a discussion of our own experience in using technological forecasting in the planning of research and development on manufacturing equipment and processes as an aid to the design of greatly improved manufacturing equipment and processes. Some ten years ago, the writer's company began to realize that the situation developing in today's world, as outlined in the opening section of this paper, made it imperative to forecast the technological future of manufacturing in more than a cursory intuitive way in the planning of such research and development. Otherwise, that research and development might turn out to have little value and to do little to advance real-world manufacturing technology. However, technological forecasting was very new then, with no recognized techniques and methods yet well defined. Therefore, it was decided to start informally with simple and easily used approaches to forecasting to determine the validity and usefulness of the process. The initial methods chosen were those now known as trend extrapolation and survey of expert opinion, although at that time they had not been so formalized.

Some of the trend extrapolation results of that early effort were reported by the writer ten years ago in a paper on the future of metalworking manufacturing [1], so it is now possible in this present paper to look back and make some assessment of the validity and usefulness of this method of forecasting.

The four most pertinent trend extrapolations of that earlier paper are shown in Exhibits 1 through 4. To test the accuracy of their predictions, these have now been updated on the basis of current manufacturing performance capabilities. The results are shown in Exhibits 5 through 8. In each case new advances in technology have occurred during the past two years and the performance associated with these has been plotted on the earlier curves. It can be seen that each of these lies on or remarkably close to the earlier extrapolated trend line. This clearly demonstrates the applicability of this method for forecasting future performance in well-established fields of manufacturing engineering.

As is evident, use of the trend extrapolation method alone gives reliable quantitative but not qualitative information about the future. Thus it accurately predicted the amount by which performance would improve in each of the four fields illustrated but told nothing about the nature of the new developments which would make this possible. Information generated by this method obviously needed to be supplemented by qualitative information about the nature of future technological events before specific decision making could take place. Since at that time the Delphi method did not exist, face-to-face expert opinion was the supplementary method used.

Leading research and development experts in the field of manufacturing equipment and processes were interviewed, and their opinion asked as to what would happen in their field in the next ten to fifteen years. However, since experts in a given technology are not necessarily experts at forecasting, their opinions were taken as input only and then further studied and analyzed to arrive at conclusions concerning the likely qualitative nature of new developments which could give the quantitative performance predicted by the trend extrapolation.

On the basis of the forecast developed ten years ago from those "trend extrapolation" and "survey of expert opinion" activities described above, company machine tool product and process research and development programs were examined and appropriately modified. In addition, some of the forecast findings were shared with U.S. Air Force manufacturing technology R&D planners through the Materials Advisory Board of the National Academy of Sciences. While all of the tangible results of the use of those forecasts obviously cannot as yet be revealed, some of them are now being told here to document the applicability and value of using technological forecasting as an aid to planning equipment and process research and development.

Concerning automation of metal removal, the quantitative potential for obtaining further significant reductions in relative man-hour requirements in machining through variable program type automation was evident from Exhibit 1. Supplementing this with expert opinion about the future of automation revealed that the type of automation having most potential to realize the performance capability forecast by Exhibit 1 was that which is now known as adaptive control. Therefore, a company research program to accomplish adaptive control of machine tools was launched, resulting in the actual improvement in performance capability shown in Exhibit 5.

Concerning the improvement of accuracy of metal removal (Exhibit 2), increased attention was paid to in-process control of accuracy, contributing to, among other things, the incorporation of capability for such control in the adaptive control system developed. Further, the findings of Exhibit 2, coupled with the results of expert opinion on this subject contributed to the company's immediate adoption of laser interferometer equipment for machine tool accuracy checks when such became available (Exhibit 6).

Concerning the improvement in tool materials (Exhibit 3), the forecast findings in this field which were shared with the Air Force contributed to their supporting development of such materials, resulting in the actual improvement in performance capability shown in Exhibit 7.

Concerning the improvement of the grinding process (Exhibit 4), the forecast findings caused increased company emphasis to be placed on grinding process, grinding machine, and grinding wheel research, contributing to the development of the combination of grinding machines of higher rigidity and higher speed and grinding wheels of uniform and improved characteristics needed to accomplish safely and effectively the improved high-speed grinding process whose performance capability is shown in Exhibit 8.

Some of the benefits of the initial effort, combining trend extrapolation and expert opinion, began to become evident quite early. Therefore, we proceeded to apply these on a broader scale to define some of the long-range qualitative trends and needs of manufacturing and to begin to develop a scenario of its future. Some of the results of this study were published by the writer [2] in 1966, defining five such important trends and needs and setting forth an initial scenario. The trends and needs described there may be summarized as follows:

1. Even in the face of growing use of mass production for manufacture of certain classes of products, industry is faced with rapid proliferation of numbers and varieties of products (spawned by today's burgeoning technology), requiring average manufacturing lot sizes to decrease. Thus, there is a growing need for more economical production of small lots.

2. Under the impetus of severe demands for increased performance in manufactured products, requirements for closer dimensional tolerances are growing. Thus there is a need for more economical means of working to higher accuracies.

3. Again, under the stimulus of demands for increased performance in products, there is a growing requirement for working increased varieties of materials with increasingly diverse properties. Thus there is a growing need for more economical means to work less conventional and higher strength materials.

4. As a result of growing efficiency in the utilization of labor in manufacturing, the cost of materials is becoming an increasingly larger part of the total cost

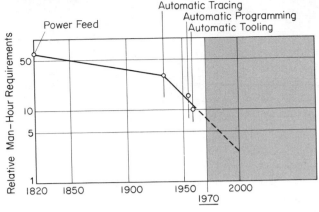

Exhibit 1. Improvement of Automation of Metal Removal.

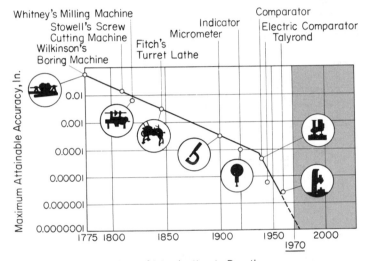

Exhibit 2. Improvement of Accuracy of Metal Removal.

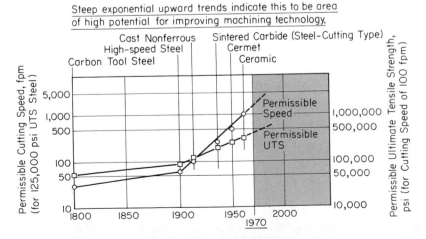

Exhibit 3. Improvement of Tool Materials.

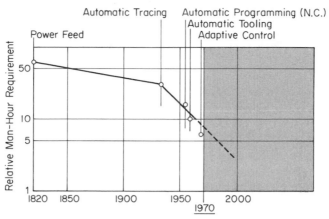

Exhibit 5. Improvement of Automation of Metal Removal.

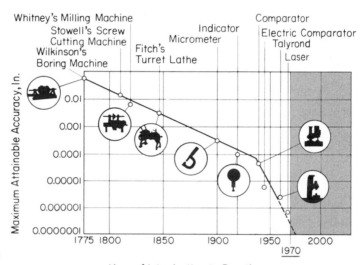

Exhibit 6. Improvement of Accuracy of Metal Removal.

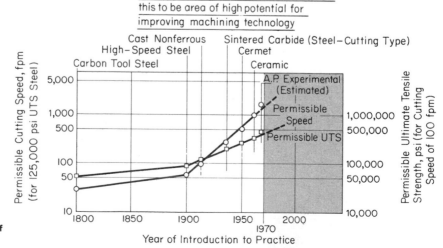

Exhibit 7. Improvement of Tool Materials.

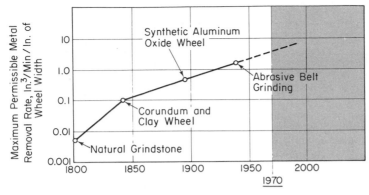

Year of Introduction to Practice

of the product. Thus there is a growing need for manufacturing processes to be less wasteful of material.

5. Under the pressure of all of the foregoing factors, ability to design products to achieve required performance at minimum cost is becoming increasingly dependent on the capabilities of the manufacturing process. Thus there is a rapidly growing need for improved communication and feedback between the manufacturing process and the design process, integrating them into a single system capable of being optimized as a whole.

Experience to date indicates that these trends and needs are quite valid. Continuation of this effort led to the further development of the scenario, resulting in the publication in 1968 [3] of an elementary scenario of one plausible and internally consistent future for the manufacturing system. However, these findings, along with those coming from the earlier trend extrapolation and expert opinion survey efforts, were the result of primarily exploratory forecasting.

By 1968 the Delphi method was becoming well known and its advantages over use of face-to-face expert opinion recognized. In addition it offered increased capability for normative forecasting. Further, published results of certain Delphi exercises had already contained some forecasts pertinent to manufacturing engineering and some opinion could therefore be formed as to its validity and usefulness for this field. Table 1 lists such forecasts by panels of experts derived from two exercises [4, 5] of a few years back. For those events which were forecast to have happened by now or a year or two hence, the accuracy is quite satisfying, since they have actually happened or are indeed imminent. Further, many of the other events listed take on a more plausible ring today than they must have had at the time the forecasts were made. Thus, confidence in the use of the Delphi method for forecasting events in the field of manufacturing has grown steadily and we are now making use of it ourselves.

As a result of the foregoing experience, technological forecasting is now a formalized activity in the writer's company, participated in by its technical

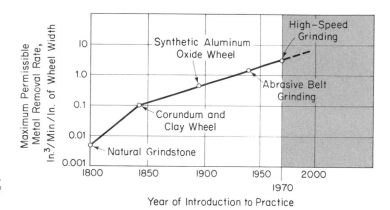

Exhibit 8. Improvement of the Grinding Process (precision grinding of steel).

and managerial staff. Further, because of the need described in the opening portion of this paper, similar formalized technological forecasting is now standard procedure in the technical activities of most progressive companies today.

Table 1. Some Delphi Forecast Results Relevant to Manufacturing Engineering.

Event	Year in Which There Is 50% Probability of Occurrence		
Forecasts Made in 1963–64 [4]	Lower Quartile	Median	Upper Quartile
1. Development of new synthetic materials for ultralight construction	1970	1971	1978
2. Operation of a central data storage facility with wide access for general or specialized information retrieval	1971	1980	1991
3. Man-machine symbiosis, enabling man to extend his intelligence by direct electromechanical interaction between his brain and a computing machine	1990 1985	2020 *Second panel* 2010	Never 2028
4. Increase by a factor of ten in capital investment in computers used for automated process control	1970	1973	1975
5. Automation of office work and services, leading to displacement of 25% of current work force	1970	1975	1975
6. Widespread use of automatic decision making at management level for industrial and national planning	1977	1979	1997
7. Evolution of a universal language for automated communication	1980	2000	Never
8. Manufacturing of propellant and raw material on the moon	1980	1990	2020

Table 1 (cont.).

Event Forecasts Made in 1966 [5]	Year in Which There is 50% Probability of Occurrence Median
1. The size of precision castings and forgings will continue to increase and will be very large compared to today's product	1968
2. There will be substantial use of composite materials employing "whisker" technology in gas turbines and jet engine airflow	1970
3. "Red hot" dies will be used to make forgings that are large, thin-walled, and free from conventional flash and parting lines	1971

REFERENCES

1. M. E. MERCHANT, "10 Years Ahead—What's in it for Metalworking," *American Machinist* (May 18, 1959), pp. 142–46; also M. E. Merchant, "Future Trends in Materials Removal Techniques," Paper No. T40, S.A.E. National Aeronautic Meeting, April, 1960.

2. ———, "The Future of Manufacturing Technology," *Frontiers in Manufacturing Technology,* University of Michigan, (1966), pp. 1–9.

3. ———, "Progress and Problems in the Application of New Optimization Technology in Manufacturing," *CIRP Annals,* Vol. 16, (1968), pp. 151–61.

4. T. GORDON AND O. HELMER, "Report on a Long-Range Forecasting Study," *Social Technology* (New York: Basic Books, 1966), pp. 44–96.

5. H. G. NORTH AND D. L. PYKE, "Technology, the Chicken—Corporate Goals, the Egg," *Technological Forecasting for Industry and Government—Methods and Applications,* (Englewood Cliffs, N. J.: Prentice-Hall, Inc, 1968), pp. 412–25.

Impact of an
Emerging Technology
on Company Operations

KARLE S. PACKARD

How will a new technology affect a given institution? Success in meeting these challenges depends on a realistic assessment of the ways in which this new technology will influence the product or service supplied, the capital equipment needed, the talents and training of personnel, and the impact on the operating structure of the company. Therefore, technological forecasting forms the basis for all other phases of such an assessment. The author outlines his application of technological forecasting to such a situation in his firm. The specific technology was that comprising the various means for integrating electronic circuits into an inseparable unit.

The introduction of integrated electronic circuits during the past decade has been widely recognized as being a major influence in the spreading applications of electronic technology. Reduction in cost, improved reliability, and the growth of the information processing branches of electronics are all closely related to it. Much has been written, and spoken, regarding the growth of integrated circuits and the potential influence on traditional industry relationships, but it is also clear that, despite the wide recognition of the importance of this technology, we have only witnessed the earliest beginning of its ultimate scope of application. An appreciation of the overall trend is not enough, however; the translation of the general technological impact of this development into the specific detailed factors that are of importance to a particular company is a very difficult task. An approach to this problem and the preliminary results obtained will be described below.

The major steps in the process of analysis used in this study are:

1. Forecast the changing technology.
2. Relate the technological forecast to the technological foundation of the company.
3. Analyze qualitatively and quantitatively the dependence of company functional areas on this technological foundation.
4. Forecast the derived impact on these functional areas.
5. Examine alternative means for coping with the forecasted changes.
6. Determine what technological development is required within the company.
7. Determine what personnel training or possible changes in types of personnel background are needed.

Each of these steps will be developed in the course of this discussion.

One of the starting points for the analysis is a knowledge of the technological foundation of the company which can be related to the overall functional operation of the company and a means for expressing this knowledge in a way that will provide for a changing technology. The method that has been developed at AIL for the past several years, takes the form of a technological relevance matrix shown in Exhibit 1. This matrix displays, for each major business area or product line, those specific technical specialties that relate to that business area. A wide variety of similar matrices have been developed by others and described in the literature using other coordinates and other measures, but this particular form has been found very useful. One measure used in this matrix, is that of *importance*, and this has usually been based on a simple three-level quantization: major, minor, or none. This is the measure shown in Exhibit 1. Another measure that is used, although it will not be discussed in this paper, is that of *capability*, also having a three-level quantization little, partial, or adequate. These two matrices have been used to guide our development program. For the purposes of this discussion, however, a third measure, relating to the technological characteristics of circuit design within these detailed technical areas, was used and will be discussed in further detail below.

Considering any one business area or product line in the matrix of Exhibit 1, its column of applicable technical areas may be considered as a vector that is descriptive of the technological base of that business area. If we consider changes in the characteristics of each of the detailed technical areas resulting from a technological change, such as that represented by integrated electronic circuits, then we can develop a transformed technology vector that will represent the technological base at some future time.

	Business Area (Product Line)					
Technology	A	B	C	D	E	F
EM receivers:						
Low–noise receivers	O	O	O	O	O	
Parametric amplifiers	O	O			o	
Scanning receivers	O	O	O	O	o	
Wide–open receivers	o	O	O	O	o	
Radiometers	O					

Some possible measures:

1. Importance $\begin{cases} \text{O} & \text{Major} \\ \text{o} & \text{Minor} \end{cases}$

2. Capability

3. Technical characteristics

Exhibit 1. Product/Technology Matrix.

The evolution of electronic circuit technology can be characterized, in one sense, by the types of *active* devices that have been dominant in applications of this technology. From the time of the invention of the Audion by DeForest in 1906 until the invention of the transistor in 1947, the vacuum tube was the dominant active device and determined the characteristics of most of the other component parts and even the approach to circuit design. By the mid 1950s the transistor had begun to replace the vacuum tube in new circuit designs, and at the present time vacuum tubes are no longer an important device for general use but are limited to those applications where their unique characteristics cannot be duplicated by semiconductor devices. Early in the 1960s integrated circuits, based largely on the semiconductor technology developed for transistors, began to see application, and although these applications are still relatively small in terms of either number of units or dollar value, it is generally conceded that integrated circuits will largely replace discrete transistors. As a concurrent trend factor, the number of individual functional elements included within a single integrated circuit device is increasing. It is expected that this latter integration trend will reach the level in the near future where a reasonably complete function will be contained on a single chip of semiconductor, that is, we will achieve integrated electronic components.

In order to quantify the forecast of this changing technology, it is helpful to base it on a reasonable forecast of future economic conditions. Exhibit 2 shows the history of total sales for the electronics industry [1]. This is extended by the author's forecast through the year 1980. This forecast was developed in considerable detail for a purpose other than this study; it is based on the latest forecast of the U.S. economy by The National Planning Association [2]. Displayed on the same graph is the past growth and forecasted future use of active elec-

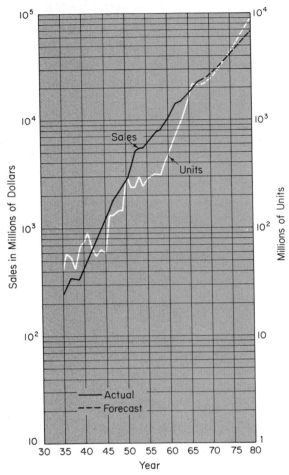

Exhibit 2. Electronic Sales Forecast and Active Electronic Devices.

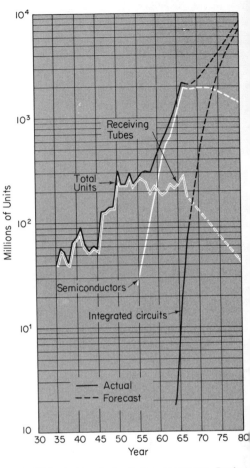

Exhibit 3. Trends in Active Electronic Device Usage (original equipment).

tronic devices. This forecast is used as a "top down" bound on the forecasted growth of the several types of active electronic devices. It should be noted that high-power vacuum tubes have been excluded from the total of active electronic devices. Their displacement by semiconductor devices will take considerably longer than for other types.

The changing relative usage of vacuum tubes, transistors, and integrated circuits is shown in Exhibit 3. Only devices used in original equipment are included; devices sold for replacement purposes are excluded. These trends are also shown in the form of substitution curves in Exhibit 4, and, in fact, these substitution curves are the basis for the trend curves shown in Exhibit 3. It is seen, that while integrated circuits presently represent only a few percent of

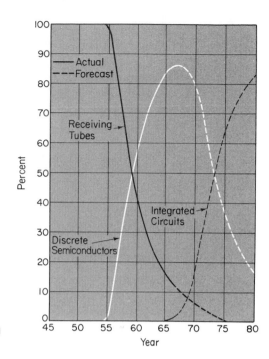

Exhibit 4. Relative Use of Active Electronic Devices (original equipment).

the total, it is forecast that they will represent fifty percent by 1973, and this share will continue to increase until it represents the bulk of the active electronic devices used. There will remain certain applications where discrete semiconductors will continue to offer the most practical characteristics, but it is expected that by the late 1970s receiving type vacuum tubes will have virtually vanished from the scene. While this forecast is centered on active electronic devices, it must also be pointed out that the change to integrated circuits will affect the use of passive electronic devices in the same way that it affects discrete semiconductors.

The above forecast is applicable to the electronics industry as a whole and could be misleading when applied to some specific portion of the industry. It also was derived mechanically, chiefly based on the use of substitution curves and corroborated by trend extrapolation. It is necessary to test the applicability of such a forecast to the future business of the particular company being examined. In order to do this a simplified Delphi experiment was performed. This experiment was independent of the other forecast and, in fact, actually preceded it, so that the respondents had no knowledge of the forecast results. Qualified technical personnel in each of the technical areas of importance to the company, were queried regarding the present and future extent of usage of integrated electronic circuits within their technical specialties and applicable to company business areas. Using the technological relevance matrix, their estimates were weighted by the relative share of total company business to obtain

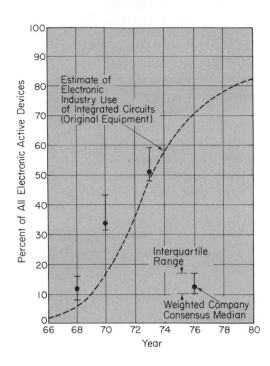

Exhibit 5. Comparison of Estimates of Penetration of Integration in the Electronics Industry to Within Company.

an overall estimate of total company applicability. Estimates were made for the present (1968) and for future years 1970 and 1973. These estimates are shown in Exhibit 5 where they are compared with the relative usage forecast of Exhibit 4. The results show very good agreement, particularly in view of the fact that AIL is a high technology company and would be expected to lead the industry average in the use of advanced technology. Thus the forecast was corroborated and judged applicable to company business.

An important dimension in the future of integrated electronic circuits is the scale of integration, that is, the number of individual circuits contained on one semiconductor chip. As the number of circuits can be increased, the complexity level of the device rises to the point where it may represent a complete functional component, such as a radio receiver or a computer memory. This trend has profound implications for both the traditional vendor/user relationships in the electronics industry and the role of individual technical specialists at all levels in the industry. For example, the background knowledge of the circuit designer will have to include, or be included with, knowledge of the physical properties of materials, heretofore a speciality of the electronic parts designer.

A forecast of the growth in the number of electronic circuits per chip, as well as the actual growth in the past, is shown in Exhibit 6. This is a straight-forward trend extrapolation, but it is corroborated by a prior company forecast based on a Delphi experiment made in early 1968. It is on this forecast that

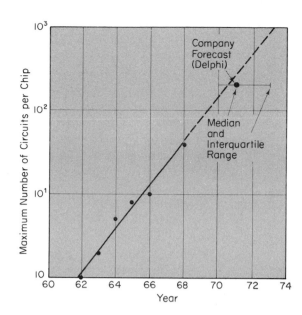

Exhibit 6. Growth in Scale of Integration.

he importance of integrated components in the evolutionary picture was based;
hat is, with the rise in level of integration above 100 circuits per chip, starting
n the early 1970s, it is expected that integrated components will be an im-
ortant factor in the industry. A potential change in the traditional roles of
nanufacturers, resulting from this trend in integrated circuits, is displayed in
xhibit 7.

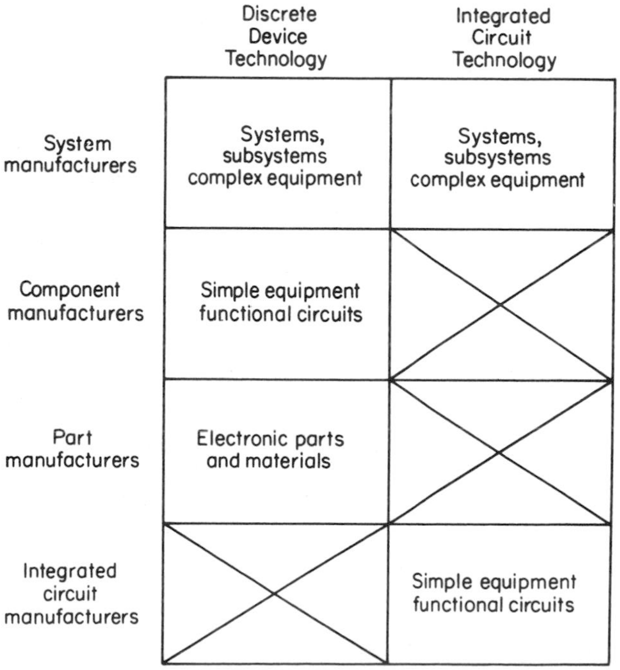

	Discrete Device Technology	Integrated Circuit Technology
System manufacturers	Systems, subsystems complex equipment	Systems, subsystems complex equipment
Component manufacturers	Simple equipment functional circuits	✕
Part manufacturers	Electronic parts and materials	✕
Integrated circuit manufacturers	✕	Simple equipment functional circuits

Exhibit 7. Industry Integration Resulting from Integrated Circuit Technology.

The implications of the forecasted technological changes for the future of the company must be examined. We wish to know how the engineering and manufacturing functions will be affected in terms of people and equipment and what might be the competitive considerations.

Another measure used with the technological relevance matrix illustrated in Exhibit 1 is the kind of engineering involved, that is, whether or not the engineering level is system engineering, component engineering, or device engineering. The relative content of these various levels of engineering within the company can also be translated into requirements for manufacturing and the purchase of materials and services. The distribution of these various types of effort will vary from company to company within even a small portion of the industry and will probably vary within a particular company with time. However, a technological change, such as that represented by integrated electronic circuits, might well change the relationships within many specific companies to a very considerable extent. A representative distribution for an electronics company of moderate size with a substantial volume of government business is shown in Exhibit 8. Here the distribution for the present is based on the assumption that the real impact of integrated electronic circuits is relatively minor even though their usage may be high. For example, the company might not have a significant integrated circuit application capability but may well use integrated circuits supplied by a traditional parts-supplying portion of the industry in a considerable fraction of their products.

As the degree of integration rises, as shown in the distributions for 1973 and 1980, there will be a substantial decrease in component engineering, component manufacturing, and the purchase of electronic parts, while the purchase of other materials, specifically integrated components, rises. These distributions assume that the in-house capability for fabricating integrated electronic circuits is not significant. Translated into value added, these distributions result in the situation shown in Exhibit 9. Here the typical industry average of sixty-five percent value added would decline to about forty-eight percent by 1980 with no in-house integrated circuit capability. On the other hand, with a full capability for manufacturing integrated circuits, the value added would actually be increased because the amount spent for electronic parts would decline without a fully compensating increase in the cost of materials. An added test of realism of this forecast was obtained by evaluating the degree of value added by the present integrated circuit capability, including the trend over the past several years. This is shown also in Exhibit 9, and although this contribution has been relatively small, it would appear that over the next few years critical decisions are indicated with respect to expansion of this capability. The identification of this need for major decisions is the primary result of this study.

Without a significant increase in integrated circuit fabrication capability it would seem that the company would tend to assume the role of system engineer relying chiefly on other manufacturers to supply the required hardware. On the other hand, if a significant integrated circuit capability is to be

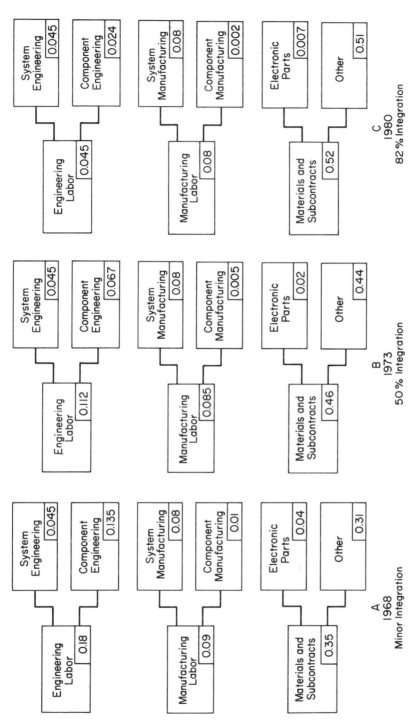

Exhibit 8. Effect of Integration on Distribution of Cost Elements.

Note: Numbers in corner boxes indicate fraction of sales represented by this cost element,

309

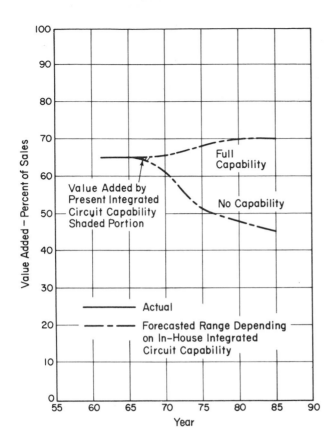

Exhibit 9. Impact of Integration on Value Added

developed, then additional technological capability and capital equipment will be required. The first alternative requires an intensive, continuing development of systems engineering capability and alignment of the marketing function to this role, since the company would have little competitive advantage based on equipment superiority. Further pursuit of this alternative would require a high level technological forecast of potential customer needs and feasible future levels of performance. While this should be done in any event, it assumes greater importance in a system engineering company, as such foresight concerning operational factors constitutes its chief competitive advantage. Pursuit of the second alternative, that of an extensive integrated circuit fabrication capability, requires further analysis and forecasting of the specific details of this technology.

Within the general field of integrated circuit technology, there are a number of detailed characteristics that are changing in time and influence the kind of preparation one should make for the future. For example, one form of integrated circuit is typified by the monolithic silicon circuit, based on semiconductor technology, wherein impurities are diffused into a restricted portion of block of silicon to produce desired electrical properties in these restricted areas. The treated areas are then interconnected to provide a more complex electrical function. The other major integrated circuit technology is based on extremely thin films evaporated onto a substrate with the electrical properties of the film and its topology used to control the electrical properties of the resulting circuit. These two technologies, in addition to having differing electrical properties, have differing degrees of cost effectiveness depending upon the application. The monolithic circuit has a much lower cost when used in high quantities but has a high cost if custom designed. It is more restricted in terms of frequency range and power handling capability than is the thin film technology. It is, however, inherently more compact and more reliable. Because of these factors, monolithic circuits are generally more applicable where the function to be performed is digital in nature, whereas thin film circuits are more usually applied in analog circuits.

Trends with regard to these two types of integrated circuits are shown in Exhibit 10, together with the trends in application of integrated circuits to analog and digital functions. The growth in analog applications despite the decline of thin film usage, is indicative of improved characteristics of monolithic circuits. The rise in analog applications is, however, probably temporary as

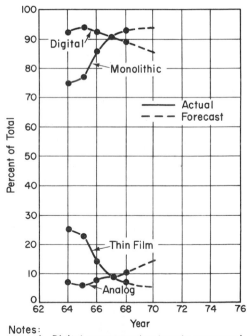

Notes:
1. Digital versus analog trends expressed as percent of total units.
2. Monolithic versus thin film comparison is based on total dollars.

Exhibit 10. Relative Use of Integrated Circuit Types.

there is an overriding trend to replace analog functions with digital ones. Reliable data for this trend are not readily available, but it should be considered before reaching final conclusions.

A specific example of the potential replacement of thin film circuits by monolithic circuits, is shown in the trend of IF amplifier miniaturization displayed in Exhibit 11. It is seen that substantial improvement in size was obtained by going from discrete semiconductor circuits to thin film circuits with continuing improvement over a succession of projects. It is also seen, however, that at the present time it would be feasible to gain almost an order of magnitude improvement by going to monolithic circuits. It should be pointed out that this trend in size reduction is a realistic one including all the required mounting hardware and not merely a theoretical one based on the volume of the component parts.

The trends in type of integrated circuit, shown in Exhibit 10, are trends for the industry as a whole, and it is necessary to examine the applicability to the specific needs of the company. This can be done by again returning to the

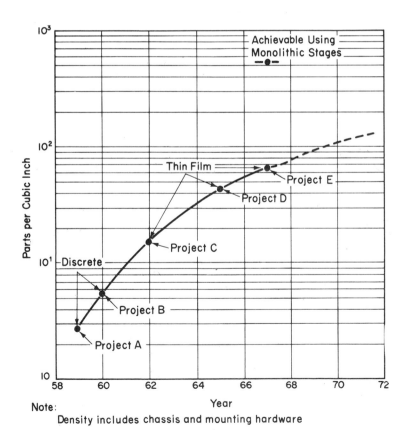

Note:
Density includes chassis and mounting hardware

Exhibit 11. IF Amplifier Miniaturization.

technology vectors illustrated in Exhibit 1. The individual technical areas must be examined with regard to the specific type of integrated circuit that is likely to dominate during the time period of interest. This was done, and some examples are shown in Exhibit 12. These specific technical areas can then be weighted by the importance to each business area and then in turn weighted by the relative importance of each business area to total company volume. By using different time frames, changes in the relative importance of the different type of integrated circuit technology can be determined. An appropriate plan can be prepared for the development of the type of integrated circuit facility needed and the technical specialties required, including a realistic, meaningful schedule for its accomplishment.

Exhibit 12. Expected Circuit Design Trends.

	Dominant Technique	
Technology Field	1968	1973
Digital computing and data handling	Monolithic IC'S	Monolithic IC's
Audio and video frequency amplification	a) Military-monolithic b) Commercial-discrete parts	} All monolithic IC's
RF amplification	Discrete parts	Hybrid IC's
Transmitter (low power, phased array)	Discrete parts	Hybrid IC's
Filtering, lumped constant (narrow band)	Discrete parts	Hybrid IC's

Notice that a variety of technological forecasting techniques were employed in this study, including:

1. Horizontal decision matrices.
2. Trend extrapolation.
3. Substitution curves.
4. Delphi.

These techniques were used selectively to develop definite phases of the analysis, to cross check each other, and to evaluate overall industry trends in relation to the particular needs of the company.

We concluded that the techniques of technological forecasting can be used to provide a practical basis for company planning. Although no single technique or combination of techniques should be considered as universally applicable, judicious choice of selected techniques to fit specific problems can provide a practical framework for investigation.

ACKNOWLEDGEMENT

The author wishes to express his appreciation for the valuable assistance of F. X. Driscoll and F. H. Williams in carrying out this study, and for the contributions of the many associates who provided the experienced opinions without which the study would not have been possible.

REFERENCES

1. ELECTRONIC INDUSTRIES ASSOCIATION, *Year Book 1968.* (Washington, D.C.: E.I.A., 1968).

2. A. AL-SAMARRIE, M. COHEN, AND T. HORI, *National Economic Projections to 1978/79.* National Planning Association Report No. 68-N-1, (January, 1969).

Computer Memory Market— Application of Technological Forecasting in Business Planning

DAVID W. BROWN

JAMES L. BURKHARDT

Can technological forecasting aid the firm in choosing the best technological-economic niche in a rapidly changing, high technology market?

The authors describe a major attempt to integrate technological forecasting into exploratory business planning. Their central concern is with computer memories. The specific query is on the types and volume of memory devices that will be developed, produced, and marketed in the five years following the date of their study (1967). Published forecasting efforts like this are rare because of the unusual mix of explicit cost-volume-technology projections.

One of the great aircraft designers in the era prior to World War II was a man by the name of William Bushnell Stout. He was, among other things, the designer of the famed Ford "Tin Goose." Stout dreamed of the day when every man would have his own airplane and, by the middle 1930s, it began to trouble him that this day was so slow in coming. Finally, he concluded that the real obstacle lay in the engines which were then available. "What we need," he announced in a speech to leaders of the aviation industry, "is an engine which develops 100 horsepower, weighs 100 pounds, and costs $100." Instantly, a member of the audience leaped to his feet and exclaimed, "And you shall have it, Mr. Stout—in 100 years!"

This story emphasizes a fundamental truth about technological forecasting. The most farsighted technical prophet will be without honor unless and until his advice appears to make good business sense. In no occupation is this principle more apparent than our own. TMA (Technical Marketing Associates) is a management consulting firm, devoted primarily to product line planning and market studies and working almost exclusively with clients whose businesses are technically oriented. What the client expects from TMA is information, interpretation, and advice which will assist him in reaching sound decisions regarding business problems. Although technological forecasting plays a part in almost every assignment which TMA performs, it must be regarded as a research tool, one of many tools available to help the client. With this in mind, what we would like to relate here is not technological forecasting per se but one modest example of how technological forecasting has been combined with other techniques to assist a client in understanding a complex market opportunity.

In the case which we are about to discuss, the client had been developing thin magnetic films which were potentially applicable as rapid-access memory elements for computers. By 1967 this work had been proceeding on a limited scale for about three years. Although the computer industry as a whole had progressed spectacularly during this period, the client saw no evidence of any large-scale trend away from the traditional ferrite core memories toward the use of thin magnetic films. Thus the client considered the further pursuit of this development open to question. Was the thin film development leading to a dead end or were there market prospects for thin film technology which justified its continuance? If, indeed, there was a future market for thin film technology in the computer industry, how attractive would that market be, and what steps would the client have to take to capitalize on it?

To provide a frame of reference for the study, a projection was needed of the future growth of the computer industry as a whole. We did not think it necessary for TMA to make such a projection on its own, since there were available many forecasts published by persons considered to be authorities on the industry. We selected as our basic projections the two curves shown in Exhibit 1. The upper curve in Exhibit 1 is a forecast of dollar sales of the data processing business as a whole, including not only computer main frames but also peripheral equipment, software, consulting services, service bureau operations, and the like. The lower curve is a forecast of dollar sales of computer "main frames" only. The main frame, of course, is the center of the computer which performs the calculating operations. It is of particular interest to us because at its heart is the subject of the study: the random access main memory.

A peculiarity of the computer industry is the lopsided distribution of the main frame business among the principal manufacturers, as shown in Exhibit 2. This tabulation highlights a problem which handicaps anyone making forecasts in the computer industry. One company, IBM, comprises more than two thirds of the industry, and IBM has very strict policies which prohibit its employees from discussing future developments with outsiders, no matter how

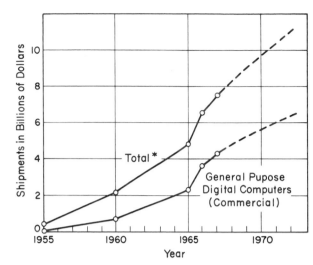

Exhibit 1. Estimated Annual Value of Computer Equipment Shipped by United States Manufacturers.

* Total includes general- and special-purpose digital computers, independent peripheral equipment, software and consulting services, service bureau operations, and supplies and supporting services.

Exhibit 2. Computer Shipments by Company, 1966.

	Amount ($ Million)	Percent of Total
IBM	2,500	68.3
Honeywell	270	7.4
Control Data	200	5.5
Univac	195	5.3
General Electric	190	5.2
National Cash Register	95	2.6
Radio Corp. of America	95	2.6
Burroughs	60	1.6
Scientific Data System	30	0.8
Others	25	0.7
	3,660	100.0

Source: *E/D/P Industry and Marketing Report.*

legitimate their interest. Consequently the future plans of two thirds of the industry can only be deduced through such bits of secondhand intelligence as may filter out through IBM's security blanket. Nevertheless the same laws of physics apply in Endicott, New York, as apply elsewhere, and we have had to content ourselves with the conviction that technological advances which make sense to the remainder of the industry will probably ultimately make sense to IBM as well.

Narrowing our outlook to the rapid access memory market, we undertook

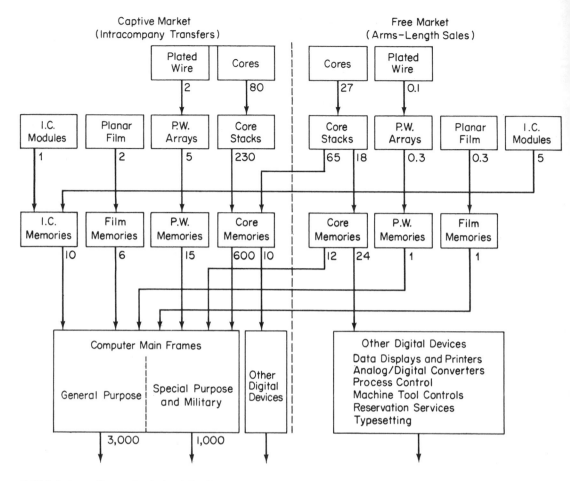

Exhibit 3. Input-Output Analysis of the Computer Memory Market Forecasted 1967 Sales and Transfers in Millions of Dollars.

next to construct for the client a chart illustrating its size and operation. This is shown in Exhibit 3. Since information as detailed as this is not available from any published source, it was necessary to compile the data shown in this exhibit by extensive interrogation of persons active in the memory industry itself.

Computer memories proved to be a big industry. The value of all the main memories made in 1967 was thought to be about $680 million. This is TMA's estimate of the total value which all the memories would have had, had they been sold on the open market. The majority, of course, were not sold at all. They were transferred within the captive portion of the market, shown on the left side of Exhibit 3, which was composed of the computer manufacturers themselves and the companies which they control. It was in the comparatively small free market, composed of independent manufacturers of memory systems, memory components and other devices, that the client's product would have to compete.

The numerals below the blocks in Exhibit 3 represent the estimated value in millions of dollars of the merchandise sold or transferred between each of the segments of the memory market in 1967. It may be seen that although the captive computer industry made most of its own memory systems and memory components, a certain portion was bought at each level from the free market. In fact, although the estimated total sales volume of the free segment was about $155 million, only about $24 million of it was finding its way into products other than those made by the captive market.

The memory market was also depicted as having several levels, each a supplier of components to the level below it. Total sales at the memory system level in 1967 amounted to about fifteen percent of the total value of the main frames. Over the prior five years this percentage had been growing. The cost effects of demands for more capacity and more speed had been largely offset, in the calculating sections of the computer, by improved design and lower prices in electronic components. Not so in memories. Cost reduction in core memory systems had not kept pace, and increasing premiums had had to be paid for advances in memory size and cycle time. It appeared that by 1972, a typical memory would probably represent twenty-five to thirty percent of the total cost of the computer. Desire to keep this percentage from growing even further motivated the intense interest in new memory technology which permeated the industry.

At each level, the market structure in Exhibit 3 could be further sub-divided on a technological basis. A glance at the figures shows that by far the greatest portion of all memories manufactured in either the captive or the free market used conventional ferrite core technology. Magnetic thin film technology was used in two types of memory systems: those using plated wire and those using planar film. Together these technologies made up no more than three percent of the total market. The only other significant technology was that using integrated circuits, and it was an even smaller factor in the market. Were our client to think of himself as a potential manufacturer of magnetic thin film memory components, it can be seen that in 1967 the total value of the market for which he might compete was only about $700,000. Small wonder, then, that the client was concerned about whether or not he should continue development work in this area.

Over the next five years, how were thin film memories likely to fare in competition with other types? This was the heart of our problem. Memories, we learned, compete with one another the same way that computers do, on the basis of data handling capacity per dollar of cost. The key measures of strength, then, are:

1. Capacity (measured in bits).
2. Speed (the inverse of cycle time in seconds).
3. Price.

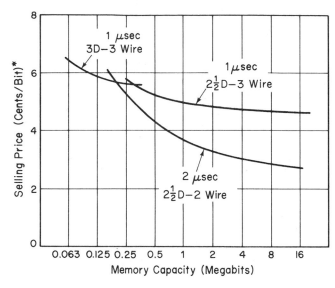

Exhibit 4. Memory Cycle Time Versus Date of Introduction for Selected "Fast" Computers.

As one memory designer put it, "We're looking for the most bits per second per buck." Using this standard we discovered that there are many good reasons why the magnetic core memory, despite many published predictions of its imminent demise, continued to dominate the memory market.

By 1967, for example, one could no longer talk about "the" ferrite core memory. There was no longer one such memory system; there were several. Ferrite core memories can now be organized in several different ways, each of which gives different performance specifications, and different cost trade-offs. The intelligent engineer, therefore, will choose between these various organizations and select the least expensive one which is suited to his requirements. A study by a leading computer designer was adapted to yield Exhibit 4, in which we can see the relationships between the cost per bit, the speed, and the size of the memory system for each of three organizational schemes which are presently popular. These newer memory organizations offer more formidable economic competition to a new technology, like magnetic thin films, than did the 3D-4 wire core memory system which was the universal standard several years ago.

Furthermore, core memories have been getting faster. This can be seen by reference to Exhibit 5, in which are plotted the decreasing memory cycle times of a series of computers which have been introduced over the last fifteen years, based on their published specifications. The newest computer with a core memory is more than ten times as fast as the computer introduced in 1953. It is interesting to observe that the computers using thin film memories have extended the speed trend of core memories but have not substantially broken

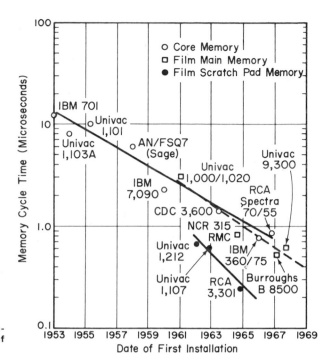

Exhibit 5. Estimated 1967 Selling Price of Complete Core Memory Systems as a Function of Capacity.

away from it. (The memories in the lower section of the chart are "scratch pad" memories, which are not directly comparable with main memories because they are not bound by the same economic constrictions.)

As core memory systems have become faster they also have become less expensive. Using figures provided by interviewees, we established the curve shown in Exhibit 6, showing the trend in price per bit of the fastest memory system of constant size in ordinary production. It may be seen that the cost per

Exhibit 6. Estimated Price/Bit of Complete 1-Million-Bit Memory Systems of Fastest Speed in Ordinary Production.

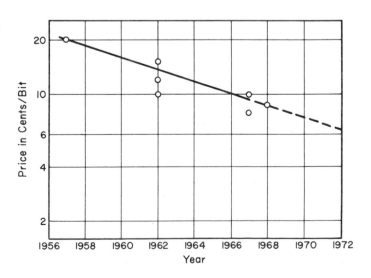

bit had been roughly cut in half, from about 20 cents in 1956 to about 10 cents in 1966, and we felt reasonably safe in extrapolating this trend out to about 6 cents per bit in 1972. This of course is the price per bit for the fastest memory in regular production; the price per bit of an average memory would be nearer to 5 cents per bit in 1966 and possibly 3 cents in 1972.

Finally, we found it instructive to examine the relationship between memory cost and memory speed at the time of our study. Exhibit 7, based on manufacturers' quotations, quantifies this relationship for a memory system of one million bits. Note that cost per bit increases very sharply for short cycle times: in fact, as the cycle time goes from 1 microsecond to 500 nanoseconds, the cost per bit doubles.

Some hard physical facts underlie this simple looking curve. Every core memory system has an electronic section and a magnetic section. The electronic section includes a switching system which puts each bit of information into the proper place in the memory and then retrieves it again when wanted, an amplifying section which makes the memory's readout signal strong enough to be used elsewhere in the computer, a "clock," and a power supply. The magnetic section consists of thousands of little ferrite doughnuts, strung on wires which carry the signals in and out of storage. Basically, the speed-limiting factor in the memory has always been the magnetic section. Engineers have discovered that they can make the ferrite cores respond more quickly by making them smaller, and over the years the diameters of these tiny doughnuts have shrunk,

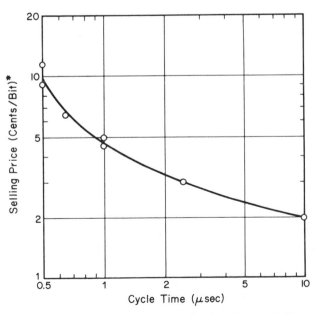

Exhibit 7. Estimated 1967 Selling Price of 1-Million-Bit Complete Core Memory Systems as a Function of Cycle Time.

* Systems include address registers, decoders, drivers, switching, core stacks, sense amplifiers, input-output registers, and timing.

bit by bit, from more than .05 inch down to as little as .012 inch. Nevertheless, these miniature doughnuts still have to be fabricated, tested, and mounted on frames, each with two or three carefully strung wires passing through its center. Over the years, the cost of making the magnetic section of the memory has actually increased. The overall reductions in memory cost have come about through drastic decreases in the cost of the surrounding electronics. In our interviews it became clear that more and more engineers were doubtful that these trends could be extended much further. It appeared then that the sharp rise in the left hand side of the curve in Exhibit 7 constituted a cost umbrella under which other technologies might take root and develop. If memories using thin film technology would operate effectively at cycle times of around 500 nanoseconds, and if there were indeed a need for memories operating that fast or faster, it seemed quite likely that they could be sold competitively at a price per bit of around 10 cents, even though quite possibly they could not be built to sell for a price as low as 5 cents per bit. This, then, was the thesis which it appeared we must investigate.

We now set about to test this thesis in the environment of the computer market of the future. Bearing in mind that our client was at least two years away from commercial sale of memory components, it appeared reasonable to try to forecast the market over a span of about five years. This, we felt, would define the competitive climate into which his new products would have to be launched.

We began by attempting to define the performance specifications of computers of the future. There was already no doubt that computer memories were becoming larger and faster. In Exhibit 8, for example, we can see the trend in memory cycle time of computers shipped over a period of three years. Simple observation indicates that average cycle time has decreased considerably. In Exhibit 9 we may make a similar comparison for maximum internal memory capacity, in which it can be seen that capacity is growing larger. We next undertook to extend these trends out to the year 1972.

It is a common observation that any product has a life cycle. The product is introduced, sales increase as it "catches on" in the marketplace, sales finally reach peak, and then, as it loses favor to other more advanced products, sales diminish until finally it disappears from the market altogether. Although computers are still so new that we do not have very much life history on them, sales statistics suggest that computer models undergo life histories of this type and that a particular model will survive on the marketplace from five to ten years, depending upon its popularity. We would suggest that the same kind of behavior characterizes whole classes of computers if they are categorized with respect to memory size and memory speed. For example, computers with memories of a given cycle time first appear in a given year, increase in usage as they outsell slower competitors, finally reach a peak in sales, and then are themselves superseded by even faster computers until, seven to ten years after their introduction, having become obsolete, they pass from the picture.

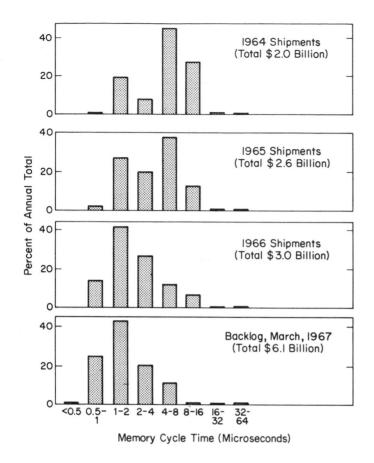

Exhibit 8. Digital Computers—Annual Shipments by Selected United States Manufacturers, Classified by Internal Memory Cycle Time.

With this thought in mind we prepared Exhibits 10 and 11. The upper curve in each exhibit is a repetition of our previous forecast of total shipments of computer main frames. The lower curves represent the past and predicted division of the main frame business on the basis of memory cycle time. Thus the sum of the ordinates of the lower curves always equals the upper curve. We may see, for example, that computers with 8 to 16 microsecond memories were already obsolescent in 1964 and that they disappeared in 1967. Computers with 2 to 4 microsecond memories appear to have reached their peak in the year 1967 and are expected to decline in popularity, dying out completely in 1971. Computers with 0.5 to 1 microsecond memories have been increasing in popularity since their introduction in 1964 and are not expected to reach their peak until around 1971. If we take a "slice" through this chart in any given year, we can obtain an approximation of the value of the computers shipped, broken down by memory cycle time. Exhibit 11 would permit us to perform the same exercise with respect to maximum internal memory capacity.

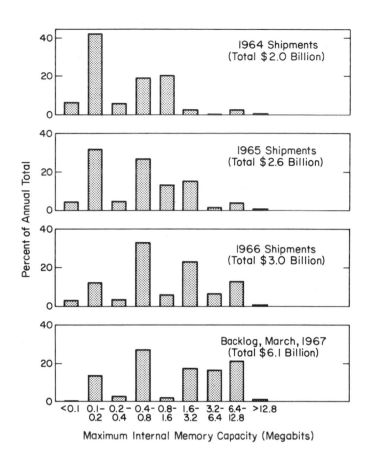

Exhibit 9. Digital Computers—Annual Shipments by Selected United States Manufacturers, Classified by Maximum Internal Memory Capacity.

By this means, we performed such a breakdown for the year 1972, and the results are displayed in Exhibit 12 using the same type of presentation that was used in Exhibits 8 and 9. In effect, what we did here was to forecast that the bulk of the computer memory market in 1972 would call for cycle times in the range of 0.25 to 1 microsecond and for memory capacities in the range of 3 to 12 megabits. This forecast, plus our previous forecast concerning memory costs, served to define the requirements which the memory designer of 1972 would be expected to meet.

It then became our task to estimate the extent to which the various memory technologies would be used to fill these requirements. In our report to the client, we disposed briefly of several memory technologies which we felt would not be of commercial significance during the period from 1967 to 1972 and concentrated our attention on the four technologies which we felt would be important: ferrite core memories, plated wire memories, thin planar film memories, and integrated circuit memories. It was obvious at the time of our study

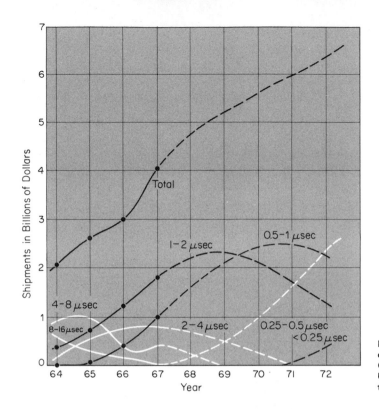

Exhibit 10. Estimated Annual Dollar Value of Shipments of General-Purpose Digital Computer Main Frames by Selected United States Manufacturers as a Function of Memory Cycle Time.

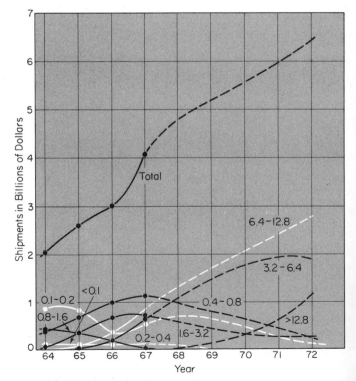

Exhibit 11. Estimated Annual Dollar Value of Shipments of General-Purpose Digital Computer Main Frames by Selected United States Manufacturers as a Function of Maximum Memory Capacity.

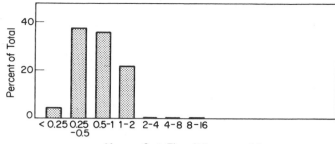

Memory Cycle Time (Microseconds)

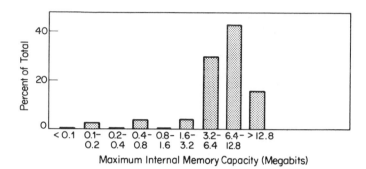

Exhibit 12. Predicted Distribution of 1972 Digital Computer Sales by Memory Cycle Time and Maximum Memory Capacity.

Maximum Internal Memory Capacity (Megabits)

that each of these technologies had vigorous adherents, and that each was receiving a wealth of R&D support. In fact, to say that one technology would outstrip all the others was to say that more than one major United States corporation had placed sizeable wagers on the wrong horse. For example, General Electric and Digital Equipment Corporation appeared to be confident that ferrite cores would be the most practical computer memory system for some time to come. Univac and National Cash Register had already committed themselves to the production of plated wire memories. Burroughs and Control Data Corporation indicated that they had bet on thin planar films. RCA appeared to be leaning toward integrated circuit memories. Only a few companies, such as Honeywell, appeared hesitant to place final bets.

This much, however, appeared to be certain: in 1972 each of these four major memory technologies would have some significant part to play in the computer industry. In Exhibit 13 TMA attempted to summarize what each of these parts was likely to be.

FERRITE CORES

Despite ten years of advances in speed and cost, the barriers which stood in the way of continued development of ferrite core memories appeared to be hard and fundamental. Further improvements in speed through the use of smaller

Exhibit 13. Present Evaluation of Principal Competitive Memory Technologies.

Performance	Ferrite Cores	Integrated Circuits	Plated Wire	Planar Film
Minimum cycle time	About 350 ns.	Below 50 ns.	About 100 ns.	About 100 ns.
Capable of NDRO	Generally not	Yes	Yes	Yes, but difficult
Physical size	Relatively large	Small, but needs cooling	Intermediate	Small
Volatility problems	None	Inherent	None	None
Aging effects	None	None	Appreciable	None
Shock resistance	Fair	Excellent	Poor	Excellent
Signal/noise problems	Negligible	Negligible	Moderate	Relatively severe
Manufacturing Cost				
Typical '67 cost/bit	$.06	$.50	$.06	$.15
Effect of increasing size	Cost/bit declines up to 10 MM bits	Cost/bit remains flat at all sizes	Cost/bit may tend to increase above 1 MM bits	Cost/bit may tend to increase above 500 K bits
Effect of increasing speed	Cost increases sharply below 1 μs. cycle time	Cost relatively independent of speed	Little increase down to 200 ns.	Little increase down to 200 ns.
Manufacturing yield	High	Low	Fairly high	Low
Cost of related electronics	Relatively low	Minimum	Intermediate	Relatively high
Investment Required				
Manufacturing facilities	Not needed	Very high	Lowest	High
Engineering and development	Not needed	Very high	Moderate	High
Adapt data processing system	Not needed	Very high	Moderate	Moderate
Capacity for Further Improvement				
Performance	Improvement in speed doubtful	Simplification and reduction in heating are likely	Size reduction likely	Increase in signal possible
Reliability	Already acceptable	No change likely	Reduction in aging effects likely	Little change likely
Manufacturing cost	Cost expected to remain about constant because of higher labor content, balanced against decreasing circuit costs.	Considerable reductions likely in all types.	Improvements likely in overall cost, through higher yield in wire mfg. and lower cost of supporting electronics.	Considerable cost reduction possible, through simplified electronics and one-process mfg. of film and electronics.

cores would require cores smaller than .012 inch diameter, which seemed to be prohibitively expensive to string. Labor cost reductions would require that plants be located in areas where skilled labor is even cheaper than that in Hong Kong or Taiwan, and it is doubtful that such a place exists. Hence, we concluded that ferrite core memory technology in 1972 would be little improved over that which we found in 1967.

Nevertheless, TMA also concluded that the majority of memories made in 1972 would still use ferrite cores. The most significant factor working to perpetuate ferrite core memories was inertia. With only a few exceptions, ferrite core memories are today's standard throughout the industry. Engineers understand them. Components, or whole systems, using them are available at competitive prices. Replacement parts and field service for them are readily obtained. Virtually all current software programs are written around core memory logic. It appeared safe to say that even if every company in the industry began at once a program to replace ferrite core memories with some other type, it would not be economically possible to get all the core memory computers out of production by 1972.

INTEGRATED CIRCUITS

Although memories built around monolithic integrated circuits and MOS arrays can potentially outperform all other types, it was not expected that they would have a major share in the memory market by 1972. Cost was the principal problem. In 1967, such memory systems carried a typical price of about 50 cents per bit. Industry observers predicted that this figure might be reduced to 10 cents per bit by 1972. A look at Exhibit 7 shows, however, that such a price would not be competitive with core memories even in 1967, except for cycle times faster than 500 nanoseconds. Since, by 1972, lower cost memories of other types were expected in the 100 to 500 nanosecond speed range, it appeared that integrated circuit memories would continue to be restricted to special purpose applications.

What sorts of applications? Two types were found: (1) Even with 1967 technology, integrated circuit memories could easily achieve cycle times of 100 nanoseconds or less. For the relatively small fraction of the computer market which would demand memory speed as high as this, the integrated circuit memory would be a strong competitor. (2) The price per bit of the integrated circuit memory remains relatively constant in all sizes; whereas, for all types of magnetic memories, the cost per bit increases sharply as the size becomes smaller. Thus somewhere there should always be a bit capacity below which the integrated circuit memory will become desirable simply because it costs less. TMA's informants predicted that in 1972 this capacity would fall somewhere well below 250,000 bits; the most likely figure was taken as 100,000 bits.

Clearly, however, if integrated circuit memories were to be limited to cycle times under 100 nanoseconds and capacities under 100,000 bits, they could claim only a very small share of the market in 1972.

PLANAR FILM

Planar film in 1967 was an "almost but not quite" technology. Adherents of planar film claimed that potentially it would do anything that plated wire would do and do it better and cheaper. Skeptics pointed out that this had yet to be demonstrated in practice. Nevertheless, the technical promise of planar film was so great that it was expected to have earned an important place in the memory market of 1972 for several reasons.

Planar film memories would offer outstanding performance, with cycle speed as low as 100 nanoseconds, combined with nondestructive readout. The memories would be unusually small, consume little power, and have no inherent limitations in capacity. Because of their inherent mechanical rigidity, and resistance to temperature changes, planar film memories would be favored for military use and for severe industrial requirements. On the negative side techniques would have to be found to limit the effects of nonuniform magnetics and of interference from stray fields, and these precautionary measures would have an adverse effect on cost.

It was not performance but inability to achieve competitive manufacturing costs which had, up through 1967, kept planar films "out of the big time." Over the next five years considerable refinement was expected in the methods for producing the magnetic film itself. By 1972, in fact, the present vacuum deposition processes might well be supplemented by other techniques which would promise a much higher yield of acceptable film planes. Present planar film memories, however, suffer a cost disadvantage in electronics as well as magnetics. The amount of circuitry needed to achieve satisfactory high-speed operation is greater than for a plated wire memory, and much greater than for a conventional core memory. Although the cost of electronics was expected to fall rapidly, it was not yet certain that the overall cost disadvantage of planar film technology could be fully overcome by 1972.

Because of the elaborate facilities needed to evaporate film planes and the painfully long programs involved in their development, the price of admission into the planar film memory business is relatively high. This was the principal factor which was expected to lead most computer makers to prefer the development of memories using plated wire techniques.

PLATED WIRE

No new technology could lay a stronger claim to an improved place in the memory market of 1972 than plated wire. The supporting evidence, while not wholly favorable, was convincing.

Two computer manufacturers had already committed themselves to the use of plated wire in 1967; the plans of at least one more (Honeywell) were in an advanced stage; and one supplier of cores and stacks (Indiana General) was expected to supplement his product line with plated wire.

Plated wire performance blanketed the expected needs of the computer market in 1972. Cycle times of 200 to 600 nanoseconds were practical and economical for plated wire, and this speed range covered about half of the market depicted in Exhibit 12. Plated wire memories could be built in any capacity, were fairly compact, and introduced no heating problems.

The reliability of plated wire memories had been demonstrated through years of experimental testing. Although they did not withstand the "shake, rattle and roll" requirements of military specifications as sturdily as planar films, plated wire memories were generally dependable and reasonably rugged. Performance was currently limited by a tendency to lose signal strength with increasing age, but it appeared probable that new materials and refined construction techniques would have overcome this weakness by 1972.

Plated wire memories seemed to offer significant potential savings in manufacturing costs. The companies which already used them claimed in 1967 that plated wire memories cost less than ferrite core memories when both were operated at cycle times of around 500 nanoseconds. As further development takes place, this cost advantage was almost certain to increase. Honeywell indicated that it would change from core memories to plated wire memories only if it was assured that the use of plated wire would cut its memory cost per bit in half, but the company seemed convinced this is likely, since it was proceeding with detailed plans to build a plant to manufacture complete plated wire memories.

TMA's interviewees seemed to agree that the investment needed to establish a facility for manufacturing plated wire was considerably less than that for any of the other new techniques discussed here. Furthermore, the engineering and development costs to adopt a plated wire system, with respect to both hardware and software, were said to be modest.

The foregoing set of opinions, derived from more than fifty lengthy interviews with engineering leaders of the computer industry, does not lend itself to precise analysis. Instead, like a typical business decision, it requires the weighing and evaluation of a large number of interrelated variables. Nevertheless, using the facts at hand, a limited amount of quantitative analysis is possible. The interviews did seem to make it clear that the various memory technologies have certain practical limits of application so far as memory size and memory speed are concerned. In Exhibit 14 these limitations were approximated. Core memories are seen to be dominant for most applications with cycle times over 500 nanoseconds. Integrated circuits dominate applications under 150 nanoseconds, provided that the capacity of the memory is less than 100,000 bits. Thin films dominate the region between cycle times of 150 and

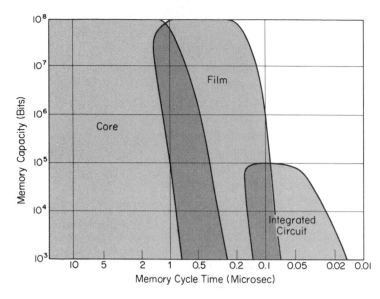

Exhibit 14. Predicted Performance Chart for Memory Systems in 1972.

Note: Shaded areas indicate operating regions in which the three memory types are expected to predominate because of performance capabilities and/or economic advantages.

500 nanoseconds. No distinction is made between planar film and plated wire, there being little to choose between them on a performance basis.

The plane on which Exhibit 14 is drawn, however, is not a plane of uniform economic opportunity. We had already forecasted in Exhibit 12 that memories of certain sizes and speeds would be much in demand in 1972, whereas other specifications would be little used. Consequently, we could now combine the predictions made in Exhibits 12 and 14 to arrive at a simulated three-dimensional picture of comparative opportunities in memory technology for 1972. In Exhibit 15 the vertical scale measures frequency of usage, while the total volume under the curved surface represents the total value of the memory market.

Having gained some technical insight into the memory market of 1972, we were now in a position to predict the economic structure of that market. This was done in Exhibit 16 using the same format which was used to describe the market of 1967 in Exhibit 3. The client's interest focused on the predicted value of the free market for thin film components: for example, plain plated wire, plated wire arrays, and planar film arrays. It can be seen that the total value of these market segments, which was about $700,000 in 1967, was expected to grow to $120 million in 1972. Such a startling increase is, of course, the result of the concurrence of several favorable factors: the rapid growth of the computer industry itself, the fact that memories will make up an ever increasing portion of the value of these computers, and the fact that thin film technology will, as we have seen, have come "out of nowhere" to occupy a

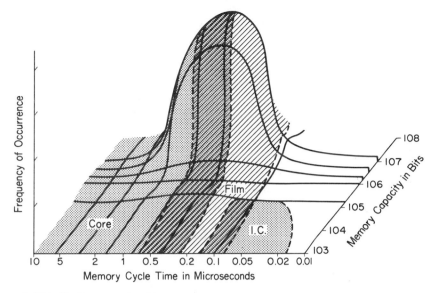

Exhibit 15. Forecasts Combined in Isometric Presentation: Frequency of Occurrence of Memory Specifications and Areas of Application for Selected Technologies for the Year 1972.

significant place in the memory market. Certainly, then, it would appear that our client had been putting his money on the right horse after all.

At this point, it may be appropriate to pause and ask, in retrospect, how accurate we feel these forecasts were. Certainly they were based on painstaking research; it is doubtful whether the doubling of our budget would have provided any significant amount of additional useful information. At the same time, it is equally obvious that we were obliged to make many assumptions, approximations, and oversimplifications in order to arrive at any conclusions at all. Also we observed, as noted above, that some large corporations have invested very substantial sums of money based on conclusions different from our own. Nevertheless, we believe that the relationships forecast for the client are, in their general outlines, correct and that the dollar values indicated on Exhibit 16 will probably prove to be accurate within a factor of approximately two.

If this proves to be so, we believe that our forecast will have served its purpose. What our client wanted to know, after all, was whether or not to continue his development program and, if he continued it, toward what end. It would take a very substantial alteration of the specific forecasted numbers to distort the general picture which the study presented.

This brings us to the point of the study: what action should the client take? It turned out that our recommendation to the client was not what might be expected. Although our picture of the future market for thin film memory components was certainly very attractive, our recommendation of thin film components as a new product venture for the client was very heavily qualified.

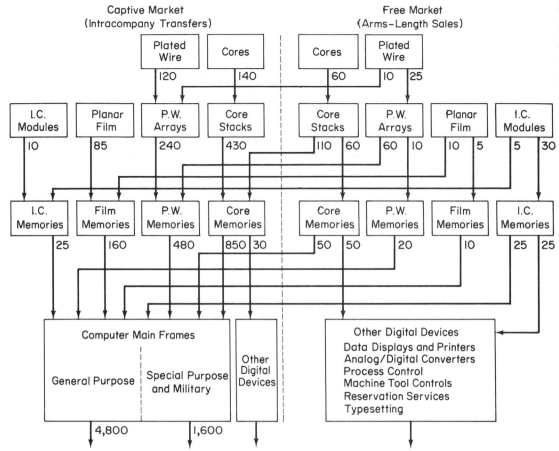

Exhibit 16. Input-Output Analysis of the Computer Memory Market Forecasted 1972 Sales and Transfers in Millions of Dollars.

Our reservations were based largely on factors not directly involved in the market forecast. It seemed to us that the client was technically far behind the competition, and that at his going rate of development he was falling further behind all the time. Only a major acquisition appeared likely to make him competitive from the technical standpoint and to give him the kind of marketing organization necessary to handle products of this complexity. In essence, we felt that the computer memory industry was like a high-stakes poker game. To be sure, the pots were expected to be large. On the other hand, the chips were expected to be very expensive, the other players had all the experience; and we were by no means sure that the client could, or should, ante up enough money to see the game through at that kind of a table.

Fortunately, the client seems to have decided to shift his attention to another aspect of memory technology to which his entire organization appears to be much better suited. Although this new approach is currently under study, it appears to us that his competitive prospects here may be more favorable.

A Multi-industry Experience in Forecasting Optimum Technological Approach

FREDERICK P. GLAZIER

What is the most desirable technology choice for a future need? The author shows how he adapted the Delphi technique to develop a multi-industry, multi-firm approach to a future technical requirement.

Basically, this example of normative forecasting assumes a future condition and then enlists interested parties to appraise the "best" technological solution. It is one of the few examples of an actual industrial use of goal-oriented analysis to choose between technological approaches. Obviously, goal-oriented forecasting is, by definition, planning. The semantics are not as important as appreciation of how this technique can be helpful in identifying the most desirable future technology.

Technological change does not just happen. It is caused in the process of responding to identified needs. The direction in which the response mechanism is developed can be influenced by the way in which the forecasts are structured and the needs identified. Thus, the environment in which the change evolves has a large effect upon how, when, and even if the change occurs.

Many have commented upon the fact that technological forecasts are not conceived in the sterile, limited environment of technology alone but that they are further nurtured in a pool of vibrant life-giving fluid supplied by politics, society, and economics. Forecasts that do not recognize the needs or pressures which these factors bring into play do not relate adequately to the future scenario. In particular, the experience which this discussion will relate was brought into being not only because of many technological forecasts, but also because the economic related forecasts (as well as the political and social) showed that the future needs could not be satisfied if technological developments would not reduce the costs of the overall system.

Our experience concerns just one finite development effort in the many efforts which must be successfully completed before a viable Mach 2.7 Supersonic Transport (SST) can be finalized. It concerns the largest single commodity in the life of the SST and touches upon the worldwide capabilities and interests of many different industries and governing bodies. It concerns the availability/price/performance of the hydrocarbon fuel required to fuel the SST.

NEED FOR IMPROVEMENT

I shall not delve too deeply into all the ramifications of the problem. To set the stage, however, it is necessary to review the changed scenario which the SST could bring into being vis-a-vis the conventional nonsupersonic commercial jet. First, the amount of fuel which the SST will require is substantially more per passenger mile than that of the subsonic jet. In fact, its fuel cost as a percent of the aircraft's direct operating cost is projected to be about fifty percent greater than that of the current commercial jet. Second, the SST fuel is exposed to a different physical environment. Its exposure to the higher temperatures caused by the adiabatic compression of the air through which the plane is moving means that the fuel will have a greater tendency to decompose and form deposits harmful to the aircraft. Third, a worldwide system of fuel manufacture, distribution, and service exists for the current jet aircraft. Duplication of these facilities to service the SST with a more oxidatively stable hydrocarbon fuel in parallel to the current product would be expensive and could lead to improper fuel being served to the aircraft through human failure. All such factors led some of the more concerned members of the various industries and governmental agencies involved to reach the following concensus as to the optimal solution:

> To not only design the SST engine and airframe to minimize the increase in the fuel's temperature as it goes from the wing tanks through the fuel system into the combustion chamber, but also to optimize fuel management to reduce maximum temperature and time of exposure. In addition, upgrade the oxidative stability (and other pertinent characteristics) of the currently used commercial jet fuel as little as necessary, so that only one fuel need be distributed to satisfy all commercial jet aircraft.

The story depicted in the following exhibits graphically gives an overview of the problem (Exhibits 1-4).

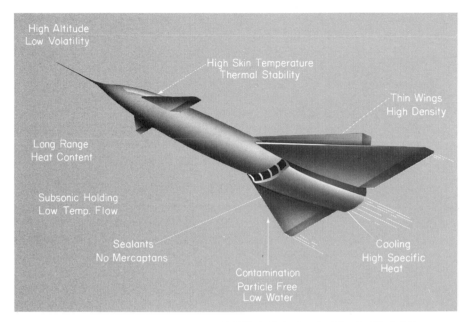

Exhibit 1. Tailoring Fuel for the Mach 3 Transport.

Exhibit 2. Turbojet Stresses Combustion Performance of Fuel.

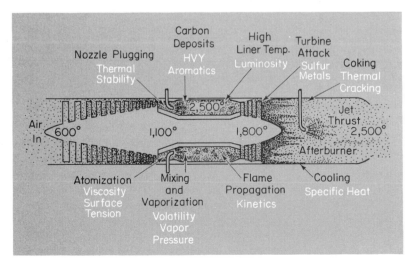

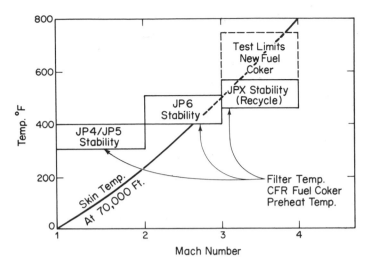

Exhibit 3. Thermal Stability Requirements of Supersonic Flight.

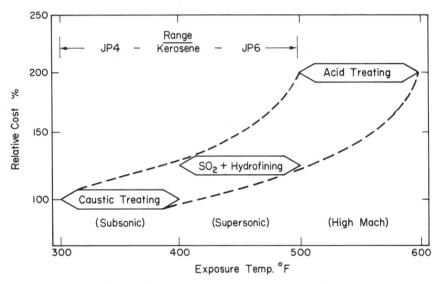

Exhibit 4. Costs for Thermal Stability Depend on Severity Level of Fuel Exposure.

Higher flight speeds lead to higher aircraft and fuel temperatures which lead to the need for more expensive refining and/or reduction in the quantity of fuel suitable for safe use. What is desired is a less critical fuel environment (through design and aircraft fuel management) and more fuel of adequate quality without additional refining or distribution facilities. If this could be attained, then the supply of the necessary quantities of SST fuel can be made at a minimal cost increment (if any) and the competitive economics may be such that the aircraft can be made viable. And if, ideally, the fuel's environment in the SST can be made mild enough so that the presently manufactured commercial aviation jet fuel performs satisfactorily for thousands of aircraft and engine hours, then present fuels can be used and any changes in the fuel's price would reflect factors other than changes in availability, quality, and facilities.

The problem is no one really knows how to define accurately the minimum acceptable performance qualities of jet fuel. Such a definition requires extensive flight tests under controlled conditions, a procedure so expensive as to be impractical. The alternative is to utilize full-scale equipment under simulated operation in the laboratory with which to gain confidence in the design (Exhibit 5). Extensive cycling tests for hundreds or thousands of hours will permit such rational design judgments, but they will not provide the petroleum refiner with a tool which he can use to show that each of his fuel batches meets the minimum acceptable quality requirements of the practical specification. He needs a test procedure which can be run in a few hours (or less) so that his fuel batch can be released in less than a few days' time. It is economically prohibitive to consider storing a refinery's production while a 200- or 300-hour QC (quality control) test is run. Thus, a practical fuel specification defines a fuel in terms of a "believable" short-term QC tool which correlates well with the full-scale laboratory or field equipment. If one considers this concept of believability as a yardstick, the chemical test of oxidative stability would be at the left of the one-inch mark, the short-time laboratory QC test using a simulated fuel system would be near the six-inch mark, the full-scale long-time research simulator near the eighteen-inch mark while the right end of the thirty-six-inch mark would reflect several years satisfactory field experience (Exhibit 6). A believable QC test must be as far to the right as is practical for the specific situation and requirement. The QC test for a one-time Apollo flight item would be further to the right than the QC test for 100,000 barrels of jet fuel.

The most suitable QC test is one that is economically accurate, one that correlates with the full-scale performance requirement, and one which has kept everyone out of the performance problems for which it is designed when the specification limits have been set at the minimum. In the absence of such a defined suitable QC tool, when a field service problem occurs, the following steps can occur:

1. Panic in all directions.
2. Quick design and test development.
3. New fuel specification requirement.

Usually, the result is the development of a test procedure which is not as economically accurate as fits the long-term needs of the industry and a specifi-

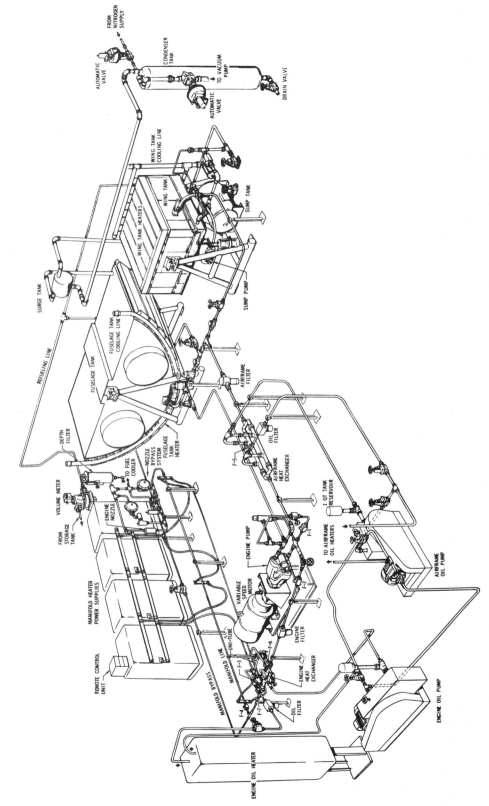

Exhibit 5. Advanced Aircraft Fuel System Simulator.

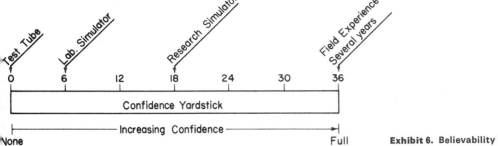

Confidence Yardstick

Increasing Confidence

None   Full

Exhibit 6. Believability in Testing.

cation performance limit which is set artificially high so as to "create" confidence.

This leads the petroleum supplier to provide fuel of greater than required performance, to give away quality (Exhibit 7), with a subsequent decrease in the quantity of fuel available to satisfy the actual performance requirements. This leads to higher costs and a reduced ability to satisfy demand.

This nonoptimum situation can be found in many areas within the society in which we participate. The experience which this paper describes has a

Exhibit 7. Fuels Are Better Than Required.

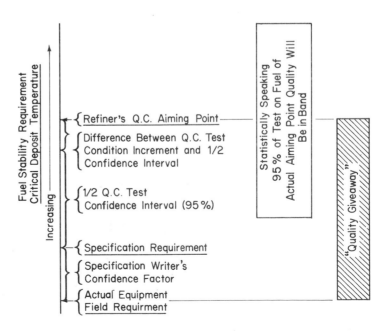

solution which is yet to evolve. One of the paths to solution has been identified by means of technological forecasting methods.

ENVIRONMENT IN WHICH EXPERIENCE OCCURS

The multi-industry cooperative environment comes about as a result of the need for competitive companies within an industry to be able to cooperate technically in establishing realistic standards for the products which will be used by other industries and governmental bodies, all without any taint of collusion. In such a group (society) members of one industry can meet with members of another industry in conjunction with the members of other consuming industries (government, for example) to generate realistic standards which will be of benefit to all factions. Such a group had, in fact, published a Jet Fuel Oxidation Stability Test Procedure, a fuel coking test in use today. In the published procedure, the group (Group A) had defined the fuel coker test confidence limits. These limits (at the ninety-five percent confidence interval) were approximately $\pm60°F$ at the 300° to 350°F temperature level. The members of Group A who felt this might adversely affect fuel availability and cost in the forecast future environment scoured the industry and other sources for other test methods of merit, found none with the desired characteristics and decided that enough interest existed to request that a more useful and more precise means of measuring the thermal/oxidation stability of jet fuel be developed.

Enter a research group (Group C) whose members come from two industries, in general the fuels industry and the equipment manufacturing industries. Group C's primary mission is to explore the interrelationships of the design and performance parameters of petroleum and the equipment used and to develop research techniques which may be of use and interest.

Group C agreed to attempt to develop research techniques which would evaluate the thermal/oxidative stability of aviation turbine fuels and do it with significantly improved precision and hopefully better credibility.

In industry cooperative ventures, the stress on the word cooperative must be emphasized and the word defined more fully. It usually means: get the work done but don't spend out-of-pocket money, and don't try to have one contributor do too much or too little. By necessity the rate of progress is tied to the consensus of the workers who can produce the most for industry's overall benefit within their own company oriented environment. When these super workers or super "producers" become identified, they find they have many needs identified for them. However, in a cooperative and democratic group, every individual must have the opportunity to contribute to the limit of his ability. Thus, a few people who "produce" the work have to justify every step of the way to "everyone" who "cooperates," even if it is only to the extent of approval. Progress becomes slow and the necessity of detailing approval and consensus is specific.

The steps taken to achieve the results in this program were as follows:

1. Identification of optimal industry requirements.
2. Conceptual design and proposal.
3. Subjective evaluation of competitive designs.
   a. Determination of the value of performance parameters.
   b. Assessment of competing designs and procedure.
4. Objective evaluation of competing designs.
5. Final development of selected design.

Our experience indicated that it is quite difficult in a cooperative activity to reach a believable consensus on any specific design objective apparently because not enough contributors are fully knowledgeable of the related data, and because those that propose a design requirement usually have become convinced (through association) of the persuasiveness of their own suggestion. Thus, those items which are identified as desirable in some form or another are included on the list because of a lack of specific antagonistic data. In this program various desired design objectives which may not have been fully justified but which represented the best industry consensus (compilation) possible were identified.

The "objectives" identified, the next step was to publicize them and request proposals. Two proposals were received, one from Manufacturer R, and the other from Manufacturer L. Manufacturer R had been active in the area of rig manufacture and test development for twenty years and, in fact, had been manufacturing a standard fuel coker for that period of time. Its proposal contained similarities to that previous standard fuel coker. Manufacturer L had, likewise, been active in the thermal/oxidative fuel stability area for twenty years. Its efforts had been to provide different types of rigs and procedures so as to take over the market enjoyed by Manufacturer R. Its proposal showed fewer similarities to that standard fuel coker. Thus, it could be inferred that the stage was set so that each manufacturer was well known to the industries involved and each had its protagonists. It was obvious that some means of accurately defining one proposal's superiority over the other must be devised, a means which would be impartial and as free from bias as possible. Supporting data would have to be provided to defend the decision from any charge of partiality.

The advantages of the Delphi system of forecasting were considered ideal for securing a consensus which would be as free as possible from bias. Moreover, this consensus would be developed in a manner which would "guarantee" that partiality did not permit the winner to be selected to the detriment of the industry. It has been pointed out that operation of the Delphi system must be performed with the following three requirements in mind:

1. The experts must be selected wisely.
2. The proper conditions under which they will perform must be created.
3. Effort must be protracted so that a reasonable consensus can be obtained.

Two distinct steps in detailing the differences between the designs were used. The Delphi technique was used to secure an industry consensus as to how much weight each parameter of worth should have in the subsequent scoring of the relative desirability of each design. The experts were selected as wisely as possible; namely, in the existing cooperative environment, it was felt politically inexpedient to rate each expert according to his degree of expertise. Since no one wanted to cast the first stone, all personnel who had been concerned and active in the area of thermal/oxidative fuel testing for several years were welcomed to the fold and invited to participate as equals by constitutional definition.

A questionnaire identifying forty-one performance parameters common to the three competing designs, including the standard fuel coker, was circulated (Exhibit 8). The questionnaire may have been devised without proper attention to clarity and definition. In retrospect, should such an exercise be undertaken again, it would be recommended that more attention be paid to the proper design of the questionnaire and the definition of terms. Strangely

Exhibit 8. Performance Parameters for Subjective Evaluation of Fuel Cokers.

Parameter of Subjective Value		Relative Value, %	
Major Group	Individual Member	Column A	Column B
A. Design parameters: between group relative value =		Within Group Relative Value	_____
	1. Heater tube surface temp. control, method of	_____	
	2. Temp. control, precision of	_____	
	3. Elimination of pump effects	_____	
	4. Method of flow control	_____	
	5. Improved filter precision (porosity)	_____	
	6. Heater tube design	_____	
	7. Filter positioning and design	_____	
	8. Heater tube temp. profile, flatness of	_____	
	9. Capability of 400 to 750°F temp. op.	_____	
	10. Impression of overall durability	_____	
	TOTAL OF COLUMN A	100%	

Exhibit 8. (Cont.)

Parameter of Subjective Value		Relative Value, %	
Major Group	Individual Member	Column A	Column B

B. Operating parameters: between group relative value =

			_____
	1. Sample size	_____	
	2. Uniformity of deposits formed	_____	
	3. Capability of automation, value of and	_____	
	4. Filter bypass or rupture capability	_____	
	5. Operator foul-up capability	_____	
	6. Operator skills required	_____	
	7. Operator judgment required	_____	
	8. Laboratory services required	_____	
	TOTAL OF COLUMN A	100%	

C. Economic parameters: between group relative value =

			_____
	1. Cost/test of each run in parts	_____	
	2. Economics of new unit cost	_____	
	3. Economics of conversion unit cost	_____	
	4. Overall economics	_____	
	TOTAL OF COLUMN A	100%	

D. Ratability of tube parameters: between group relative value =

			_____
	1. Ease of tube rating	_____	
	2. Other	_____	
	TOTAL OF COLUMN A	100%	

E. Manufacturing parameters: between group relative value =

			_____
	1. Production control of filter	_____	
	2. Production control of heater and tube	_____	
	3. Overall control	_____	
	TOTAL OF COLUMN A		

F. Check-out capability parameters: between group relative value =

			_____
	1. Check-out of temp. meas. device	_____	
	2. Check-out of flow control (oper. check)	_____	
	3. Check-out of gas contamination	_____	
	4. Check-out of tube temp. profile	_____	
	5. Straightness of tube	_____	
	6. Other	_____	
	TOTAL OF COLUMN A	100%	

G. Preparation parameters: between group relative value =

			_____
	1. Possible sample contamination	_____	
	H_2O _____		
	Dirt _____		
	N_2 _____ TOTAL 100%		
	2. Policy of used vs. new parts reqd,	_____	
	New parts _____		
	Tube clean. proc. _____ TOTAL 100%		
	3. Preparation of fuel before going in test section	_____	
	Aeration _____		
	Filtration _____ TOTAL 100%		
	4. Ease & capability of total device clean. proc.	_____	
	5. Ease of assembly & dis. &/or prep. of tests		
		100%	100%

enough, however, the objective of obtaining a subjective determination of the value index which should be assigned to the parameters used in evaluating the competitive proposals vis-a-vis the standard fuel coker was satisfied in spite of this handicap.

The reiterative procedure inherent in the Delphi method provided the understanding and the rationale necessary to cause the value index of the parameter to converge. As one of the participants in the Delphi system exer-

Exhibit 9. Sample Round 2 Questionnaire in Subjective Evaluation of Fuel Cokers.

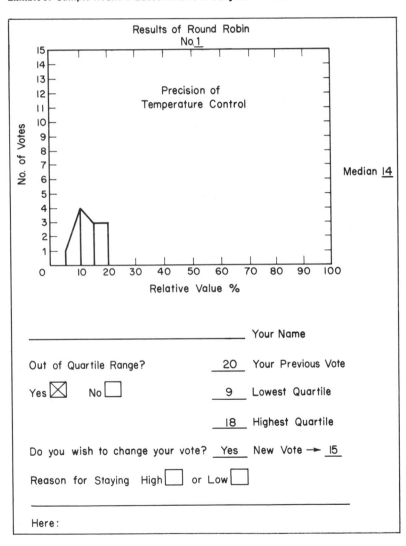

cise said, "Never have I had my arm twisted in as nice and as effective a manner." The iterative procedure was performed by mail. Results of the first questionnaire were plotted by percentage distribution (the questions as to the relative value of the parameters were to be answered on a 0 to 100 percent basis) and were returned to the participants with the *reported* derived interquartile range and the request: "Do you want to change your vote? Are you in or out of the I.Q. range? If out, and you desire to stay where you are, please give your reasons for staying high or low and please respond in two or three weeks" (Exhibit 9). The second iteration results were analyzed similarly and the reasons for staying high or low were typed on the questionnaire and returned for the third iteration with a similar set of questions.

The rules of optimizing Delphi were followed as befitted the situation. The experts were selected wisely, meaning we knew of no others. The proper conditions may not have been established for their noncommittee interaction, but, in fact, these experts had all been exposed to the available information over the past few years and resided in their own environment and near their own reference libraries. Unfortunately, there is no way to guarantee that the home office is the proper environment for such an exercise, and comments were received to the effect that it took too long to fill out each of the questionnaires. Obviously, this shows that the experts were diligent and were attempting to apply more than a minimum amount of subjectivity to the establishment of the relative value of the parameter. It was observed, however, that the loudest voice in committee sessions carried no more than his share of weight in this exercise and that the expert with the greatest reputation was no more prone to be considered correct than any other (Exhibit 10).

No more than the three mentioned iterations were conducted. Three were considered adequate because (1) either a satisfactory consensus for the purpose of obtaining a median value index had been reached or, (2) as mentioned earlier, it was apparent that no further significant convergence of opinion would be forthcoming, and (3) the interest in continuing was waning, and (4) the need for the summation was imminent.

Not only had the parameter's value indices been established, but also the point had been reached in the cooperative venture at which the relative value of the proposed designs should be determined so that the industry would not be saddled with the necessity of testing a design which obviously had no promise. A concept scoring session was thus scheduled before the objective test program would begin so that any proposed design not considered better than the standard fuel coker would not be tested.

It had been pointed out earlier that participation in a program which was forecasting change in a multi-industry cooperative environment required that every step be conducted in an established and well-documented atmosphere. This is readily apparent when it is realized that not only are the members of the participating industries interested in evolving a test and/or procedure

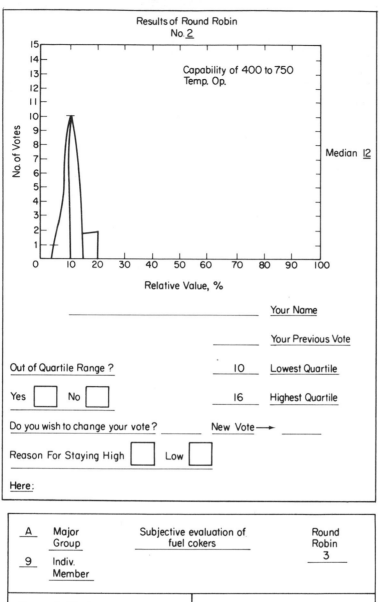

Results of Round Robin
No. 2

Capability of 400 to 750
Temp. Op.

Median 12

No. of Votes

Relative Value, %

_____   Your Name

_____   Your Previous Vote

Out of Quartile Range ?

Yes [ ]   No [ ]

10   Lowest Quartile

16   Highest Quartile

Do you wish to change your vote? _____   New Vote ⟶ _____

Reason For Staying High [ ]   Low [ ]

Here:

A	Major Group	Subjective evaluation of fuel cokers	Round Robin
9	Indiv. Member		3

Reason for Staying: High.	Low.
Ability to cover range important for special military fuels.

May need high temp. range for some future fuels. May be able to trade temperature for time. | Parameter only something desired for military applications, cannot agree with median of 11% of all other design parameters. Should be considered ~5% |

Exhibit 10. Sample Round 3 Questionnaire in Subjective Evaluation of Fuel Cokers.

Notes: 400 to 750°F operating temperature should be attained for new coker design.

which optimally fits industry's needs but also when it is realized that both representatives of Manufacturer R and Manufacturer L are participating in all activities save for the actual pencil contributions. Thus, the concept scoring session must be prepared for in a most meticulous way. As in the case of selecting the Delphi experts, it becomes important to have only participants in the concept scoring sessions who are indeed experts in the field.

The problem of defining expertise here was somewhat easier than might have been possible under other circumstances. In the competitive drive to present its own rig and test procedure in the most favorable light, both Manufacturer L and Manufacturer R had constructed three rigs, developed a test procedure, and made them available for operation. Thus, since it was possible to find many operators (and even some executives) who were familiar with and had operated a standard fuel coker, it was stipulated that only those "experts" who were familiar with (and had operated) a standard fuel coker and who had also operated each of the candidate test devices (or had observed an operator in a full test operation) would be considered qualified to participate in the concept scoring session. Again, it was considered politically inexpedient to attempt to classify the experts by their actual or even apparent degree of expertise.

An open concept scoring session was held. All of the qualified experts were gathered in a room; the performance parameter was identified; its value index as had been developed in the Delphi exercise was read and an interpretation of the meaning of the performance parameter given by the session moderator. In addition, a discussion of the desirable features of the proposed designs, as prepared by the manufacturers, was read. Each expert in the concept scoring session was given an opportunity to express his interpretation, ask for other interpretations, and argue out a consensus. When all present (and in this type of session the manufacturers' representatives from either Manufacturer L or Manufacturer R were excluded) expressed satisfaction, the raters were asked to assign a relative value to each rig. Thus the three competing designs (including the standard) were rated on each parameter at the same time and with the same relative scale; in this case, 10=Excellent, 5=Average, and 0=Poor (Exhibits 11-13). In this way, it was felt that each rig could be compared to the other two by rater, parameter, and scale, so that the differences between the various rater's scoring severities could be normalized by calculating the differences between the rig ratings on a percentage basis. Such turned out to be the case.

Immediately after this first concept scoring session had been completed, a two minute check rating by major parameter groups was made in order to check the bias of the qualified expert (Exhibit 14). This was done to see if a quick general subjective rating would generate significantly different results from the detailed step-by-step approach described above. It did not.

The results of the rig scoring session are tabulated below for the total ranking as well as the major parameter grouping. Individual parameter scorings

Exhibit 11. Subjective Evaluation of Fuel Cokers: Rated Value of Test RIG S (Standard).

Parameter of Subjective Value		Column A	Column B	Column C	
Major Group	Individual Member	Value Factor	Merit Value	Total Value	Sub-totals
A. Design parameters:	1. Heater tube surface temp. control, method of	.0475	0	_____	
	2. Temp. control, precision of	.0450	0	_____	
	3. Elimination of pump effects	.0350	0	_____	
	4. Method of flow control	.0150	3	.0450	
	5. Improved filter precision (porosity)	.0125	5	.0625	
	6. Heater tube design	.0225	2	.0450	
	7. Filter positioning & design	.0075	2	.0150	
	8. Heater tube temp. profile, flatness of	.0250	0	_____	
	9. Capability of 400–750°F temp. op.	.0275	0	_____	
	10. Impression of overall dura.	.0125	0	_____	.1675
B. Operating parameters:	1. Sample size	.0360	5	.1800	
	2. Uniformity of deposits formed	.0432	0	_____	
	3. Cap. of automation, value of &	.0108	3	.0324	
	4. Filter bypass or rupture capa.	.0090	0	_____	
	5. Operator foul-up capability	.0234	5	.1170	
	6. Operator skills required	.0234	5	.1170	
	7. Operator judgment required	.0252	5	.1260	
	8. Laboratory services required	.0090	5	.0450	.6174
C. Economic parameters:	1. Cost/test of each run in parts	.0198	5	.0990	
	2. Economics of new unit cost	.0102	5	.0510	
	3. Economics of conversion unit cost	.0090	10	.0900	
	4. Overall economics	.0210	5	.1050	.3450
D. Ratability of tube parameters:	1. Ease of tube rating	.1520	5	.7600	
	2. Other	.0380	5	.1900	.9500
E. Manufacturing parameters:	1. Production control of filter	.0182	5	.0910	
	2. Production control of heater and tube	.0287	5	.1435	
	3. Overall control	.0231	5	.1150	.3495
F. Check-out capability parameters:	1. Check-out of temp. meas. device	.0510	0	_____	
	2. Check-out of flow control (oper. check)	.0240	5	.1200	

Exhibit 11. (Cont.)

Parameter of Subjective Value		Column A	Column B	Column C	
Major Group	Individual Member	Value Factor	Merit Value	Total Value	Sub-totals
	3. Check-out of gas contamination	.0135	0		
	4. Check-out of tube temp. profile	.0420	0		
	5. Straightness of tube	.0195	5	.0975	
	6. Other	.0000	0		.2175
G. Preparation parameters:	1. Possible sample contamination	.0240	5	.1200	
	2. Policy of used vs. new parts required	.0110	5	.0550	
	3. Preparation of fuel before going in test section	.0180	5	.0900	
	4. Ease & capability of total device cleaning procedure	.0230	5	.1150	
	5. Ease of assembly & disass. and/or preparation of tests	.0240	5	.1200	.5000
	TOTAL	1.000		3.1469	3.1469

Exhibit 12. Subjective Evaluation of Fuel Cokers: Rated Value of Test RIG R (Manufacturer R)

Parameter of Subjective Value		Column A	Column B	Column C	
Major Group	Individual Member	Value Factor	Merit Value	Total Value	Sub-totals
A. Design parameters:	1. Heater tube surface temp. control, method of	.0475	5	.2375	
	2. Temp. control, precision of	.0450	8	.3600	
	3. Elimination of pump effects	.0350	10	.3500	
	4. Method of flow control	.0150	3	.0450	
	5. Improved filter precision (porosity)	.0125	5	.0625	
	6. Heater tube design	.0225	5	.1125	
	7. Filter positioning & design	.0075	5	.0375	
	8. Heater tube temp. profile, flatness of	.0250	5	.1250	
	9. Capability of 400–750°F temp. op.	.0275	5	.1375	
	10. Impression of overall dura.	.0125	5	.0625	1.5300

Exhibit 12. (Cont.)

Parameter of Subjective Value		Column A	Column B	Column C	
Major Group	Individual Member	Value Factor	Merit Value	Total Value	totals
B. Operating parameters:	1. Sample size	.0360	5	.1800	
	2. Uniformity of deposits formed	.0432	5	.2160	
	3. Cap. of automation, value of &	.0108	3	.0324	
	4. Filter bypass or rupture capa.	.0090	5	.0450	
	5. Operator foul-up capability	.0234	4	.0936	
	6. Operator skills required	.0234	5	.1170	
	7. Operator judgment required	.0252	4	.1008	
	8. Laboratory services required	.0090	5	.0450	.8298
C. Economic parameters:	1. Cost/test of each run in parts	.0198	5	.0990	
	2. Economics of new unit cost	.0102	5	.0510	
	3. Economics of conversion unit cost	.0090	5	.0450	
	4. Overall economics	.0210	5	.1050	.3000
D. Ratability of tube parameters:	1. Ease of tube rating	.1520	5	.7600	
	2. Other	.0380	5	.1900	.9500
E. Manufacturing parameters:	1. Production control of filter	.0182	5	.0910	
	2. Production control of heater and tube	.0287	4	.1148	
	3. Overall control	.0231	5	.1155	.3213
F. Check-out capability parameters:	1. Check-out of temp. meas. device	.0510	5	.2550	
	2. Check-out of flow control (oper. check)	.0240	8	.1920	
	3. Check-out of gas contamination	.0135	5	.0675	
	4. Check-out of tube temp. profile	.0420	5	.2100	
	5. Straightness of tube	.0195	5	.0975	
	6. Other	.0000	0		.8220
G. Preparation parameters:	1. Possible sample contamination	.0240	3	.0720	
	2. Policy of used vs. new parts required	.0110	5	.0550	
	3. Preparation of fuel before going in test section	.0180	5	.0900	
	4. Ease & capability of total device cleaning procedure	.0230	0		
	5. Ease of assembly & disass. and/or preparation of tests	.0240	3	.0720	.2890
	TOTAL	1.000	—	5.0421	5.0421

Exhibit 13. Subjective Evaluation of Fuel Cokers: Rated value of Test RIG L (Manufacturer L).

Parameter of Subjective Value		Column A	Column B	Column C	
Major Group	Individual Member	Value Factor	Merit Value	Total Value	Sub-totals
A. Design parameters:	1. Heater tube surface temp. control, method of	.0475	8	.3800	
	2. Temp. control, precision of	.0450	6	.2700	
	3. Elimination of pump effects	.0350	10	.3500	
	4. Method of flow control	.0150	8	.1200	
	5. Improved filter precision (porosity)	.0125	5	.0625	
	6. Heater tube design	.0225	9	.2025	
	7. Filter positioning & design	.0075	5	.0375	
	8. Heater tube temp. profile, flatness of	.0250	8	.2000	
	9. Capability of 400–750°F temp. op.	.0275	5	.1375	
	10. Impression of overall	.0125	8	.1000	1.8600
B. Operating parameters:	1. Sample size	.0360	8	.2880	
	2. Uniformity of deposits formed	.0432	8	.3456	
	3. Cap. of automation, value of &	.0108	8	.0864	
	4. Filter bypass or rupture capa.	.0090	5	.0450	
	5. Operator foul-up capability	.0234	8	.1872	
	6. Operator skills required	.0234	8	.1872	
	7. Operator judgment required	.0252	8	.2016	
	8. Laboratory services required	.0090	5	.0450	1.3860
C. Economic parameters:	1. Cost/test of each run in parts	.0198	5	.0990	
	2. Economics of new unit cost	.0102	5	.0510	
	3. Economics of conversion unit cost	.0090	5	.0450	
	4. Overall economics	.0210	5	.1050	.3000
D. Ratability of tube parameters:	1. Ease of tube rating	.1520	5	.7600	
	2. Other	.0380	7	.2660	1.0260
E. Manufacturing parameters:	1. Production control of filter	.0182	5	.0910	
	2. Production control of heater and tube	.0287	8	.2296	
	3. Overall control	.0231	8	.1848	.5054
F. Check-out capability parameters:	1. Check-out of temp. meas. device	.0510	8	.4080	
	2. Check-out of flow control (oper. check)	.0240	5	.1200	
	3. Check-out of gas contamination	.0135	5	.0675	

Exhibit 13 (Cont.)

Parameter of Subjective Value		Column A	Column B	Column C	
Major Group	Individual Member	Value Factor	Merit Value	Total Value	Sub-totals
	4. Check-out of tube temp. profile	.0420	9	.3780	
	5. Straightness of tube	.0195	5	.0975	
	6. Other	.0000	0		1.0710
G. Preparation parameters:	1. Possible sample contamination	.0240	8	.1920	
	2. Policy of used vs. new parts required	.0110	8	.0880	
	3. Preparation of fuel before going in test section	.0180	5	.0900	
	4. Ease & capability of total device cleaning procedure	.0230	8	.1840	
	5. Ease of assembly & disass. and/or preparation of tests	.0240	8	.1920	.7460
	TOTAL	1.000		6.8944	6.8944

Exhibit 14. Subjective Evaluation of Fuel Cokers: Sample Rating Form to Check Results by Major Parameter Groups.

Date_____

Name_____

Rig_____

Parameter of Subjective Value	Column A	Column B	Column C
	Value Factor	Merit Value	Total Value
Major group			
A. Design parameters:	25	____	____
B. Operating parameters:	18	____	____
C. Economic parameters:	6	____	____
D. Ratability of tube parameters:	19	____	____
E. Manufacturing parameters:	7	____	____
F. Check-out capability parameters:	15	____	____
G. Preparation parameters:	10	____	____
TOTAL	100%	____	____

were tabulated in Table 1 with a summary by major parameter groups as follows:

Subjective Evaluation of Thermal Stability Devices, % Rating.

	Subjective Evaluation of Cokers		
Industry Members—January, 1969	Standard	Man. R	Man. L.
Total	21	34	45
Design parameters	11	41	48
Operating parameters	20	33	47
Economic parameters	30	34	36
Ratability parameters	34	31	35
Manufacturing parameters	26	31	43
Check-out capability parameters	18	30	52
Preparation parameters	22	31	47

A study of these results indicates that both candidate device designs were considered superior to the present fuel coker. In fact, Manufacturer L's design likewise could be considered superior to the design of Manufacturer R in all major areas. Perhaps in certain business environments this would be considered as suitable justification for Manufacturer R to be eliminated from further consideration. However, in some multiindustry cooperative environments, it is considered that a subjective scoring result should not be used when the means for securing objective data have already been arranged for on a cooperative basis. Such was the case and it provided an objective means for comparing actual objective performance versus expert subjective anticipation.

A limited test program to determine rig/test repeatability had been designed by statisticians. The data would be used to determine the superiority of either rig design and would be compared to similar results on the standard fuel coker. Both rig designs would be run at two different laboratories for at least sixty times. Since the two different laboratories were very familiar with the operation of a standard fuel coker, it was considered worthwhile to conduct a concept scoring session with each laboratory's personnel after the testing portion of the program had been completed. It was hoped that this would provide insights into differences between promises, judgments, and actualities. And, as was discussed in obvious jest, it would provide two additional comparative subjective scorings of the two designs, so that a selection decision could be made in case the two rigs/test procedures were not significantly different on a statistical performance basis.

Thus, similar concept scoring sessions were held at the two different laboratories about eight months after the first session. The operators of the rigs (as well as some supervisory personnel) were questioned. The results on a total and major parameter basis are shown below; the individual parameter rating values are shown in Tables 2 and 3.

Table 1. Results by Industry Members.

	Subjective Evaluation of Thermal Stability Devices, % Rating		
	Standard	Man. R	Man. L
Design			
1. Heater tube surface temp.	2	40	58
2. Precision of temperature control	15	46	39
3. Elimination of pump effects	4	44	52
4. Method of flow control	26	31	43
5. Improved filter precision	21	28	51
6. Heater tube design	18	32	50
7. Filter positioning and design	23	38	39
8. Flatness of temperature profile	9	51	40
9. Capability of 400–750°F temp.	4	43	53
10. Impression of overall design	16	35	49
Operating parameters			
1. Sample size	9	34	57
2. Uniformity of deposits formed	16	33	51
3. Value of capability of automation	19	30	51
4. Filter bypass rupture capability	30	34	35
5. Operation foul-up capability	25	30	45
6. Operator skills required	27	34	39
7. Operator judgment required	26	33	41
8. Laboratory services required	28	33	39
Economic parameters			
1. Cost/test of each run in parts	30	29	40
2. Economics of new unit cost	33	33	34
3. Economics of conversion unit cost	28	47	25
4. Overall economics	28	31	40
Ratability of tube ratings			
1. Ease of tube ratings	31	34	36
2. Other	30	32	37
Manufacturing parameters			
1. Production control of filter	25	28	47
2. Production control of heater & tube	26	32	42
3. Overall control	29	33	38
Check-out capability parameters			
1. Check-out of temp. measuring device	19	33	49
2. Check-out of flow control	26	35	39
3. Check-out of gas contamination	10	30	60
4. Check-out of tube temp. profile	0	17	83
5. Straightness of tube	37	32	31
6. Other	—	—	—
Preparation parameters			
1. Possible sample contamination	26	28	46
2. Policy of used vs. new parts reqd.	24	25	51
3. Prep. of fuel before test section	30	32	38
4. Ease of capab. of total device cleaning	19	23	58
5. Ease of assembly and disassembly for test	24	23	53

Table 2. Results for **Lab. W,** Subjective Evaluation of Thermal Stability Devices, % Rating.

	Standard	Man. R	Man. L
Design			
1. Heater tube surface temp.	17	28	55
2. Precision of temp. control	12	47	40
3. Elimination of pump effects	4	36	51
4. Method of flow control	21	25	54
5. Improved filter revision	18	38	45
6. Heater tube design	11	29	60
7. Filter positioning & design	16	37	46
8. Flatness of temperature profile	41	21	37
9. Capability of 400–750°F temp.	7	44	50
10. Impression of overall design	12	31	57
Operating parameters			
1. Sample size	8	35	56
2. Uniformity of deposits formed	18	21	61
3. Value of capability of automation	12	30	58
4. Filter bypass or rupture capab.	34	19	47
5. Operating foul-up capability	11	26	63
6. Operator skills required	32	32	37
7. Operator judgment required	28	20	52
8. Laboratory services required	22	29	49
Economic parameters			
1. Cost/test of each run in parts	25	31	44
2. Economics of new unit cost	33	33	33
3. Economics of conversion unit cost	35	31	34
4. Overall economics	23	33	44
Ratability of tube ratings			
1. Ease of tube ratings	35	29	36
2. Other	21	28	51
Manufacturing parameters			
1. Production control of filter	26	33	40
2. Production control of heater tube	25	32	43
3. Overall control	27	34	39
Check-out capability parameters			
1. Check-out of temp. measuring device	13	44	42
2. Check-out of flow control	31	21	49
3. Check-out of gas contamination	33	27	40
4. Check-out of tube temp. profile	19	30	52
5. Straightness of tube	32	34	34
6. Other	—	—	—
Preparation parameters			
1. Possible sample contamination	17	35	49
2. Policy of used vs. new parts reqd.	18	20	62
3. Prep. of fuel before test section	25	30	45
4. Ease & capability of total device clean.	24	25	51
5. Ease of assembly & disassembly for test	22	22	55

Table 3. Results for Lab. P, Subjective Evaluation of Thermal Stability Devices, % Rating.

	Standard	Man. R	Man. L
Design			
1. Heater tube surface temp.	21	41	38
2. Precision of temp. control	19	38	42
3. Elimination of pump effects	8	50	42
4. Method of flow control	21	21	58
5. Improved filter precision	33	33	34
6. Heater tube design	25	40	35
7. Filter positioning & design	27	35	38
8. Flatness of temp. profile	27	43	30
9. Capability of 400–750°F temp.	9	43	47
10. Impression of overall design	26	32	42
Operating parameters			
1. Sample size	10	30	60
2. Uniformity of deposits formed	27	36	37
3. Value of capability of automation	36	20	44
4. Filter bypass or rupture capab.	33	33	33
5. Operator foul-up capability	31	32	37
6 Operator skills required	30	32	39
7. Operator judgment required	26	33	41
8. Laboratory services required	30	39	31
Economic parameters			
1. Cost/test of each run in parts	34	31	34
2. Economics of new unit cost	42	31	28
3. Economics of conversion unit cost	52	30	17
4. Overall economics	37	35	28
Ratability of tube ratings			
1. Ease of tube ratings	31	37	32
2. Other	32	33	35
Manufacturing parameters			
1. Production control of filter	30	35	35
2. Production control of heater & tube	32	27	42
3. Overall control	30	34	36
Check-out capability parameters			
1. Check-out temp. measuring device	29	35	36
2. Check-out of flow control	26	22	52
3. Check-out of gas contamination	30	35	35
4. Check-out of tube temp. profile	7	29	64
5. Straightness of tube	30	49	21
6. Other	—	—	—
Preparation parameters			
1. Possible sample contamination	16	36	48
2. Policy of used vs. new parts reqd.	29	32	40
3. Prep. of fuel before test section	26	31	43
4. Ease & capab. of total device cleaning	20	40	40
5. Ease of assembly & disassembly for test	19	26	55

Subjective Evaluation of Thermal Stability Devices, % Rating

	Subjective Evaluation of Cokers		
	Standard	Man. R	Man. L
Lab. W—September, 1969			
Total	22	31	47
Design parameters	12	36	52
Operating parameters	20	30	50
Economic parameters	27	31	42
Ratability parameters	32	29	39
Manufacturing parameters	26	33	41
Check-out capability parameters	22	34	44
Preparation parameters	22	27	51
Lab. P—September, 1969			
Total	27	34	39
Design parameters	20	41	40
Operating parameters	26	33	41
Economic parameters	39	32	29
Ratability parameters	31	36	33
Manufacturing parameters	31	31	38
Check-out capability parameters	24	34	42
Preparation parameters	21	33	46

A study of these data shows some differences from the scoring made previously. This should not really surprise anyone, since hindsight is 20/20 and the promises of a vendor usually exceed his performance. It would be unfair, however, to postulate as above without also vaguely inferring that it is possible that prior institutional loyalties and experience affected to some extent the ratings assigned to the competing rigs by the personnel at the two laboratories. For example, suppose one of the laboratories had been deeply involved in the initial design and development of the standard fuel coker. One would assume that the advantages of this design approach would be considered apparent in the evaluation of a design having similarities. On the other hand, if one of the laboratories had been deeply involved in the initial application and use of the standard fuel coker and had had some difficulty in convincing itself that the right rating device was in use, one could assume that the advantages of a design having few similarities to the current fuel coker would be apparent. The procedures of the Delphi system in minimizing possible bias and also in reaching a consensus might be useful in future concept scoring sessions where the results need to be more closely resolved.

In this experience, as shown in the following table, each individual expert who scored the three rigs scored them in similar order, although not all to the same degree.

Subjective Evaluation of Thermal Stability Devices, % Rating

Rater	Relative Value, %			
	Standard	Man. R	Man. L	Total
1	21	31	48	100
2	21	35	44	100
3	22	37	41	100
4	25	30	45	100
5	24	34	42	100
6	24	31	45	100
7	22	35	43	100
8	20	34	46	100
9	21	33	46	100
10	17	36	47	100
11	21	31	48	100
12	17	38	45	100
13	20	30	50	100
14	25	33	42	100
15	23	33	44	100
16	22	35	43	100
17	29	34	37	100
18	22	38	40	100
19	25	37	38	100

The differences in such scoring cannot be ascribed to prejudicial bias, only to the fact that engineering judgment (or economic, legal, political, or social judgment) is based upon the factors experienced during an individual's history, thus the difference in such scoring only reflects the bias built in by such experience.

Group C, a multi-industry organization dedicated to the development of research techniques, thus had at this point identified two test techniques which had been rated subjectively as better than the current technique, and had completed an objective test program on the two competing designs to identify that design which should be developed further into a tool suitable for general industry use. A statistical analysis of the objective data would permit this selection, unless, of course, the analysis would not permit one device to be identified as significantly more precise than the other. The results showed that two research techniques had been identified as showing promise of being more precise than the current technique and worthy of further development and as having equivalent precision.

The hope that the concept scoring results could be used in selecting which of the two devices should be further developed was not realized when it was ruled that the conclusions and recommendations of the overall report could refer only to the objective data obtained during the physical development of the data. Thus, no recommendations based upon the subjective evaluation were made.

It must be recognized that the morphological sequence of iterations involved in the Delphi system and in the operation of a series of concept scoring sessions provided an ideal mechanism for communicating with all parties in an informed manner. The results of such subjective evaluations and the contributing judgments made during the evolution of the two-step consensus procedure described are widely disseminated in a participating and credible atmosphere.

The ultimate subjective judgment is made when money is committed. In this case, approximately ten companies (beside the manufacturers) have agreed cooperatively to participate in the further development of the rig and procedure proposed by Manufacturer L, none in the Manufacturer R program. The final development and proving of Manufacturer L's proposed design and procedure is underway.

CRITIQUE

Several observations emerged from this experience which might be useful to one who faces a similar requirement in any one of many environments.

1. Before starting, always check to see whether or not the rules of the game will permit the results of such a structured approach to be utilized.
2. If not, then devote your energies to changing the rules so an open, step-by-step, well-ordered procedure can be used to make judgment selections.
3. If so, then:
   a. Optimize the choice of the design objectives; time spent here will aid in the following steps and minimize disagreement nuclei from forming at the moment of truth.
   b. Minimize the number of performance parameters which will be subjected to Delphi convergence, but word them in as clear (and succinct) a manner as possible.
   c. Use as broad a spectrum of experts as possible (multidiscipline) in reaching a consensus as to a parameter's value index.
   d. Be sure that a reasonable population of experts (multi-interests) contribute to the scoring of the design-device concept so as to prevent a stacked deck.
   e. Score all concepts at the same session (sequentially by parameter) and treat all results, rater by rater, on a percentage relative value basis so as to minimize differences in rating severities.

Adherence to these principles should permit a creditable and credible forecast to be made. Using a structured and step by step procedure in this adherence increases the possibility that the forecast will be normative. Technological change does not just happen.

ACKNOWLEDGEMENT

Previously published descriptive illustrations have been used in this discussion. The kind permission of W. G. Dukek of Esso Research and Engineering, H. R. Lander of the United States Air Force, and R. M. Schirmer of Phillips Petroleum Company to use them is gratefully acknowledged.

A Three-dimensional Model for Assessing the Impact of Future Technology on Navy Business Management

ROBERT L. HANEY

How will future technology affect logistics activities in the Navy? The unique model developed in this chapter divides managerial activities along dimensions of level, function, and resources. Using Navy logistic operations as the focus of study, the author reports the correlation of this 3-D matrix to an information system based on COSATI (Committee on Scientific and Technical Information of the Federal Council on Science and Technology) field classifications. Having gathered and filed data according to their model, the study group was able to develop future scenarios efficiently and in moderate detail.

> Management now is generally coming to understand that the forces of change will either destroy our system or that we shall shape these forces so that business can continue to be private and effective and productive.[1]

This statement by G. William Miller, President of Textron Inc. in 1969 might well have been made by the leadership of the Supply Corps of the Navy

[1] From an address to Associated Industries of New York State, "The Challenge to Management in a New Era," January, 1969.

in 1966 with only slight paraphrase of "business managers of the Navy" for "management" and the Supply Corps for "business." Restated, the Supply Corps' top management recognized in 1966 that forces of change were threatening to destroy the established methods of supplying the U.S. Navy, and the Supply Corps was in need of guidelines that would permit it to shape the forces of change such that the Corps could continue to be effective and productive. Few people in or out of the Department of Defense had attempted to catalog the forces of change and relate them to a specific organization for the purpose of deriving long-term objectives or guidelines for the organization. Management & Economic Research, Inc., was selected to perform the research partially because several members of its staff had experience in long-range forecasting and were familiar with the Supply Corps.

Officially the study was titled "The Impact of Future Technology on Naval Business Management." Eight months were alloted to achieve the study objective of creating a set of parameters within which the flag officers of the Corps could select fifteen-year goals for the Corps. A steering committee was appointed with representatives from the various specialty groups within the Naval Supply Systems to assist the outside contractor. My role, as a Senior Industrial Economist with Management & Economics Research, Inc., was to develop a methodology for compiling and assessing technological forecasts likely to affect the management of logistics and to assess the impact on the supply system managers. The study was to include a review of the socioeconomic and political environment, analysis of the present skills employed in logistics management, a determination of the skills to be required in the next fifteen years, and the role of the Supply Corps in the total logistics management of the Department of Defense. This paper reports the methodology developed to relate forecasts to goals for this government organization.

Our study team was truly interdisciplinary; it was composed of a professional educator and consultant to educational institutions, an expert in government research policy and organization, three Supply Corps officers who had completed their first year of the Harvard Business School and were on loan for the three summer months to the study, me, and several other consultants in technological forecasting and technical fields.

It became obvious that a framework was essential for coordinating and integrating the output of this Tower of Babel resulting from a multidisciplinary team. Our first problem was to attempt to define logistics management and the relationship of the Supply Corps to the other groups comprising the logistics managers of the Department of Defense. Semantic difficulties are nothing short of traumatic in systems work. We found that we needed more than two dimensions to define logistics, so that we could relate the forecasts to the people and activities loosely described as "those who supply the men who fire the guns." We knew that any framework must permit us to relate forecasted innovations to levels at which men worked, to functions performed, and to the resources used. The best solution turned out to be a three dimensional cube conceived by Admiral Ensey some years ago. We modified it to reflect the levels, functions, and resources of 1966.

Before going into the detail of this matrix, I should point out that parallel effort was taking place in the design of a cataloging system for technological forecasts and the collection of those forecasts themselves. In addition, extensive interviews with key members of the Defense Department were scheduled to give us a view of the defense environment and the benefit of the Defense Department leaders' thinking on the role of innovation in defense management.

Now to the matrix description of logistics (which is most appropriate also to describe a business enterprise). We found that our definition of logistics must be sufficiently broad to permit consideration of alternative roles for the Supply Corps. Exhibit 1 represents this three-sided cube that we found to be the keystone of our effort. On the top face, A, we described the levels at which managers and other workers perform logistics functions. This dimension of the matrix some call the scope of actions taken. Our definition chose three levels ranging from the lowest level termed operations to the highest level termed management of the enterprise. In between we find the management of operations where most of the managers function in any organization. More specifically, these levels are:

1. *Operations* has been defined as the physical or intellectual activity required to perform assigned work under supervision, for example, the unloading of a cargo ship by longshoremen or the stocking of shelves in a warehouse. The efforts of our study team in developing a framework is an example of an intellectual operation.

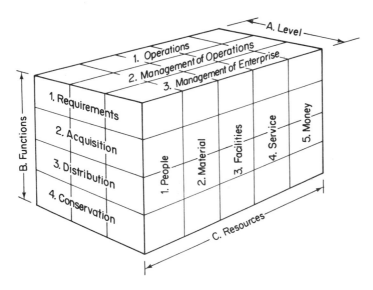

Exhibit 1. Matrix of Logistics Management Functions.

2. *Management of operations* is best described as supervising those who are performing operations in accordance with prescribed policies and procedures. Even so-called senior management finds itself doing this level of work more often than perhaps it should.

3. *Management of the enterprise* is described as the level at which strategic and tactical planning and goal setting takes place. At this top level policy is made, and integration of logistics, tactics, and strategy in the military or naval organization is carried out. In industry this would be the policy making, goal setting, and strategic planning level of action.

Now we move on to the second dimension of the logistics definition labeled the "functions." On this face of the cube, B, are shown four broad functions required to carry out the logistics activities.

1. *Requirements* include those actions required to determine resources and schedules required to support the tactics or strategy of a military organization. This corresponds to the planning function of a business enterprise and may take place at any of the three levels of action described above.

2. *Acquisition* includes those actions that must be taken to obtain the resources and ranges from the early research and development phase to the testing and inspection.
   a. Research and development and design and development of prototypes of hardware or procedures (may be conceptual or may deal with hardware).
   b. Purchase, hire, or lease of materials and services is more frequently described as procurement or purchasing activities.
   c. Fabrication is the actual manufacturing and assembly of hardware or the training of people required to perform certain skills, for example.
   d. Testing and inspection is a check for quality and meeting of specifications of things and the testing and qualification of people.

3. *Distribution* represents those specific actions of transferring objects or people but does not include the transportation or physical movement which is considered a service or resource used in the process. In the military establishment this would include:
   a. Receipt is the various inventory record keeping actions.
   b. Storage is both short- and long-term placement for access.
   c. Packing is the physical actions and the intellectual process such as determination of economic load sizes.
   d. Issue, assign, and allocate are those actions that put the object or service at the disposal of the user.

4. *Conservation*, a functional group, describes those actions designed to preserve and restore objects or people or money to insure efficient use of resources. Such actions as maintenance, repair, modification, salvage, and accounting related to audit and conservation would fall into this group. Personnel development and health care are examples of conservation applied to the people resource.

The final dimension to our description of logistics is that of the "Resources" used in the functions just described and at the levels described earlier. On face C of the cube we show five resources commonly found in the military organization or a business enterprise:

1. *People,* in the case being described, were classified as either military or civilian.
2. *Material* includes all material things but excludes facilities.
3. *Facilities* includes structures, buildings, and various installations of permanent or temporary construction.
4. *Services* are supporting systems such as transportation, subsistence or food service, data processing, utilities, and retailing such as PX stores.
5. *Money* includes all the financial resources of the enterprise or military organization.

Perhaps the most useful aspect of this matrix was its capability to describe any activity that took place in logistics and thereby provided a noncontroversial definition of the organization and activities for which future technology was to be assessed. During the summer of 1969, this matrix was adapted to describe the activities of some of my present company's subsidiaries and it appears to be quite viable. In a multiservice company like Transamerica Corporation, our subsidiaries range from the extremes of United Artists and Liberty Records to more conservative industries like Occidental Insurance and Transamerica Financial Corporation. Naturally none of them like to be considered as "like" any others. The matrix appears to be a solution to this semantics problem.

Concurrent with the above matrix development, we were compiling forecasts of technological change that had some chance of being relevant to logistics systems and management. Frankly, there were few data bits thrown out because almost any forecast had some potentially significant impacts on logistics. We set up a cross-indexing system with a unique vector process for selecting data for retrieval. A consultant on such problems assisted us in designing this manual system that met four criteria as follows:

1. Accept any form of data such as reports, media clippings, periodicals, magnetic tape, interviews and so on.
2. Permit abstracting where possible (and it often was desirable due to limited availability of documents or extreme length).
3. Permit any number of references to a specific document or data element stored in one place.
4. Provide selectivity in retrieval of forecasting data.

Each bit of information received was assigned a single number starting with one and numbered consecutively. Abstracts made by the study team were numbered this way also. If the document was a periodical, we added a suffix, P, to indicate a separate file drawer for these bulky documents. Other odd size information units could be given a suffix like T for magnetic tape and C for computer printouts of literature searches to indicate separate storage locations as needed. After assigning this simple file location number, the information was given one or more codes by the technology analysts and research assistants to identify the field and subfield of technology and the logistics levels, functions, and resources that may be relevant.

After examining many proposed coding or classification schemes designed to index technological data, we decided to stay with the COSATI list (promulgated by the Committee on Scientific and Technical Information of the Federal Council on Science and Technology) now in common use by the Federal government. COSATI includes twenty-two broad fields ranging from Aeronautics to Space Technology. Under these twenty-two broad groups or fields, there are 188 subfields or groups, for example: Field 10—Energy Conversion (nonpropulsive) had three subfields of conversion techniques, power sources, and energy storage; Field 06—Biological and Medical Sciences had twenty-one subfields. Although it may have some controversial groupings, we found it eminently useful. Its best feature was its capability to be expanded into subfields as a second level of detail and down to a third level if needed. For a broad survey such as this study, however, we never had to go below the subfields provided by the COSATI list. (This list was being revised in 1968 and varies slightly from that used in this study in 1966–67.)

Each forecast data bit was coded with the major field and the subfield. For example, a report forecasting the introduction of the C–5A aircraft and its cargo carrying capabilities would have been coded 01–03 for Aeronautics aircraft and also under 13–04 for Mechnical, Industrial, Civil, and Marine Engineering with the subfield −04 of containers and packaging and subfield −06 for ground transportation equipment and possibly for subfield −08 for industrial processes such as new concepts for loading and unloading such new transportation vehicles.

In addition to the coding for field of science and technology, a second code was added for levels, functions, or resources likely to be affected. For example, projections of the development of electronic money would have impact on the two levels, 1, operations, and 2, management of operations, of face A of the cube matrix. It would also have impact on all functions 1 thru 4 of face B and on the money resource of face C, 5. This report or forecast would have been coded 115, 125, 135, 145, 215, 225, 235, and 245 in addition to the COSATI numbers.

To permit easy, manual indexing of the documents and other data catalogued, a 3 × 5 card was set up for each COSATI major field and each subfield under the major field if needed. An index card was also set up for each of the sixty cubes in our matrix using numerical coding such as 111 for the operation of determining requirements for people. The format of the cards was designed to allow random entry of document numbers by last digit order. The format used was similar to this:

COSATI Field	05							Behavioral & Social Sciences	
Sub field	01							Administration and Management	
			Document Serial Number (By Last Digit)						

01	02	03	04	05	06	07	08	09	00
11	22		04	35	76	467	228	109	300
301	452		464	75					
321			714						

With these file cards established for each field or subfield of the COSATI list and for each combination of level, function, and resource, we had created a vector system for retrieving forecasted bits of data. To assess impacts of electronics on the operational level for all functions using the money resource, for example, only five cards would be pulled to search for documents: the card for the electronic field (or one subfield) and the index cards for 115, 125, 135, and 145 matrix areas. By scanning the five index cards, the researcher would examine all five (by column) for the same document serial number. Sometimes several dozen documents would turn up as relevant to the field of technology and function-resource being examined. If no "hit" appeared, that is no document serial numbers common to both the technology field and cube description, then the researcher could (1) abandon the search or (2) examine all documents on all cards related to the search terms.

This vector scheme using document serial numbers permitted us to move through the analysis of impact areas much more rapidly and also eliminated the need to consider all documents in the file on each step of the analysis. Computerization was avoided in 1966 as software and hardware costs were not as reasonable as in the 1970's. Such a scheme is only as effective as the judgment, knowledge, and imagination of the researcher coding and retrieving the documents and cross-reference index cards. An author file was set up also but was soon abandoned due to rare use.

About 500 documents and abstracts were collected over four months and entered into our technology files. Sources varied from computerized literature searches made by the Defense Documentation Center to the very current articles in the *Wall Street Journal* on such devices as the Gunn device for communications or laser applications. In retrospect we found that books were the least useful because the delay in getting technical information published often resulted in it becoming stale by the time it was published in book form. Periodicals presented a useful but time consuming source and the most current, if not detailed source, was the *Wall Street Journal*. Tapes of the *New York Times* and the *Journal* would be a prime source for current announcements. Of course, we used the better known reports such as the Army Technological Forecasts, the "Technology For Tomorrow" prepared by the Air Force, the various Rand reports, "The World of 1975" by Stanford Research Institute, Kahn's *The Year 2000,* and so on.

At this point, the study team needed a systematic way to relate these bits and pieces of forecasted changes to the logistics activities described in the matrix. The two-dimensional lattice shown in Exhibit 2 provided this system for relating the technology to the organization and activities. The twenty-two major fields of science and technology are listed down the left margin of the lattice. Although each of the subfields under each major field was examined, only the aggregated effect on the activities was shown for the major field on this summary table. Across the top are displayed the three levels at which the logistics managers function, and under these three levels are shown the twenty combinations of functions and resources that may be affected by technological change. For example, the small cube within the large matrix numbered 111 would be the operation of determining requirements for people and those fields of technology that would offer potential impact on this function are shown in the column marked 111. Another example would be the column 335 representing the management of the enterprise (strategy and tactics and logistics) of the distribution function relating to the monetary resource. We tested this matrix and lattice combination for several days with every conceivable logistics area and found it to be quite suitable for this broad review of technological impact.

Now we come to the exciting part of the research, the examination of the forecasts and determination of the impact of the forecasted innovation on the sixty combination cubes of the matrix. To ensure completeness and to permit a degree of dynamics in this step, two teams went through the twenty-two major fields with their 188 total subfields for each column or sixty combinations of levels, functions, and resources. Simple mathematics indicates that such a process could mean 11,280 iterations and naturally this was not done. Fortunately for the researchers, if not for the report, we had no useful forecasts in a number of the 188 subfields of science and technology. Conversely, we were swamped with many forecasts of electronic information technology (often conflicting). Obviously the judgment and intuition of the research teams had to be relied upon at this point. Each of us had a broad knowledge of the sixty elements of the logistics matrix and several of us had special knowledge of some elements. None of us professed to be experts in any one field of technology but all of us had a general knowledge and varying degrees of specialized knowledge in several major fields of technology. The bias of any one scientific discipline was avoided consequently.

After the two teams went through the lattice both examining the fields one at a time across the sixty areas and conversely taking each of the sixty areas independently and running down the twenty-two fields, the composite view was recorded. Degrees of impact were marked on the lattice and file items were tagged for use in preparing scenarios described later. A number of weighting

The table below uses the following symbol key:
● Major Impact · ⊗ Minor Impact · ○ Probable Impact

Operations columns: 111, 112, 113, 114, 115, 121, 122, 123, 124, 125, 131, 132, 133, 134, 135, 141, 142, 143, 144, 145

Field	111	112	113	114	115	121	122	123	124	125	131	132	133	134	135	141	142	143	144	145
Aeronautics							●	●				⊗		●			⊗	⊗	⊗	
Agriculture							⊗	⊗												
Astronomy and astrophysics																				
Atmospheric sciences						○	⊗	⊗				⊗	⊗				⊗			
Behavior and social science	⊗	⊗	⊗	⊗	⊗	●	●	●	●	●	●	●	●	●	●	●	⊗	⊗	⊗	⊗
Biology and medical science							●	⊗				⊗	●			●	⊗			
Chemistry							●	⊗												
Earth sciences and oceanography						○	●										○			
Electronics	●	⊗	⊗	○		⊗	⊗	●	●	●	⊗	●	●	⊗	●	●				
Energy conversion							●	⊗												
Materials							●													
Mathematical science	⊗	⊗			⊗	⊗	●			⊗		●				○			○	
Mechanical engineering, etc.							●	●	⊗			●					●			
Methods and equipment	⊗	⊗	⊗	⊗	⊗	●	●	●	●	●	⊗	●	●	⊗	●	●	⊗	●	⊗	⊗
Military science	⊗	⊗	⊗	⊗	⊗	⊗	●	⊗	●	⊗	●	●	●	⊗	●	●				
Missile technology			⊗		⊗	⊗	●	●	⊗			●	⊗				⊗			
Nav. comm. detection	⊗	●	○	●	⊗		●	⊗	⊗	⊗		●	⊗	⊗	●	●				
Nuclear science							○	○				○								
Ordnance												⊗								
Physics							●					⊗								
Propulsion fuels							●	●			●	○					⊗			
Space technology												⊗								

Management of Operations columns: 211, 212, 213, 214, 215, 221, 222, 223, 224, 225, 231, 232, 233, 234, 235, 241, 242, 243, 244, 245

Field	211	212	213	214	215	221	222	223	224	225	231	232	233	234	235	241	242	243	244	245
Behavior and social science	●	●	●	●	●	●	●	●	●	●	●	●	●	●	●	●	●	●	●	●
Earth sciences and oceanography	●	●	●	●	●	●	●	●	●	●	●	●	●	●	●	●	●	●	●	●
Mathematical science	⊗	●	⊗	●	⊗	●	○	○	⊗	○	⊗	○	○	○	○	○	●			
Methods and equipment	⊗	●	⊗	●	○		●	⊗	●		●	⊗	●		●	⊗	●	⊗	●	●
Military science						●	●													
Nav. comm. detection						○	●	⊗	⊗		⊗	●						⊗	⊗	

Management of Enterprise columns: 311, 312, 313, 314, 315, 321, 322, 323, 324, 325, 331, 332, 333, 334, 335, 341, 342, 343, 344, 345

Field	311	312	313	314	315	321	322	323	324	325	331	332	333	334	335	341	342	343	344	345
Behavior and social science	●	●	●	●	●	●	●	●	●	●	●	●	●	●	●	●	●	●	●	●
Earth sciences and oceanography	●	●	●	●	●	●	●	●	●	●	●	●	●	●	●	●	●	●	●	●
Mathematical science	●	⊗		⊗	●	○	⊗	○	●	●	⊗	●	○	○		○				⊗
Methods and equipment	●	●	⊗	●	●	●	⊗	●	●	⊗	●	●	●	●	●	●	●	●	●	●
Nav. comm. detection	⊗	⊗	⊗			○			○	⊗		⊗		⊗	⊗	⊗	⊗

● Major Impact
⊗ Minor Impact
○ Probable Impact

Exhibit 2. Lattice of Technology Impact on Logistics.

schemes were examined to record the degree of impact on each of the logistic areas and a very simple one was selected. We wanted an indicator that could be understood by a layman with no special knowledge of the fields of technology. The purpose of the entire study was to give a steering committee, composed of civilian and military personnel with various backgrounds, a set of guidelines showing where the technology would have impact. If this committee had to be educated in all fields in order to understand or have confidence in the degree of impact we found likely, we would have failed to meet our study objectives.

The three degrees of impact used were "major," "minor," and "probable." Our definition of major was determined by a consensus that at least one subfield of the broad field would create major changes in the method and resources as well as skills required to perform the activity so marked. The term minor impact was defined as one or more of the subfields or the broad or major field affecting the marked activity in such a way that some changes might have to be made in the methods, resources, or skills required. Finally, the term probable impact was used to indicate that there was a good chance that the activity would be affected but the study team could not identify specific functions or methods or skills. From the pattern of impacts that emerged (Exhibit 2), it is clear that the forecasts we had access to in 1966 indicated that the operations level would be affected by change in more functions and in more resource areas than the two management levels. Heaviest impact was shown in the acquisition, distribution, and conservation functional areas rather than in the requirements area. Viewed from a different perspective, six fields of science and technology would have major or minor impact on almost all logistics activities. These six are:

1. Behavioral and social sciences.
2. Electronics.
3. Mathematical sciences.
4. Methods and equipment.
5. Military sciences.
6. Communications.

Numerous other analyses can be drawn from this lattice. Detailed forecasts by field were fully compiled in the second volume of the report. Using the lattice as an index, one could go to the forecast volume and read in detail the innovations related to logistics activities for the major field and its subfields. Only forecasts that were selected by the study teams as relevant to the logistics subject and within the fifteen- to thirty-year time frame were included in the hundred page volume. In preparing this volume, I attempted to relate the forecast to specific matrix terms. (It became a guide to long-range investment opportunities as well as a guide for the leaders of the Naval Supply System and the Supply Corps.)

From this volume of selected forecasts (and using the lattice for a road-map), two members of the original study team sifted and synthesized scenarios of how several basic functions managed by the Supply Corps would be performed in the 1980's based upon the selected forecasts. It was here that the most interesting cross fertilization from one field to another became possible. Two examples of these scenarios are summarized here to show how this emerged.

SCENARIO OF FOOD SERVICE (SUBSISTENCE)— 1980

Environmental Trends

The trend toward standardization of methods and procedures in the Department of Defense would result in a standard ration (both cost and quality) for all services but with individual service menus possible. Procurement of food would be centralized and standard facilities and equipment would be required. In the socioeconomic and political spheres, the world food problem would have impact on the Navy's food service just as the changing eating habits of Americans would. Our higher standard of living, demographic changes, worldwide travel, and increased leisure are resulting in a shift in eating habits. Cost of producing and distributing food within the naval service would rise despite the advances in preparation, packaging, and transportation methods. (When this was written in 1966, we should have been forewarned about food costs in the 1970's!) Increased awareness of nutrition in the 1970's would result in noticeable changes in menus.

Technological Trends

New additives will permit freeze-drying and other long shelf life processes to be applied to food with little loss in taste quality. New sources such as spun protein or soybean derivatives will replace, to a limited extent, very high cost foods such as beef. Beef, for example, will be "factory" grown and processed in a production line mode. Even so, the demand for beef will increase much faster than supply resulting in the need to process the bone, hair, and skin together with the meat that will be constituted in new "cuts" to keep the price down during the next thirty years. Processing of most food stuffs will be revolutionized with freeze-dried and irradiated foods replacing many fresh and frozen items. Nature's' perfect package, the egg, has been altered by cooking the white and yolk in a cylinder and cylindrical core shapes, respectively, reassembling the egg in this cylinder shape, and processing it so that it will have a shelf life of two to three weeks if wrapped in an airtight film. This permits all perfect slices for serving. (It is interesting to note here that a number of processes we identified in 1966 have now become household items such as freeze-dried coffee and fruits in breakfast cereals. Beef is being slaughtered and cut into package size at factories in Iowa now, all within fifty miles of where it was raised.)

Mechanical and chemical advances in the preparation of food will permit full automation of food service on a naval ship or shore station in the 1980's by use of improved microwave cooking units and pouched vegetables fed into integrated storage and preparation units. Physical distribution of food will change significantly as a result. Computerized inventory planning and food ordering will replace the present methods. Facilities will be smaller and longer shelf life food will permit ships to be less dependent on replenishment of stores at sea or at shore stations. Coupled with nuclear power, this lengthened replenishment cycle extends cruising ranges and schedules.

The evening watch will not have to be "dogged" to permit all hands to eat dinner in the customary 1600 to 2000 hours. In fact, chow time will be randomly selected by crews at their pleasure reflecting the societal shift of our younger people to eating lighter meals more frequently during the day. Food services in universities have already been faced with this shift in demand. KP duty will disappear due to the preprocessing of raw foods. One forecast stated that the lowly potato would be too costly to distribute in its raw form within a few years. The weight and space saved by dehydrating and shipping the potato in that form are economically unbeatable. Concurrently, the taste of the reconstituted potato has improved immensely in the past few years. These new food processing techniques coupled with packaging breakthroughs are expected to reduce shipping weights from twenty-five to fifty percent.

Finally, the entire concept of food service on ships and shore facilities will have to be redesigned. One of the special roles of the Supply Officer on a ship in the past has been the procurement and management of food service or subsistence. The advances outlined in the study will call for less of his time expended in procurement, inventory, and record keeping and supervision of the food service itself. The crew of a ship or shore facility will have a wider choice of food available at all hours through automated galleys.

The retailing of food at PX's will be affected also; supermarket items stocked will increase from about 6,000 items today to about 10,000 or 12,000 within the next several decades. One food technologist forecast that the supermarket may disappear and be replaced by prepackaged modules selected from a wide choice by the housewife and delivered by truck to her home. There, the module would be plugged into the wall of her kitchen and entire meals could be ordered and served in several minutes from the meal planning control panel.

Another example of the scenario approach to relating forecasts from various fields to a particular function is in inventory management.

SCENARIO OF INVENTORY MANAGEMENT

Environmental Trends

The three major trends in government toward centralization, standardization, and consolidation of functions will be extended further in inventory management. Extension of planning programming budgeting systems (PPBS) for planning and control, systems analysis approach for all three services, and national standards for nomenclature and measurement (metric system) are expected in the 1970's.

Technological Trends

Automation's still sleeping giant in production and warehousing will come alive in the 1970's due to advances at least in several fields: microelectronics, packaging concepts and materials, and transportation. Electronic parts and equipment used by the Navy will be designed with logistic self-support features built into the components thereby reducing by orders of magnitude the number and quantity of spare parts. A tenfold increase in the reliability of electronic components by 1972 was stated in this scenario as was the proliferation of components in factory-assembled form ready to be plugged in. In addition, a fifty percent drop in cost of these components was expected.

Dramatic changes in intermodal transport and speed of response were called out based on the advent of the C–5A aircraft and new concepts of prepositioned logistics stores on LDS (logistics deployment ship) type craft or perhaps floating, anchored barges throughout certain areas of the world. One stop delivery of supplies from manufacturer or supplier to the consuming naval unit will be possible with concept for airdrops to ships at sea. This could mean a drastic reduction in staging points such as naval supply depots. By use of centralized and on-line inventory control using worldwide satellite communications (circuit time will be quite inexpensive compared to 1966 costs), a ship at sea can be supplied within hours after the requirement is known. More likely, sophisticated techniques for predicting mean-time-between-failure and automatic requisitioning will "drive" the naval supply pipeline. Automated factories with capabilities to increase production by orders of magnitude on short notice will be on line to the centralized supply control centers. Packaging concepts will reduce weight by up to fifty percent with reduced loss due to handling and transport. In fact the whole concept of operating and managing the naval supply system will shift from one of individual item-by-item to an integrated continuous flow concept with fine tuning of "valves." Environmental control concepts of stores will shift from warehouses to containers which become temporary warehouses providing self-contained humidity, temperature, tests for readiness, and theft control systems.

Other findings resulted from the study and are too numerous to list here. The impact on the skill requirements for Supply Corps officers occurs in many areas. A blurring of what used to be discrete functions in the logistics world is already obvious and will accelerate. These once independent functions such as transportation, subsistence, and retailing are becoming very dependent on other functions and cannot be examined or changed in a vacuum. The managers of these functions cannot be allowed to optimize their function to the exclusion of the effect on interfaced functions. We suggested that two new "superfunctions" be used in place of the many older ones: distribution management and resource management. (This was still new in concept for much of the Department of Defense in 1966 when the study was done.) The managers of these superfunctions should not be trained in just one or two of the formerly independent functions of procurement, warehousing, or electronic data processing. None of the above statements are intended to imply that the Navy had not been moving in this direction in the 1960s; our point was simply that the rate and depth of changes were accelerating to a degree that retraining of present logistics managers and entirely new educational programs for officers entering the Corps are essential.

Our study of the impact of future technology on naval business management identified specific impacts and their source from twenty-one of the twenty-two fields of science and technology. Eighteen fields will offer major impact and seven fields will affect more than half of the functions at the operational level. Major changes will occur in six major fields that will affect all three levels of logistics; behavioral and social sciences, biological and medical, electronics, military sciences, communications, and mathematical sciences. The old relationship of the men firing the guns being more important than those supplying the guns and ammunition is reversing. Within the supply establishment, some functions such as retailing (PX) and subsistence are candidates for extensive automation with management functions moved up from the Supply Officer of a unit to a more centralized, consolidated location. New opportunities are opening to the Supply Corps in the superfunctions of distribution management and resource management. Both the nature of warfare (both cold and hot) and the rapid developments in communications, information systems, and systems analysis are creating a growing dependence of traditional functions upon one another.

The question posed by technology is not one of specialists versus generalists, but rather it is a case of the need for the logistics manager who has broad-gauged knowledge and a variety of highly developed skills in designing and operating the systems that provide the fast response required in the next fifteen years. These skills in communications, distribution, quantitative management, and systems analysis will permit man and machine to compliment each other.

Man's intuitive judgment when coupled with the computerized information systems of the 1970's will be the heart of the logistics system of the future.

Changes in the circuitry and organization of computers have been occurring at a logarithmic rate, and from 1966 to 1976 it is projected that cost per computer operation will drop by a factor of 200 from 1966 levels. Throughput speeds will increase from an average of 4 million additions per second in 1966 to 900 million additions per second, for example. Weight of a typical computer will drop by a factor of 300 to 400 in the 1966 to 1976 period. Space required by a typical computer will diminish by an order of magnitude of 1,000 between 1966 and 1976, and the power requirements will be reduced by a factor of 500. (Already we are seeing this forecast come true in the computers used on the Apollo flights and in the navigational computers on the Boeing 747.) A 1975 type computer that would offer the computational power of the monsters of 1953 would weigh 1/20 of a pound, take .1 cubic foot of space and use 1/2 watt of power. In our study we quoted a prediction of the fourth major breakthrough in man's ability to communicate that will occur between 1970 and 1985. It will most likely be based upon these phenomenal advances in computers and some breakthrough in mode of communications such as laser power to transmit analog and digital data in digital form.

There are many aspects of this study I cannot go into for various reasons. Much quantitative data was prepared and reported. In addition, the role of the Supply Corps in the overall logistics management of the Department of Defense was examined and recommendations were developed for the Corps leaders to consider.

I hope this summary of the techniques we developed to collect, assimilate and assess technological change may offer some useful ideas in attempting to relate forecasts to planning. I would like to pass on a quotation that sums up my view of where we are today in evolving techniques for assimilating forecasts of technology today:

> There are ranges of mountains so vast—the Atlas, the Sierra, the Himalayan, the Andes—that even when we have negotiated the passes and crossed to the other side, we are not aware that we have, in truth, crossed them. In our time now—in the years immediately ahead—the human species is crossing such a divide. We are beginning to move from a mechanically fragmented world, a thing of bits and pieces, into a process world, where the wave is more important than the drops that make it up. Like "Alice Through The Looking Glass," we have begun to immerse ourselves in an electromagnetic environment—where the White Rabbit is always late (because there is no time) and where the White Queen is ever running faster, just to stay in one place. When we annihilated time and space through the human use of the electromagnetic spectrum, we began to cross The Great Divide.[2]

2 Kaiser Aluminum News, Vol. III, 1966.

Application of
Technological Forecasting
to the Computer Industry

KENNETH E. KNIGHT

One can hardly mention progress in technology without arousing thoughts of the most powerful innovation of our day—the computer. It is interesting, though, that we have very few studies in depth on the evolution of this technology. Knight's chapter, based on his doctoral dissertation at Stanford, is unquestionably the most exhaustive quantitative examination of changes in the key parameters of the computer.[1]

The author develops a complex model that embraces many computer attributes. He then plots performance and cost curves by years. The data can be used to project computational power into the future. The rigorous, detailed information on 214 computers makes this chapter one of the few truly quantitative case histories of advances in a technological capability.

The first twenty plus years of the computer industry have been hectic ones. Great strides have been taken to provide reliable and inexpensive com-

---

[1] Aspects of the research reported in this article have appeared in the following two articles. K. E. Knight, "Changes in Computer Performance" *Datamation,* Vol. 12, No. 9 (September, 1966), pp. 40–54, and K. E. Knight, "Evolving Computer Performance 1963–67" *Datamation* Vol. 14, No. 1 (January, 1968), pp. 31–35.

putation capability. To obtain a clearer picture we will see where we have been and how fast we have moved to get to where we are today.

1. We generate a performance description for 310 general-purpose computer systems. The performance description estimates the overall capabilities of each computer system based upon its hardware features and basic elementary scientific and commercial computation problems.

2. Using the performance descriptions for the computers introduced in any one year, we generate a technology curve for that year. The technology curve describes the theoretical performance that can be purchased for different monthly rental expenditures.

3. The economics of the scale for any one year is the relation between computing power and system cost which is approximately as follows: Computing power $= (C \times \text{system cost})^2$; $C = \text{constant}$.

4. Technological improvement in number of operations per dollar between 1950 and 1962 has been at an average rate of almost 100 percent.

5. The major historical findings, economies of scale and rate of technological change, become the basis for technological forecasts—trend extrapolation.

FUNCTIONAL DESCRIPTION OF GP COMPUTERS

The capability of each system to perform its computing tasks represents the functional description (or evaluation) of that system. For our purposes we will look at only two aspects of computer performance: (1) computing power, indicated by the number of standard operations performed per second (P); (2) cost of the computing equipment, which equals the number of seconds of system operations per dollar of equipment cost (C).

Computing power (P) evaluates the rate at which the system performs information processing, the number of operations performed per second. Two machines solve specific problems with different internal operations because of their individual equipment features. P will therefore describe operations of equivalent problem solving value to provide the desired measure of a computer's performance. We will estimate P from structure. In order to do this, we first must understand which structural factors influence computing capability. Then we determine the manner in which the structural factors interact to develop the functional model. P consists of three main components: (1) the internal calculating speed of the computer's central processor (t_c); (2) the time the central processor is idle and waiting for information input or output $(t_{I/o})$; and (3) the memory capacity of the computer (M). These factors are the important performance measures needed to determine (P). We define t_c as the time (in microseconds) needed to perform 1 million operations, and $t_{I/o}$ as the nonoverlapped input-output time (microseconds) necessary for these one million operations. Therefore, the computer performs $10^{12}/(t_c + t_{I/o})$ operations per second. The computer's memory has a strong influence on P. We found that the memory factor interacts with internal operating time to determine computing power as follows:[2]

$$M \times \frac{10^{12}}{t_c + t_{I/O}} = P$$

The internal speed of the central processor, t_c, is the time taken by the computer to perform its information processing tasks. The speed equals the internal operation times of each computer, multiplied by the frequency with which each operation is used. To determine the internal speed, therefore, it is necessary to measure the frequency with which the various operations are performed in a typical problem. For scientific computation we considered approximately 15 million operations of an IBM 704 and IBM 7090 from a mix of over 100 problems. In the analysis of the operations used in this "problem mix" the instructions were grouped into five categories:

1. Fixed add (and subtract) and compare instructions performed.
2. Floating add (and subtract) instructions required.
3. Multiply instructions.
4. Divide instructions.
5. Other manipulation and logic instructions—this category combines a large number of branch, shift, logic, and load-register instructions.

The relative frequency with which each of the five types were used in the scientific programs we traced is presented in Exhibit 1.

To determine the frequency with which the different operations were used in commercial computation, nine programs were analyzed in detail (two inventory control, three general accounting, one billing, one payroll, and two production planning). All nine problems were run on an IBM 705, representing over one million operations. We analyzed the nine programs using the same five instruction categories selected for scientific computation. The relative frequency with which each of the five types of instructions were used in commercial computation is presented in Exhibit 1. The time the central processor stands idle waiting for information input or output, $t_{I/O}$, is a function of the amount of information that must be taken into the computer, the amount of information that must be sent out of the computer, the rate at which information is transferred in and out of the computer, and the degree to which input and/or output can take place while the central processor is operating.

---

[2] A more detailed description of the development of the functional model is presented in K. E. Knight, "A Study of Technological Innovation—The Evolution of Digital Computers," (Ph.D. Dissertation, Carnegie Institute of Technology, 1963).

$$P = \frac{10^{12}\left[\dfrac{(L-7)(T)(WF)}{32,000(36-7)}\right]^{i}}{t_c + t_{I/O}}$$

$$t_c = 10^4\left[C_1 A_{F_1} + C_2 A_{F_2} + C_3 M + C_4 D + C_5 \mu\right]$$

$$t_{I/O} = P \times OL_1\ 10^6\left[(W_{I_1} \times B \times 1/K_{I_1}) + (W_{O_1} \times B \times 1/K_{O_1}) + N(S_1 + H_1)\right] R_1$$
$$+ (1-P)\ OL_2\ 10^6\left[(W_{I_2} \times B \times 1/K_{I_2}) + (W_{O_2} \times B \times 1/K_{O_2}) + N(S_2 + H_2)\right] R_2$$

Variables—Attributes of each Computing System

P = the computing power of the nth computing system

L = the word lengths (in bits)

T = the total number of words in memory

t_c = the time for the Central Processing Unit to perform one million operations

$t_{I/O}$ = the time the Central Processing Unit stands idle waiting for I/O to take place

A_{F_1} = the time for the Central Processing Unit to perform one fixed point addition

A_{F_2} = the time for the Central Processing Unit to perform one floating point addition

M = the time for the Central Processing Unit to perform one multiply

D = the time for the Central Processing Unit to perform one divide

μ = the time for the Central Processing Unit to perform one logic operation

B = the number of characters of I/O in each word

K_{I_1} = the input transfer rate (characters per second) of the primary I/O system

K_{O_1} = the output transfer rate (characters per second) of the primary I/O system

K_{I_2} = the input transfer rate (characters per second) of the secondary I/O system

K_{O_2} = the output transfer rate (characters per second) of the secondary I/O system

S_1 = the start time of the primary I/O system not overlapped with compute

H_1 = the stop time of the primary I/O system not overlapped with compute

S_2 = the start time of the secondary I/O system not overlapped with compute

H_2 = the stop time of the secondary I/O system not overlapped with compute

R_1 = 1 + the fraction of the useful primary I/O time that is required for nonoverlap rewind time

R_2 = same as R_1 for secondary I/O system

When we studied the input-output requirements we were unable to count the actual number of pieces (or number of words) read or written. Instead, the time the computer's central processing unit (1) operated alone, (2) operated concurrently with I/O, and (3) idled, waiting for information input-output to take place, was measured. From the actual input-output times and published input-output rates, it was possible to estimate the number of words read and written. The following computing systems were studied to estimate $t_{I/O}$; IBM 704, 705, 650, 7070, 7090, and 1401; Philco 211; and Bendix G15. The figures for the 7090 were accurately obtained, using the system's clock for single channel I/O, double channel I/O, and double channel I/O with program

Exhibit 1 (Cont.)

Semiconstant Factors		Values	
Symbol	Description	Scientific Computation	Commercial Computation
F	the word factor a. fixed word length memory b. variable word length memory	1 2	1 2
W_{I1}	weighting factor representing the percentage of the fixed add operations a. computers without index registers or indirect addressing b. computers with index registers or indirect addressing	 10 25	 25 45
W_2	weighting factor that indicates the percentage of floating additions	10	0
W_3	weighting factor that indicates the percentage of multiply operations	6	1
W_4	weighting factor that indicates the percentage of divide operations	2	0
W_5	weighting factor that indicates the percentage of logic operations	72	74
	percentage of the I/O that uses the primary I/O system a. systems with only a primary I/O system b. systems with a primary and secondary I/O system	 1.0 variable	 1.0 variable
W_{I1}	number of intput words per million internal operations using the primary I/O system a. magnetic tape I/O system b. other I/O systems	the values are the same as those given above for W_{I1} 20,000 2,000	 100,000 10,000
W_{O1}	number of output words per million internal operations using the primary I/O system		
W_{I2}/W_{O2}	number of input/output words per million internal operations using the secondary I/O system	the values are the same as those given above for W_{I1}	
U	number of times separate data is read into or out of the computer per million operations	4	20
OL_1	overlap factor 1—the fraction of the primary I/O system's time not overlapped with compute a. no overlap—no buffer b. read or write with compute—single buffer c. read, write and compute—single buffer d. multiple read, write and compute—several buffers e. multiple read, write and compute with program interrupt—several buffers	 1 .85 .7 .60 .55	 1. .85 .7 .60 .55
OL_2	overlap factor 2—the fraction of the secondary I/O system's time not overlapped with compute	values are the same as those given above for OL_1, a through e	
	the exponential memory weighting factor	.5	.333

interrupt. The other figures were obtained by less precise counting methods. The results obtained from the precise 7090 measures, and from the other systems, were very similar and are presented in Exhibit 1.

The memory capacity (M) of a computing system greatly influences its computing ability. Increased memory markedly improves the processing of very large problems which would otherwise be split into subproblems. There are also advantages to larger memories when performing smaller problems because they allow the use of compiling routines, subroutines, and so on. Recently, with the advent of multiple input-output capability and multiple program operation with executive and interrupt routines, larger memories provide additional advantages for all sizes and types of problems.

We were unable to find a feasible means to measure analytically the influence which memory has upon a computer's performance capability. Our best alternative was to obtain the opinions of the individuals who were most familiar with computers. A total of forty-three engineers, programmers, and other knowledgeable people were contacted and asked to evaluate the influence of computing memory upon performance. While their opinions varied, their answers were analogous enough to construct the functional model that estimates the effect memory has upon computer performance. The results of our inquiry are presented in Exhibit 1.

MACHINES COVERED

The two characteristics of the functional description for each computer which this study considers are calculated for the general purpose computers in the United States known to the author. The list of computers was obtained through a detailed search of the computing literature. All the systems that did not have structural elements to satisfy the functional model (specifically P) were deleted from the list. Computers which are not in the class of functionally similar products defined by the functional model are those that were built and used to perform a set of specialized information processing tasks. As a result these systems contained limited and specialized input-output equipment or limited internal arithmetic capabilities and are not included in our sample.

Most of the recent general purpose computers have been manufactured in quantities from tens to thousands. With quantity production the manufacturers have offered a large number of alternative system configurations. For these computers one functional description does not fully describe the computer. Many of the computers offer over eight memory sizes, three input-output systems, four input-output channel configurations, and four arithmetic and control extras. This represents over $(8 \times 3 \times 4 \times 4)$ 384 different computing systems. Although only a few configurations eventually are produced, the modern systems potentially consist of several hundred alternatives. It would be impossible to calculate P for even a few alternatives of each system. We must therefore settle on one configuration for each computer.

There appears to be a good method for selecting the configurations and that is to consider the most typical configuration of the computer. Where structural changes have been made, we have used the equipment which was available when the system was first introduced. In a few cases where important modifications have been introduced at a later date, these modifications are considered as separate computers and are treated as such in the study. The values of P and C for both scientific and commercial computation for the 218 computers introduced between 1944 and 1962 are presented in Table 1.[3]

Table 1 Computing Systems.

Computer		Date	Scientific Computation		Commercial Computation	
			P	C	P	C
No.	Name	Introduced	Ops/Sec	Secs/$	Ops/Sec	Secs/$
1	Harvard Mark I	1944	.0379	50.94	0.406	50.94
2	Bell Lab Computer Model IV	March 1945	.0068	509.4	0.035	509.4
3	Eniac	1946	7.448	31.81	44.65	31.81
4	Bell Computer Model V	Late 1947	.0674	84.83	0.296	84.83
5	Harvard Mark II	Sept. 1948	.1712	50.94	0.774	50.94
6	Binac	Aug. 1949	21.75	127.2	11.70	127.2
7	IBM CPC	1949	2.126	207.8	14.37	207.8
8	Bell Computer Model III	1949	.0674	102.2	0.296	102.2
9	SEAC	May 1950	102.8	50.94	253.8	50.94
10	Whirlwind I	Dec. 1950	110.7	31.81	45.57	31.18
11	Univac 1101 Era 1101	Dec. 1950	682.5	50.94	301.8	50.94
12	IBM 607	1950	5.666	479.6	34.06	479.6
13	Avdiac	1950	108.5	84.83	51.20	84.83
14	Adec	Jan. 1951	54.26	42.42	57.16	42.42
15	Burroughs Lab Calculator	Jan. 1951	5.605	254.5	7.718	254.5
16	SWAC	March 1951	632.2	50.94	324.7	50.94
17	Univac I	March 1951	140.1	24.94	271.4	24.94
18	ONR Relay Computer	May 1951	.2937	127.2	1.050	127.2
19	Fairchild Computer	June 1951	2.000	127.2	4.539	127.2
20	National 102	Jan. 1952	1.260	848.3	2.998	848.3
21	IAS	March 1952	467.0	84.83	305.0	84.83
22	Maniac I	March 1952	302.7	101.9	163.4	101.9
23	Ordvac	March 1952	268.8	72.76	127.8	72.76
24	Edvac	April 1952	31.56	54.22	14.86	54.22
25	Teleregister Special Purpose Digital Data Handling	June 1952	12.16	78.93	26.43	78.93
26	Illiac	Sept. 1952	123.1	72.76	50.43	72.76
27	Elcom 100	Dec. 1952	1.278	424.2	3.241	424.2
28	Harvard Mark IV	1952	63.99	42.42	64.95	42.42
29	Alwac II	Feb. 1953	10.17	509.4	12.08	509.4
30	Logistics Era	March 1953	52.85	72.00	39.01	72.0
31	Oarac	April 1953	24.38	141.4	35.71	141.4
32	ABC	May 1953	29.88	212.1	11.66	212.1

[3] Only computers through 1962 were considered in the original study. (See Knight, "A Study of Technological Innovation," *op. cit.*)

Table 1 (Cont.)

Computer		Date	Scientific Computation		Commercial Computation	
			P	C	P	C
No.	Name	Introduced	Ops/Sec	Secs/$	Ops/Sec	Secs/$
33	Raydac	July 1953	171.3	8.483	244.6	8.483
34	Whirlwind II	July 1953	233.4	21.21	95.96	21.21
35	National 102A	Summer 1953	4.089	116.5	8.400	116.5
36	Consolidated Eng. Corp. Model 36–101	Summer 1953	38.31	181.8	21.07	181.8
37	Jaincomp C	Aug. 1953	4.745	103.9	3.375	103.9
38	Flac	Sept. 1953	61.55	50.94	107.9	50.94
39	Oracle	Sept. 1953	1,002.	31.81	563.4	31.81
40	Univac 1103	Sept. 1953	749.0	28.34	666.2	28.34
41	Univac 1102	Dec. 1953	460.3	50.94	240.0	50.94
42	Udec I	Dec. 1953	16.38	72.67	21.93	72.67
43	NCR 107	1953	16.99	254.5	34.44	254.5
44	Miniac	Dec. 1953	10.91	267.6	9.545	267.6
45	IBM 701	1953	992.7	18.34	615.7	18.34
46	IBM 604	1953	2.766	974.2	20.19	974.3
47	AN/UJQ-2(YA-1)	1953	21.48	84.83	56.16	84.83
48	Johnniac	March 1954	319.2	84.83	284.9	84.83
49	Dyseac	April 1954	72.18	50.90	172.4	50.90
50	Elcom 120	May 1954	5.471	261.9	6.456	262.0
51	Circle	June 1954	14.04	318.1	10.59	318.1
52	Burroughs 204 & 205	July 1954	80.84	77.94	187.3	77.94
53	Modac 5014	July 1954	6.238	299.8	10.09	299.8
54	Ordfiac	July 1954	2.607	92.51	6.011	92.51
55	Datatron	Aug. 1954	113.7	113.2	243.1	113.2
56	Modac 404	Sept. 1954	7.116	254.5	15.29	254.5
57	Lincoln Memory Test	Dec. 1954	1,925.	9.285	768.7	9.285
58	TIM II	Dec. 1954	7.414	848.3	7.439	848.3
59	Caldic	1954	23.99	203.8	41.34	203.8
60	Univac 60 & 120	Nov. 1954	.0924	356.3	1.473	356.3
61	IBM 650	Nov. 1954	110.8	155.9	291.1	155.9
62	WISC	1954	7.736	145.7	6.413	145.7
63	NCR 303	1954	3.491	117.6	8.281	117.6
64	Mellon Inst. Digital Computer	1954	14.23	169.9	10.55	169.9
65	IBM 610	1954	.1408	519.6	0.437	519.6
66	Alwac III	1954	44.80	302.7	91.42	302.7
67	IBM 702	Feb. 1955	394.4	20.78	1,063.	20.78
68	Monrobot III	Feb. 1955	.3743	299.8	1.188	299.8
69	Norc	Feb. 1955	545.8	10.17	268.2	10.17
70	Miniac II	March 1955	11.76	267.6	17.44	267.6
71	Monrobot V	March 1955	.4678	295.5	1.607	295.5
72	Udec II	Oct. 1955	7.244	84.83	10.65	84.83
73	RCA BIZMAC I & II	Nov. 1955	285.6	5.668	967.9	5.668
74	Pennstac	Nov. 1955	26.75	212.1	22.98	212.1
75	Technitral 180	1955	110.0	46.19	190.1	46.19
76	National 102D	1955	7.317	112.3	14.20	112.3
77	Monrobot VI	1955	.3293	222.7	0.966	222.7
78	Modac 410	1955	24.18	203.8	51.84	169.9
79	Midac	1955	101.6	169.9	29.00	169.9
80	Elcom 125	1955	31.24	164.1	29.01	164.1

Table 1 (Cont.)

Computer		Date	Scientific Computation		Commercial Computation	
			P	C	P	C
No.	Name	Introduced	Ops/Sec	Secs/$	Ops/Sec	Secs/$
81	Burroughs E 101	1955	.6898	580.0	2.319	580.0
82	Bendix G15	Aug. 1955	57.34	419.9	30.25	419.9
83	Alwac III E	Nov. 1955	41.50	249.4	90.15	249.4
84	Readix	Feb. 1956	80.63	194.9	87.99	194.9
85	IBM 705, I, II	March 1956	734.0	13.27	2,087.	13.27
86	Univac 1103 A	March 1956	2,295.	19.49	1,460.	19.49
87	AF CRC	April 1956	81.66	31.81	28.97	31.81
88	Guidance Function	April 1956	5.246	461.9	7.744	461.9
89	IBM 704	April 1956	10,670.	13.18	3,785.	13.18
90	IBM 701 (CORE)	1956	2,378.	17.81	1,807.	17.81
91	Narec	July 1956	444.8	25.45	190.6	25.45
92	LGP 30	Sept. 1956	41.94	479.6	32.75	479.6
93	Modac 414	Oct. 1956	28.26	169.9	42.94	169.9
94	Elecom 50	1956	.5990	139.2	1.776	1,039.
95	Udec III	March 1957	25.11	72.76	20.85	72.76
96	George I	Sept. 1957	1,538.	50.94	571.9	50.94
97	Univac File O	Sept. 1957	35.20	41.02	73.17	41.02
98	Lincoln TXO	Fall 1957	1,471.	10.19	359.6	10.19
99	Univac II	Nov. 1957	1,155.	22.27	2,363.	22.27
100	IBM 705 III	Late 1957	2,379.	13.27	7,473.	13.27
101	Teleregister Telefile	Late 1957	286.0	65.98	935.9	65.98
102	Recomp I	Late 1957	25.76	363.8	16.14	363.8
103	IBM 608	1957	15.21	389.7	60.69	389.7
104	Mistic	1957	64.28	101.9	24.50	101.9
105	Maniac II	1957	1,491.	72.84	1,421.	72.84
106	IBM 609	1957	18.19	530.7	75.21	530.7
107	IBM 305	Dec. 1957	94.47	163.0	96.47	163.0
108	Corbin	1957	1,794.	50.90	2,407.	50.90
109	Burroughs E 103	1957	.6736	551.8	2.286	551.8
110	AN/FSQ 7 & 8	1957	36,730.	2.834	15,560.	2.834
111	Alwac 880	1957	2,198.	50.90	959.7	50.90
112	Univac File I	Jan. 1958	42.49	41.05	92.04	41.05
113	Lincoln CG24	May 1958	6,394.	21.21	5,933.	21.21
114	IBM 709	Aug. 1958	1,869.	8.882	10,230.	8.882
115	Univac 1105	Sept. 1958	4,433.	14.50	5,527.	14.50
116	Lincoln TX2	Fall 1958	82,050.	8.483	34,000.	8.483
117	Philco 2000–210	Nov. 1958	29,970.	17.81	28,740.	17.81
118	Recomp II	Dec. 1958	41.36	249.4	28.03	249.4
119	Burroughs 220	Dec. 1958	810.2	79.94	1,616.	79.94
120	Mobidic	1958–1960	8,741.	10.19	12,250.	10.19
121	Philco CXPO	1958	2,622.	15.91	1,576.	15.91
122	Monrobot IX	1958	.4598	2,545.	1.334	2,545.
123	GE 210	June 1959	1,884.	44.54	5,085.	44.54
124	Cyclone	July 1959	234.6	215.0	119.6	215.0
125	IBM 1620	Oct. 1959	94.79	331.7	47.20	331.7
126	NCR 304	Nov. 1959	1,136.	40.23	2,445.	40.23
127	IBM 7090	Nov. 1959	97,350.	9.742	45,470.	9.742
128	RCA 501	Nov. 1959	638.7	38.97	1,877.	38.97
129	RW 300	Nov. 1959	218.6	45.58	534.3	45.78
130	RPC 9000	1959	14.50	138.6	9,521.	138.6

Table 1 (Cont.)

No.	Computer Name	Date Introduced	Scientific Computation P Ops/Sec	C Secs/$	Commercial Computation P Ops/Sec	C Secs/$
131	Librascope Air Traffic	1959	3,043.	16.94	6,130.	16.94
132	Jukebox	1959	16.56	338.9	18.66	338.9
133	Datamatic 1000	1959	480.8	13.44	1,455.	13.44
134	CCC Real Time	1959	393.8	77.17	280.3	77.17
135	Burroughs E 102	1959	.6670	580.0	1.847	580.0
136	Burroughs D 204	1959	2,354.	68.00	1,183.	68.00
137	AN/TYK 6V BASICPAC	1959	1,365.	50.90	493.0	50.90
138	CDC 1604	Jan. 1960	58,290.	18.34	20,390.	18.34
139	Librascope 3000	Jan. 1960	5,177.	12.47	25,320.	12.47
140	Univac Solid State 80/90 I	Jan. 1960	329.1	124.7	489.6	124.7
141	Philco 2000–211	March 1960	105,844.	14.845	55,740.	14.85
142	Univac Larc	May 1960	142,600.	4.619	40,450.	4.619
143	Libratrol 500	May 1960	21.07	286.0	20.38	286.0
144	Monrobot XI	May 1960	4.839	890.7	10.30	890.7
145	IBM 7070	June 1960	2,813.	23.98	5,139.	23.98
146	CDC 160	July 1960	119.3	354.3	49.63	354.2
147	IBM 1401 (Mag. Tape)	Sept. 1960	496.7	83.14	1,626.	83.14
148	AN/FSQ 31 & 32	Sept. 1960	172,200.	6.235.	48,360.	6.285
149	Merlin	Sept. 1960	8,306.	42.42	2,925.	42.42
150	IBM 1401 (Card)	Sept. 1960	340.9	215.0	967.8	215.0
151	Mobidic B	Fall 1960	5,251.	12.72	8,630.	12.72
152	RPC 4000	Nov. 1960	89.91	249.4	54.11	249.4
153	PDP-1 (M.T.)	Nov. 1960	4,455.	41.57	2,173.3	41.6
154	PDP-1 (P.T.)	Nov. 1960	166.6	215.	57.16	215.0
155	Packard Bell 250 (PT)	Dec. 1960	62.23	506.9	22.21	506.9
156	Honeywell 800	Dec. 1960	28,790.	14.85	23,760.	14.85
157	General Mills AD/ECW-57	Dec. 1960	143.9	141.7	44.03	141.7
158	Philco 3000	Late 1960	102.2	155.9	66.13	155.8
159	Maniac III	Late 1960	11,140.	25.45	4723.	25.45
160	Sylvania S9400	Late 1960	62,510.	9.306	49,550.	9.306
161	Target Intercept	Late 1960	16,800.	33.89	16,070.	33.89
162	Westinghouse Airbourne	1960	10,950.	12.47	4806.	12.47
163	RCA 300	1960	1,466.	25.98	687.7	25.98
164	Mobidic CD & 7A AN/MYK	1960	12,410.	10.39	15,430.	10.39
165	Litton C7000	1960	18,200.	11.34	5,323.	11.34
166	Libratrol 1000	1960	84.16	254.5	50.85	254.5
167	GE 312	1960	122.0	299.8	47.12	299.8
168	Diana	1960	102.1	127.2	48.85	127.2
169	DE 60	Feb. 1960	.6384	1,155.	1.855	1,155.
170	Burroughs D107	1960	311.8	63.62	73.95	63.62
171	AN/USQ 20	1960	22,390.	20.78	23,670.	20.78
172	AN/TYK 4V Compac	1960	1,610.	41.57	616.1	41.57
173	General Mills Apsac Jan.	Jan. 1961	16.22	424.2	7.084	424.2
174	Univac Solid State 80/90 II	Jan. 1961	3,199.	69.28	3,044.	69.28
175	Bendix G20 & 21	Feb. 1961	37,260.	33.17	17,060.	33.17

Table 1 (Cont.)

Computer		Date	Scientific Computation		Commericial Computation	
			P	C	P	C
No.	Name	Introduced	Ops/Sec	Secs/$	Ops/Sec	Secs/$
176	RCA 301	Feb. 1961	323.0	113.4	1,055.	113.4
177	BRLESC	March 1961	47,240.	12.72	28,550.	12.72
178	GE 225	March 1961	6,566.	77.94	7,131.	77.94
179	CCC-DDP 19 (Card)	May 1961	5,159.	138.6	3,027.	138.6
180	CCC-DDP 19 (MT)	May 1961	7,908.	59.38	8,073.	59.38
181	IBM Stretch (7030)	May 1961	371,700.	2.078	631,200.	2.078
182	NCR 390	May 1961	2.034	328.2	10.43	328.2
183	Honeywell 290	June 1961	354.3	207.8	182.8	207.8
184	Recomp III	June 1961	48.28	311.8	35.76	311.8
185	CDC 160A	July 1961	1,015.	138.6	1,780.	138.6
186	IBM 7080	Aug. 1961	27,090.	11.34	30,860.	11.34
187	RW 530	Aug. 1961	13,460.	59.38	5086.	59.38
188	IBM 7074	Nov. 1961	41,990.	19.49	31,650.	19.49
189	IBM 1410	Nov. 1961	1,673.	62.35	4,638.	62.35
190	Honeywell 400	Dec. 1961	1,354.	71.67	2,752.	71.67
191	Rice Univ.	Dec. 1961	7,295.	50.90	2378.	50.90
192	Univac 490	Dec. 1961	17,770.	24.94	15,050.	24.94
193	AN/TYK 7V	1961	4,713.	41.57	9,077.	41.57
194	Univac 1206	1961	20,990.	42.42	17,700.	42.42
195	Univac 1000 & 1020	1961	3,861.	66.33	3,292.	66.33
196	ITT Bank Loan Process	1961	492.6	34.64	1,916.	34.64
197	George II	1961	298.	31.81	675.1	31.81
198	Oklahoma Univ.	Early 1962	7,723.	50.90	2,616.	50.90
199	NCR 315	Jan. 1962	3,408.	65.63	11,460.	65.63
200	NCR 315 CRAM	Jan. 1962	3,364.	73.36	9,896.	73.36
201	Univac File II	Jan. 1962	33.46	38.97	94.49	38.97
202	HRB-Singer Sema	Jan. 1962	129.2	890.7	56.94	890.7
203	Univac 1004	Feb. 1962	1.789	479.6	25.29	479.6
204	ASI 210	April 1962	8,868.	135.5	4,114.	135.5
205	Univac III	June 1962	22,720.	27.11	22,790.	27.11
206	Burroughs B200 Series-B270 & 280	July 1962	163.3	95.93	615.3	95.93
207	SDS 910	Aug. 1962	4,841.	249.4	2,355.	249.4
208	SDS 920	Sept. 1962	9,244.	65.63	4,964.	65.63
209	PDP-4	Sept. 1962	220.2	479.6	75.97	479.6
210	Univac 1107	Oct. 1962	138,700.	12.47	76,050.	12.47
211	IBM 7094	Nov. 1962	175,900.	8.782	95,900.	8.781
212	IBM 7072	Nov. 1963	22,710.	34.64	8,694.	34.64
213	IBM 1620 MOD III	Dec. 1962	214.8	259.8	56.89	259.8
214	Burroughs B5000	Dec. 1962	43,000.	32.82	15,910.	32.82
215	ASI 420	Dec. 1962	27,790.	44.54	11,090	44.54
216	Burroughs B200 Series—Card Sys	Dec. 1962	114.3	160.1	437.2	164.1
217	RW 400 (AN/FSQ 27)	1962	7,437.	12.47	11,240.	12.47
218	CDC 3600	Dec. 1962	315,900.	11.34	74,900.	11.34

Table 1 also contains date of introduction for each of the 218 computers considered. For our study we define the date of introduction as the delivery date of an operating system to the first user. Where the computer is manufactured and used by the same organization, the date of introduction is defined as that when the completed computer passes a minimal acceptance test.

STATISTICAL AVERAGING TECHNIQUE

The procedures used to calculate the computing power (P) and computing cost (C) represent a statistical averaging technique. The calculated numbers for a particular machine should not be taken as the "measure" for that particular machine. In making the calculations we used only one configuration for our average set of problems. The configuration selected was the one that was representative of the earlier systems. It should be emphasized that no attempt was made to optimize either throughput (number of calculations per second) or cost (number of calculations per dollar) for the machine.

The calculations of P (operations/second) and C (seconds/dollar rental) are intended to provide overall comparisons between machines of various sizes and between machines introduced in different years. From this data we determine the advances in computing power over time and investigate the differences between small and large computers. Because of the averaging technique used to calculate P and C, our data do not provide direct comparisons between two machines for the specific set of user needs.

TECHNOLOGY CURVES

Since the functional descriptions consist of two attributes, we can display them on a two-dimensional graph. Exhibit 2 contains points obtained when operations/second (P) is plotted against seconds/dollar (C) for computers performing scientific computation. Because of the tremendous range of P and C, Exhibit 2 is drawn on log-log graph paper. The number next to each point identifies the corresponding computer as listed in Table 1.

From an initial glance at Exhibit 2, it is apparent that the early systems generally fall on the lower left portion of the graph and the newer ones on the upper right. The graph shows how much computing power is obtained at different costs; here are high cost systems (few seconds per dollar) and low cost ones (many seconds per dollar). In any year, an expensive computer has greater computing power (higher number of operations per second) than a less expensive one. It is also apparent from Exhibit 2 that for a constant C we obtain greater P over time.

A curve that connects the functional descriptions of the computers in a single year describes the computing technology for that year. Improved per-

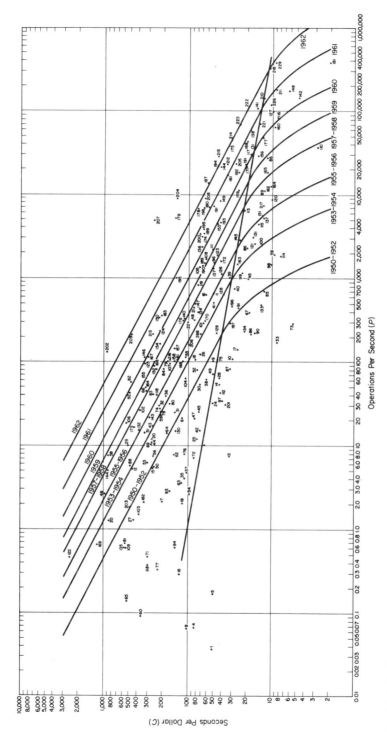

Exhibit 2. Technology Curves for Scientific Computers. (The numbers in this graph used to identify each computer correspond to the numbers in Table 1.)

formance consists of a continual shift over time, enabling an increased number of operations per second to be performed for a given cost.

We now wish to use our data to develop the technology curves. Unfortunately, the points for a particular year in Exhibit 2 do not form smooth parallel curves. For any one year considerable scattering occurs because (1) not all systems are equally technically advanced, and (2) there are errors in the estimates of P and C.

The first reason for the scatter of points needs little explanation. In the computing industry, there have been many systems introduced, and these have resulted in a wide range of performance from improved to poorer. Some systems will make significant improvements and fall far to the right of the other points. Alternatively, many systems will not match the capabilities of existing computers and will lie in the range of the industry's previous know-how.

The second reason for the scatter is the expected variance in the estimates of the functional descriptions. P was obtained by means of the functional model which estimated each system's actual performance. There are differences both in the pricing policies of the manufacturers and in our ability to determine what equipment constitutes each particular system that creates a variance in C. In the calculation we performed, many small errors could have entered into the estimates of P and C to produce random error, even if all the systems came from an identical level of technological knowledge.

Recognizing that variance exists, it is necessary to use a curve-fitting technique to estimate the desired technology lines. For this study we have used least square regression analysis. From a visual analysis of Exhibit 2 it appears that the technology curves for the different years are approximately the same in shape, with a shift to the right over time. Thus, the data were fitted to the following equation:

$$(1) \quad ln(C) = \alpha_0 + \alpha_1 \, ln(P) + \alpha_2 \, [ln(P)]^2 + \beta_1 S_1 + \beta_2 S_2 + \ldots + \beta_7 S_7$$

The α's and β's represent the regression coefficients to be determined by the least squares analysis. The $S_1, \ldots, S_7$ represent dummy variables (or shift parameters) for the different years considered. To fit the curve, the data were grouped into eight time periods (for example, 1962, 1961, 1960, 1959, 1957–58, 1955–56, 1953–54, and 1950–51–52). The earlier years were combined because of the small number of systems introduced in each of these years. The dummy variables were used in the following manner: for 1962, $S_1, \ldots, S_7$ were all set equal to 0; for 1961, $S_1 = 1$ and $S_2 = S_3 = \ldots = S_7 = 0$; for 1960, $S_2 = 1$ and $S_1 = S_3 = \ldots = S_7 = 0$; $\ldots$ and finally for 1950–51–52, $S_7 = 1$ and $S_1 = S_2 = \ldots = S_6 = 0$. ln P and $[ln\,(P)]^2$ were both initially included in the equation since a visual analysis of the lines made them appear curved.

After the initial regression estimate, all points that were more than one half a standard deviation below and to the left of the curve for their year were omitted. Eliminating points in this manner provides a distinct procedure for determining which points we will include in the final determination of technology curves and forces the technology curves to the right to provide a more accurate picture of the performance limits.

The regression analysis, using the data for computer performance in scientific computation with Equation 1, showed that $[ln\ (P)]^2$ term was not significant. The least squares technique was then used to fit Equation 2 to the data.

$$(2) \qquad ln(C) = \alpha_0 + \alpha_1 ln(P) + \beta_1 S_1 + \beta_2 S_2 + \ldots \beta_7 S_7$$

For the linear equation, all the terms were significant and the correlation coefficient was $r = +.9596$. Since the correlation coefficient equaled only $+.9596$ with Equation 1, it appeared most reasonable to use the simpler linear equation to plot the technology curves. In the calculation of both the polynomial and the linear equation, over 120 observations were used so that the sample sizes would be adequate. The equation for the scientific computation technology curves is as follows:[4]

$$(3) \qquad ln(C) = 8.9704 - .51034\ ln(P) - .3650(S_1) - .7874(S_2)$$
$$- 1.0724(S_3) - 1.3028(S_4 - 1.6639(S_5)$$
$$- 1.9859(S_6) - 2.5013(S_7)$$

The eight curves described by Equation 3 are drawn in Exhibit 2.

We now perform a similar analysis for commercial computation. The results of the calculation of the technology curves for systems performing commercial computation are shown in Equation 4.[5]

$$(4) \qquad ln(C) = 8.1672 - .459\ ln(P) - .3643(S_1) - .6294(S_2)$$
$$- 8561(S_3) - .9011(S_4) - 1.187(S_5)$$
$$- 1.454(S_6) - 2.164(S_7)$$

The eight curves drawn from Equation 4 are shown in Exhibit 3.

[4] For this equation the following list contains the standard error and the test of significance (student's t test) for each regression coefficient.

Regression Coefficient	Standard Error	t Value
β_1	0.0171	−30.41
		− 2.112
β_2	0.1608	− 4.897
β_3	0.1887	− 5.682
β_4	0.1687	− 7.723
β_5	0.1992	− 8.349
β_6	0.1750	−11.34
β_7	0.1943	−12.87

[5] The correlation coefficient, r, for the linear equation (Equation 2) was $+.8543$. The curves using Equation 1 and Equation 2 were similar and the correlation coefficients almost equal so that the simple linear equation was used to construct the technology curves.

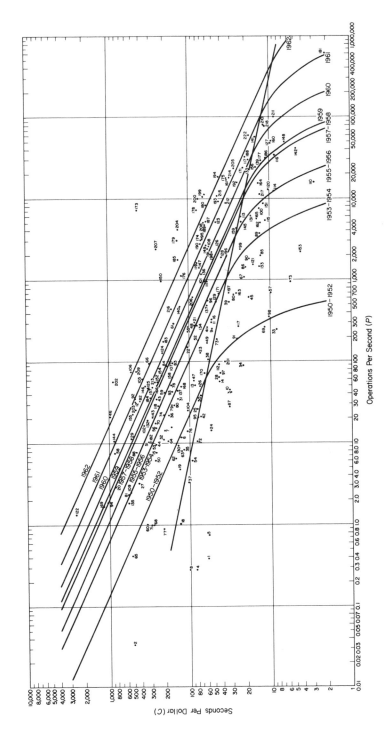

Exhibit 3. Technology Curves for Commercial Computation. (The numbers used in this graph to identify each computer correspond to the numbers in Table 1.)

We analyze the meaning of the technology curves by first restating the general equation for the curves:

(5) $$C = (\alpha_0)(P^{\alpha_1})(e^{\beta_1})(e^{\beta_2}) \ldots (e^{\beta_7})$$

where
and $\alpha_0, \alpha_1, \beta_1, \beta_2, \ldots, \beta_7$ are the values calculated with the least square regression analysis.

From Equation 5 we obtain the following:

(6) $$\frac{\text{seconds}}{\text{dollar}} = k \text{ (Shift parameter to adjust for years)} \left(\frac{\text{operations}}{\text{sec.}}\right)^{\alpha_1}$$

For any particular year we can combine the constant, k, and the shift parameter into a new constant K. If we, therefore, set $a_1 = -\alpha_1$ Equation 6 now becomes:

(7) $$\frac{\text{dollars}}{\text{second}} = \frac{1}{K} \text{ (operations/sec.)} \, a_1$$

For scientific computation the value for $\alpha_1 = -.519$ so that a_1 equals .519. For commercial computation $\alpha_1 = -.459$ so that a_1 equals .459. We can therefore assume that a_1 is approximately equal to .5 and rewrite Equation 7 as follows:

(8) $$\text{System Cost} = \frac{1}{K} \sqrt[2]{\text{Computing Power}}$$

This represents a very interesting result because it indicates that within the limits of the computing technology one can construct four times as powerful a computer at only twice the cost.

(9) $$\text{Computing power} = (K \text{ system cost})^2$$

That computing power increases as the square of cost was proposed in the late 1940s by Herb Grosch. Since that time the relationship expressed in Equation 9 has been referred to as Grosch's Law. We have seen the industry develop a sense of humor over its twenty-year life with frequent jokes being made in reference to Grosch's Law. In a recent article by Charles W. Adams, "Grosch's Law Repealed," the author proposes to "replace the square (Grosch's Law) by the square root."[6] Grosch's Law has received much attention because of its implications about economies of scale, yet, has never been supported with adequate quantitative data. We still need to question whether the law (Computing power = Constant (Cost)^2) is true or if it is the artifact of the computer companies' pricing policy. The popularity of the law and the difficulty in setting prices leads us to suspect the possibility of some bias in our data.

We must express another word of caution before we attach too much significance to Grosch's Law. In calculating the technology curves, we were able to use the systems actually built. The equations derived are, therefore, applicable within the limited range of computers studied. Special consideration has to be given to the fact that there are definite limits to the maximum computing power that can be obtained at any one time. As the bounds of technological knowledge are reached, additional computing power is purchased at a very high price. For high values of P the technology curve will not remain a straight line but will curve downward to show an ever increasing negative slope. The reason that this did not show up in the regression analysis is that only a few computers came close to the maximum limits of computing power. Three noticeable ones are the AN/FSQ 7 and 8 (the Sage computers), the Univac Larc, and the IBM Stretch. These computers each obtained a new high evaluation for absolute computing power but at considerably lower number of operations per dollar. Grosch's Law did not hold for these three machines because the increases in power were obtained at less than the squared or even a 1 to 1 relationship with Cost. In other words the slope of the curve, or a, is less than -1. We cannot build larger and larger computers at reasonable costs since at any point in time there are absolute limits to the size and speed obtainable. This fact needs to be kept in mind when talking about Equation 8. The most powerful computing systems we could possibly build today or tomorrow would not be the most economical.

In order to estimate where the turning point occurs we use the computers that have had, at one time, the maximum absolute efficiency. For scientific computation there are eight systems and for commercial computation ten. We add to Exhibits 2 and 3 lines of maximum efficiency through these points. The point where the line crosses the technology curves for each year provides an estimate of where the technology curves start to slope downward to yield diminishing marginal returns for systems with greater computing power. The latter curves are drawn freehand on Exhibits 2 and 3 to show their approximate shape.

PERFORMANCE IMPROVEMENTS FROM 1950 TO 1962

The continual stream of performance improvements appears to result from the dynamic nature of the industry itself. Most people in the computing field

[6] C. W. Adams, "Grosch's Law Repealed," *Datamation* (July, 1962), pp. 38–39.

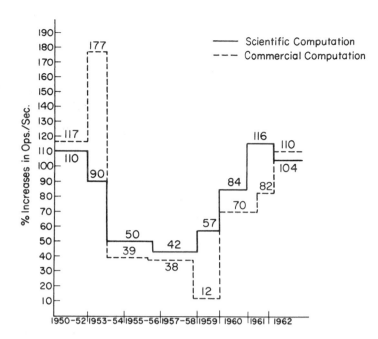

Exhibit 4. Average Yearly Shift of the Technology Curves.

search conscientiously for faster and more economical machines. However, most of these individuals have a limited idea of what has happened over the past twenty years. For instance, they greatly underestimate the number of innovation systems produced and the amount of performance advances.[7] From 1950 through 1962 the technology curves have an average improvement of eighty-one percent per year for scientific computation and eighty-seven percent per year for commercial computation. It is seen from Exhibit 4 that there has been some variance in yearly percent improvement. The improvements in both scientific and commercial computation have been fairly similar, with the first five years, 1950–54, and the last three years, 1960–62, showing greater improvement than the years, 1955–59. The large commercial computation improvement in 1953–54 that we mentioned earlier as being significantly above the mean can be explained by the great increase in speed and the number of machines using magnetic tape units. Since commercial computation relies more upon input-output capability than does scientific computation, the improvements and increased utilization of magnetic tapes aided this category more than they did the other. No simple explanation has been found for the other variations shown in Exhibit 4.

---

[7] The shift in the technology curves could be measured as either: (a) for constant cost (C), the percent increase in performance (P), or; (b) for constant power (P), the percent change in cost (C). For our analysis we have looked at alternative (a). Therefore our numbers will be larger than if we had used alternative (b).

TECHNOLOGICAL FORECAST

The conclusions reached on the rate of technological change and the economies of scale represent the basic data with which to make a technological forecast for the computer industry. Because of the strength of the conclusions in the study one could predict that the same relationships will hold in the future. After the original study was distributed, men in business, nonprofit and governmental groups as well as academics, started using the results as a technological forecast. The work helped to illustrate the tremendous decrease in cost (C) for computing power that could be expected which would make it economically feasible to use computers and, at the same time, point out a growing machine obsolescence problem. The results also show the continual increase in the power (P) of the largest machine and predict a continued rapid increase.

After being contacted by hundreds of organizations who were using the results of the research to make technological forecasts, we decided to update the study to include machines introduced from 1963–66. There were many of the people making the inquiries betting on both sides: pro, the results would still hold, con, the rate of change and economies of scale would no longer be found. Using the identical procedure, we calculated computing power (P) and equipment cost (C) for the computers introduced between 1963 and 1066 (Table 2).

Table 2. Computing Systems

Computer No.	Name	Date Introduced	Scientific P (Ops/Sec)	Commercial P (Ops/Sec)	C (Sec/$)
219	IBM 7040	4/63	21,420	9,079	44.54
220	IBM 7044	7/63	67,660	23,420	23.98
221	RCA 601	1/63	68,690	58,880	13.86
222	Honeywell 1800	11/63	110,600	57,750	17.81
223	Philco 1000	6/63	6,811	10,440	65.63
224	Philco 2000–212	2/63	369,800	84,230	9.169
225	Librascope L 3055	12/63	114,000	30,620	10.39
226	H. W. Electronics 15K	2/63	119.6	50.98	1,247
227	GE 215	6/63	5,246	6,924	89.07
228	DDP-24	6/63	580.4	632.7	124.7
229	CDC 3600	6/63	459,065	156,375	12.47
230	UNIVAC 1050	9/63	12,028	19,675	113.4
231	UNIVAC 1004	9/63	97.12	1,473	415.7
232	PDP-5	10/63	6,338	12,519	311.8
233	IBM 1460	10/63	1,611	7,200	69.28
234	IBM 1440	11/63	1,412	5,559	183.40
235	Honeywell 1400	12/63	1,770	6,821	41.57
236	ASI 2100	12/63	24,628	10,241	178.2
237	SDS-9300	12/63	43,876	10,646	89.07
238	Burroughs 273	1/64	714.6	3,467	87.82

Table 2 (Cont.)

Computer No.	Name	Date Introduced	Scientific P (Ops/Sec)	Commerical P (Ops/Sec)	C (Sec/$)
239	GE-235	1/64	28,557	22,244	51.96
240	IBM 7010	1/64	5,729	11,537	31.18
241	Burroughs B 160–180	4/64	295.5	1,599	145
242	CDC 160G	4/64	54,065	20,278	89.07
243	IBM 7094 II	4/64	217,108	95,146	8.20
244	CDC 3200	5/64	195,256	87,510	51.96
245	GE 415	5/64	7,472	15,668	77.94
246	UNIVAC 1004 II, III	6/64	79.16	1,878	283.4
247	SDS-930	6/64	73,181	21,035	103.9
248	GE 425	6/64	11,485	22,160	62.35
249	GE 205	7/64	1,775	6,188	311.8
250	Hcneywell 200	7/64	1,148	7,027	103.9
251	RCA 3301	7/64	126,761	58,359	44.54
252	PDP-6	7/64	46,359	32,803	51.96
253	CDC 6600	9/64	7,021,619	4,091,293	8.31
254	UNIVAC 418	9/64	58,767	166,564	62.35
255	NCR 315–100	11/64	6,164	17,251	155.9
256	GE 635	11/64	338,958	253,898	11.34
257	CDC 3400	11/64	269,859	157,202	29.69
258	Burroughs B5500	11/64	376,275	544,201	20.78
259	SDS 925	2/65	92,692	150,102	155.9
260	SDS 92	2/65	19,140	79,065	239.8
261	CDC 3100	2/65	118,462	74,391	77.94
262	ASI 6020	3/65	28,160	13,161	178.1
263	DDP-224	3/65	52,330	81,492	103.9
264	DDP-116	4/65	2,176	4,023	677.7
265	GE 625	4/65	224,374	118,154	15.20
266	PDP-8	4/65	1,768	990.5	230.9
267	PDP-7	4/65	68,497	29,571	103.9
268	IBM 360/40	5/65	33,438	50,073	54.08
269	IBM 360/30	5/65	7,942	17,104	72.88
270	NCR 315 RMC	7/65	132,060	153,770	62.35
271	UNIVAC 1108 II	8/65	2,075,181	2,088,142	10.39
272	GE 435	8/65	24,803	56,623	41.57
273	IBM 360/50	9/65	187,488	148,967	27.47
174	IBM 1130	9/65	16.38	56.76	692.8
275	NCR 590	9/65	4.288	21.76	519.6
276	ASI 6240	10/65	33,177	13,232	155.9
277	UNIVAC 491 & 492	10/65	4,929	48,490	36.68
278	RCA Spectra 70/15	10/65	1,837	16,586	164.1
279	Raytheon 520	10/65	29,118	13,427	207.8
280	IBM 360/75	11/65	3,560,854	1,437,806	11.81
281	Honeywell 2200	12/65	12,222	14,332	77.94
282	CDC 3800	12/65	690,510	150,726	12.47
283	RCA Spectra 70/25	12/65	4,818	36,366	103.9
284	Friden 6010	1/66	1.66	48.66	1,039
285	CDC 6400	1/66	696,086	193,785	12.47
286	DDP-124	1/66	5,812	7,618	249.4
287	Honeywell 1200	1/66	2,130	10,907	115.5
288	IBM 360/20	1/66	1,932	4,497	239.8

Table 2 (Cont.)

Computer No.	Name	Date Introduced	Scientific P (Ops/Sec)	Commericial P (Ops/Sec)	C (Sec/$)
289	UNIVAC 1005 II, III	2/66	88.25	1,677	259.8
290	UNIVAC 10051	2/66	71.73	1,186	366.8
291	Honeywell 120	2/66	2,108	9,526	190
292	IBM 360/65	3/66	1,385,573	809,738	13.86
293	UNIVAC 494	3/66	1,291,740	1,527,140	24.94
294	SDS 940	4/66	289,444	301,365	34.64
295	RCA Spectra 70/55	7/66	1,341,132	1,224,010	19.48
296	RCA Spectra 70/45	7/66	211,610	290,493	41.57
297	RCA Spectra 70/35	7/66	61,186	126,391	77.94
298	Philco 200–213	10/66	6,251,118	4,307,061	7.793
299	IBM 360/44	10/66	1,025,941	858,520	62.35
300	Honeywell 4200	5/67	45,569	32,270	31.18
301	SDS Sigma 7	12/66	894,566	554,280	41.57
302	PDP-8/S	9/66	1,595	8,546	1,247.
303	PDP-9	12/66	107,672	352,534	1,247.
304	SDS Sigma 2	1/67	118,152	101,079	155.8
305	Burroughs B 2500	2/67	22,153	28,791	124.7
306	Burroughs B 3500	5/67	154,842	130,251	69.31
307	UNIVAC 9300	6/67	4,350	18,424	138.6
308	UNIVAC 9200	6/67	1,592	7,458	415.7
309	Burroughs B 6500	2/67	3,127,266	2,755,760	15.59
310	CDC 3500	9/67	1,086,342	1,021,365	29.69

PERFORMANCE IMPROVEMENT 1963 TO 1966

Data for commercial and scientific computation (Table 2) are plotted in Exhibits 5 and 6. As in the earlier article a regression technique has been used to describe the changes in computer performance from year to year and also to compare the computer performance per dollar of computer cost. The equation fitted is the same as the one used for the period 1950–62:

$$(10) \qquad ln(C) = \alpha_0 + \alpha_1 \, ln(P) + \beta_1 S_1 + \beta_2 S_2 + \beta_3 S_3 + \beta_4 S_4$$

In this equation the α's and β's represent the regression coefficients to be determined by the least squares analysis. The S_1, S_2, S_3, and S_4 represent dummy variables (or shift parameters) for the different years considered (1963–66). For the regression we will include 1962 as the base year. We will also consider all the systems from 1963–66 in the regression analysis.

The results of the regression calculation using 111 computers introduced between 1962 and 1966 are as follows.

For scientific computation:

(11) $ln(C) = + 6.823 - .322\ ln(P)$
$+ 0.000\ (1962)\qquad + 0.822\ (1965)$
$+ 0.272\ (1963)\qquad + 0.988\ (1966)$
$+ 0.415\ (1964)$

For commercial computation:

(12) $ln(C) = + 7.441 - .404\ ln(P)$
$+ 0.000\ (1962)$
$+ 0.385\ (1963)$
$+ 0.723\ (1964)$
$+ 1.186\ (1965)$
$+ 1.550\ (1966)$

The plots of Equations 11 and 12 are shown in Exhibits 5 and 6.[8, 9] We see that the most striking observation once again is the rapid advance in computer performance. For scientific computation the average improvement in performance over the previous year, holding cost constant, is determined by the 115 percent increase per year in computer capability for equal cost. The previous result shows for the years 1950 through 1962, an average of 81 percent per year. For commercial computation we find an average of about 160 percent per year for 1963 through 1966 against the earlier 87 percent for the years 1950 through 1962. As shown in Exhibit 7 each year 1963–66, that is each of the

[8] The technology curves shown in Exhibits 5 and 6 are not comparable with those in Exhibits 2 and 3. The curves shown in the current exhibits were calculated using all the general purpose computers introduced in the period 1962–66. The earlier article used a procedure that eliminated some of the systems that were technology inferior, those that fell far below and to the left of the technology line.

[9] Note that the slopes of the technology curves shown in Exhibits 2 and 3, 1950–62, were steeper than the current curves, 1962–66. This results in the appearance that the base 1962 curves in Exhibits 5 and 6 have been rotated, and therefore they are not identical to the ones shown earlier, that is:

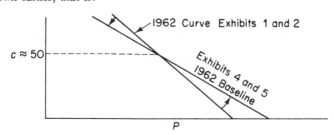

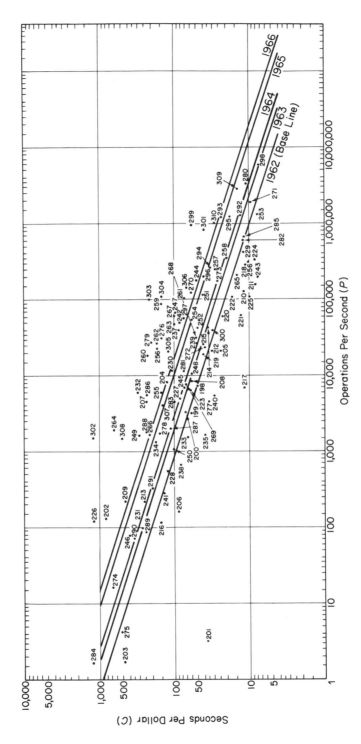

Exhibit 5. Plot of Equation 11: Regression Calculation for Scientific Computation.

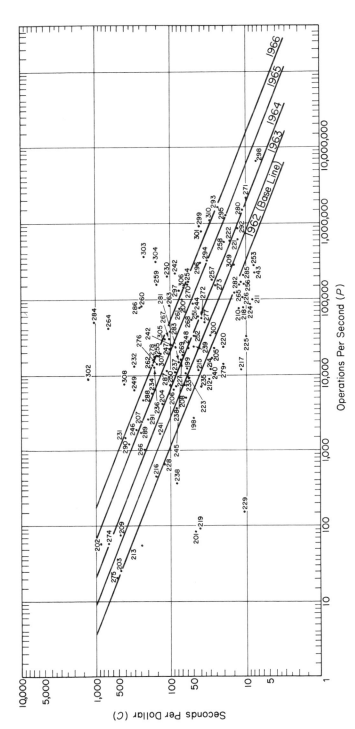

Exhibit 6. Plot of Equation 12: Regression Calculation for Commercial Computation.

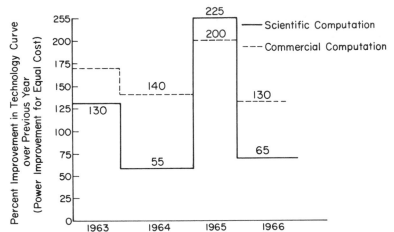

Exhibit 7. Average Yearly Shift of the Technology Curves (power, _P_, improvement for constant cost, _C_).

four years, had a significant improvement in both scientific and commercial computation capability.[10]

From our regression equation we obtain the new calculation of the economies of scale for the years 1962 through 1966. Rewriting Equation 11 we get

$$(C) = K(P)^{\alpha_1}, \text{ or}$$
$$(Cost) = K(Power)^{\alpha_1}$$

K is a constant which represents a combination of α_0 and the yearly shift parameter. Our earlier finding was that computer power increased approximately as a function of cost squared; for twice the cost you get four times as much computing power: $-\alpha_1 = 0.5$.

Our regression equations yield $-\alpha_1 = +.404$ for commercial computation, and $-\alpha = +.322$ for scientific computation. Both of these indicate that the returns to scale are greater than those we had found in the period 1950–62. For the 1950–62 period the results were that $-\alpha_1 = +.519$ for scientific computation, and $-\alpha_1 = +.459$ for commercial computation. We now find that during the years 1962–66 our equation for scientific computation is $P = K(C)^{2.5}$, and for commercial computation $P = K(C)^{3.1}$. These returns to scale are greater than those found during the period 1950–62.

Our use of a detailed historical analysis of the general purpose digital computers was a good short-term technological forecast. We did find that the tremendous rate of improvement in computing power for fixed cost that w

---

10 Again we must note that our economies of scale are calculated as percent change (C) for a constant (P). The numbers would have been smaller if we had calculated percent change in (P) for constant (C), since we have the economies of scale included in our calculations.

observed between 1950 and 1962 has continued and possibly slightly accelerated from 1963 through 1966 with the introduction of the third generation computers. We also find that the economies of scale are still present and that today there appear to be even greater economies of scale, with larger machines providing equivalent computation at much less cost.

INSIGHT INTO TECHNOLOGICAL FORECASTING

Our technological history of computers has been profitably used to generate technological forecasts. While many simplifying assumptions went into the constructions of the determination of computing power (P) and cost (C), the results provide a good overview of the development of this industry and the important technological and economic relationship of its elements. The research has had some success in predicting, but more importantly it is a first step that will allow others to improve upon our crude technical estimates.

The most difficult aspect of the research was to obtain the data on the various computers that were used in the functional model. The data were obtained through a lengthy and tedious search of the published literature, examination of the manufacturers equipment specifications and many helpful friends. The tremendous reception that the research has received has more than rewarded the efforts.

Normex Forecasting of Jet Engine Characteristics

A. WADE BLACKMAN, JR.

The normex forecasting technique is developed and applied to examine the historical dynamics of the commercial jet aircraft engine market and to forecast future market characteristics as a function of selected engine performance parameters. It is shown that lognormal distributions can be used to represent the variations of market share with various engine performance parameters and that market positions are highly susceptible to technological changes in engine performance. Forecasts of market share as a function of selected engine performance parameters are made for the 1975 time period.

INTRODUCTION

Improved techniques for predicting the characteristics of future markets for technology-intensive products are increasingly in demand because of the increasing capital investments needed to respond to market demands, reduced lead times, high risks, competitive pressures, and so on.

Two generally accepted forecasting techniques currently widely used to predict future technological capabilities are the exploratory forecasting and normative forecasting techniques which are discussed in detail by Jantsch [1]

Exploratory technological forecasting is a technique in which key performance parameters characterizing a particular area of technology are identified and future progress is predicted by extrapolating existing rates of change of the performance parameters to indicate future capabilities. Normative technological forecasting techniques seek to identify future sociological, geopolitical, and economic environments along with their concomitant demand patterns and to indicate the rate of technological progress required in a given field to meet the future needs.

A problem of compatibility sometimes exists when these two approaches are interfaced, because the normative forecast may indicate a need for a particular level of technology which cannot be met unless existing rates of technological change (as identified by the exploratory forecasts) are accelerated from those which can be achieved with current resource allocations. Alternatively, the demand for a technology may have a limited lifetime which may no longer exist when the required technological level is achieved. These concepts are illustrated schematically in Exhibit 1.

In the case of technology-intensive products having sales characteristics which are largely dependent on the technology content of the product, the interfacing problem between the two forecasting techniques can be minimized and an improved indication of future demand may be obtained by relating the market characteristics to technological performance parameters. By predicting future market characteristics, the future technological characteristics which will be demanded by the market in a future time period can then be identified. Such an approach has been utilized by Brown and Burkhardt [2] to indicate the future characteristics of the computer memory market. Because this approach applies exploratory (trend extrapolation) techniques to obtain a normative (that is, need or market-demand oriented) type of forecast, the technique will be termed "normex" forecasting. To fully realize the potential of normex forecasting, a need exists for the development of general mathematical procedures which can be used to predict the future relationships expected to exist between market characteristics and technological performance parameters.

The purpose of the work presented herein is to develop mathematical procedures applicable to the normex forecasting technique and to apply these procedures to indicate past trends in and future characteristics of the commercial jet aircraft engine market.

ANALYSIS

The point of departure for the application of normex forecasting is an analysis of the salient characteristics of as large a segment of the total market as possible; ideally it should be based on annual world sales. This analysis should include a determination of the time variation of product market shares and a

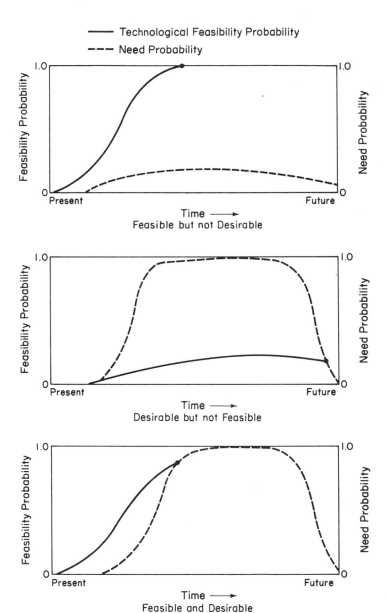

Exhibit 1. Temporal Relationships between Technological Feasibility and Demand.

delineation of the pertinent technological characteristics of the products making up the market.

The next problem which must be faced in application of the normex procedure is that of finding a mathematical function which will provide an adequate approximation to the frequency distributions of sales (or market shares) as functions of pertinent technological performance parameters selected on the basis of their influence on sales.

For the work described herein, the engine performance parameters of specific fuel consumption (defined as the pounds per hour of fuel flow divided by the engine thrust) and engine specific weight (defined as the engine weight in pounds divided by the engine thrust in pounds) were selected for consideration. The influence of these performance parameters on the economics of aircraft propulsion is discussed by Sens and Meyer [3]. Specific fuel consumption (SFC) and engine specific weight are shown to have an important influence on aircraft direct operating cost. For example, a 10% change in SFC or engine thrust-to-weight ratio (the inverse of specific weight) produced a 4.9% and 1.0% change, respectively, in the direct operating cost of a long-range subsonic transport.

Histograms (to be presented later) relating the selected engine performance parameters to market shares in a given year revealed a general skewness in the distributions with the long tail extending to the right. It appeared that a normal distribution would not be likely to provide an adequate approximation but that a lognormal distribution might provide the desired characteristics. It would not be implausible to assume that market shares in a given period of time might be determined by the product rather than the sum of performance factors as would be the case with a lognormal distribution. The logarithm of the performance parameter would be the sum of independent random factors, and as the number of these random factors increased, the distribution of the logarithm of the performance parameter would approach the normal distribution according to the central limit theorem. These concepts are discussed further by Holt, *et al.* [4].

The characteristics of the lognormal distribution are very appealing. As discussed by Pessemier [5] and Aitchison and Brown [6], it is a distribution which can be made sufficiently symmetric or asymmetric to yield a good fit to a wide variety of distributions encountered in practical applications. Also, the parameters of the distribution can be easily estimated using either graphic or nongraphic procedures [5]. The graphic procedure may be applied by measuring the points on a cumulative frequency plot on lognormal probability paper at which the fitted straight line crosses the 0.5 line and the 0.84 line on the probability scale, and the shape parameters of the underlying normal distribution can be estimated from the following relationships:

(1)
$$\mu = \ln X_{0.5}$$

(2)
$$\sigma = \ln \frac{X_{0.84}}{X_{0.5}}$$

μ = mean of underlying normal distribution
σ = standard deviation of underlying normal distribution
X = variable

When the shape parameters (μ and σ) are known, a lognormal frequency curve may be fitted to histograms (for example, constructed by plotting market shares on the ordinate and selected engine performance parameters on the abscissa) utilizing the following equation (see Croxton and Cowden [7]).

$$(3) \qquad \Upsilon = \frac{0.398\ Ni}{X\sigma} \exp\left[\frac{-(\ln X - \mu)^2}{2\sigma^2}\right]$$

where

$\Upsilon =$ value of ordinate
$X =$ value of abscissa
$N =$ number of items in sample (for example total units sold in a given year)
$i =$ class interval
$\mu =$ mean of underlying normal distribution
$\sigma =$ standard deviation of underlying normal distribution

Note that the constant in Equation (3) differs from that given in Croxton and Cowden [7] because in Equation (3) natural logarithms are used rather than logarithms with a base of ten.

The analytical procedures utilized in this study may be summarized as follows:

1. Historical data were assembled on annual world sales and on the technological performance characteristics of the products which contributed to annual sales.
2. For selected years, histograms were constructed of the frequency distributions of sales as functions of engine performance parameters.
3. For each of the above histograms, cumulative frequency plots were constructed on lognormal probability paper and the mean and standard deviation of the underlying normal distribution were estimated graphically.
4. Values of the mean and standard deviation of the underlying normal distributions as determined for the selected years were plotted versus time and extrapolated into the future. Similar plots were constructed of annual unit sales and were also extrapolated into the future.
5. Using the extrapolated values obtained above in Step 4, along with extrapolated values of future world sales obtained above in Step 1, frequency distributions of sales as a function of technological performance parameters were calculated utilizing Equation (3) for future years of interest.

RESULTS AND DISCUSSION

Past and Current Market Characteristics

The historical dynamic characteristics of the commercial jet aircraft engine market as affected by changing technology are illustrated by the isometric plot in Exhibit 2. The early engines which dominated the market in the 1958 to

1962 time period were pure turbojets which had SFC values of approximately 0.75–0.76.[1] The first generation of turbofan engines began their market penetration in approximately 1960. As a class, the first generation turbofans had SFC values which showed a greater degree of variation (approximately 0.51 to 0.69) because of varying fan bypass ratios, and so on. Their better specific fuel consumption as well as other desirable characteristics, gave them an economic advantage vis-à-vis the turbojet and led to their market dominance by approximately 1966. Currently, this dominance is beginning to be eroded by the second generation of turbofans which have performance capabilities superior to those of the first generation designs (SFCs of 0.33). It is clear from this plot that technological improvements have an exceedingly important effect on market dynamics. There appears to be relatively little market reluctance to the acceptance of engines having improved performance capabilities. This ready acceptance results from the strong influence of engine performance on operating economies and overall return on investment.

Exhibit 3 shows the variation of specific fuel consumption with time and gives an indication of the rate of technological progress. To obtain these plots,

[1] The SFC values used throughout are values at takeoff conditions and are generally lower than values at cruise conditions (cf., refs. [8] and [9]).

Exhibit 2. Commercial Jet Aircraft Engine Market Dynamics.

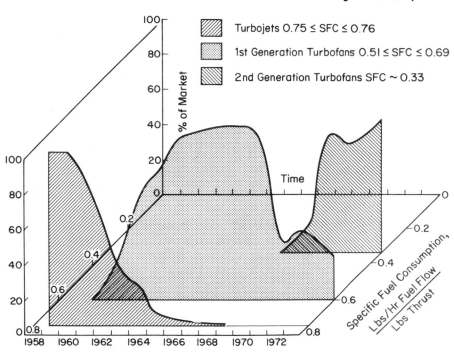

Turbojets 0.75 ≤ SFC ≤ 0.76

1st Generation Turbofans 0.51 ≤ SFC ≤ 0.69

2nd Generation Turbofans SFC ~ 0.33

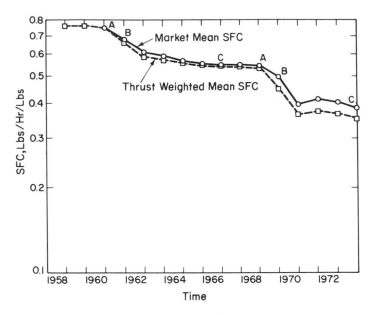

A — Introduction of New Generation

B — Rapid Penetration of New Generation

C — Nearly Complete Dominance of Market by New
Generation

Exhibit 3. Variation of Market Mean and Thrust Weighted Mean SFC with Time.

the SFC of the various engines making up the market were weighted according to either their market share or the engine thrust produced by a given engine type as a percentage of the total thrust of all engines in the market. It can be seen that the mean values obtained were relatively independent of the averaging method. Similar plots of the variation with time of the market mean SFC and specific weight are presented in Exhibits 4 and 5, respectively. Shown along with the mean performance values are the plus and minus one-standard-deviation values which give an indication of the dispersion of the data.

Histograms and cumulative frequency plots showing market characteristics as a function of the performance parameters of SFC and specific weight are presented in Exhibits 6 through 13 for selected years. The circular data points shown on these figures indicate the market shares captured by a particular model jet engine which produced the performance indicated by the abscissa value. Shown also in these figures are theoretical frequency distributions superposed on the histograms. These distributions were calculated using the shape parameters (determined graphically) in conjunction with the computational procedure discussed previously.

In general, good agreement was obtained between the histogram data and the theoretical lognormal distributions. It should be noted that because the

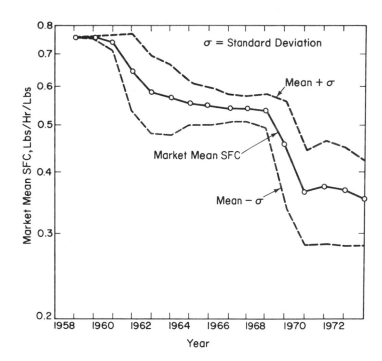

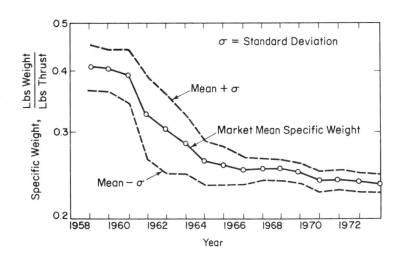

lognormal distribution is asymmetric, the mean of the distribution occurs slightly to the right of the peak frequency values.

The cumulative frequency plots which are also presented in Exhibits 6 through 13 were obtained by plotting on the ordinate the accumulated values of market shares versus the performance parameter value on the abscissa. As log-

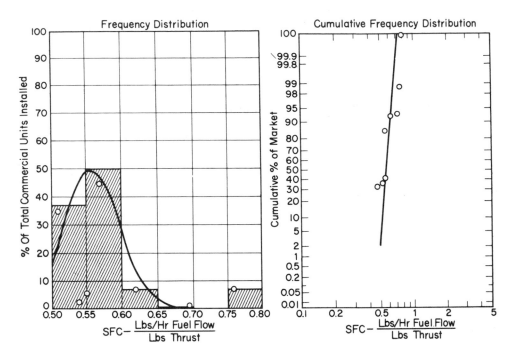

Exhibit 6. SFC Variations among Commercial Installations in 1964.

Exhibit 7. SFC Variations among Commercial Installations in 1966.

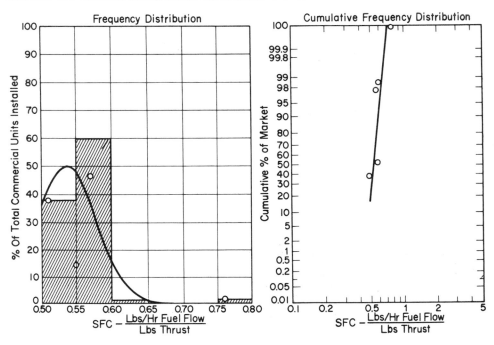

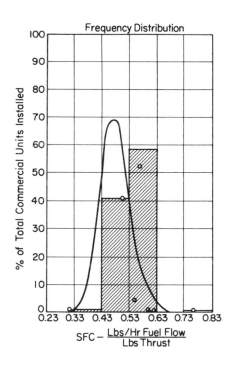

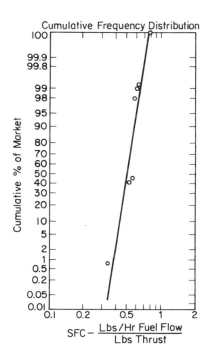

Exhibit 8. SFC Variations among Commercial Installations in 1968.

Exhibit 9. SFC Variations among Commercial Installations in 1970 (estimated).

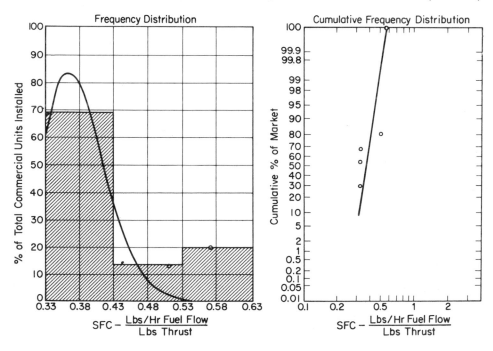

normal probability graph paper was used, a straight line would indicate a lognormal distribution. The ordinate indicates the portion of the market made up of engines having performance values less than or equal to the abscissa value. It can be seen that good agreement between the data and straight-line plots was obtained, thus indicating that the lognormal distribution hypothesis is valid.

Both the mean and shape of the distributions can be seen to vary with time. Variations in the mean are caused by improvements in the technological state-of-the-art resulting from research and development efforts. Changes in the shape of the distributions are probably related to the dynamics of the replacement of existing equipment by improved designs and would be expected to be a function of the rate of obsolescence, lag times involved in engine change-overs, the investment required to replace engines, capital costs, degree of amortization, and so on.

It is significant that at a given point in time, the most advanced technological capability makes up only a very small portion of the market. Hence, if one relies on exploratory extrapolation of maximum performance parameters (Jantsch [1]) to forecast future product performance and no consideration of the market characteristics is introduced, it is possible to be misled. Furthermore, a market-oriented forecast can serve to identify trade-offs which may be oc-

Exhibit 10. Specific Weight Variations among Commercial Installations in 1964.

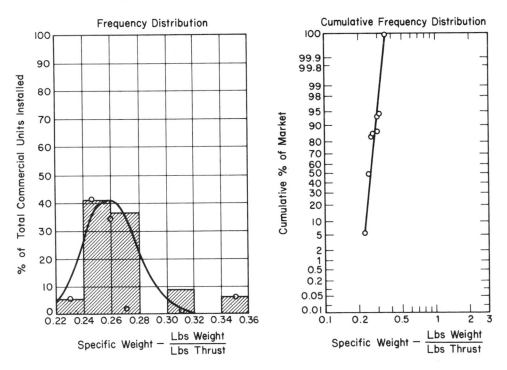

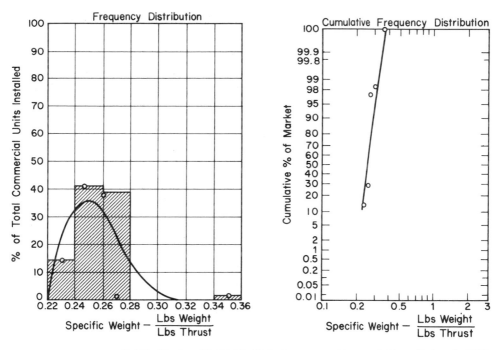

Exhibit 11. Specific Weight Variations among Commercial Installations in 1966.

Exhibit 12. Specific Weight Variations among Commercial Installations in 1968.

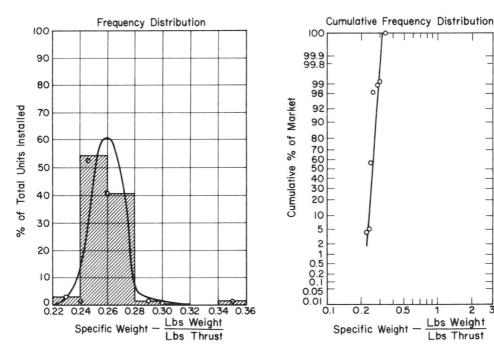

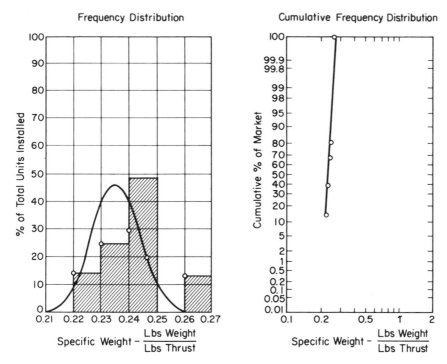

Exhibit 13. Specific Weight Variations among Commercial Installations in 1970 (estimated).

curring between performance parameters; for example, noise limitations may impose thrust and/or specific fuel consumption operating requirements which are less favorable than the levels which are technologically possible.

Future Market Characteristics

Having gained some insight into the past and current characteristics of the commercial jet aircraft engine market, the future characteristics of this market will be forecast utilizing the assumption that existing trends will continue through the 1975 time period.

Plots of the variation with time of the shape parameters (μ and σ) are presented in Exhibits 14 through 17 for the performance parameters of SFC and specific weight, respectively.

In most cases, a linear variation with time was approximated when the shape parameters were plotted on semilogarithmic graph paper. The curves of μ, the mean of the underlying normal distribution, generally have less scatter than the curves of σ, the standard deviation of the underlying normal distribution. When new engine types having improved performance characteristics are first introduced to the market, some time is required before the new engine sales

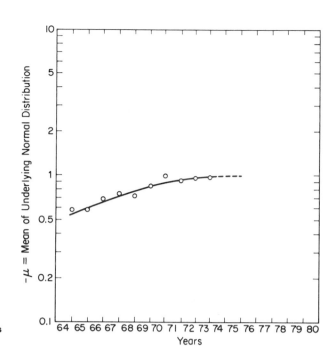

Exhibit 14. Variation of SFC Shape Parameters with Time.

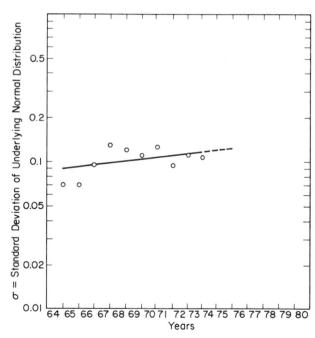

Exhibit 15. Variation of SFC Shape Parameters with Time.

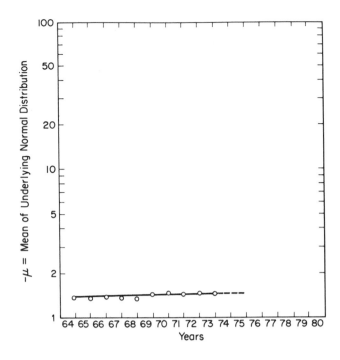

Exhibit 16. Variation of Specific Weight Shape Parameters with Time.

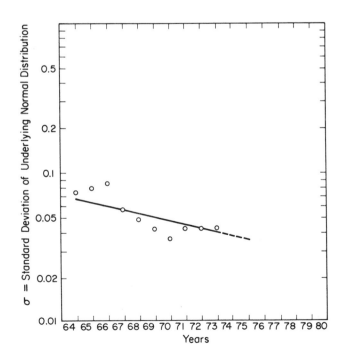

Exhibit 17. Variation of Specific Weight Shape Parameters with Time.

have an appreciable impact on the market. During this time period, the new engine's introduction would have little influence on the market mean because the new engine's market share is low; however, the dispersion of the distribution would be affected, and a scatter in the σ time series greater than in the μ time series would result.

By extrapolating the distribution shape parameters (from Exhibits 14 through 17) and estimating unit sales for future time periods, it is possible to make forecasts of the future market characteristics using the calculation procedures previously discussed. Such forecasts for the 1975 time period are presented in Exhibits 18 and 19 for performance parameters of SFC and specific weight, respectively. Relative to current market mean performance levels, the forecasts show reduced SFC and reduced specific weight as would be expected from the extrapolation of existing technological progress directed toward fulfilling the market requirements of lower aircraft direct operating costs and improved return on investment.

At this point, some remarks relative to the accuracy of these forecasts would appear to be appropriate. It is recognized that uncertainty is the hallmark of

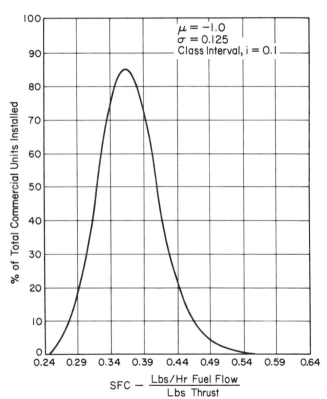

Exhibit 18. Forecast of Distribution of Jet Engine SFC in Commercial Market of 1975.

$$SFC - \frac{Lbs/Hr\ Fuel\ Flow}{Lbs\ Thrust}$$

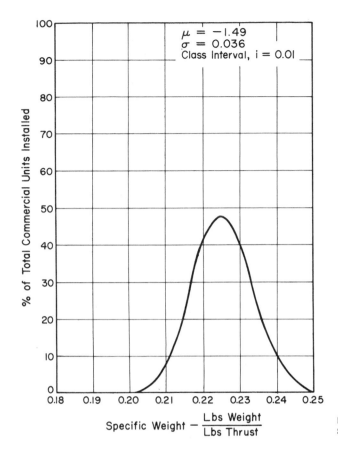

$$\mu = -1.49$$
$$\sigma = 0.036$$
Class Interval, $i = 0.01$

Exhibit 19. Forecast of Distribution of Jet Engine Specific Weight in Commercial Market of 1975.

practically all forecasting techniques and in this respect the normex technique is no exception. Normex forecasts are based on the assumption that currently existing trends will continue in the future. To the extent that this hypothesis is invalid, error will be introduced. Also, as previously pointed out, the normex technique is predicated on the assumption that there is a strong relationship between a product's sales potential and its technological performance parameters; that is, its application (within the context of its current formulation) is largely restricted to technologically intensive products. However, the methodology is general and should be applicable to products whose sales depend primarily on other characteristics such as marketing, advertising budgets, production, and so on, provided that the product's sales are related to the correct parameters. Through its ability to forecast both the mean and variance associated with future technological performance parameters, the normex technique provides the capability for evaluating the level of uncertainty associated with a forecast, and hence a measure of the extent of possible error is presented along with the forecast.

In general, it would appear that forecast errors would be minimized by making a number of different individual forecasts utilizing many different analytical techniques applicable to the same time period and basing the final forecast on those forecasts having the greatest degree of internal consistency. It would appear that the normex technique would be most useful in this context; that is, it should serve as an adjunct to existing forecasting techniques to provide useful consistency checks.

The primary advantages of the normex technique are summarized below:

1. The results give an indication of the technical requirements for future product designs and can serve to establish goals and priorities for future research programs necessary to achieve the technology requirements demanded by the market in a specified future time period.
2. The technique can indicate market oriented trade-offs between technological performance parameters and can serve to indicate the desired interface between technological capability and market demand.
3. The normex technique produces an evaluation of both the mean and the variance of technological parameter forecasts and thereby provides a clear indication of the uncertainty associated with future forecasts.

REFERENCES

1. JANTSCH, E., *Technical Forecasting in Perspective*. Organization for Economic Co-operation and Development, Paris, 1967.
2. BROWN, D. W., and J. L. BURKHARDT, The Computer Memory Market: An Example of the Application of Technological Forecasting in Business Planning. Paper presented at the Second Session of the Second Annual Technological Forecasting Conference, Lake Placid, New York, June 9–13, 1968.
3. SENS, W. H., and R. M. MEYER, New Generation Engines—The Engine Manufacturer's Outlook. SAE Paper No. 680278 presented at the Air Transportation Meeting, New York, April 29—May 2, 1968.
4. HOLT, C., *et al.*, *Planning Production, Inventories, and Work Force,* pp. 272–98. Englewood Cliffs, New Jersey: Prentice-Hall, 1970.
5. PESSEMIER, E. A., *New Product Decisions,* pp. 169–74. New York: McGraw-Hill, 1966.
6. AITCHISON, J., and J. A. C. BROWN, *The Lognormal Distribution.* Cambridge Univ. Press, 1966.
7. CROXTON, F. E., and D. J. COWDEN, *Applied General Statistics,* pp. 613–19, 2nd ed. Englewood Cliffs, New Jersey: Prentice-Hall, 1955.
8. *Gas Turbine International,* **10** (January/February 1969).
9. WILKINSON, P. H., *Aircraft Engines of the World* 1966/1967. Washington, D. C.: Wilkinson, 1967.

Forecasting Errors in the Selection of R&D Projects

DENNIS L. MEADOWS

Almost every firm frequently and implicitly does short-term technological forecasting when R&D personnel and managers make estimates and commit resources to technological goals. The recent cataclysmic time and cost errors in Rolls Royce and Lockheed development projects received worldwide publicity, yet they are simply grand examples of smaller technological forecasting errors that occur in many firms and government laboratories. Hopefully, more attention will now be paid to short-term technological forecasting in development work.

Meadows has studied laboratory records on numerous development projects in a dozen firms to compare forecasted results with actual outcomes. By statistical analysis of project histories and through models of information flow, he traces the sources and nature of errors in management estimates of time, cost, and manpower requirements, as well as probabilities of technical and commercial success.

INTRODUCTION

Management's typical reaction to any proposal for a formal technological forecasting system is that the accuracy of the forecasts would not justify the cost of the system. Forecasts are inaccurate. There is little quantitative research into forecasting errors, but the few studies which have been conducted, together with a great deal of managerial experience, suggest that errors are often large. Occasionally they are catastrophic. Proponents of integrated long-range technological forecasting systems cannot deny this. They can only suggest that systematic consideration of alternatives, automated handling of the vast quantities of relevant data, and comprehensive consideration of important interrelationships together offer an improvement over management's more intuitive and ad hoc approach to forecasting and planning. That reasoning has obvious merit, but it will lack real imperative until accompanied by a more systematic investigation of the errors actually inherent in forecasts and of their implications for the forecasting techniques currently proposed.

The low usage of quantitative procedures for forecasting the value of alternative development projects illustrates this point. Formal project selection models were the first technological forecasting procedures to receive wide attention in the management science literature. A 1964 survey of project selection techniques discovered over eighty different proposed forecasting methods [1]. The research revealed, however, that there was little implementation of any technique, primarily because none of the procedures satisfactorily treated the problem of forecast accuracy. Other evidence supports this conclusion [2].

More recent proposals for project selection formulas and for other technological forecasting techniques still have not considered the effects of the errors in the forecasts upon which they are based. A few studies have explicitly treated the problem of forecasting accuracy [3, 4]. They have, however, compiled only post mortems by calculating differences between initial estimates and final outcomes. Implicit in such "before-after" research is the assumption that forecasts and outcomes do not interact in the process of technological development and transfer. Exhibit 1 is a more realistic representation.

R&D management is a dynamic process, adjusting more or less continuously to differences between what was forecast and what has apparently been achieved through the efforts of the technical staff. Decisions based on the forecasts lead to the allocation of the laboratory's resources among alternative projects. After delays inherent in the development process, this expenditure results in technical achievements. Management attempts to monitor this progress, but there are

Exhibit 1. Decision Feedback Loop.

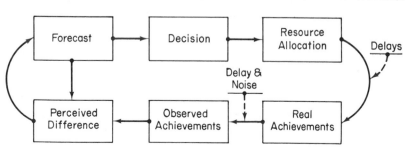

delays, noise, and deliberate bias in both the formal and the informal communication channels of the laboratory. These factors together with initial errors in the forecasts lead to differences between management's observations and its expectations. Those differences typically lead to revised forecasts and new decisions designed to decrease the difference between perceived and forecasted or desired results. Most research managers have personally witnessed this interaction between forecast and error in some project where the expected and actual cost leapfrogged each other upwards, while the project's technical solution appeared always just around the corner. Obviously a before-after analysis of forecasting error does not fully characterize the errors in such a system as that illustrated above.

The continuous sequence of forecast-decision-results-errors-forecast constitutes a feedback loop, a fundamental structure inherent in the control of all physical and social systems. It is through a plexus of feedback loop structures like that in Exhibit 1 that forecasting techniques and their associated errors affect the performance of the firm. If we are to understand the implications of errors, they must be viewed in the context of this feedback loop structure.

Several members of the M.I.T. Research Program on the Management of Science and Technology have begun to study the characteristics and the dynamic implications of errors in the forecasts used to select among alternative development projects. Two aspects of that work are: data on the errors which actually occur in industrial-development-oriented forecasts and results from computer simulations which trace the effects of such forecasting errors as they are propagated through the decision loop networks of the laboratory.

The specific characteristics of laboratory decision loops will vary from one firm to another, but the decision feedback loop structure underlying industrial development activity is quite general. Individual laboratory differences can easily be accommodated by the statistical and simulation techniques I will present. It is thus already possible to generalize some results of the research, and I will point out several significant conclusions. I must emphasize, however, that the real importance lies with this approach to forecasting. Thus I will stress the procedures as much as the conclusions which have so far resulted from the work. Before the potential of technological forecasting systems can be fully realized, each firm must study the process by which it plans and allocates its own resources in the development of new technology.

ACTUAL ERROR CHARACTERISTICS

The before-after approach to estimate error does not completely assess the characteristics of errors in a firm's forecasting procedures. It does, however, indicate the magnitude of the errors which forecasting procedures must accommodate. If there are sufficient data one may also find statistical relations between error magnitude and the characteristics of the firm's organization and technology. Both of these are useful inputs to any dynamic model of error and project performance. Thus we have included this approach as the initial phase in our study. We have studied four firms in detail. Their forecasting performance will be analyzed here. Data from a fifth firm, studied by Edwin Mansfield, will be included to check the generality of our findings [5].

Three of the laboratories are in chemical companies engaged primarily in the development, production, and sale of industrial intermediates. Two of the chemical firms, A and B, have between $100 and $300 million annual sales. The third is one of the largest firms in the nation. The fourth laboratory, called here the Electronics Laboratory, is maintained by an instrumentation company to engage in work of interest to NASA and the Department of Defense. The fifth set of project data comes from the Mansfield study of the central research unit in a prominent equipment manufacturer.

Data from each of the five laboratories were obtained from three sources. The initial estimates of project cost, probability of technical success, and probability of commercial success, when available, were obtained from project evaluation forms completed at the time each project was first approved for funding. Accounting records supplied the actual costs, and the leader of each project indicated whether the project had eventually been technically and/or commercially successful.

Data were obtained on 144 completed projects, and these were divided into four mutually exclusive categories according to their outcomes.

> Miscellaneous failure: The project was closed out because of manpower shortages, changes in the market objectives of the firm, or other nontechnical reasons. The project did not result in a product.
>
> Technical failure: The project was closed out when unforeseen technical difficulties prevented the development program from producing the desired product.
>
> Commercial failure: The project did produce the product initially desired, but the product was not sold.
>
> Commercial success: The project was technically successful, and the product did produce sales income for the firm.

With the exception of Chemical Laboratory B, no attempt was made to rank projects on the basis of their technical or commercial performance. The definitions above do not apply to the projects in Chemical Laboratory C where only the estimated and actual costs for one year's development on each project were available.

The data from each laboratory were analyzed to determine the magnitude of the forecasting errors associated with each project and to discover significant relations between those errors and other attributes of the laboratory. The conclusions resulting from this work are listed below and then followed by a detailed presentation of the data and the analyses which lead to them.

Conclusions

1. There is a very low correlation between the actual and estimated values of cost, probability of technical success, and probability of commercial success figures employed in selecting among development projects.

2. There is a pronounced tendency for technically and commercially unsuccessful projects as a group to incur greater cost overruns than commercially successful projects.

3. The average magnitude of cost overruns and the rate of commercial success may differ depending upon the sources of the project idea.

4. Technically and commercially unsuccessful projects cost more on the average than commercially successful projects.

5. Forecasting error typically leads management to expend more than fifty percent of a firm's development resources on projects which do not produce commercially successful products.

6. Forecasting errors are influenced by the laboratory's organizational form.

Discussion

1. There is a very low correlation between the actual and estimated values of cost, probability of technical success, and probability of commercial success figures employed in selecting among development projects.

Cost Estimates. In Chemistry Laboratories A and B, the cost of successfully meeting the technical requirements of a project is predicted before deciding whether it should be funded. The actual costs of projects which were technically successful can thus be compared with the initial estimates to obtain one measure of the error in those estimates. Exhibits 2 and 3 compare initial estimates with actual costs for each of the two organizations. The actual costs are generally found to differ substantially from those initially forecast. Where there are deviations of this type, it has been suggested that a linear formula in the form,

$$\text{Actual Cost} = a + (b) \text{ (Estimated Cost)}$$

be employed to modify the initial estimates, correcting for the errors inherent in them [6]. It is easy to determine statistically the "best" linear formula for any given set of historical data, but many factors not explicitly considered can limit the predictive ability of such an equation by causing some deviation of the actual costs from the modified estimates. The correlation coefficient, r, measures this deviation. The quantity r^2 indicates what percent of the variance in actual costs is explained by the modified estimates. If r (consequently r^2) is low, the estimates will be of little use in predicting actual costs. The correlation coefficient for each set of projects is given on the corresponding figure. The most accurate estimates, those in Chemical Laboratory A, explain only twenty-five percent of the variance in the actual project costs even when these estimates

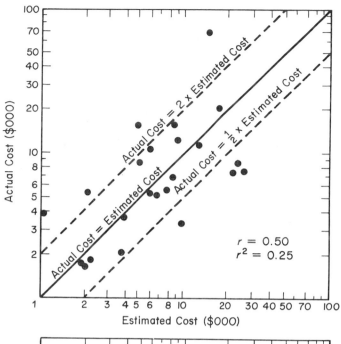

Exhibit 2. Estimated Versus Actual Costs for 23 Technically Successful Projects, Chemical Laboratory A.

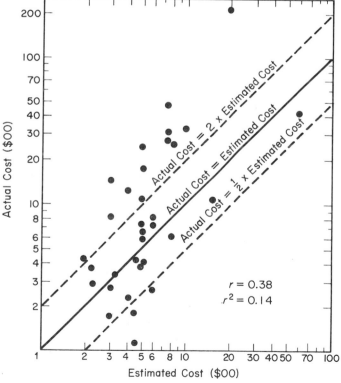

Exhibit 3. Estimated Versus Actual Costs for 33 Technically Successful Projects, Chemical Laboratory B.

are modified with the best linear correction formula for that particular set of data.

In Chemistry Laboratory C project costs are predicted for only one year at a time. The uncertainty in annual cost estimates should be substantially less than that in the initial estimates of total project costs. The estimates were found, however, to account for no more than fifty-seven to seventy-four percent of the variance in actual costs.

Historical data on cost forecasting accuracy may also be expressed in the form of cumulative frequency distributions. Exhibits 4 and 5 present the cumulative probability curves of cost overruns among the projects studied in Chemical Laboratory A and B respectively. In Laboratory B, as indicated by dotted line 1, thirty-eight percent of the projects were completed without incurring cost overruns. For those projects the ratio (Actual Cost/Estimated Cost) was less than or equal to 1. Similarly, seventy-five percent of the same projects were completed at less than or equal to 2.4 times the initially forecast development cost (dotted line 2). This representation of historical errors has particular value in evaluating the utility of forecasting models. Its use in this role will be discussed later.

Probability Estimates. The staffs of Chemical Laboratory A and the Equipment Laboratory estimate the probability of technical success and the prob-

Exhibit 4. Cumulative Distribution of Cost Overruns for 23 Technically Successful Projects, Chemical Laboratory A.

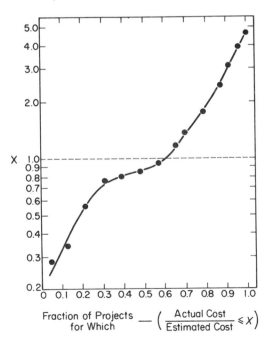

Fraction of Projects for Which $\left(\dfrac{\text{Actual Cost}}{\text{Estimated Cost}} \leqslant x \right)$

Exhibit 5. Cumulative Distribution of Cost Overruns for 33 Technically Successful Projects, Chemical Laboratory B.

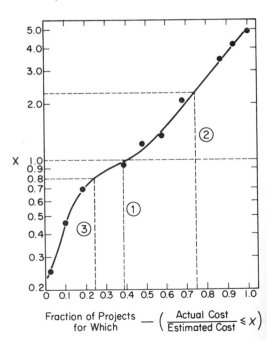

Fraction of Projects for Which $\left(\dfrac{\text{Actual Cost}}{\text{Estimated Cost}} \leqslant x \right)$

ability of commercial success of each development project before deciding whether it should be funded. The estimated values are employed by management as if they characterized a binomial process. For example, of all the projects which receive the initial estimate PTS = 0.80, 80 percent are expected to be technically successful.

Probability estimate errors are more difficult to measure than cost estimate errors. A numerical overrun ratio can be employed to indicate the error in an estimated cost figure. There is no equivalent measure of probability estimate error. When a coin lands with one side up on eight of ten tosses, it does not necessarily disprove the initial estimate that the probability of that side landing up is 0.50. Similarly, a deviation between the fraction of projects which actually succeed and the probability of technical success initially assigned a set of projects does not necessarily prove the initial estimate to have been in error. A deviation of the fraction actually successful from that expected on the basis of the initial forecast is particularly likely to occur when there are only a few projects in the group. Thus only statistical measures of probability estimate error are useful.

To measure the accuracy of the probability of technical success estimates in Chemistry Laboratory A, all projects were divided into groups according to the probability of technical success values initially assigned. The estimated probability of technical success values are indicated on the horizontal axis of Exhibit 6. The fraction of each group which actually succeeded was determined from the data, and these fractions are indicated by vertical white bars.

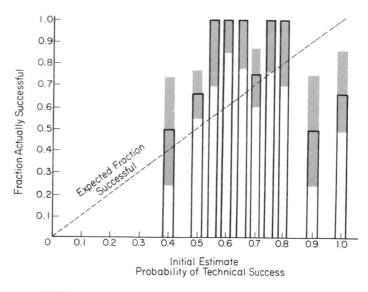

Exhibit 6. Accuracy—Estimated Probability of Technical Success in 30 Projects, Chemical Laboratory A.

50 Percent Confidence Interval for True Probability of Success

Finally, that fraction actually technically successful was used to calculate limiting values of the true probability of success which the staff attempted initially to forecast. From the fraction actually successful in each set of projects, an interval was found which has a fifty percent probability of including the true but unknown probability of technical success. These fifty percent confidence intervals are shown in Exhibit 6 as shaded bands. Although the procedure is complicated, it is not difficult to interpret the results. If the estimates are accurate, about half of the shaded bands should cross the line indicating the expected fraction successful.

The measurement of probability of commercial success differs from the above in only one respect. Each estimate of probability of commercial success was based on the assumption that the project would be technically successful, that is, that it would result in a marketable product. Thus only the twenty-three technically successful projects should be included in the analysis shown in Exhibit 7. Missing data on three projects permit the accuracy to be measured for only twenty projects. It is apparent that neither the probability of technical success nor the probability of commercial success forecasts distinguish usefully among the more and the less successful projects. The results also tend to confirm the laboratory managements' belief that market estimates are less reliable than technical estimates.[1]

Mansfield assigned a different meaning to the Equipment Laboratory probability of technical success estimates. Thus his analysis of estimate error differs from that described above. Nevertheless, he too concluded that, "although there is a direct relationship between the estimated probability of

1 Seiler [7], page 177.

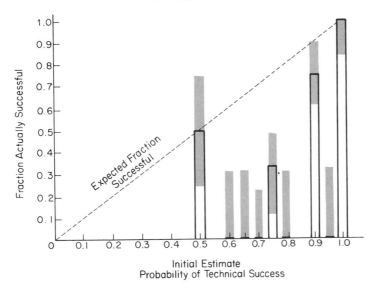

Exhibit 7. Accuracy—Estimated Probability of Commercial Success in 20 Projects, Chemical Laboratory A.

50 Percent Confidence Interval for True Probability of Success

success and the outcome of the project, it is too weak to permit very accurate predictions."[2]

2. There is a pronounced tendency for technically and commercially unsuccessful projects as a group to incur greater cost overruns than commercially successful projects.

Cost Overruns. When projects are grouped according to their outcomes, the ratio of total actual to total estimated costs for the unsuccessful projects of each laboratory is found to be greater than for the corresponding group of successful projects. This finding applies to the three laboratories in which the necessary data were available, and it appears to be independent of the average size of the projects undertaken by the laboratory. This phenomenon will be explained later in terms of the decision feedback loop structure underlying the allocation of resources to development projects. Table 1 shows the average cost overrun for each category of project in the three laboratories.

[2] Mansfield and Brandenberg [5], page 462.

Table 1. Average Cost Overrun by Project Outcome.

Project Outcome	Laboratory		
	Chemical Lab. A	Chemical Lab. B	Equipment Lab.
Miscellaneous failure			0.96
	2.55*	3.84*	
Technical failure			1.25
Commercial failure	1.28	4.25	
			0.96*
Commercial success	0.94	1.27	

* Two categories were combined in collecting the data.

3. The average magnitude of cost overruns and the rate of commercial success may differ depending upon the sources of the project idea.

Cost Overrun. In Chemistry Laboratory B the source of the project idea was recorded. It is thus possible to determine the average cost overrun of projects initiated by the laboratory, the marketing staff, or by the customer. The results are summarized in Table 2.

Table 2. Average Cost Overrun by Project Source, Chemical Laboratory B.

Source of Project Idea	Average $\frac{\text{Actual Cost}}{\text{Estimated Cost}}$
Laboratory	2.20
Marketing department	2.02
Customer	1.27

The different overruns in Table 2 probably correspond to the magnitude of technological advance attempted in each group of projects. Customers would tend to request only product modifications, while laboratory personnel are inclined to suggest more difficult technical problems. The marketing staff would presumably fall between the two extremes. This interpretation is supported by a study of military development projects conducted by Marshall and Meckling. Twenty-six programs were divided into three categories according to the magnitude of the technological advance each attempted. The cost overrun factors were 1.4 for small, 1.7 for medium, and 3.4 for large technological advances.[3]

Rate of Success. Projects in Chemistry Laboratory B which were initiated at the suggestion of customers have a much greater probability of commercial success than projects originating in either the marketing department or the laboratory. Table 3 gives the percent of each project group which resulted in no sales, small, medium, or large sales. The expenditure invested in projects from each source is also included in the table. Taken together, Tables 2 and 3 suggest that the magnitude of uncertainty associated with a development project may be related to the project source.

[3] Marshall and Meckling [3], page 472.

Table 3. Commercial Outcome by Project Source, Chemical Laboratory B.

Incremental Sales	Source of Project Idea		
	Laboratory (%)	Marketing (%)	Customer (%)
None	66	58	33
Small	17	14	33
Medium	17	14	13
Large	0	14	21
	100	100	100
Percent of total budget expended	40	40	20

4. Technically and commercially unsuccessful projects cost more on the average than commercially successful projects.

Average Costs. Data presented above indicated that unsuccessful projects tend to incur greater cost overruns (actual cost/estimated cost). Table 4 illustrates the tendency of unsuccessful projects to cost more in absolute terms as well.

5. Forecasting errors lead management to expend more than fifty percent of the firm's development resources on projects which do not produce commercially successful products.

Cost of Unsuccessful Projects. Projects which are closed out before they result in a product may contribute some new knowledge to future projects, but

Table 4. Average Project Cost by Project Outcome.

	Average Project Cost*			
Project Outcome	Chem. Lab. A	Chem. Lab. B	Equip. Lab.	Elec. Lab.
Miscell. failure	3.21†	2.51†	0.60	0.67
Technical failure			1.16	2.28
Commercial failure	2.21	2.36	1.00†	1.05
Commercial success	1.00	1.00		1.00

* Costs are expressed as a multiple of the average project cost for commercially successful projects in the respective laboratory.
† Two categories were combined in collecting the project data.

they produce no direct sales revenue or profit for the firm. The errors in current forecasts lead most of the firm's development expenses to be invested in this type of project. Table 5 exhibits the allocation of money among the four project categories actually found in the study.

Table 5. Laboratory Investment in Unsuccessful Projects

	Investment as % of Total Cost			
Project Outcome	Chem. Lab. A	Chem. Lab. B	Equip. Lab.	Elec. Lab.
Miscell. failure	45*	18*	27	10
Technical failure			20	25
Commercial failure	37	51	51†	18
Commercial success	18	31		47
	100	100	98	100

* Project leaders responded to the question, "Did the project achieve its technical objectives?" Thus no distinction was made between technical and miscellaneous failures.
† Mansfield did not obtain information on the commercial outcome of the projects in his study. Thus it is impossible to distinguish between those projects which were commercially successful and those which succeeded only technically. Rounding errors prevent percent total actual cost from equaling 100 percent.

6. *Forecasting errors are influenced by the laboratory's organizational form.*

Organization and Error. It was suggested above without any objective support that forecasting errors and the decision structure of the laboratory interact throughout the conduct of a development project. The conclusion was initially derived from qualitative consideration of the way in which development work is actually carried out, but it is supported by the research results available on forecasting errors.

Chemical Laboratory C underwent a thorough reorganization during the course of our study. Initially the laboratory was organized along scientific disciplines. All work was channeled through a steering committee composed of top laboratory and corporate personnel. The reorganization placed control of most development decisions under the jurisdiction of product-oriented business teams composed of technical, production, and marketing representatives. Data are available on the forecasting accuracy of the laboratory staff two years before and two years after the reorganization. Although the data represent cost forecasts by the same laboratory personnel for roughly comparable projects, there is a significant improvement in the accuracy of forecasts after the reorganization.

A second study by Marquis and Rubin of government sponsored development projects also indicated the relation between forecasting accuracy and laboratory organization [8]. The cost and schedule overruns in thirty-seven development projects, each over one million dollars in total cost, were measured. Work on twenty of the projects was controlled with PERT procedures. Seventeen laboratories used some other form of decision procedure in allocating resources among the groups involved in the project effort. Seventy-one percent of the laboratories not using PERT incurred overruns; only forty percent of the PERT-organized laboratories incurred similar overruns.

That completes the presentation of the more important results from our study of errors in the forecasts employed in selecting among development projects. After a few words about the generality of these errors, we will turn to the implication of this information for the performance of a laboratory and the selection of its forecasting techniques.

We are accumulating data from other laboratories, but those included in the sample above are considered technical leaders by their competitors. Each company is profitable and growing. One cannot dismiss the seemingly poor estimating performance as the result of inept technical staff or poor management. Neither can the errors be attributed to the relatively poor predictability of the particular parameters studied. Probability of technical success and project cost are the two parameters which laboratory managers feel they can forecast most accurately.[4] Thus it is expected that similar errors will be found to characterize the preponderance of industry's shorter term technological forecasts. The data above are taken from four different industries, they represent projects which differ greatly both in size and in the amount of technological advance attempted. It appears likely therefore that most firms are confronted with similar errors in the technological forecasts upon which their development decisions must be based.

IMPLICATIONS OF FORECASTING ERRORS

Information on the characteristics of errors in short-run forecasts is useful only to the extent that it is possible to determine its implications for the performance of the firm and, indirectly, for the technological forecasting process

4 Seiler [7], page 177.

dures which will be useful to the firm's management. The implications of any errors will depend upon the way in which the forecasts are employed. The project cost and product sales forecasts with which we have been chiefly concerned typically are used in two ways: (1) to select among alternative projects on the basis of their forecast profit and (2) to schedule the effort of the laboratory personnel, for example, to determine how many projects may be undertaken and to assign the available personnel among them. It is quite easy to determine the effects of errors when forecasts are used in the first role. To measure the impact of inaccuracy on scheduling decisions, however, we must return to the notion of the decision feedback loop structure underlying laboratory management.

In forecasting profitability, the effects of errors are quite straightforward. Simple arithmetic will reveal for those projects undertaken the decrease in profits which stems from development cost overruns, overoptimistic sales forecasts, or projects closed out before they produce a marketable product. As we have discovered, elimination of the errors in predicting technical and commercial success would permit the firm to cut its development expenditures by fifty percent or more without any decrease in the generation of new products.

It is important in such analyses to distinguish between contract- and market-oriented development. In fixed price or incentive contract research with no production follow-on, sales forecasts are not relevant. Cost estimate errors will, however, be an important determinant of the firm's profit. The study has found cost forecasts in error by as much as 400 percent; these would have tremendous impact on the contract-oriented laboratory.

In commercial development programs, the situation is reversed. Costing errors are relatively unimportant. Development costs are typically a small percentage of the profits from a successful product, and cost overruns of even 100 or 200 percent will not generally be important in comparison to the profit. However, even small percentage errors in the sales forecasts can make important differences in the profit of the firm.

When using a formula or other quantitative procedure to forecast the profit of a proposed project, it is important to know the confidence which may be placed in the forecast. Error research similar to that presented above, when employed by the firm with the records of development projects it has recently completed, will permit management to calculate approximate confidence intervals for the forecasts it currently is employing. The information required for these calculations can most usefully be supplied in the form of the cumulative probability curves introduced above.

Exhibits 5 and 6 above give the cumulative frequency distributions of different cost forecasting errors among the projects already completed in Chemical Laboratory A and B. Similar curves could be constructed for sales estimates or any other quantitative forecasts for which the actual outcomes can be presently measured. To the extent that projects under consideration are similar to those completed by the firm, the historical cumulative frequency curves may be taken as good indications of the errors likely to be inherent in current forecasts. Under these circumstances it would become possible to construct history-based confidence intervals about the current forecasts.

One simple example may illustrate the use of cumulative frequency data in this way. The director of Chemical Laboratory B has been given an estimate of $10,000 for the development cost of a project currently under consideration for funding. If it were undertaken, the project would be conducted by the same personnel who were involved in the completed projects represented in Exhibit 5. The new product concept is also based on the same general fund of technical expertise as those completed projects. Thus the manager expects the forecast cost figure to be about as accurate as those he has received in the past.

We have already shown in Exhibit 5 that 25 percent of the completed projects in Chemical Laboratory B exceeded their initially estimated cost by 240 percent (dotted line 2). The manager would thus assign a probability of 0.25 to the chance that the new project would ultimately cost more than $24,000. Similarly, the manager would assign a probability of 0.38 to the current project's cost being less than the $10,000 forecast (dotted line 1) and a probability of 0.25 to its costing less than $8,000 (dotted line 3). The chances of other errors may also be determined from the cumulative overrun curve.

This technique may be extended to formulas which incorporate more than one potentially inaccurate forecast. For example, the commonly used Expected Profit Ratio,

$$\text{Expected Profit Ratio} = (\text{PTS})(\text{PCS})(\text{Net Profit})/(\text{Dev. Cost})$$

PCS = Probability of Commercial Success
PTS = Probability of Technical Success

may be treated in the same manner. It is only necessary that cumulative error probability curves have been determined for each of the forecasts used in the calculation and that the forecast and project be similar to those for which data are available. The second condition makes it necessary for the utility of the procedure to be determined in each laboratory by trial. In general, however the calculation of history-based confidence intervals will provide useful information on the implications of errors in the forecasts used for predicting the profitability of projects.

It is not as easy to determine the implications of errors inherent in the forecasts which are used to schedule the laboratory's resources, for here the feedback loop structure of laboratory decision making becomes especially important. The simple decision loop in Exhibit 1 may be elaborated to represent the information feedback structure underlying the scheduling of resources in development projects. Exhibit 8 represents a development program in which man-hours are the principal input and cost.

Personnel are initially allocated to the project on the basis of the desired completion date and the estimated total man-hour requirements of the tech

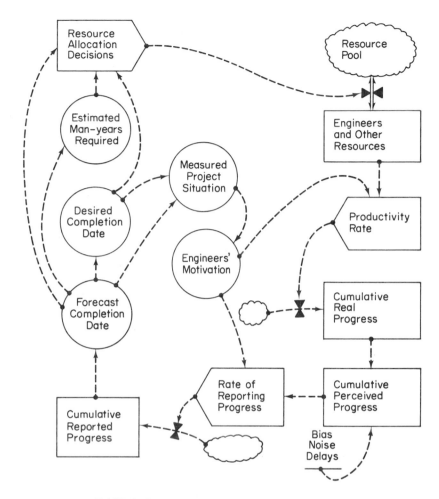

Exhibit 8. Feedback Loop Structure of Project Management Decisions.

nical objectives. Because the desired and the forecast completion date are initially the same, the engineers' motivation and hence their productivity are somewhat less than that obtainable under conditions of extreme schedule pressure. There typically will be more time spent in leisure and time spent on unnecessary engineering refinements than in a crash program.

The number of engineers assigned to the project and their level of motivation together determine the rate of progress toward the technical objectives of the project. Technical accomplishments are generally difficult to quantify, however, and all social communication systems have inherent delays and errors. The engineers will therefore often be lead to an inaccurate perception of their progress. Additional errors and bias are introduced by the attitudes and the motivations of the engineers and their immediate supervisors. The technical personnel know the current schedule and the rewards and penalties associated with positive and negative deviations from the schedule. This information together with their perceptions of the actual progress will influence the problems and the achievements which they report upward to the laboratory management. Whether the cumulative progress is monitored informally or through some mechanism like PERT, management will generally be forced by the accumulated errors to revise the forecast completion date. Faced with a schedule overrun, management may revise the desired completion date, assign more men to the project, or exert more pressure on the current engineering staff. Whatever management's response, the effects of its decisions will be propagated through the system of loops to result eventually in new inputs for further decisions.

Some results from the statistical analyses of project histories may be explained in terms of this feedback loop structure. To avoid penalties associated with technical failures, the laboratory personnel may delay or bias their reports of technical difficulties. Occasionally this may yield sufficient funds and time to eliminate the difficulties. When it does not, however, especially great overruns and high project costs will result. PERT techniques or the profit motivation of business teams may decrease schedule overruns by lowering the ability or incentives of the technical staff to bias the information upon which management's allocation decisions are based.

In this context simple calculations or sensitivity analyses are no longer adequate for understanding the implications of forecasting errors. Inaccuracies may lead to many different kinds of costs depending upon the nature of the decision structure in the laboratory. Technical manpower resources are relatively fixed in the short run. Men precipitously added to one project must therefore be taken from another. Then the program with lower priority will itself experience schedule overruns or, often, be dropped altogether. Such projects appeared above in the category of miscellaneous failures. Table 5 indicates that the money wasted on this type of project amounted to twenty-seven and ten percent respectively of the total development budgets in the Equipment and the Electronics Laboratories.

Another cost stems from the time inevitably lost whenever men are transferred from one project to another. The project effort is disrupted while the new personnel become familiar with the work and are assimilated into the working groups. If management responds to schedule slippage by increasing the pressure on the technical staff, the laboratory personnel may conceivably be demoralized by what seems to be an impossible completion goal. Costs then would result from decreased efficiency. Finally, there may be contract or market penalties associated with schedule overruns.

The magnitude of all these costs will depend upon the size and the direction of the initial error, the rate at which it is detected and accommodated, the efficiency with which resources can be transferred between projects, the reaction of the technical staff to increased schedule pressure, and the implications of schedule overruns. The specific values of these and other parameters may vary from one laboratory to another, but the structure itself is a general representation of all project-organized development programs. It thus includes important determinants of the project performance in each of your laboratories, and it can be used in each to determine some implications of forecasting errors.

It is actually possible to fit the parameters of such a model to a particular laboratory. The management science and organizational psychology literature, the experience of corporate and laboratory administrators, and empirical research similar to that discussed above all can provide information on the elements and the relationships which comprise the important feedback loops in any particular laboratory. In practice, of course, the description will be much more detailed than that in Exhibit 8. We have found that laboratory simulation models must contain about thirty elements before they begin to exhibit the behaviour modes characteristic of real organizations. Some of the models which have been used to explore laboratory performance are an order of magnitude more complex than that.

While it is possible for managers to describe accurately the decision feed-back loop structure which determines the performance of their laboratories, the complexity of any useful model makes it impossible to intuitively predict the performance which will result from different forecasting errors. Thus the laboratory description must be converted into a set of equations which can be employed in computer simulation experiments.

System Dynamics is a methodology which enables one to identify the information feedback structure which determines the performance of social and economic systems. Included in the methodology is the necessary computer software for converting the structure into a computer simulation model. Several published references give specific instructions in System Dynamics and describe its application to industrial development programs [9–11]. However, I want merely to suggest the utility of the technique in studying the implications of forecasting errors. Thus I will limit my remarks to a discussion of one particular study. It illustrates the use of simulation models in studying the performance of development projects, and its results appear relevant to some types of development programs. Anyone interested in pursuing this approach to technological forecasting may gain additional information on the methodology from the references above.

Scambos and his co-workers were specifically interested in the schedule slippages which result when initial forecasts of the effort required to complete a project are in error. To explore the implications of inaccurate forecasts they elaborated on the basic feedback structure in Exhibit 8. Several important features were incorporated into their simulation model [12].

1. Management's initial forecasts of the total effort required by the project were in error. The magnitude of the error, its direction, and the rate at which it became apparent to management were varied in the simulations.
2. Two engineering groups were involved in the conduct of the project. Their

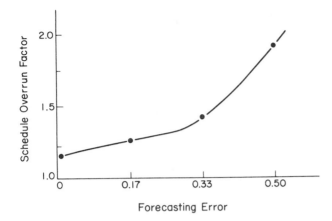

Exhibit 9. Schedule Overruns Associated with Different Forecasting Errors.

activities were interdependent. Whenever one group slipped too far behind the other, the second group was held up in its work as technical information and hardware necessary for its work were not available from the laggard group. The productivity of each group was influenced by its perceptions of the other's progress. Although management is usually content to merely meet minimum standards, engineers generally prefer to refine the product technically. Thus each group in the simulation model paces itself to finish just behind the other. That goal prevents it from delaying the project, yet it provides the most time for technical refinements.

3. Technical competence was held constant during the course of the project. More detailed models have included the dependence of competence upon management's hiring, firing, and training policies and upon the past work experience of the laboratory staff.

Two results of the model simulations are important here. First, it was more costly in terms of schedule slippages to underestimate than to overestimate the total effort required in the project. Underestimating causes large schedule slippages. The minimum time required to complete the simulated project was thirty-six months. Estimates only fifty percent in error could increase that time to fifty-eight months. Exhibit 9 indicates the slippage resulting from different forecasting errors. Overruns increase exponentially with the magnitude of the forecasting error.

SUMMARY

Errors in the forecasts used to select among development projects are costly in terms of wasted resources and schedule overruns. Inaccuracy cannot be much improved through simple correction equations. Instead management must quantify the error and its implications through research. Only then will it be possible to design the optimal forecasting and planning procedures. Research into forecasting accuracy is currently being conducted both at M.I.T. and in several large industrial laboratories. While the results above must yet be confirmed and extended, there is no longer any doubt that both research tools, simulation and historical analyses, can provide useful insights into the forecasting performance of industrial laboratories. This paper was designed to encourage such research and to present some details of the techniques which make it possible. The design of improved forecasting procedures is a difficult technical problem, but it will yield to research. The potential rewards of the improved forecasting techniques which will result from this research are enormous.

REFERENCES

1. N. R. BAKER and W. H. POUND, "R and D Project Selection: Where We Stand." *IEEE Transactions on Engineering Management,* EM-11, No. 4, 123–34, December, 1964.

2. DENNIS L. MEADOWS, "Estimate Accuracy and Project Selection in Industrial Research" *Industrial Management Review,* Massachusetts Institute of Technology, Cambridge, Mass., pp. 105–20, Spring, 1968.

3. A. W. MARSHALL and W. H. MECKLING, "Predictability of Costs, Time and Success of Development" in National Bureau of Economic Research (Ed.), *The Rate and Direction of Inventive Activity: Economic and Social Factors,* Princeton University Press, Princeton, N.J., 1962, pp, 461–75.

4. M. J. PECK and F. M. SCHERER, *The Weapons Acquisition Process, An Economic Analysis,* Harvard University Press, Boston, Mass., 1962.

5. E. MANSFIELD and R. BRANDENBURG, "The Allocation, Characteristics, and Outcome of the Firm's R and D Portfolio: A Case Study" *The Journal of Business,* 39:447–64, 1966.

6. G. H. FISHER, "The Problems of Uncertainty," in J. P. Large (Ed.) *Concepts and Procedures of Cost Analysis,* The RAND Corporation, RM-3589-PR, Santa Monica, Calif., June, 1963.

7. R. E. SEILER, *Improving the Effectiveness of Research and Development,* McGraw-Hill, New York, 1965.

8. IRWIN RUBIN, "Factors in the Performance of R and D Projects" in *Proceedings of the Twentieth National Conference on the Administration of Research,* Denver Research Institute, University of Denver, Colorado, 1967, pp. 67–71.

9. J. W. FORRESTER, *Industrial Dynamics,* John Wiley & Sons, Inc., New York, 1961.

10. EDWARD B. ROBERTS, *The Dynamics of Research and Development,* Harper & Row, New York, 1964.

11. ROBERT S. SPENCER, "Modelling Strategies for Corporate Growth" presented at the Society for General Systems Research session of the conference of the American Association for the Advancement of Science, Washington, D. C, December 26, 1966.

12. W. BEARD, B. QVALE, and T. SCAMBOS, "Application of Industrial Dynamics to Certain Aspects of R&D" unpublished research paper, Sloan School of Management, M.I.T., Cambridge, Mass., May 20, 1966.

Morphological Analysis:
A Method for Creativity

LUCIEN GERARDIN

By systematically exploring a field of possibilities, morphological analysis reveals many potential technical possibilities that have been neglected or that are strange at first glance. This stimulates the imagination and causes the creative technical work that we had hoped for. Computerization of the written description of each solution in a form easily grasped by the mind has noticeably increased this stimulation.

Although morphological analysis is beautifully logical, applications in industry are very difficult. The author suggests how he and his colleagues are applying this technique.

HISTORY OF MORPHOLOGICAL ANALYSIS

Although the term *morphological analysis* was coined by F. Zwicky, the method is in fact very old. One can trace it back to the Majorcan logician and mystic monk Ramon Lull (1235–1315). Lull had an idea which he called the "Great Art." By systematically combining a very small number of principles one would have the possibility of solving all the problems of philosophy and metaphysics. But the practical means that he had at his disposal were insufficient. His principles were materialized by boxes on circles rotating around the others as we can see in Exhibit 1. Unhappily R. Lull did not have a computer in order to enumerate all the combinations of principles.

Exhibit 1. Lull's Diagram—the "Great Art."

Descartes had studied the art of Lull and had seen the dangers of thought mechanization. He said in his *Discours de la Methode,* a few lines before explaining the famous "Quatre Regles":

> *"I have studied a little being young between philosophy and logic...but upon examining them I take care that, for the logic, its syllogisms...serve rather for explaining to others the things that one knows or even, as in the Art of Lull, to speak without judgment of those that one ignores and must learn."*

Leibnitz was a strong partisan of the art of Lull and he spoke of it with praise in his *Dissertatio de Arte Combinatoria.* Moreover, the method had been generalized. In effect R. Lull had limited the number of principles to be combined to nine as we can see in the table reproduced in Exhibit 2 from a

Exhibit 2. The Nine Principles of Lull.

1. Essence.
2. Vnité.
3. Perfection.

		A.	B.	C.	D.	E.	F.	G.	H.	I.	K.
Predicats	Abso-luts.		Bonté.	Magni-tude.	Eternité.	Puif-fance.	Sageffe.	Volonté.	Vertu.	Verité.	Gloire.
	1. Re-latifs.		Diffe-rence.	Concor-dance.	Cōtrarieté Duration.	Prin-cipe.	Milieu.	Fin.	Maiori-té.	Equa-lité.	Minorité.
Alpha-bet, ou principes de cét art.	M.Queſtions		Sça-uoir?	Qui eſt?	Dequoy?	Pour-quoy?	Quant?	Quel?	Quand?	Où?	Cōme quoy ou cōment?
	N. Suiets.		Dieu.	Ange.	Ciel.	Hom-me.	Imagi-natiue.	Senſiti-ue.	Vegeta-tiue.	Ele-men. tatiue.	Inſtrumē-tiue.
	O. Vertus.		Iuſti-ce.	Pruden-ce.	Force.	Tempe-rance.	Foy.	Eſpe-rance.	Chari-té.	Patien-ce.	Pieté.
	P. Vices.		Aua-rice.	Glouton-nie.	Luxure.	Super-bité.	Pareſſe.	Enuie.	Ire.	Men-ſonge?	Inconſtan-ce.

book of the seventeenth century. But certain people have understood that it would be possible to break through that restriction.

Exhibit 3 is excerpted from a book by the famous Jesuit, A. Kircher (the inventor of the magic lantern, therefore the great-grandfather of today's movies), and it is a very clear example of morphological analysis.

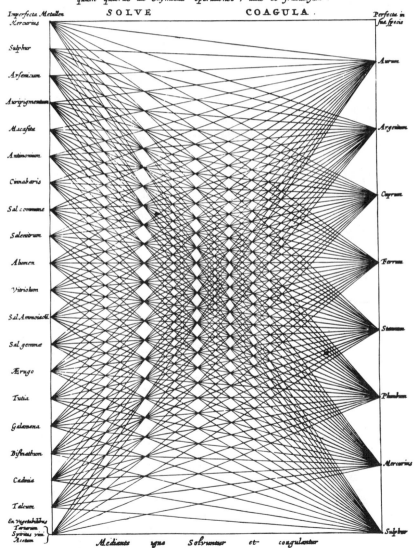

TABULA COMBINATORIA,
Qua
Tom.II.303.

Quicquid in tota Alchimia continetur, veluti in Synopsi anacephalæotica ob oculos pom:
tur Curiosi Lectoris. Neqꝫ extra hanc, sive lapidis fabricam spectes, sive Metallurgi:
cam artem, sive mixturas Metallicorum corporum, sive denuo eorum geneses spectes, quic:
quam quæras ad Chymicas operationes, utile et fructuosum.

Exhibit 3. Kircher's Example of Morphological Analysis.

Unhappily, this method brings within itself a poison, that which Descartes has pointed out—it is the risk of thought mechanization. Some Lullist fellows of the sixteenth and the seventeenth centuries gave full rein to this. It was easy to criticize this as the humorist Swift has done with high spirits (and also a little mischief) in his *Gulliver's Travels*. He describes a contrivance invented by a savant of Laputa that randomly combines letters and in this way "permitting the most ignorant person to write books without the least help of genius or study." Thus, the art of Lull fell little by little into oblivion despite all its potential value.

The criticism of Swift is accurate; in particular it permits clearly pointing out what is and what is not morphological analysis. There is never a miracle or an *amplification of ideas*. A computer is useful for rapidly enumerating all the solutions coming from a combinatorial process, but its role is purely mechanical and nothing more. The value of solutions found is related to the value of the analysis; the solutions must be studied and exploited intelligently. In fact, morphological analysis is a *method for creativity*, more exactly a *systematic aid to creativity*. It absolutely does not eliminate the creative work of man, but it stimulates and amplifies it by making the imagination work on a greater number of ideas than that which was possible by a more classical approach.

Therefore it would be better to substitute for the very ambitious definition of F. Zwicky the following: *"morphological analysis is a method for systematically exploring all the possible solutions to a given problem."* Despite its apparent simplicity this definition is also ambitious.

BASIC PHILOSOPHY

What is the basic general philosophy of morphological analysis? According to the definition given above, we want to study the problem to be solved from a viewpoint as general as possible in order to be sure (or more exactly, almost sure) of exploring all the possible solutions. Starting from the known, or from the plausible, we can then hope to discover the new. First, we always know some solutions to the problem to be solved as schematized in Exhibit 4.

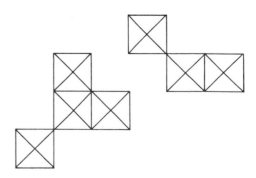

Exhibit 4. The Known.

By starting from that known, we define and structure the problem in an abstract form, specifying and limiting it as schematized in Exhibit 5.

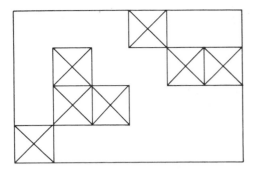

Exhibit 5. The Structured.

By enumerating all the different possibilities that are inside the abstract structure thus elaborated, we explore the field of all the possible solutions, which is the goal aimed for. Among all the new solutions discovered, a few of them can upon examination appear to be of special interest (Exhibit 6).

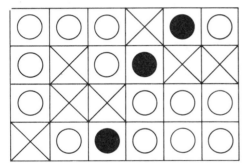

Exhibit 6. The New.

How can we translate this general philosophy into practice? This is very simple and can be summarized in a few rules. First, it is necessary *to clearly define the problem and delineate it,* something that is not particular to this method since it is the universal base for any rational method. Once this is done we know, as we have said, some particular solutions to the given problem. Starting from the problem definition and using the description of particular known solutions, *we determine the main parameters that characterize the problem,* also the main values to assign to each of these parameters. Value is to be taken in a very broad meaning generally more qualitative than quantitative. In this way we can construct what F. Zwicky called *the morphological box* (Exhibit 7).

	Values			
Parameter A	A_1	A_2	A_3	
Parameter B	B_1	B_2		
Parameter C	C_1	C_2		
- - - - - - - - - - -	- - - - -	- - -		
Parameter H	H_1	H_2	H_3	H_4

Exhibit 7. The Morphological Box.

Finally we *enumerate all the possible associations for the different values of the different parameters* and find in this way all the different possible solutions to the studied problem (Exhibit 8). It is useful to use an electronic computer for that, but this computer is by no means imperative.

	Values			
Parameter A	A_1			
Parameter B		B_2		
Parameter C		C_2		
Parameter H				H_4

Exhibit 8. Example of a Particular Situation.

TO MANAGE RESULTS

The main difficulty of morphological analysis lies in its too great richness: combinatorial process proliferates in geometric progression and we quickly arrive at tens of thousands of solutions. As an example, if we use a morphological box of eight parameters of four values each we will have 65,536 different solutions. If it only requires half an hour to study each of these 65,536 solutions (some will be easily studied but others will conceal more difficulties and it may quite well be that these are the most interesting), the required work represents roughly fifteen man-years. This is out of practical possibilities. But on the other hand, if the number of possible solutions is too small because the analysis had to be oversimplified, we risk finding only self-evidences. One of the main problems, if not *the problem* to practice morphological analysis is then the following: *how to study each of all the solutions given by a morphological box.*

Some people proposed using the idea of *abstract distance between solutions.* If two different solutions differ only by one single parameter value, they are spaced by one; if they differ by two values, the distance will be two and so on. Around a given solution (for example a known solution), *we can cluster the solutions* given by the combinatorial process in solutions at distance one, distance two, and so on.

Then we can study in more detail only the solutions closer to the starting known solution or on the contrary, the solutions that are further apart. This idea seems very useful; it is true that abstract distances can aid in studying the results but only if this is not too systemized. It may be that some solutions differing by one will be as distant as solutions differing by four or five and the reverse.

Another method is *to randomly choose a small set of solutions* from the different possibilities given by the combinatorial process, for example, 1 solution out of 10 or 1 out of 100. In this case we do not use the full possibilities that the method could give, but if this randomly chosen set permits finding new ideas, we reach the goal we have aimed for and that is sufficient.

We can combine the two methods above by *using proximity clustering starting from solutions chosen at random which have been judged to be interesting.* By using the concept of distance in order to organize the exploration around particular solutions, this method is a genuine daughter of the method of solution clustering by distance criteria.

Finally, we can also *study all possible solutions.* In this case it is necessary to design the morphological box in order to have a number of different solutions which correspond to a reasonable time for the work of examination. This means that we must then choose the number of parameters and the number of values for them in such a way that the combinatorial process will give only a few hundred different solutions. With 7 parameters of 3 values each there will be 2,187 solutions; with 7 parameters of 2 values each, there will be only 128. The compromise lies between.

In this case we must have the courage to simplify the complexity, so that the abstract structure taken as a base for the morphological box will satisfy the above requirements. This simplification must be done by respecting the relevance of the description: parameters must each have almost the same importance, the same for the values given to each parameter. This is very interesting in that it forces us to make a deep analysis of the problem being studied.

In every case, we can try to reduce the number of different solutions given by the combinatorial process by *seeking the restrictions which might exist on the values of different parameters* (two, three, or more). These restrictions can reflect genuine impossibilities (either physical or logical) or combinations that seem to be unrealistic for the time scale being considered. This pruning work is not mandatory; its only purpose is to simplify the study of final results by reducing the number of possible different solutions.

DETAILED CASE STUDY

In order to better show how to put morphological analysis into practice, we will describe an actual application made by the Look-Out Studies Group of Thomson-CSF regarding the problem of communication by mass media. The purpose of this study was the definition of a research program. It was therefore necessary to explore the field of mass media from the standpoint of needs and products in order to find:

1. *New needs for known products.*
2. *New products for known needs.*
3. *New needs to be satisfied by new products.*
4. *New products for new needs.*

The first step is to choose the analysis viewpoint. Needs and products belong to different levels which are:

1. *Operational level* (needs and situations).
2. *Technological level* (products).

The interaction between these levels is generally not a direct one because it is very improbable that a single product will satisfy a given need. More often a *system* is required, and that system is organized in such a way that it has several *functions*. Therefore the idea of using morphological analysis at the *functional level* (systems) makes the morphological box the metastructure which links needs and products (Exhibit 9).

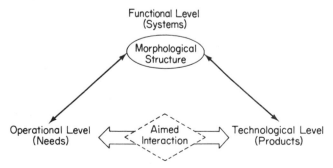

Exhibit 9. Structure as an Aid to Link Needs and Products.

It was therefore necessary to describe in functional terms the structure of mass media communication between human beings. The problem is so complicated that this description is not unique. We have to remind ourselves of that which has been stated before regarding morphological analysis: it is not possible to describe the problem as a whole and in all its implications; we must choose a viewpoint which is relevant to the goal aimed for, here, the definition of a research program especially technological research. We are not interested in communication semantics and in the influence upon those between whom the communication is exchanged.

After many discussions we have adopted the system organization given in Exhibit 10. We notice that some blocks of the diagram are drawn with solid lines while others are drawn with dotted lines. The viewpoint adopted here excludes the contents of a message and its use. We have limited our morphological description to only the functions outlined by a solid line, therefore, from the transmitting interface(s) to the receiving interface(s). It is absolutely necessary to clearly and accurately define the functions shown in that diagram. These definitions have to be written in order to ensure language coherence throughout the study. If we satisfied ourselves with intuitive definitions, we would be sure to arrive very quickly at total confusion because these intuitive definitions would be different from one man to the other. Each one would easily believe that the others understand him while all the languages would be different.

Therefore, we have begun with about sixty known situations. Each situation has also been the subject of a written description (in a few lines) always for reasons of mutual understanding.

From the known, in agreement with the diagram in Exhibit 10, we have built the morphological structure by determining the parameters and their values. We must avoid *intuitive genius*: experience has shown us that it is absolutely mandatory to work in a group because nobody can trust the truth on a given subject. Choice of parameters has been made by respecting the rules given above:

1. Mutual independence of the parameters.
2. Equal relative weight of the parameters and of the values of parameters.
3. Relevance according to viewpoint chosen for the analysis.

In order to see that these rules have been fulfilled, we have used the known solutions by verifying that each one of them corresponds to one and only one abstract morphological solution.

We have also randomly chosen abstract solutions and tried to translate them on operational situations to see if selected parameters and values describe in an adequate manner the problem of communication. As we had decided to use systematic exploration of all the possible solutions, it was also necessary to limit the number of parameters and values. Seven morphological boxes had been tried before we decided on those in Exhibit 11.

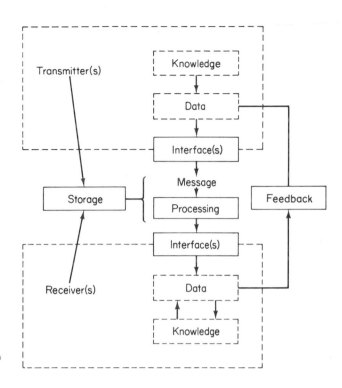

Exhibit 10. System of Communication between Human Beings.

Exhibit 11. Morphological Box for Mass-Media Communication.

Parameter	Value	Numerical Indicator
Interface T	Single Multiple	1 2
Interface R	Single Multiple	0 2
Feedback	Null Partial Total	0 4 5
Message	Sound Graphic Audiovisual	0 12 24
Processing	None Transpose Transform	0 36 72
Simultaneity	Yes No	0 108
Availability	Existing Nonexisting	0 216

The numbers in the right-hand column when added together define a number associated with each solution. For example, in Exhibit 12 the solution description shown would have a numerical value of 7 associated with it.

Exhibit 12. Example of Solution.

Parameter	Value	Numerical Indicator
Interface T	Single	1
Interface R	Multiple	2
Feedback	Partial	4
Message	Sound	0
Processing	None	0
Simultaneity	Yes	0
Availability	Exists	0
Total		7

There are 432 different solutions, but we have found four logical restrictions which result in infeasible combinations of parameters, two involving two parameter values and two involving three parameter values. (Exhibit 13).

Exhibit 13. Four Restrictions of Mass-Media Communication.

Parameter	Value
Feedback Simultaneity	Partial No
Feedback Simultaneity	Total No
Interf. Trans. Interf. Recep. Feedback	Single Multiple Total
Interf. Trans. Interf. Recep. Feedback	Multiple Single Total

These infeasible combinations reduce the number of different solutions to 252 instead of 432. Putting these restrictions into operation has been automated in the following way. A given solution corresponds to vector which for the above example (solution number 7) is:

$$1, \quad 2, \quad 2, \quad 1, \quad 1, \quad 1, \quad 1.$$

The indexing follows the order of the parameters of the morphological box by giving to each parameter the number of its value (for example, 1 or 2, or 1, 2, or 3 in the present case). To each restriction we make a similar vector correspond. For example, for the infeasible combination first quoted in the above table, we have the vector,

$$0, \quad 0, \quad 2, \quad 0, \quad 0, \quad 2, \quad 0$$

obtained by putting the flag 0 for the parameters which are not taken into consideration by the restriction. By masking the vector corresponding to a solution by the vector corresponding to each restriction, we study the appearance of zeros. If the number of these zeros corresponds with the degree of the restriction (that is, the number of non-zero elements in the masking vector), we reject that solution and do not print it out. In the example given only *one* zero will appear (in the third position) while there are *two* 2's in the masking vector and hence, this solution is at least feasible.

OUTPUT COMPUTERIZATION

As we have said, morphological analysis is a method to aid creativity. Therefore, it is necessary that *the way a solution is presented will stimulate the imagination.* Everybody will agree that an abstract description such as the one given above carries little stimulation to the imagination; it is too abstract. We have already noticed that; even though, this presentation has already permitted us to discover a few new ideas.

Since we want to discover new ideas of needs and/or products, why should we not try to obtain a more striking description if possible by mechanical means. In other words, can we, according to the ideas of R. Leclerc [19], *axiomatize* in some way the problem? This is what we have tried to do first at the operational level of the needs. To each abstract functional value of a parameter, we have written an operational commentary. Can we go further by making a commentary correspond to the combination of two parameters? We must remember at this point one of the main characteristics for parameters of a morphological box: the parameters must be mutually independent. This means that *the commentary corresponding to the combination of two parameters must not contain more than the logical sum of the two commentaries corresponding to the isolated parameters.* This in itself is an excellent way to verify the independence of the parameters retained to constitute the morphological box. This is something we have found in the present case, with only a single exception (that is, the two parameters of simultaneity and availability are two functional aspects that technology reunifies in message storage).

It is not of interest to systematically study all the combinations corresponding to the sets of three parameters because *the number of threefold combinations is too great.* There are almost as many combinations as there are solutions, and it would not be profitable to systematically study all these combinations. But nothing prohibits taking into account a commentary corresponding to a three-fold combination that has been judged to be of interest.

In the present case the commentary table has been that shown in Exhibit 14.

Exhibit 14. Example for Operational Commentary.

Parameter	Value	Commentary
Interface T	Multiple	Face to face/Conference Network
Interface R	Multiple	Mail, Phone, Broadcasting
Feedback	Null	Unilateral Communication
Feedback	Partial	Return Circuit(s)
Feedback	Total	Bilateral Communication
Simultaneity	Yes	Direct
Simultaneity	No	Delayed
Availability	Exists	Recorded at Transmission and/or Reception
Simultaneity Availability	Yes Exists	Computer Aided

As for the restrictions, each commentary is given a vector which by masking permits a subprogram to select the set of comments relative to a given solution (Exhibit 15).

Exhibit 15. Example of Presentation with Commentaries.

Parameter	Value
Interface T	Multiple
Interface R	Multiple
Feedback	Partial
Message Sent	Sound
Processing	None
Simultaneity	Yes
Availability	Exists
Face to face/Conference Network	
Phone, Broadcasting, Mail	
Return Circuit(s)	
Message	Sound
Processing	None
Recorded at Emission and (or) Reception	
Computer Aided	

The stimulation is clearly improved. It is then straightforward to go much further in this way by also using technological commentaries. As an example, some of them are given in Exhibit 16.

Exhibit 16. Example of Technological Commentaries.

Parameter	Value	Commentary
Message	Sound	Micro/Synthetic Voice
Message	Graphic	Keyboard/Camera/Designation
Message	Sound/Image	Micro/Synthetic Voice
Message Processing	Sound Transform	Visual Display
Message Processing	Graphic None	Visual Display
Message Processing	Graphic Transpose	Visual Display
Message Processing	Sound None	Loud Speaker/Headphone
Message Processing	Sound Transpose	Loud Speaker/Headphone
Message Processing	Graphic Transform	Loud Speaker/Headphone
Message Processing	Sound/Image Transform	Visual Display/Loudspeaker
Message Availability	Sound Exists	Tape Recorder
Message Processing Availability	Graphic Transform Exists	Tape Recorder
Message Processing Availability	Graphic None Exists	Printer

It is necessary to avoid a lengthy description if we want to have some overall effect when we see the description. Such an overall effect enhances the production of new ideas. Therefore, we decided not to print the abstract description and finally present only the commentaries as shown in Exhibit 17.

Exhibit 17. Example of Commented Description.

Case No. 51	Single
Interface T	
Mail, Phone, Broadcasting	
Unilateral Communications	
Message	Graphic
Processing	Transpose
Direct Transmission	
Recording at Transmission and (or) Reception	
Computer Aided	
Keyboard/Camera/Optical Designation/Synthetic Image	
Storage	
Visual Display	

This evidently stimulates the imagination more than the corresponding abstract form shown in Exhibit 18.

Exhibit 18. Example of Abstract Form.

Parameter	Value	Case No. 51
Interface T	Single	
Interface R	Multiple	
Feedback	Null	
Message	Graphic	
Processing	Transpose	
Simultaneity	Yes	
Availability	Exists	

The practice has shown us that the stimulation is now very strong; the imagination works much more and finds many new ideas.

One of the advantages of this computerization of results must be emphasized. If we have an idea of something new, either operational or technological, that novelty can be described in the form of a commentary. It is then very easy to use the combination program by indicating to it to print only the solutions including the value of the parameter (or combinations of the parameter values) corresponding to the new commentary. In this way we *only stimulate the imagination where it must work and everywhere it must work*. Therefore a morphological box is not exploited once and for all; we can return to it several times later on to find new ideas each time.

REFERENCES

1. Fritz Zwicky, "Morphology and Nomenclature of Jet Engines," *Aeron. Eng. Review* (June, 1947).

2. ———, "The Morphological Method of Analysis and Construction," in *COURANT,* Anniversary Volume (New York: Intersciences Publish., 1948), p. 461–470.

3. ———, *Morphological Astronomy* (Berlin: Springer Verlag, 1957).

4. ———, *Morphology of Propulsive Power,* Monographs on Morphological Research, No. 1 (Pasadena, California: Society for Morphological Research), 1962.

5. K. W. Norris, "The Morphological Approach to Engineering Design," in *Conference on Design Methods,* (eds.) J. C. Jones and D. G. Thornley (Elmsford, N.Y.: Pergamon Press, Inc., 1963), p. 115–140.

6. Fritz Zwicky, *Entdecken, Erfinden, Forschen in Morphologische Weltbild* (Munich and Zurich: Droemer/Kname, 1966). English translation: *Discovery, Invention, Research, Through the Morphological Approach* (The MacMillan Co., 1968).

7. A. V. Bridgewater, "Long Range Process Design and Morphological Analysis," *The Chemical Engineer* (April, 1968), p. CE 75–CE 81.

8. ———, "Morphological Methods—Principles and Practice," in *Technological*

Forecasting, (ed.) R.V. Arnfield (Conference on Technological Forecasting, University of Strathclyde, 1968) (Edinburgh: University Press, 1969), p. 241–252.

9.. E. P. HAWTHORNE and R. J. WILLS, "Forecasting the Market for Certain Machine Tools, 1974–1999," in *Technological Forecasting,* (ed.) R.V. Arnfield (Conference on Technological Forecasting, University of Strathclyde, 1968) (Edinburgh: University Press, 1969), p. 241–252.

10. ARTHUR D. HALL, "Three-Dimensional Morphology of Systems Engineering," *IEEE Transactions on Systems Science and Cybernetics* (April, 1969), p. 156–160.

11. ROBERT U. AYRES, "Morphological Analysis," in *Technological Forecasting and Long Range Planning* (New York: McGraw-Hill, Inc., 1969), Chap. 5, p. 72–93.

12. R. J. WILLS and E. P. HAWTHONE, "Morphological Methods Applied to Metalworking Processes", *Technological Forecasting, Some Techniques,* Symposium at Aston University, Birmingham, 9/10 September 1969.

13. S. A. GREGORY, "Morphological Methods: Antecedents and Associates," *Technological Forecasting, Some Techniques,* Symposium at Aston University, Birmingham, 9/10 September 1969.

14. G. ROYSTON, "Morphological Analysis and the Development of the Brewing Process," *Technological Forecasting, Some Techniques,* Symposium at Aston University, Birmingham, 9/10 September 1969.

15. R. D. WATTS, "Some Theoretical Principles in Morphological Analysis," *Technological Forecasting, Some Techniques,* Symposium at Aston University, Birmingham, 9/10 September 1969.

16. J. H. McPHERSON, *Structured Approaches to Creativity,* LRPS Research Reports (private circulation) (Palo Alto, Calif.: Stanford Research Institute, 1969).

17. A. KAUFMANN, M. FUSTIER, and A. DREVET, *L'Inventique, Nouvelles Methods de Creativite* (Paris, 1970), a) techniques to manage the results (p. 144–166), b) two case studies (p. 188–202).

18. R. FOSTER, E. LEONARD, and R. REA, "*A Scenario Generating Methodology,*" Abt Associates Report, Cambridge, Mass., June 1970.

19. RENE LECLERCQ, "The Use of Generalized Logic in Forecasting," *Technological Forecasting and Social Change,* II, No. 2 (1970), p. 193–194.

XPANDING THE SCOPE OF TECHNOLOGICAL FORECASTING– THE INTERFACE WITH OTHER ENVIROMENTS

PART FOUR

Anyone who seriously approaches technological forecasting soon becomes aware that technology is even less autonomous than generally assumed. Its dependency on other elements of the environment seems to be growing, and the fate of DDT, the SST, and phosphate detergents are cases in point.

It follows that sound technology forecasts will have to embrace consideration of more than technical factors. This section includes chapters on economic input-output analysis, environmental constraints, a survey of political forecasting, pollution control policy impact on raw materials, and international postures (gross national product, population, and so on).

In addition, the reader is introduced to the virtually unknown yet extremely significant subject of *values* in the future. All of us are aware that value systems have undergone major change and possibly lie behind the demise of the SST program and the downgrading of the space program. But what are "values"? How are they formed, and what will they be in the future? Professor Nicholas Rescher, a leading student of values, introduces this subject.

How can one anticipate value changes or at least societal responses to proposals for change? Researchers from The Institute for the Future describe an experiment conducted for the state of Connecticut, *simulation gaming,* to test reaction to future policies.

The Environment,
A Growing Constraint
on Technological Change

JOHN R. EHRENFELD

Unquestionably, the environment interacts with technology. Ecological problems of current social concern, such as air and water pollution, have been caused in large measure by man's drive to provide more and better means for carrying out his life's work. However, as the author aptly points out in this paper, the environment and its elements are becoming increasingly more important constraints on technological progress. Thus forecasts of technological change must accommodate these constraints. Sound technology forecasts will have to provide a view of the future in which environmental impacts and reactions are properly balanced.

One problem in dealing with a topic of exceptionally high current interest and exposure is the looseness of definition of subject and treatment that often accompanies wide public discussion. Because TF (technological forecasting) like most systems oriented processes should be inclusive in order to reduce the likelihood of a priori exclusion of meaningful (but unfortunately obscure) inputs, let me adopt a very broad definition of first, the *physical environment* as: everything that was here before man arrived on the scene or would be here if man was removed now.

This definition includes properly, from our context today, such things as:

air, water, natural resources and minerals, land, and plant and animal life. Whereas much attention is directed to air, water and, to a lesser degree, land pollution, considerations of changes in all of these quantities enter and affect the TF process in parallel ways and are best examined together.

There is a second more difficult to define environment, which I shall call the *sensory environment*, that depends entirely on the presence of man within the physical environment and is best defined by examples: noise; aesthetic-perspective, clutter, trash, bill boards, and urban sprawl; visibility; and smell.

The difficulty of developing a clean-cut definition for this class can be illustrated by the classic paradox that goes: if a tree falls in a forest, and no one is around, does it make a *noise* when it hits the ground.

Since we are to deal principally with changes and their effect, let me further define two kinds of changes. One I will call *real* changes, which can be measured by some physical means. For example, we can speak of changes in air quality in terms of the measurable concentration of certain chemical species, or in water by the change in biological oxygen demand or temperature, in noise by the decibel level, and so forth. There is a second kind of change equally important right now although perhaps not so much in the future. These kinds of change I will call *apparent* changes and are the consequence of a change in social, economic, political, or moral *values*, even in the absence of any *real* change.

Today much of our interest and involvement in ecology and in the environment derives from apparent changes. Not many years ago belching smoke stacks were looked upon with great pride by city officialdom. Today the same men look out on the same stacks, whose emissions probably have been greatly reduced by the forced installation of control measures, and complain about the insult to our environment that is being produced. Recognition and evaluation of such apparent changes may have more impact on the technological forecasting process than a concomitant real change in the environmental quantities being examined. Before moving to discuss how these changes affect the TF process, here are several examples of both kinds related to the present time frame.

The chemical composition of the major life supportive elements of our physical environment have been changing since man appeared upon spaceship earth. Carbon dioxide (not normally considered a pollutant but equally a part of the physical environment) concentration on a global scale has continuously increased as both animal life (through exhalation of CO_2) and use of fossil fuel for energy have increased. The tremendous increase in energy production in this century has produced such measurable changes in our lifetime that one school of atmospheric scientists predicts a significant raising of the global temperature (by the greenhouse effect) to a point where the ice caps would melt and flood all the coastal regions of the earth. This global kind of change, its effects, and projections are now the subject of serious studies. For example, MIT, under the egis of many agencies, organized a summer study during which this and other global environmental problems related to the consequences of present practices were examined.

As an aside, fortunately, for those of us who live within earshot of the oceans, there is another school which claims that the observed, continuing increase of dusts in the upper atmosphere are reflecting more of the sun's energy back into space, resulting in a cooling trend. The two effects appear to have worked in opposition; there has been a very slight decrease in global temperature in recent years. The level of pollutants, defined as undesirable elements of the environment, has increased tremendously in the past several decades, as a result of growing population, affluence, economic growth, and a more or less laissez-faire attitude about environmental problems. The past several decades have seen as a result the death of 3,000 in a London smog, destruction of an ancient monument (Cleopatra's Needle), general decay in the proliferation of unswimmable and undrinkable waterways, and the blotting out of urban perspective by smog.

These are of course only a few examples of the constant real changes in the environment. These changes generally are continuous, well-behaved functions of complex sets of demographic and economic variables, which characteristic permits application of many formal techniques for estimation. On the other hand apparent changes seem to take place in quantum steps. The passage of omnibus air and water pollution control acts by Congress in recent years is an example of a political change which already has had tremendous impact on technology and will continue to exert even more influence in the future. The Clean Air Act of 1967, amended substantially in 1970, provides for setting air quality standards. By setting such standards we instantly quantify environmental levels with regard to acceptability and at the same time identify the need for corrective courses of action. Second, the Act requires that such courses of action (implementation plans) be developed and put into practice through whatever legislative and administrative means are necessary.

The emergence of economic concepts which treat air or water or land and, in some parts of the world, the mineral resources as public commodity is producing apparent changes. We are seeing, right now, an aroused public demanding action. The real situation outside has changed only incrementally in the last few years, while public demand for action has taken a tremendous jump. This more than any real change is responsible for the high position of environment on the list of important issues in the United States today. All these changes will influence the process of technological forecasting. I hope to show, in a general way, how consideration of environmental change must be included.

In examining the consequences of new technology, environmental factors can act as either *positive* or *negative* constraints. We have seen numerous examples of both recently, with the negative side getting more public play.

One of the major new technology developments in the United States is the SST (supersonic transport). In the early analysis of the consequences of the development of such an aircraft, little attention was paid to environmental factors, either to noise or to high altitude atmospheric pollution. In the years that have passed since the program started these factors continue to affect the project, and I believe in the long run will shut it down.

Public acceptability of sonic booms was tested in a controversial series of flights in the southwest. It was determined that the public would or could be made to live with the noise. Since then general concern with noise has grown tremendously. A bill to prohibit the SST from landing or taking off within Massachusetts has been brought before the Massachusetts legislature. It did not pass but was seriously considered.

Unwillingness to accept the *uncertainty of the effects of environmental changes* is also increasing. In the past the burden of proof most often rested on an individual or the public to show that the action of an agency would infringe on his rights in order to gain injunctive action (and was very hard put to succeed). Now more and more, the burden is shifting; agencies must demonstrate that no *harmful* consequences will accrue from its actions (as, for example, in the current pure food and drug regulations). The promoters of the SST debunk the arguments of those who claim that the SST exhaust would pollute the upper atmosphere and cause profound changes in global weather, simply by stating that there is no *clear cut evidence* that this is true. Those promoters are indeed correct. There is, however, no question that changes would occur, and there is considerable presumption that the consequences would be unfavorable. I would not be at all surprised to see the concern about noise and atmospheric pollution translated into a social value of such negative magnitude to outweigh the positive economic value claimed by the proponents of the SST. When this occurs, public support will undoubtedly end, bringing about collapse of the entire project.

Automotive development is now being profoundly affected by environmental considerations; certainly any and all planning that was done without considering environmental consequences has been upset. One might have thought some years ago that lead tetraethyl would meet its demise when increases in atmospheric lead content reached levels considered dangerous. It now appears that if lead tetraethyl does go (and I would bet on its disappearance), it will be as a result of its negative effects on the ability of present internal combustion engines, and attached pollution control devices, to meet emission standards for nitrogen oxides and hydrocarbons.

The prohibition of lead tetraethyl would not only dim the future of the several companies in the business but would change basic petroleum refining engineering and economics. Even considering engine modifications using lower octane gasoline (resulting from lead removal), a significant shift in gasoline composition toward aromatics will be necessary. Refinery optimization techniques and continuing present research on catalytic processes would have to shift in parallel.

The technological future of the internal combustion engine is clouded by environmental considerations. The long American romance with horsepower is being broken up by increasing concern and growing *regulations,* that is, environmental consequences. One can predict that if the population of humans and vehicles continues to grow at present rates, that even with the low emission engine configurations required to meet the federal regulatory timetable, the emissions will creep back above tolerable limits within several decades. Note these comments made some time ago by Dr. John Middleton, Commissioner of the National Air Pollution Control Administration (Dec. 1969):

> There is no question that the emission standards now on the books will significantly reduce the amount of pollutants discharged to the atmosphere from the automobile. As these standards are applied successively to each new generation of automobiles, and as the older generations of uncontrolled cars are replaced, we can expect that year by year there will be a gradual reduction in emissions from motor vehicles until the year 1980, when virtually all cars on the road will be under partial control. However, this increasing control of automobile emissions is offset to a considerable degree by the projected increase in the number of vehicle miles that will be driven in our cities. Further, it is our experience that in the hands of the motoring public, and under the care of the typical garage, emission control systems on cars in actual use do not reflect the sustained degree of control that has been predicted from the performance of test cars in the laboratory. Instead, the control systems deteriorate substantially.
>
> In my judgment, the best that we can expect from the standards now in effect is that hydrocarbon and carbon monoxide emissions will by 1980 dip to approximately sixty percent of current emissions, or roughly what they were in 1953. And after 1980, when these standards have passed the saturation point of their effectiveness, as vehicle use continues to increase, the levels of pollution will resume their upward climb.

This estimate becomes both a tremendous negative constraint in examining the technological future of internal combustion engines, and also a powerful positive constraint in considering alternate solutions to vehicle power plants.

Until recently, steam cars were as technologically obsolete as the proverbial buggy whip; now look at the spate of activity in that field. If any of the several competitors are successful (some have already dropped out after disastrous experiences), it will be the direct result of environmental pressures on technology.

The whole utilities industry provides many examples of the importance of environmental considerations. Power consumption in the United States has about doubled every ten years, seriously straining our production capacity. Primary attention over the past several decades has been devoted to means to meet the demand to improve productivity. The first environmental constraint that entered the planning forecasting process was consideration of fossil-fuel supply limitations. Nuclear energy was hailed as the means to reduce the demand on a finite resource and to produce cheaper power. Environmental considerations concerned with safety have dictated many factors in design and siting of nuclear reactors and have certainly markedly influenced the development process.

The situations concerning demand continue today, but environmental factors have *upset* many aspects. Nuclear plants are being attacked not on the basis of safety, once the bugbear, but on the basis of thermal pollution. Conventional fossil-fuel burning plants are being threatened with shutdown by an angry public who, characteristically in these hectic times, simultaneously flail the industry for failing to provide enough power for all their newfangled conveniences. Per capita use of power has risen from about 2,500 kilowatt hours in 1950 to about 8,000 now and is estimated to reach 22,000 kilowatt hours in 1990.

The utilities, in supporting R&D during the last several decades, allocated insignificant funds to environmental problems, not out of callousness but for failure to foresee the changes that have occurred. Fuel producers cannot easily supply current demands for low-sulfur fuels. Current technology cannot easily nor economically provide solutions for treating effluent gases to remove deleterious pollutants.

These are things that have already happened; the future promises even more to upset the environmentally unconstrained growth of utility and energy technology. Pressures on the domestic natural gas supply from users who cannot practically reduce emissions by application of processes may force a national allocation policy limiting utility use. With the development of nuclear plants slowed by siting and technological problems, coal, the most plentiful resource, would have to be substituted. This in turn would promote the need for economic means for gasification and sulfur removal, and for the large-scale flue-gas treatment processes. And so the problems multiply and interact all because of changes in the environment; in this case a real increase in air and water pollution amplified by changing social values. These events are coupled with a real change in fossil-fuel resources which are coupled, in turn, with begrudging acceptance of their finiteness.

Before moving on to the last part of this discussion, let me simply list other examples of changes in our environment and their influence on technology.

Product or Process	Environmental Effect
Detergents	Eutrophication-algae blooms
Basic oxygen steelmaking	Particulate emissions
Fuel cells/Magneto-hydrodynamics	Thermal pollution
Mass transportation	Land abuse and air pollution
Many chemical processes	Stream pollution
Paper making	Air and water pollution
Alkali production	Increased mercury pollution
Insecticides, herbicides	Gross ecological changes

I am not familiar with the whole area of technological forecasting or of the analysis of technological change. However, having examined several treatises on the subject of technical change, I was intrigued by the development of models relating change to economic growth and other factors. (For example, see *The Economics of Technological Change*, E. Mansfield, Norton, 1968, or *Invention and Economic Growth*, J. Schmookler, Harvard Press, 1966). Almost no mention of environmental constraints was made in any of them. I believe that the validity of these models, as developed on historical bases, must be seriously questioned and that modification to include environmental factors must be made. The relationship between economics and technological change has been based on air and water at zero cost, vastly undervalued disposal costs, little or no social cost inputs, and, in general, classical elastic economic theory for mineral resources.

No such assumptions are realistic today. Environmental constraints as measured by some economic equivalent may soon dominate model behavior patterns. If so, then how does one provide *estimates* for the magnitude of these environmental equivalents. In many ways the overall projection/prediction problem parallels that of technological forecasting itself. The methods can be roughly divided in three types:

1. Trend extrapolation.
2. Models of causal relationships.
3. Expert opinion.

Our company has been working with each of these in several current programs looking at air pollution from stationary fossil-fuel combustion equipment under contract to the National Air Pollution Control Agency, now part of the Environmental Protection Agency. One key phase of the job has been to estimate present levels of emissions of sulfur and nitrogen oxides and particulates and project the estimates to 1990. The main task has been to develop estimates of equipment capacity. With these data it is rather simple first to estimate fuel consumption and, then, emissions by application of well-known emission factors. The methods used include both time-series extrapolation and more general econometric correlations. To the skilled econometrist the application is almost trivial; to the environmental planner, this exercise provides a reasonable basis for analysis and decision making.

We have been able to compare the results of air projections against several other bench marks. Residential and the small types of combustion equipment correlate well with residential construction, which over a period of years correlates well with population growth. Since we can estimate population growth reasonably accurately, we would expect these estimates to be quite realistic.

The sales of large fire tube boilers (over 500,000 pounds of steam per hour or over 50 megawatts electric equivalent) match new construction plans of the utilities amazingly closely for the five years following our base projection. This industry, which published all of its statistical data, such as plants under construction or planned, has served as an excellent case on which we have both developed and checked correlational methods. The trend projection method which seems to work well for the several categories mentioned, fails when applied to a smaller size range of water tube boilers. I mention this here to point to the dangers of unrestrained reliance on the results of such calculations. In cases where classical models are inapplicable, it is sometimes possible, as we did, to superpose subjective judgment or some nonlinear function to explain the anomalous behavior and to generate estimates that seem reasonable.

Such formal methods are available to project real environmental changes. For example, extensive regional input-output models have been developed for water resource management. With knowledge of individual productivity indices for each element in an input-output matrix, it is possible to estimate to the gross generation or emission of environmental quantities (including waste heat). Then with *phenomenological models* these quantities can be transformed into significant terms. For example, important measures of air and water pollution are concentrations of pollutants. Transport, diffusion, and lifetime models are used to effect the transformation. We, as other companies, have, for example, developed a series of atmospheric diffusion models for single and multiple emission sources to predict ground level concentrations, under a wide variety of meteorological conditions.

As briefly described, the combination of econometric and phenomenological models constitutes the only analytic methodology for estimating real changes with which I am familiar. The estimation of apparent changes is considerably more difficult. Changes in economic value particularly in the introduction of social cost elements, generally have been discontinuous functions. The determination of reasonable numerical values to use is difficult, although methods for estimating social costs are now developing.

Political factors are equally difficult to predict. In certain cases, the present laws provide timetables. For example, the Clean Air Act of 1967 with the 1970 amendments lays out, as noted above, a complete schedule of the actions that are to be taken. One cannot predict the content of the actions individual regions will take, for example, specific values for annual mean concentrations, but, at least, one knows when the decision will be made.

In closing, I will emphasize several of what I believe are the key points in this discussion:

1. The environment is changing as a consequence of technological and economic development.
2. These changes are manifest both as *real* and as *apparent* changes.
3. The evaluation of technological forecasts *must consider* environmental factors
4. The problems of (and methods for) forecasting environmental changes are similar to those for forecasting technological changes.
5. The problem is exceedingly complex.

SELECTED BIBLIOGRAPHY

BOWER, BLAIRE T., ET AL, *Waste Management,* N.Y., Regional Planning Association, 1967.

Duprey, R. L., *Compilation of Air Pollutant Emission Factors,* U.S. Public Health Service, P.H.S. Pub. #999-AP-42, Washington, D.C., U.S. Govt. Printing Office, 1968.

FWPCA, U.S. Department of Interior, *The Cost of Clean Water,* Vol. 1-3, Washington, D.C., January 1968.

FWPCA, U.S. Department of Interior, *The Cost of Clean Water and Its Economic Impact,* Washington, D.C., 1969.

FWPCA, U.S. Department of Interior, *The Economic Impact of the Capital Outlays Required to Attain the Waste Water Quality Standards of the FWPCA,* Washington, D.C., 1968.

Goldman, Marshall I. (ed.), *Controlling Pollution: The Economics of a Cleaner America,* Englewood Cliffs, N.J., Prentice-Hall, Inc., 1967.

HEW, *The Cost of Clean Air,* Report of the Secretary of the Department of Health, Education and Welfare to Congress, June, 1969 & March, 1970.

Jarrett, H., *Environmental Quality in a Growing Economy,* Baltimore, Md., Johns Hopkins Press, 1966.

Joint Committee on Atomic Energy, 91st Congress, *The Environmental Effects of Producing Electric Power* (Hearings), Washington, D.C., 1967.

Kneese, Allen V., *Economics and the Quality of the Environment,* Baltimore, Md., Johns Hopkins Press, 1969.

———— and Blaire T. Bower, *Managing Water Quality: Economics, Technology, Institutions,* Baltimore, Md., Johns Hopkins Press, 1968.

———— and O. C., Hernfindahl, *Quality of the Environment (An Economic Approach to Some Problems in Using Land, Water and Air),* Washington, D.C., Resources for the Future, Inc., 1965.

Landsberg, Hans H. and Sam H. Schurr, (Resources for the Future, Inc.), *Energy in the U.S.,* N.Y., Random House, 1968.

———— Leonard L. Fishman and Joseph L. Fisher, *Resources in America's Future: Patterns of Requirements and Availabilities 1960–2000,* Baltimore, Md., Johns Hopkins Press, 1963.

Morrison, Warren E. and Charles Readling, *An Energy Model for the United States Featuring Energy Balances for the Years 1947–1965 and Projections and Forecasts to the Years 1980–2000,* U.S. Dept. of Interior, Bureau of Mines, Information Circular #8384, July, 1968.

Ridker, R. G., *Economic Costs of Air Pollution,* N.Y., Frederick A. Praeger, 1967.

U.S. Congress, Senate, Committee on Public Works, *Hearing Before the Subcommittee on Air and Water Pollution: Thermal Pollution 1968,* 1968.

Varga, J., Jr. and H. W. Lownie, Jr. (Battelle Memorial Institute), *A Systems Analysis Study of the Integrated Iron and Steel Industry,* conducted for NAPCA under Contract #PH22-68-65, May 15, 1969.

Wolozin, H. (ed.), *The Economics of Air Pollution,* N.Y., W. W. Norton and Co., Inc., 1966.

EPA PUBLICATIONS

Air Quality Criteria for Particulate Matter, #AP-49 (Jan. 1969).
Air Quality Criteria for Sulfur Oxides, #AP-50 (Jan. 1969).
Control Techniques for Particulate Air Pollutants, #AP-51 (Jan. 1969).
Control Techniques for Sulfur Oxide Air Pollutants, #AP-52 (Jan. 1969).
Air Quality Criteria for Carbon Monoxide, #AP-62 (1970).
Air Quality Criteria for Photochemical Oxidants, #AP-63 (1970).
Air Quality Criteria for Hydrocarbons, #AP-64 (1970).

Control Techniques for Carbon Monoxide Emissions from Stationary Sources, #AP-65 (1970).

Control Techniques for Carbon Monoxide, Nitrous Oxides and Hydrocarbon Emissions from Mobile Sources, #AP-66 (1970).

Control Techniques for Nitrous Oxide Emissions from Stationary Sources, #AP-67 (1970).

Control Techniques for Hydrocarbons and Organic Solvent Emissions from Stationary Sources, #AP-68 (1970).

Political Forecasting

DAVID V. EDWARDS

The interaction between political processes and technology is increasing. There is no question that the influence of the political environment on the technological environment, and vice versa, will be a continuing reality. It follows that the technological forecaster must learn to allow for political forces that lead to technical change. Professor Edwards reviews the status of political forecasting, its role relative to technological forecasting, and its future potential. He offers his own views on a method of predicting sociopolitical forces.

We are well aware that politics depends upon technology not just because political decision is often informed through the effective use of communication, information processing, and other such obviously technological devices but also because the subject of decisions made by political figures is increasingly technological. We know also that there are many difficulties in technological forecasting, on some of which we have not yet made much progress. Probably the most fundamental limitation on technological forecasting today, although perhaps not the most obvious, is its dependence on the much less developed art of social forecasting and especially political forecasting. This dependence is clearly manifest in the case of military research and development. Another rather obvious subject for which this dependence is crucial is urbanization and especially demands arising out of the political composition and the resource problems of the city. In the case of military hardware, to take but one example, we can clearly see the reciprocal effect of technology and politics. Clearly, in this area, there is a tendency to develop and then even to use whatever we discover technologically; and further, the nature of our military establishment clearly affects international politics and domestic politics both. But in addition, and equally obviously, decisions about technological development are made in political arenas where nontechnological determinants predominate. For these reasons, there can be little question that our technological forecasting should be informed by and concerned with political forecasting.

Political forecasting, as we are all well aware, goes all the way back to the Greeks (notably the original Delphi) and the Bible. Although political forecasting is increasingly in vogue today, it often seems little better than it was in those ancient days. Of such forecasts today we demand and require both accuracy and credibility. Unfortunately, until accuracy is so good that it is itself grounds for credibility, accuracy and credibility will generally require different qualities of a forecast. Thus credibility at this time tends to depend fundamentally upon the scientistic quality of the forecast, that is, qualities of rigor, quantification, and deductive argument—all of which somehow, mysteriously, lead us to have greater confidence in arguments made about present or future. Accuracy, on the other hand, is most heavily dependent upon there being a basis in theory, or at least in empirical regularity, for the forecasts being made. The function of this theoretical base is to enable the forecaster to escape his own intuition and find good evidential reasons for the forecasts that he makes. But not until this theoretical base is widely shared by his colleagues will its explicit use tend to make his forecasts more credible. And we are still far from such a time in the social sciences.

Despite these limitations, examples of past political forecasting merit brief attention. In 1942, twenty-six professional people and public figures were asked to write statements of what they believed the political world would look like a decade from that time. These forecasters relied sheerly on intuition, at least if we can believe the account that Hans Toch later wrote of this effort [1]. Much more recently, Herman Kahn and Anthony Wiener in 1967 published a book relying fundamentally on trend extrapolation in an effort to predict the state of the economy and politics, particularly international politics, in the year 2000 [2]. Although, of course, there have been many other efforts in between these two, these are probably the most indicative of their types, and their types have long been fashionable. In the study of technological forecasting, we have found it worthwhile to examine past forecasts in an effort to discover their inaccuracies and determine the sources of those errors. We can also learn much from examining these two rather typical efforts at political forecasting. The political forecaster of the forties whose work sociologist Toch assessed was quite wrong in most of his predictions. The reasons for these errors seem to lie fundamentally in lack of adequate attention to geopolitical and subsequent ideological developments in international politics. Moreover, these errors can be attributed to an almost total lack of appreciation of the economic significance both nationally and internationally of the war in which the United States was then engaged. The drastic variances among these forecasts, as well as the variance of almost all of the forecasts from what actually took place subsequently in the world, can be explained largely in terms of these major misconceptions and hence the inadequate social and political theory on the basis of which the predictions were made. The forecasts of Kahn and Wiener cannot yet be assessed, of course, because the period over which they are predicting has not yet occurred. Nonetheless, it is impressive to notice the extent to which they, because they are fundamentally employing trend extrapolation, are forced to emphasize the quantifiable and hence trendable factors in the world and further the obvious difficulties they face and recognize in attempting to forecast transformations in economic, social, and political systems.

But if these are unimpressive efforts by and large, they are also probably representative of the best efforts of their types available. One way of discovering why this is so is to examine a catalog of social predictive methods to see what tools are or can be in the hands of forecasters when they attempt to engage in political and social prediction. There are ten such prediction or forecasting methods to which I should like to direct your attention. (I examine these in somewhat greater detail in Chapter 14 of my book, *International Political Analysis* [3].)

The first such method is also the oldest: reliance on *prophecy*. As unfashionable as prophecy now is among academicians, it is still highly favored in many quarters. The widespread and increasing interest in astrology is perhaps the best indication. The weaknesses of prophecy for our purposes are obvious and we need not discuss them.

The second method of prediction is the employment of *chance*. This method is equally or perhaps even more unimpressive than prophecy, and yet it has advocates. It consists, fundamentally, in the use of devices such as coin flipping to determine what the future will hold. It is in favor primarily with people whose other efforts at prediction have proved less than fifty percent accurate and who hence believe they can improve upon their predictions by employing coin flipping and other such devices. It suffers not only from the obvious weakness of not relating to the factual material out of which a prediction might be made, but also from the obvious difficulty of structuring one's predictive decisions in yes-or-no form.

The next major approach is perhaps the most widespread academic approach. This is, of course, the use of *intuition*. What distinguishes intuitive prediction from other types based on evidence and principles is that intuition is presumably based on evidence and principles that the predictor is either unable or unwilling to divulge or specify to those people to whom he makes his prediction. The intuitive prediction is the clearest case of a prediction which requires that its credibility be based upon the past record of the predictor, rather than upon anything inherent in the prediction itself.

The fourth type of prediction is the reliance upon *analogy*. The clearest cases of this are efforts based on historical analogy. Seeking to predict some particular development, a predictor will discern present conditions he believes relevant, attempt to find a set of similar conditions somewhere in history, and upon finding such a set will predict that the same outcome will occur as resulted from the original set of conditions in time past. The difficulties of such prediction are, of course, monumental, primarily because it is so difficult to take account of all the relevant conditions, particularly when political, social, and technological systems are subject to such rapid transformation. Hence, the reliability of any particular historical, analogical effort is called into question on these grounds.

The fifth predictive method is the reliance upon *correlation*. This approach, of course, endeavors to discover empirical regularities in linkages of happenings without inquiring into the theoretical bases for such correlations. Predictions are then made on the assumption that what has covaried in the past will continue to do so. For example, upon discovering that the phases of the moon are cyclically regular and that the menstrual cycle tends to correspond to this cycle, we might attempt to predict future developments in a menstrual cycle from knowledge about developments in the lunar cycle. The dangers of this approach should be obvious. And they seem to apply to other instances, such as the effort to predict the business cycle via the sunspot cycle.

The next major collection of methods involves *projection* from the past through the present to the future. There are three basic types of projection that might be employed. The first of these is an effort to extrapolate persistence or consistency and relies upon the discovery or at least the attribution of unchangingness to certain major features of the situation. These unchanging features are then assumed to be unchanging in the future, and, hence, predictions are made that things will continue as they have in the past. Examples of these types of prediction are not hard to find, and perhaps the most obvious are those that rely on institutional activities for the basis of their predictions. For example, we might predict with a high degree of confidence that the United States will hold a presidential election in 1972 as it did in 1968 and as it will again in 1976. However, because so few things in the world seem to be so persistent, or so consistent and so regular, the effort to predict in this way is likely to be limited to uninteresting, insignificant affairs.

The second major type of projection is the extrapolation of trends. Because this is so prominent in technological forecasting, little more need be said about its nature or its problems beyond indicating that it is difficult to find trendable, trending, or other developing situations, circumstances, or factors that can be easily predicted by trend extrapolation in the realm of the political and social.

The third type of projection is the extrapolation of trends with foreseeable trend changes built into that extrapolation. Instances of this, too, are not hard to find, although the effort is often very difficult to make in practice. A revealing example might be an effort to extrapolate political participation. We may find that secular changes in degree of participation result primarily from the spread of communications facilities and of media receptacles within the population. Thus we may extrapolate trends in voting, for example, as a function of this technological development. However, we also know that the extension of the franchise through legislation will be the major determinant of the level of political participation, but this determinant must be predicted not through such extrapolation but rather through the concentration upon legislative effort. That is, if we have reason to believe that Congress will pass a law extending the franchise to those who are not presently guaranteed it or extending civil rights and protection to people who have thus far been denied the opportunity to participate, we may insert into our extrapolated trend based on developments in communication a jump in the trend that allows for the immediate entry into the voting ranks of people who have already been mobilized by the communications facilities but who have not previously been allowed to express that mobilization politically. This approach to prediction has obvious merits where the situations and data allow or require its effective employment.

The next major category of predictive methods includes efforts at *simulation* and *gaming*, both through use of the computer and through the use of human resources directly. Some scholars find the employment of these approaches promising as predictors. There certainly are technological advantages to the employment of computer simulation, especially those made possible by the opportunity to play permutations on a situation in great number and with many controlled variations very quickly and relatively inexpensively in order to see what impact slight changes in one or another parameter might have. Nonetheless, as devices for predicting political developments, these still leave a great deal to be achieved because the mathematical models on which they are grounded are not yet satisfactory.

The next approach might be termed *invention*. By this I mean an effort to conceive of all possible futures of a given situation, circumstance, capability, or whatever and then through the use of some independent means, assign probabilities to each of those possible futures. This approach, of course, suffers from the same difficulties of technological approaches that endeavor to conceive of all possible ways of designing, for example, a jet engine, and then through one means or another determine which of these seems to be the most likely or promising way of approaching further development. Immediately one is struck with the difficulties imposed by the limitations of the analyst, including possible "tunnelvision," bias, or an inability to grasp the relevant interactions of system elements. Moreover, difficulties arise due to unforeseen events and changes in the value structure.

The next approach to prediction is what I would call *negation* or *elimination*. In this approach one begins without a clear sense of what all the possibilities for development of a situation are, but one looks for those things which clearly cannot happen and eliminates each of those in turn. It is assumed (somewhat naively perhaps) that at the end of these efforts one will have a rather clear sense of what is left in the realm of possibility and hence what may happen—and further hoping that there will prove to be one or only a few basic possibilities.

The final major approach to prediction of social events is the employment of the *dialectic*. This approach is based upon the theory that any complex social system will inevitably generate within itself contradictions—contradictory tendencies that will in their turn become more and more prominent and eventually come to combat or permute the previously dominant social or political circumstance. Ultimately, it is assumed, the situation will develop into some combination of or variation on these two tendencies. The best instances of this sort of prediction are presently to be found in the field of sociology where some scholars have desired and endeavored to use what they call "counter-system models" to determine likely antitheses to present society and social arrangements and hence to foresee the source of social transformations which may be imminent. Thus some scholars are now attempting to see beyond our present industrial society to what is often called the "postindustrial society" deriving from the spread of affluence and the increasing interest of the citizenry in removing the alienation and bureaucratization which increasingly upset social arrangements and threaten to tear the social fabric.

Of course, the ultimate method of prediction is the employment of comprehensive general explanatory theory, theory in this case of the political system and its environment or of the social system within which the political system resides. This most difficult and most promising sort of prediction I shall turn to shortly.

The key question in our effort to predict political developments is: Are the significant determinants of politics known and are they predictable as known? We may well know what it is that causes social and political change, but we may be unable to predict what will happen to those determinants in the future—at least not without a good deal of knowledge of the underlying determinants of these major determinants. The interrelationship of determinants which we have already noted and which is so obvious to us as we study any of these subjects requires not just adequate partial theory (that is, for example, theory of one small area within politics) but also a model of society as a whole that enables us to see on what other factors political developments ultimately depend.

Are these key political determinants known then? Sociopolitical theory at present—empirical sociopolitical theory—is very much underdeveloped. Nonetheless, it is clear that we are now making considerable progress in empirical theory building in the social sciences, and that we can expect a great deal of improvement in the near future, improvement that may well enable us to engage in predictive efforts much more impressive and indeed more effective and accurate. This should be clear to anyone who studies any subfield of sociology or political science.

As this work progresses, we shall have more general theory for use in political forecasting as well as in more rigorous prediction. The requirement for being able to predict generally and accurately is that we be able to explain, for the structure of both explanation and prediction is basically identical. Both involve the effort to move from general laws and statements of specific conditions to conclusions about the events that follow from these. More specifically, in either venture we begin with a comprehensive description of the conditions existing at the time, and then a statement of general principles that link various conditions existing at a given time with subsequent developments or events. By combining these statements and principles, we should be able to show that certain events follow from the conditions.

This sort of predictive venture requires theory that is general rather than particular and limited to the immediate phenomenon, for as we have seen events in the political and social realm are highly interdependent. Thus, for example, to predict American political developments we almost certainly must know a great deal about coming developments in the economy, particularly developments in affluence and poverty; and, further, we must know a great deal about coming events in society, particularly events involving social mobilization and communication, among other factors.

Without comprehensive theoretical knowledge of this sort, we can, of course, forecast by relying upon correlation to get approximations. We will be relying upon correlations because we know neither the causal connections of conditions and subsequent events nor the possible exogenous factors, factors which are unexpected or unconsidered because they lie outside of our generalizations. But we can only engage in such correlational forecasting on condition that we have available statements of empirical regularities. And inevitably such forecasts will suffer from the continual problem of credibility because we are so aware of the likelihood of spurious correlation. Unfortunately, the only way we can generally test the empirical generalizations of correlations is to predict in many different areas and then find linkages between those areas that give us independent grounds for confidence in our predictions.

Our efforts at prediction suffer in part because our models of society or of the political system are generally not really models but merely conceptual schemes or analytical frameworks. In other words, we have lists of factors that academicians believe we should look for as possible causes, but we have neither indications of the actual causal relations between these factors and outcomes, nor the parsimony which is required of models. If we had knowledge of causal relations, our predictions would be more credible and we would be better able to understand and anticipate the dynamic which results in system transformation and is hence essential to any effort to predict over a period of time. Further, if we had the parsimony we seek—that is, if we could limit the number of factors presently listed in our analytical frameworks or conceptual schemes, we would then be able actually to understand the contentions we were making and to engage in efficient prediction. (I have discussed and demonstrated at length various ways of developing such general theory in *International Political Analysis* [3].)

Once we have a general theory, we would be able to employ it in predicting, because it would consist of a great series of propositions, each of the form: Under condition ABC, the result will be XYZ. And our predictions or forecasts would be comprehensive, accurate, and credible. It is toward this goal that we social scientists are working. But it is with what partial theories and scattered insights, laced with a hefty draught of intuition, that we continue to attempt estimates of the political context of technological development.

REFERENCES

1. HANS H. TOCH, "The Perception of Future Events: Case Studies in Social Prediction," *Public Opinion Quarterly*, 22 (1958), 57–66.
2. HERMAN KAHN AND ANTHONY J. WIENER, *The Year 2000* (New York: Macmillan, 1967).
3. DAVID V. EDWARDS, *International Political Analysis* (New York: Holt, Rinehart & Winston, 1969).

Impact on
Raw Material Markets
of Environmental Policies

JARED E. HAZELTON

Environmental quality, or quality of life as it is sometimes referred to, is becoming a major concern to economists as well as to technologists. This author views production activity on a continuum in which today's resources not only produce tomorrow's products but also tomorrow's wastes, and technology is a medium for this transformation.

There is now strong social pressure suggesting that, in the future, technology must be directed towards improving the relationship between waste and resource by either making more efficient use of the limited resources to create less waste, or by recycling the waste and thereby creating a new resource.

Hazelton shows several examples of waste recovery and then points out that in some cases, such as sulphur, pollution control probably will have a cataclysmic effect on traditional raw material demand. Here is a thought provoking forecasting concept for every firm based on raw materials production.

It has been suggested that future technology is essentially the result of society's present assessment and reaction to environmental changes—political, economic, and social—as well as of technological activity. Technology is responsive to social and economic pressures, which often materialize as political decisions that affect progress. Technological forecasting, then, must of necessity consider the fact that future socioeconomic conditions will require the creation of new technological capabilities. My purpose in this paper is to examine one particular type of socioeconomic change, the increasing public awareness and concern over deterioration of the natural environment, and to analyze the implications of this change for the direction and pace of future technology.

PUBLIC CONCERN FOR ENVIRONMENTAL QUALITY

No single issue has so captivated public attention in recent months as concern over deterioration of the natural environment. Problems of air and water pollution, land use planning, population control, and resource sufficiency have come to dominate the public forums of the nation. Awareness of these problems has created increased demands for the development of public policies for controlling the environment. With the establishment of the President's Council on Environmental Quality and the enactment of legislation at both the state and federal levels strengthening programs in air and water pollution control and solid waste disposal, it appears that all levels of government are responding to public pressures for environmental protection.

I believe it would be a mistake to view concern over deterioration of the environment as a passing fad which will be satiated by proper recognition of the problem in articles in weekly magazines, political platforms, and industry advertising. This concern stems from a combination of forces that are unlikely to abate over the remaining years of this century. The source of these forces is the inevitable conflict between man's economic activities and the status of the natural environment.

The natural environment provides three important services for man: (1) it is the source of his raw materials: (2) it provides space for waste accumulation and for regeneration and assimilation of chemically and biologically active wastes; and (3) it is a principal determinant of health levels and life style. As ecologists have known for years, these services are not separable but are highly interdependent.

In the early stages of economic development, there is only a minor conflict of these roles. However, as increased population and industrialization place more pressures on the environment's ability to provide raw materials and store, dilute, and chemically degrade waste products, the capacity of the environment to provide a satisfactory basis for a high quality life is also threatened. If current trends in population and economic growth are continued for the next thirty years, it is inevitable that continued conflicts between man's economic activities and the state of the natural environment will occur. These conflicts will result in continued political pressures for environmental protection. Viewed from the standpoint of corporate management, the course taken by public policies implementing environmental controls will affect not only the future direction of technology but will also influence the firm's opportunities for profitable growth and development.

In this paper I shall first present an analytical framework for viewing the conflict between man's economic activities and his natural environment. Utilizing this framework I will then consider the problems of resource scarcity, recycling of wastes, and pollution control. In each of these problem areas, I will attempt to identify the socioeconomic forces which will affect the development of future technology. I will examine specific technological alternatives and attempt to assess their direct impact on raw material requirements evaluated in terms of existing supply and demand conditions and forecasted changes.

FRAMEWORK FOR ENVIRONMENTAL ANALYSIS

Economists have only recently begun to recognize environmental pollution and its control as a materials balance problem for the entire economy.[1] A simplified diagram depicting the relations between an economy and the natural environment is shown in Exhibit 1. Our economic system may be thought of as a throughput process: we extract ores and fossil fuels; we process these into energy and commodities which are "consumed" by the economy; in so doing, we produce negative outputs of pollutants, for example, by-products of production and consumption with negative values; and we dispose of these by-products in pollutable reservoirs in the natural environment, what Kenneth Boulding has called, "negative mines."[2]

One advantage to viewing the economy in this way is to emphasize that there is no such thing as "final consumption"—witness the piles of junk cars which dot our countryside. When we speak of consumption of certain commodities, we are actually referring to the consumption of services rendered

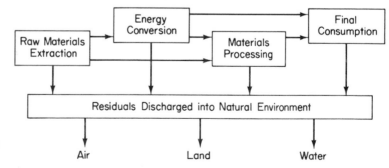

Exhibit 1. Material Flow Diagram of the Economy.

[1] Robert I. Aryes and Allen V. Kneese, "Production, Consumption, and Externalities," *American Economic Review* (June, 1969), p. 285.
[2] Kenneth Boulding, *Economics as a Science* (New York: McGraw-Hill Book Company, Inc., 1970), p. 42.

by them. Their material substance remains in existence either to be reused or discharged into the natural environment. The total tonnages of materials extracted for use by an economy, if we assume no foreign trade or capital accumulation, would be approximately equal to the tonnages of waste products generated by the economy in the long run.

The implications of this analysis for environmental control are clear. First, production of goods and services requires increasing amounts of raw materials which are subject to varying degrees of depletion. Reduction of wastes can only be accomplished by lowering the amount of material throughput. Second, the wastes from extraction, production, and consumption must inevitably reside in the air, the land, or the water. Efforts to reduce the effect of these residuals on one environmental media, such as air, can only succeed by shifting these discharges to other environmental media. As a consequence pollution cannot be properly dealt with by considering environmental media—the air, the land, or the water—in isolation. Finally, recycling of residuals will not only help conserve raw materials, but will also help alleviate disposal problems.

PROBLEM OF RESOURCE SCARCITY

Classical economists treated resources as a single entity for analysis, subsumed under the term land. This led to identification of resources with substances or tangible things. This misconception about the nature of resources was challenged in 1933 by Erich W. Zimmermann.[3] Zimmermann developed a functional concept of resources, stressing that the word "resources" does not refer to a thing or a substance but to a function which a thing or substance may perform. The function is to attain a given end, the satisfaction of man's wants. This semantic advancement had several important implications. Resources cannot be defined except in terms of known technology. Mineral deposits which are unknown, which are inaccessible, or which cannot be processed and used with known technology are not resources. An equally important implication is that resources can only be defined in relation to the needs of man. This leads to the conclusion that pollution, the deterioration in the quality of a resource, can only be defined and analyzed in the context of its effect in impairing the serviceability or usefulness of that resource to man.

Early concern over resource sufficiency concentrated attention on the stock or nonrenewable resources. The economic literature of the nineteenth century is replete with dire predictions of impending resource scarcity in such vital resources as coal. The Conservation Movement in the United States around the turn of the century emphasized the need to conserve natural resource stocks for future generations. Even as recently as 1952, the President's Materials Policy

3 Erich W. Zimmermann, *World Resources and Industries* (New York: Harper & Brothers, Publishers, Revised Edition, 1951).

Commission (The Paley Commission) expressed concern over diminishing supplies of "the evident, the cheap, the accessible" resources.

Economic research in recent years has revealed why the pessimistic predictions of those concerned with impending scarcity of exhaustible resources have failed to materialize. Harold J. Barnett and Chandler Morse studied the question of whether or not increasing scarcity of fixed resources has resulted in a secular increase in the average costs of extractive products. Their work reveals that increasing scarcity of particular resources encourages the discovery or development of alternative resources, not only equal in economic quality but often superior to those replaced. Thus they find that "relative unit costs of total extractive goods and agricultural goods, have been constant, and those of minerals have fallen; those of forestry alone have risen."[4] In summarizing the Barnett and Morse findings, John Krutilla has observed:

> Those who take an optimistic view would hold that the modern industrial economy is winning its independence from the traditional natural resources to a remarkable degree. Ultimately, the raw material inputs to industrial production may be only mass and energy.[5]

However, not all experts are so sanguine. A study released last year by the Committee on Resources and Man of the National Academy of Sciences strikes a different chord when it concludes: "It is not certain whether, in the next century or two, further industrial development based on mineral resources will be foreclosed by limitation of supply."[6]

Close examination of the likely existing reserves of the major raw materials reveals very few minerals in which a shortage appears imminent. The exceptions, which receive mention in the National Academy study, are mercury, tin, tungsten, and helium. While the short-run outlook for each of these minerals is for shortage, given the existing price structure and technology, there is good reason to believe that in the long run, changes in price will encourage the development of new sources of supply or the substitution of alternative materials for most uses.

Barnett and Morse contend that the market provides a built-in mechanism for dealing with natural resource scarcity in the form of sociotechnical change which is brought into force by an increase in the price of a raw material. This sociotechnical change may take any or all of three forms.

First, additional supplies of scarce minerals may be obtained by expansion of the resource base through the discovery of new deposits. The discovery of new deposits may be facilitated by changes in the technology of minerals exploration. If the new deposits are of similar quality and form to those pre-

[4] Harold J. Barnett and Chandler Morse, *Scarcity and Growth* (Baltimore: The Johns Hopkins Press for Resources for the Future, Inc., 1963).

[5] John Krutilla, "Conservation Reconsidered," *American Economic Review* (September, 1967), p. 778.

[6] National Academy of Sciences, Committee on Resources and Man, *Resources and Man* (San Francisco: W. H. Freeman and Company, 1969), p. 6.

sently being worked, no change in the technology of use will be required. However, it is more likely that changes in technology will make possible the profitable exploitation of new deposits which are lower in quality or different in form from those being worked at present.

Second, increase in scrap and waste recovery for both nondissipative elements recovered in original form (for example, scrap metal) and for converted elements in solutions and gases (for example, recovery of sulphur from industrial gases) will act to reduce demands for new inputs.

Third, scarcity of certain minerals may be mitigated by the substitution of more plentiful elements and a consequent reduction in demand.

Acting singly or jointly, these sociotechnical changes help maintain resource sufficiency. The tendency of most studies of resource adequacy is to extrapolate present trends of supply and demand without sufficient regard for the stimulus to technological change provided by higher raw material prices. One danger of this approach is the temptation to call for governmental controls to "conserve" the resource. The presence of government controls may act to retard the very types of technological change which are essential in maintaining resource sufficiency.

RECYCLING OF RESIDUALS

The high degree of interaction between environmental controls, new technology, and raw material markets is no where more apparent than in the area of recycling of residuals. As our framework for viewing the environment shows, residuals are an inherent part of the extraction, production, and consumption process in a modern economy. The recycling of these residuals may be expected to increase in response to pressures on a finite resource base generated by a growing population and continued economic growth. In addition concern over environmental pollution may be expected to result in legislation which will accelerate the move toward increasing recycling of residuals. In response to these forces new technology may be expected to develop which will enable recycling to increase. However, the direction and pace of this new technology will depend in great measure on the market for the recycled materials.

The raw materials that are extracted from the earth may be classified as follows: (1) those that are wasted in the process of extraction and manufacture and which may be reclaimed directly by the processing plant; (2) those that are used by society in such a way that they are dissipated and dispersed, making recycling extremely unlikely; and (3) those that are not dissipated but are manufactured or converted into durable products which may be used as a secondary source of raw materials upon termination of the original product life.

It has been estimated that of the inputs of raw materials into the United States economy, approximately three quarters of the overall weight is eventually discharged to the atmosphere as carbon (combined with atmospheric oxygen as CO or CO_2) and hydrogen (combined with atmospheric oxygen as H_2O). The remaining residuals are either gases (such as CO, NO_2, and SO_2), dry solids (like rubbish and scrap), or wet solids (such as garbage, sewage, and industrial wastes suspended or dissolved in water).[7] It is these remaining residuals which offer the best potential for recycling. Two examples of recycling, recovery of sulphur from stack gases and recycling of obsolete scrap, illustrate the close relationship between impending environmental controls, recycling technology, and raw material markets.

The United States Public Health Service estimates that each year, 28.6 million tons of sulphur oxides are emitted into the atmosphere. Sulphur oxide emissions are increasing at a rate of from six to seven percent per year. The principal sources of sulphur oxide emissions are coal- or oil-fired electric utilities and industrial plants. In the long run, it is likely that a shift from fossil fuels to nuclear power generation will reduce the magnitude of the sulphur oxide problem. However, this is not likely to occur until better means of nuclear generation can be developed which do not present the threat of thermal pollution and accidental radiation emissions. In the meantime it is likely that over the next decade, legislation, mainly at the local level, will require the sulphur oxide content of power plant fuels and stack emissions to be reduced to a level of from 0.5 to 1 percent.

Desulphurization of fuels and stack gases is technically feasible at present. Residual fuel oil which ranges in sulphur content from 1.47 percent (domestic) to 2.6 percent (Caribbean) can be treated to reduce the sulphur content to 0.5 percent.[8] Desulphurization of coal is less likely to occur. It is possible to remove from 20 to 60 percent of the sulphur contained in coal by physical means. The remaining 40 to 80 percent of the contained sulphur can only be removed by converting the coal to a gas and cleaning the gas stream. While this is technically feasible, it is more likely that utilities will switch to low sulphur coal which is in abundant supply in the United States. The main problem with switching is that low sulphur coal deposits are located in areas remote from the main areas of use. Thus the problem is more one of transportation costs than of desulphurization technology.[9]

It appears unlikely that the transportation prospects for low sulphur coal will change dramatically in time to be of assistance in meeting the stringent regulations being imposed by local governments on sulphur emissions. As a result, the principal means used to meet these requirements over the next decade, both for utilities and industrial plants, will be to desulphurize stack

[7] *American Economic Review* (June, 1969), p. 285.

[8] One problem with desulphurization of residual oil from the Caribbean which remains to be solved is handling the high metal content of the oil. Development of new catalysts for hydrodesulphurization that are not deactivated by the heavy metals in residual fuel oils.

[9] See "A Crisis That Looks Permanent," *Business Week* (October 3, 1970), pp. 14–15.

gases. First generation desulphurization facilities for stack gases are already in use. Research on second generation methods is underway.[10]

The technology and economics of removal of sulphur from stack gases will be greatly affected by the market for elemental sulphur. It has been estimated that potential sulphur from utilities alone without abatement in the year 2000 is 112 million long tons.[11] This implies removal of 79 million long tons a year just to hold the level of pollution from this source to the 1970 level of 33 million long tons. The magnitude of this potential increment to sulphur supplies is shown in Exhibit 2. Projected United States sulphur consumption for the

[10] Areas of technology in which second-generation processes might lie include: aqueous scrubbing, solid metal oxides, other inorganic solids, inorganic liquids, organic liquids, organic solids, catalytic oxidation, direct reduction to sulphur, and physical separation. See American Society, Committee on Chemistry and Public Affairs, *Cleaning Our Environment: The Chemical Basis for Action* (Washington, D.C.: The American Chemical Society, 1969), p. 71.

[11] Alfred Petrick, Jr., "The Effect of Urbanization on Mineral Demand," Paper delivered to the Western Resources Conference, University of Denver, July, 1970.

Exhibit 2. Sulphur Demand Projections Relative to Sulphur Supply from Pollution Abatement.

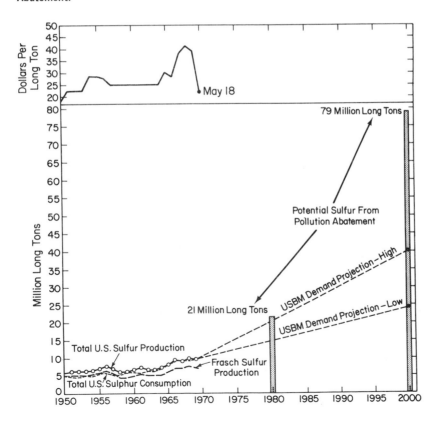

year 2000 has been estimated by the Bureau of Mines at 40 million long tons. Thus desulphurization may lead to a dramatic surplus of sulphur.

Sulphur recovery from stack gases will not only provide increased supplies of either elemental sulphur, or its principal product, sulphuric acid, but it will provide these supplies as a by-product in plants widely distributed across the nation. This will have two effects on sulphur markets. First, whereas sulphur production is at present concentrated in the Frasch deposits of the Texas-Louisiana Gulf Coast, the deposits in West Texas, and the sour gas field of the Southwest, recovered sulphur from stack gases will be available wherever sulphur bearing fuels are used, principally in industrialized urban areas. This should result in a shift to a pattern of small local markets served by local sources. Second, the price of elemental sulphur has been in large measure controlled by the two principal producers of domestic Frasch sulphur. Prices have been relatively stable over the history of the industry and profits have been relatively high.[12] However, the widespread desulphurization of stack gases will provide a means of entry into the industry and will likely disturb the stability of sulphur markets, resulting in lower prices.

Lower prices for sulphur will affect the economics of sulphur removal techniques. Lack of local markets for sulphur may result in a preference for removal techniques which do not produce by-product sulphur, such as those employing limestone and dolomite rather than catalytic conversion, alkalized alumina sorption, or chemical absorption processes.

In the long run, however, as noted by the American Chemical Society's Committee on Chemistry and Public Affairs, it does not seem likely that emission of sulphur by industrial combustion of fossil fuels can be controlled to the extent required by stack gas treatment. The sources are too numerous, too small, and too diverse.[13] It is likely that substitution of low sulphur fuels or development of new combustion processes with reduced sulphur oxides emission will be required. Given the restrictions on sulphur oxides emission expected to prevail during the next thirty years, new technology will be required.

Recycling of scrap is already a big business in the United States. The secondary materials industry includes more than 9,000 recognized establishments with total annual sales of more than $5 billion. In 1966, secondary aluminum accounted for about twenty percent of total consumption, secondary copper for about forty-two percent of consumption, secondary iron and steel for about forty-five percent, and secondary zinc for about twenty-five percent. The rubber industry in 1966 consumed some 265,000 long tons of reclaimed rubber, or about twelve percent of total consumption of new rubber of all types. Paper and textiles are among other materials that are recycled in large volume.[14]

[12] Jared E. Hazleton, *The Economics of the Sulphur Industry* (Baltimore: The Johns Hopkins Press for Resources for the Future, Inc., 1970).
[13] American Chemical Society, Committee on Chemistry and Public Affairs, *op. cit.*, p. 61.
[14] *Ibid.*, p. 183.

There are three principal sources of scrap: residuals from the processing of mineral ores (home scrap); residuals from the operations of fabricators (prompt industrial scrap); and discarded articles flowing from consumers (old or obsolete scrap). The potential for scrap recovery depends on the material's resistance to chemical and physical breakdown, the quantities of it available in the recoverable unit, its location, the price, and the ease with which it can be extracted from the primary product.

Home scrap and prompt industrial scrap present the fewest recovery problems and are, for the most part, recycled at present. Obsolete scrap presents the largest potential source of recycling. Hans Landsberg, in his book, *Resources in America's Future,* projected increases in the use of obsolete scrap for the major metals.[15] For each metal he provided a "low," "medium," and "high" projection, using 1960 as his base year. Exhibits 3 A, B, C, and D depict the projections made by Landsberg for zinc, lead, aluminum, and copper and also show the actual amounts of scrap recovery for each metal for the period from 1960 to 1968.

It is interesting to note that based on the actual trends in scrap reuse for these metals between 1960 and 1968, Landsberg's projections appear to be too high. The trend in recycling for aluminum, copper, and lead have barely approximated his "low" projection, while that for zinc has fallen considerably below this level.

Recycling of scrap faces many difficulties which restrict the potential for environmental improvement. Collection of scrap into sufficient concentrations and in the proper locations may be extremely difficult. While much of the residuals are collected by municipalities, a recent study showed that 14 percent of the trash hauled off in municipal packer trucks is incinerated. It found that 3 percent is salvaged, 5 percent goes into sanitary landfills, 0.5 percent becomes

15 Hans H. Landsberg, Leonard L. Fischman, and Joseph L. Fisher, *Resources In America's Future* (Baltimore: The Johns Hopkins Press for Resources for the Future, Inc., 1963).

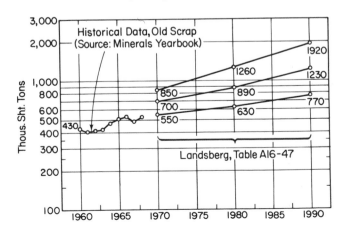

Exhibit 3a. Copper, Secondary Recovery from Obsolete Scrap.

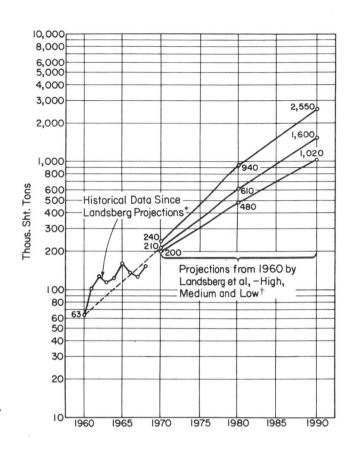

Exhibit 3b. Aluminum, Secondary Recovery from Obsolete Scrap.

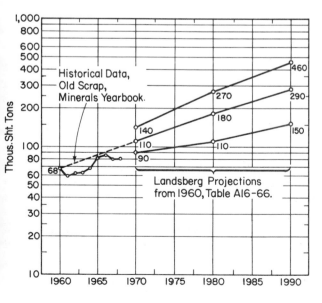

Exhibit 3c. Zinc, Secondary Recovery from Obsolete Scrap.

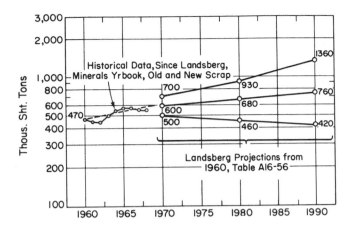

Exhibit 3d. Lead, Recovery from Obsolete Scrap, Old and New Scrap.

compost, and 77.5 percent is dropped into open dumps.[16] The technology of waste separation for municipal refuse is inadequate for present needs. Increased recycling of paper, glass, and cans may depend on the availability of voluntary collectors.

Scrap recycling is also hampered by economic considerations. The steel industry's switch from open hearth furnaces, which could take a forty-five percent scrap charge, to basic oxygen furnaces, which can take only a thirty percent scrap charge, sharply reduced the price for scrap metal. While some smaller steel firms are introducing electric melting furnaces which can take a 100 percent scrap charge, the prospects are for continued domination of the basic oxygen furnace. In many other metals the cost of the scrap reclaimed element exceeds the cost of newly mined ores.

Finally, increased recycling will probably depend on the adoption of legislation to encourage or even require re-use. Local zoning laws, air pollution regulations, urban renewal programs, and highway beautification efforts have seriously restricted the growth in secondary recovery. It appears unlikely that significant relaxation of these ordinances will occur given increased political pressures for environmental protection.

CONTROL OF EMISSIONS

Residuals are an inherent part of the economy of a modern society. If they cannot be recycled, they must be discharged into the air, the land, or the water. These, together with the living organisms which they house, constitute the renewable resources. They are interrelated in a manner that has caused them to be considered as a single entity called the biocycle. All living things draw their sustenance from air, water, and soil, in various combinations, and make their

16 See "Turning Junk and Trash Into A Resource," *Business Week* (October 10, 1970), p. 67.

home within one or more of these aspects of nature. They form the environment for the living creatures within them. Any change which occurs in these elements works a massive effect upon any life within them, even to precipitating its occasional disappearance.

It is not surprising that we are today witnessing repetition of the pessimistic resource forecasts of the nineteenth century, only with concern shifted from the nonrenewable to the renewable resources. It is commonplace to read of dire predictions of environmental disasters which will befall man if he does not change his ways. Interestingly enough, some of these predictions suffer from the same fault as prior attempts at predicting resource availability. They fail to account for the impact of market forces. It was a rise in the price of coal which led to technological change which made us less dependent upon coal. In a similar fashion, it will be an increase in the price of water which will make us take some more effective efforts to conserve our water supplies.

However, one cannot be sanguine about sole reliance on market forces to take care of the allocation of what are becoming scarce resources. For unlike the stock resources, we must be concerned not only with the *quantity* but also the *quality* of the renewable resources. For example, if iron ore of five percent concentration becomes scarce, iron ore of one percent concentration can be substituted provided certain technological changes occur in iron ore processing. However, there is no way in which one hundred cubic feet of slightly polluted water or air can be substituted for fifty or ten feet of clear air or water, at least, not for most purposes. Thus, we move from a concern over the availability of renewable resources to an interest in the availability of renewable resources of a given quality.

There are important trade-offs between the forms that residuals may take. Most efforts at pollution control will be aimed at changing the form of emission to utilize a different environmental media for disposal. As with recycling, there is a close relationship between environmental control over emissions, new technology for waste disposal, and raw material markets. A good example of this interaction is provided by current efforts to curb air pollution from the automobile.

According to the National Air Pollution Control Administration some 142 million tons of carbon monoxide, hydrocarbons, nitrous oxides, sulphur dioxide, and particulates are dumped into the air annually across our nation. The amount of such emissions and their sources are indicated in Table 1. From these figures it is apparent that air pollution control must begin with the biggest source of emissions, the automobile.

Each year, over 100 million motor vehicles in the United States discharge into the atmosphere 66 million tons of carbon monoxide, 1 million tons of sulphur oxides, 12 million tons of hydrocarbons, 6 million tons of nitrogen oxides, and 2 million tons of particulate matter plus residuals of chemicals added to improve the quality of fuel. The major sources of automobile pollution are (1) exhaust emissions, (2) emissions from the fuel tank and carburetor, and (3) emissions from the crankcase. Technology has already dealt with the

Table 1. National Air Pollution Emissions (millions of tons per year, 1965).

Source	Total	% of Total	Carbon Monoxide	Sulphur Oxides	Hydro-carbons	Nitrogen Oxides	Particles
Automobiles	86	60	66	1	12	6	1
Industry	23	17	2	9	4	2	6
Electric power plants	20	14	1	12	1	3	3
Space heating	8	6	2	3	1	1	1
Refuse disposal	5	3	1	1	1	1	1
Totals	142		72	26	19	13	12

Source: *The Sources of Air Pollution and Their Control*, Public Health Service Publication No. 1548 (Washington, D.C.: U.S. Government Printing Office. 1966).

latter two sources of emissions; future technology will have to deal with the most important source, exhaust emissions.

Auto emission can be controlled in two places: first emissions from the exhaust system can be reduced either by a thermal or catalytic muffler; and second, emissions of fuel additives may be reduced by removing lead and other additives from fuels. These two actions, additives control and emission control, are closely related. A decision on one will certainly affect the decision regarding the other. The raw materials impact of emission control in the auto industry is depicted in Exhibit 4.

Emission control seeks to reduce or eliminate the pollutants which are emitted from the exhaust. In a catalytic muffler, hydrocarbons and carbon monoxide are oxidized to yield only water vapor and harmless carbon dioxide. Some catalysts also break down nitrogen oxides into nitrogen and oxygen, although an alternative method of eliminating nitrogen oxides is to recirculate the exhaust. In experiments emission has been reduced to 50 ppm (parts per million) hydrocarbons and 0.5 percent carbon monoxide. The major problem with catalytic mufflers are catalyst life which must withstand high temperatures, mechanical durability, packaging, and cost. Despite these problems, General Motors has just announced that all of its 1975 models will be equipped with catalytic mufflers.

Metal demands to equip 10 million cars per year with the catalytic muffler would be on the order of 50 to 80 million pounds of catalysts annually. Valued at an average of $1 per pound, this is a potential market of significant scope. Twelve firms have been competing for this market. The catalyst will probably be base metal compounds such as copper chromite, alumina, vanadium pentoxide or copper oxide.[17]

The only alternative at present to the catalytic muffler is the thermal reactor. In the thermal reactor, carbon monoxide and unburned carbohydrons

[17] Robert J. Leak, John T. Brandenburg, and Milton D. Behrens, "Use of Alumina Coated Filaments in Catalytic Mufflers Testing with Single Cylinder Engine," *Environmental Science and Technology*, Vol. 2.

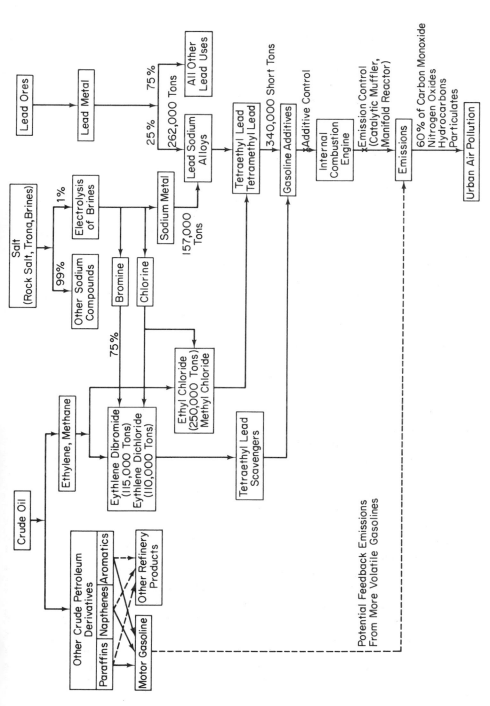

Exhibit 4. Mineral Industry Relationships to Gasoline Engine Emission Control.

are oxidized or burned before being emitted, but this is accomplished by keeping exhaust gases very hot and by adding extra oxygen. The insulated reactor, made of thin metal that heats up quickly, would replace the exhaust manifold, and outside air would be pumped in to form the fire. The thermal reactor is being developed by DuPont Company and the Ethyl Corporation and is still under active consideration by the Ford Motor Company. Since the thermal reactor uses heat and oxygen to oxidize pollutants rather than a catalyst, it performs efficiently on gases containing tetraethyl lead, a gasoline additive that causes problems in the catalytic muffler. The adoption of the thermal reactor would also affect metals demand, but commodities would be different, probably stainless steel and nickel alloys.

The choice between catalytic mufflers and thermal reactors is closely related to the question of gasoline additive control. Control of additives principally tetraethyl lead and the lead scavengers now appears certain. The additive tetraethyl lead promotes even burning of gasoline in the engine cylinder and raises the octane rating of the fuel. The scavengers remove lead from the cylinder after it has done its job.

The impact of gasoline additive control is depicted in Exhibit 4. It would have a significant and immediate impact on raw materials markets. (1) Lead demand would be reduced by about twenty-five percent, a reduction approximately equal in magnitude to a decade of growth in lead markets other than gasoline additives; (2) nearly all of the market for sodium would be eliminated since sodium is necessary to produce an intermediate in tetraethyl lead manufacture; (3) about seventy-five percent of the bromine market would be eliminated since bromine is used to produce intermediate compounds in the manufacture of lead scavengers; and (4) a significant part of the chlorine market would be affected since chlorine is used to produce intermediates in both additive and scavenger manufacture.

The petroleum industry will also be affected by the removal of gasoline additives through the elimination of the market for the petroleum derivative ethylene. However, reduction in this market would probably be offset by new markets created by the efforts of petroleum producers to duplicate present day gasoline octane ratings without the use of lead. It is estimated that $3 billion would have to be invested in refineries in addition to the $2 billion United States refineries already intend to spend over the next three years to keep pace with normal demand growth. It is likely that the race to produce higher octane nonleaded gasolines will take the form of increased use of aromatics in gasoline blends. These aromatics will be supplied by increasing catalytic reforming capacity at refineries. Catalytic reforming is a continuous process used to upgrade low-octane virgin, thermal, or heavy catalytically cracked naphthas into high octane components. Volatility is increased and sulphur content is reduced. The increase in catalytic reforming will in turn produce a boom in the demand for platinum based catalysis. It may also increase the demand for rhenium, since platinum-rhenium catalysts have proven competitive with the conventional platinum catalysis in pilot plant tests. The sole United States producer of rhenium, Cleveland Refractory Metals at Solon, Ohio, projects an increase in demand for rhenium on the basis of platinum-rhenium-catalyst demand from the present 1,000 to 2,000 pounds per year to 20,000 pounds per year by 1972.[18]

It should be noted that the increased use of aromatics in automotive gasolines could lead to higher photochemical pollution than that experienced with prior leaded fuels. In a United States Bureau of Mines study, the photochemical reactivity of automobile emissions was found to be increased by as much as twenty-five percent when the fuel was changed from a typical United States leaded gasoline to a prototype unleaded gasoline of comparative octane quality. The increase is attributed to the characteristics of blending components that were used to obtain the required octane quality without using lead.[19]

CONCLUDING COMMENT

The public concerns that are generating pressures for the enactment of environmental controls stem from the inevitable conflict between man's economic activities and the state of the natural environment. Our economic system is essentially a closed system in which raw materials are extracted from the natural environment and processed into energy and commodities which are then "consumed." Residuals from extraction, production, and consumption, which are approximately equal to the volume of inputs, must be eventually discharged into the natural environment.

The increasing production of goods and services in a modern industrialized economy makes heavy demands upon the resource base. Technology plays an important role in helping to maintain resource sufficiency through the discovery of new deposits, increased recycling of residuals, and discovery of substitute inputs. Recycling of residuals not only acts to conserve resources, but also alleviates problems created by the dumping of wastes into environmental media. Pollution control is primarily concerned with the shift of residuals from one form to another or from one type of environmental media to another. Technology can play an important role in this shifting process.

As an economist, I want to stress that the response of technology to the demands created by environmental controls will be shaped by the forces of supply and demand acting through resource markets.

---

[18] Petrick, *op. cit.*
[19] U.S. Bureau of Mines, *Comparative Emissions from Some Leaded Prototype Lead-Free Automobile Fuels*, R.I. 7390 (Washington, D.C.: Government Printing Office, 1970), p. 23.

Madness, Mediocrity, or Mastery: A Threat Analysis for a New Era

HAROLD A. LINSTONE

The defense community, government, and industry, have entered upon a period of flux not encountered since World War II. Linstone's study concentrates on the world environment of the 1970–85 period. It is an attempt to set the stage for more detailed corporate long-range planning exercises in this area by suggesting broad environmental forces and their potential impact.*

> *Interviewer:* Are you optimistic or pessimistic about the future?
> *Forecaster:* I am basically optimistic.
> *Interviewer:* Then why do you look so sad?
> *Forecaster:* I don't think my optimism is justified.

Most environmental forecasts extrapolate familiar parameters describing familiar trends from the recent past into the future. Never has this practice

*The work described in this chapter is based on studies performed for Project MIRAGE 85 of Lockheed Aircraft Corporation in 1969–1970 and reported in University of Southern California Industrial and Systems Engineering Technical Report 70–2, March 1970. All views expressed are those of the author and do not necessarily represent those of Lockheed.

been more dangerous than it is today. The two views presented in this discussion underscore the hazards and their significance.

"TRADITIONAL" LOOK INTO FUTURE

Era of Two Superpowers

For the past quarter century, the United States and the Soviet Union have been the undisputed number one and two powers. They alone have the ability to effectively destroy any nation on earth, including each other. For example, the McNamara posture statements from 1964 to 1969 indicate that in all but one of the sixteen cases he analyzed, the United States can inflict 70 to over 120 million fatalities on the Soviet Union after absorbing an enemy first strike.[1] Economically, the ratio of strength in terms of GNP (gross national product) between the first and second power is of the order of 2:1. If this trend continues, the gap will, of course, widen. A 1965 gap in GNP per capita of $2,300 can easily become a 1985 gap of $3,900.

Both nations have used their power to gain or hold influence in Eastern Europe, Western Europe, Latin America, Africa, the Middle East, and Asia.[2] Currently over one million American troops are stationed abroad. There are 2,270 United States military bases overseas (343 major ones) in twenty-nine foreign countries as well as in United States possessions.[3] They range from Greenland to Africa, Spain to Japan.

The extensive conventional military aid provided by the two superpowers to other countries has occasionally resulted in strange bedfellows: support by one superpower to two opposing nations, for example, United States to Greece and Turkey, to Israel and Jordan, and to Pakistan and India; support by both superpowers to one nation, for example, Pakistan and Iran each supplied by both superpowers.

The aims of the two superpowers brought them into both cooperation and conflict beginning with World War II in Europe. They cooperated to rescue Europe from Hitler and divided Germany. Then conflict began. Following the 1948 Berlin crisis, instigated by the Soviet Union, the formation of the United States dominated NATO pact in 1949 provided stability. This, in turn, permitted spectacular recovery of the shattered European economy on both sides of the Iron Curtain. Most importantly, the policy of "conquer and

---

[1] W. W. Kaufmann, "The Strategic Nuclear Forces," unpublished paper, 1969, p. 31. The exception is the case where the Soviet Union develops a greater than expected offensive and defensive threat and even then the number of Soviet casualties is about 50 million.

[2] Until 1950, the United States enjoyed a unique image in Europe as the only major country without the "normal" power aims. Evidence: aid to England, France, and Russia in two world wars and the Hoover and Marshall postwar aid programs.

[3] *Los Angeles Times*, June 22, 1969, p. A–27. *U.S. News and World Report*, August 4, 1969, p. 47. These figures represent the results of a cutback in bases begun in 1966–67.

divide" has effectively prevented the rise of a single dominant country (for example, Germany) in Europe which could once more threaten its neighbors.

Similarly, in the poor world the two superpowers have a history of both cooperation and conflict. Since the demise of European colonialism, a power vacuum has made this area much more the center of superpower activity than the advanced countries. Of more than fifty internationally significant local wars since World War II, only one, the Hungarian revolt, has been fought between advanced industrial states or on the territory of an advanced state.[4] Superpower cooperation is evidenced by the stand against nuclear proliferation and attempts to settle the India-Pakistan conflict.

Conflict has arisen from Soviet efforts to replace United States dominance in Cuba (in the United States sphere from 1823 to 1959), to replace British and United States influence in the Arab world (beginning in 1946 with the Soviet actions in Iran), and to vie with the United States for influence in the former French and Japanese areas in Southeast Asia (Indochina, Korea).[5] Finally, cooperation is apparent in the initiation of strategic arms limitations talks in 1969.[6]

We can readily extrapolate this era of the superpowers into the next quarter century. But there are signs that the era is drawing to a close. The United States difficulties in Vietnam, the Soviet difficulties in Czechoslovakia, as well as the impotence of the United States and Soviet Union in the Middle East are obvious indications of change.[7] The limits of power are coming into focus and the "Age of the Cold War" may well fade into the history books.

Challengers

Extrapolating current trends, a ranking of nations in terms of productivity yields the following table for 1985:

GNP Ranking (Medium Estimates)

United States
Soviet Union
Japan
West Germany
France

4 R. E. Osgood, "The Reappraisal of Limited War," *Adelphi Papers No. 54* February, 1969, p. 42.

5 The American interest in the Pacific actually started with the opening up of Japan. It may almost be viewed as a continuation of the nineteenth century drive to conquer the Western frontier on this continent. Has it reached its zenith with Korea and Vietnam?

6 It is also notable that Soviet trade with capitalist countries has increased twenty-five percent between 1958 and 1967.

7 Soviet writer Andrei Amalrik views his nation in these words: "Too authoritarian to permit everything, too weak to repress everything, the regime is tottering toward death." ("The Fall of the Soviet Empire", *Atlas,* February 1970, p. 24.)

Japan and Germany, the two principal instigators and losers of World War II, are the foremost challengers in terms of total productivity. Both have had a very favorable government-business relationship and diverted only minor resources to the military establishment. They exercise economic rather than political power in international affairs.

The 2 : 1 ratio for the United States and Soviet Union in total GNP is also reasonable for the Soviet Union and Japan. Neither challenger represents a real military threat to the Soviet Union or to the United States in the next fifteen years. However, the Land of the Rising GNP may try an economic version of its "Greater Southeast Asia Coprosperity Sphere." Herman Kahn has even suggested that the twenty-first century may become known as the Japanese century. One wonders how long Japan can maintain its current phenomenal growth rate.[8]

Germany may encourage one of several European options: a fragmented Europe dominated economically by Germany, a loosely federated Europe with no dominating country but a considerable number of common functional institutions, or a strongly integrated Europe which becomes an independent "Third Force." A Western European Union can match the Soviet Union in economic strength, that is, GNP. Medium growth estimates for 1985 give the Soviet Union a GNP of $788 billion and the combination of West Germany, France, United Kingdom, and Italy $851 billion. So far, Europe has been hampered by inadequate education of its manpower, weak management and marketing capabilities, and inefficient joint ventures. The small and rigid advanced education systems and the apprenticeship concepts are obsolete. There have been numerous instances of European leadership in technology nullified by poor exploitation due to marketing and management (examples: commercial jet aircraft, computers, steel processing). Cooperative international projects such as Euratom, the Coal and Steel Community, and ELDO have not been truly integrated operations. The vacuum has been filled by American companies operating mostly with European financial and manpower resources.[9] A strong European economy will require:

1. Formation of large industrial units with strong management in technologically advanced areas.
2. More surrender of national sovereignty than has been exhibited heretofore.
3. Major revision of the educational system.
4. Drastically changed social structures to provide organizational flexibility.

---

[8] There are some interesting contrasts between the United States and Japan. Example: In Japan young industries are supported by the government; later they move to full private control. In the United States, on the other hand, old industries are supported by the government (for example, shipbuilding and railroads).

[9] According to Servan-Schreiber [*The American Challenge* (New York: Avon, 1967), p. 14], ninety percent of the American investment in Europe is financed with European funds.

In recent years, GNP per capita has come into wide use for evaluating and comparing the status of nations. Sawyer considered 236 variables for 82 nations and concludes that "if one were to know but three characteristics of a nation, their population, GNP per capita, and political orientation would seem to be a good choice."[10]

Exhibits A–1 to A–3 (Appendix A) show the dynamics of population and GNP per capita growth for most nations to 1985 on a regional basis. The 1985 GNP per capita ranking differs notably from that shown earlier for GNP.

Ranking of Countries Moving Beyond
"Mass Consumption" Stage
(Medium Estimates)

United States
Sweden
Canada
West Germany
East Germany

Here also we find a clue to the question of continuation of the era of the superpowers. The challengers reaching for the highest level include a different breed of nation.

Poor World

Two thirds of the nations, with seventy percent of the world population, have a GNP per capita of less than $500. Exhibit A–4 (Appendix A) shows the last decade of growth and Exhibits A–1 to A–3 (Appendix A) indicate their relationship to the rest of the world in the next fifteen years. The widening gap is evident by the directions of the arrows. Compare, for example, Israel and UAR, Japan and Indonesia.

The most severe problem is in the proper management of human resources (for example, education) and capital, the key ingredients for development. We usually find one of the following situations:

1. Dictatorship of the right: The establishment is status quo oriented, usually corrupt and incompetent. Often the military establishment is the governing body. Examples: Portugal, Haiti, Paraguay.

2. Dictatorship of the left: There is a wide populist power base and engagement of the masses is emphasized. Extravagant promises and ideological trappings accompany socioeconomic changes. Often these changes are based on unsound theories and actually retard development. Here, too, the military establishment frequently provides the leadership. Corruption and incompetence also mark many of these regimes. Examples: China, Cuba, Syria, Algeria.

10 J. Sawyer, "Dimensions of Nations: Size, Wealth, and Politics," *American Journal of Sociology,* Vol. 73, No. 2 (September, 1967), p. 159.

These dictatorships are often "supported" by one of the two superpowers with its own aims (and sometimes both). Usually the United States "supports" the right, while the Soviet Union "supports" the left.

3. Moderate or compromise: In this group we see the liberal or reform leadership seeking change by evolution and widening the power base progressively. Again corruption, incompetence, and external "support" plague the government. Examples: Iran, Thailand, India.

Failure of (3) creates a shift to either (1) or (2). However, we also observe other shifts: from left to moderate, for example, Yugoslavia; and right to left, for example, Cuba.

The need is for highly capable, dedicated, and farsighted leadership which can mobilize capital effectively, educate and train people rapidly, and manage balanced growth of the nation while preventing domination by other powers—the requirement is formidable indeed. Today's advanced nations themselves could not solve the problem successfully during their transformation from an agricultural to an industrial society. Violence and revolution accompanied modernization in the United States, Soviet Union, England, Germany, and other nations.

In 1968 the rich nations spent about $6.9 billion in aid to the poor countries.[11] In terms of GNP fraction the United States aid program ranked seventh with .38 percent (France led the honors with .72 percent). It is often said that the advanced nations should increase their technical and financial aid to the poor countries. But it is debatable whether present know-how in the advanced countries is able to provide adequate guidance for balanced growth in the poor nations even if talent and money are liberally applied. Economically, the desirability of groupings of nations to strengthen their capital-human resource position is as valid for the poor as it is for Europe. Technology has already reduced the dependence of the wealthy on single crops or natural resources of the poor. Synthetic rubber, new foods, electric cars, and materials of the future will further lessen the leverage of the poor countries. The question is whether nationalistic (or tribalistic) prejudices and traditions can be submerged.

Alliances and Conflict

In a world of two superpowers, alliances are not groupings of equals but instruments placing the weak under the protection or control of the strong. The major post-World War II alliances—NATO, 1949; SEATO, 1954; the Warsaw Pact, 1955; OAS, 1956;—are all in this "umbrella" category. While still effective in relations between the two superpowers, the changing environ-

---

[11] For comparison, it is noted that the world was spending over $180 billion on arms that year.

ment can only reduce their significance in the next fifteen years. The appearance of challengers who can gain status and influence by strictly nonmilitary means and the worsening plight of the poor world tend to alter the alliance needs considerably. For one thing, "balance of power" alignments, a hallmark of nineteenth century Europe, again become possible. However, the two aging superpowers will find it most difficult to recognize and accept change. The key institutions in both countries (for example, defense and foreign affairs) undoubtedly will press for maintenance of these umbrella alliances. Their principal argument will be that any change will alter the relative positions of the United States and the Soviet Union.

"NEW" LOOK INTO FUTURE

The family and the nation have made up the most important social units in the world for many years. Now both are undergoing technology-induced stresses which may dramatically alter their future and invalidate the customary trend extrapolations. Let us first very briefly consider the family. The impact of technology is deep as two examples will show:

1. The difference in education between old and young reduces the authority of the former and hence loosens family ties.[12]

2. Communications and transportation have a similar effect. Mobility separates family clans and even separates parents from children. Communications tie in the young to the outside world early in a direct manner without the parents acting as filters.

In short, the family today finds itself weakened in a rapidly changing environment.

The reader may wonder why a defense planning study should even mention the "family" problem. The explanation will become evident after we discuss the nation in more detail. During the past fifty years, about 100 new nations have been created. At the same time, supranational concepts have had an ephemeral existence. The League of Nations failed; the United Nations organization is barely alive. The seemingly most successful example, Communism, has been bowing to nationalistic pressures. The COMINTERN failed and conflict in the Communist world is also between nations (for example, Soviet Union and Czechoslovakia, Soviet Union and China).

Success as a nation, like success as a corporation, has required a combination of capable leadership and objectives which motivate and unify its members. Common goals and values, together with competent and adaptive manage-

---

12 An illustration is afforded by mathematics. Earlier generations were taught Euclidean geometry and the decimal system as "eternal truths." Today students learn that many other geometries with opposing axioms are equally correct and that any number greater than one can serve as a basis for a number system. The students soon begin to question other ideas accepted by their elders.

ment or administration, create the self-discipline and synergistic group activities which characterize great achievements. Japan and Israel in the last decade are examples of such success. These countries also illustrate two essential points, respectively:

1. Unless a war results in annihilation, the outcome does not appear particularly significant for national success. Japan and Germany, the losers in World War II, are progressing far better than some of the victors (for example, England) if measured in terms of productivity or GNP per capita growth.

2. A motivation which serves admirably is fear of survival, represented by threat of military attack against the nation. In the case of Israel, the Arab threat of driving the Israelis into the sea unifies and drives forward a very heterogeneous society.

The meaningfulness of GNP per capita as a measure of national success in the next quarter century is open to question:

1. A country which moves beyond the mass consumption stage into a post-industrial stage where industry becomes a secondary activity can hardly consider productivity as a suitable measure of national status. New measures of individual and group achievement will be needed.

2. The famine-ridden poor nations will not consider industrial productivity central to their success. India has already recognized that in its current stage agricultural improvement is more crucial than industrialization (as recognized by Gandhi). Success of the "green revolution" cannot be measured adequately by GNP per capita.

3. Problems are mounting in both rich and poor nations which imply that "success" can no longer be couched in purely economic terms for any country. Money loses its significance for an individual who is incurably ill.

Can the government of the wealthiest nation in history be considered capable if it is unable to (1) protect its citizens against destruction (in the event of nuclear attack), (2) protect its citizens from attack in the streets, (3) efficiently stop military advances and subversion instigated by a small nation, (4) fight a successful "War on Poverty" at home, (5) assure clean air and water for its citizens, (6) lead the world in health (for example, infant mortality rate, expected life span of its men), and (7) plan the evolution of its society beyond the industrial stage? The time available to solve this nation's problems and avoid crises is clearly limited; one overview of the situation is depicted in Table 1.

The source of the global problems facing nations is the exponential growth of technology.[13] In the last fifty years, technology has drastically cut the death rate, thereby causing a population explosion (Exhibit 1). It has developed

[13] T. J. Gordon combines annual per capita power use, steel production, intercity passenger travel miles, working force of engineers and scientists, and communications (newspaper circulation, radio and TV receivers) to obtain a technology index. This index is doubling every twenty years.

Table 1. Classification of Problems and Crises by Estimated Time and Intensity (United States).

Grade	Estimated Crises Intensity (Number Affected X Degree of Effect)	Estimated Time to Crises*		
		1 to 5 Years	5 to 20 Years	20 to 50 Years
				(Solved or dead)
1.	Total annihilation	Nuclear or RCBW escalation	Nuclear or RCBW escalation	
2.	10^8 Great destruction or change (physical, biological, or political)	(Too soon)	Participatory democracy Ecological balance	Political theory and economic structure Population planning Patterns of living Education Communications Integrative philosophy
3.	10^7 Widespread almost unbearable tension	Administrative management Slums Participatory democracy Racial conflict	Pollution Poverty Law and justice	?
4.	10^6 Large-scale distress	Transportation Neighborhood ugliness Crime	Communications gap	?
5.	10^5 Tension producing responsive change	Cancer and heart Smoking and drugs Artificial organs Accidents Sonic boom Water supply Marine resources Privacy on computers	Educational inadequacy	?
6.	Other problems— important, but adequately researched	Military R & D New educational methods Mental illness Fusion power	Military R & D	
7.	Exaggerated dangers and hopes	Mind control Heart transplants Definition of death	Sperm banks Freezing bodies Unemployment from automation	Eugenics
8.	Noncrisis problems being "overstudied"	Man in space Most basic science		

Source: J. Platt, "What We Must Do", *Science*, November 28, 1969, p. 1118.
* If no major effort is made at anticipatory solution.

504

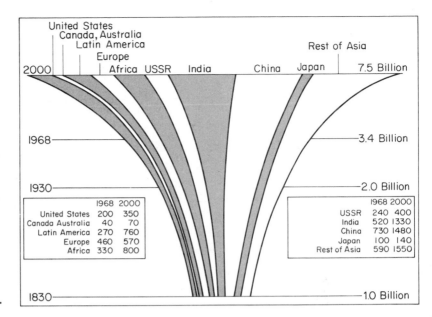

	1968	2000
United States	200	350
Canada Australia	40	70
Latin America	270	760
Europe	460	570
Africa	330	800

	1968	2000
USSR	240	400
India	520	1330
China	730	1480
Japan	100	140
Rest of Asia	590	1550

Exhibit 1. Population.

high speed global transportation, unleashed vast energy resources, and, most important of all, created near-instantaneous audiovisual communications. Person to person data flow is increasing far faster than GNP, population, energy consumption, or transportation. The speed of human transportation has increased by a factor of ten in fifty years and only one more such change can be envisioned. But computer speed has been rising (and cost declining) by the same factor every four years and the pace is likely to continue for awhile.[14] Data are not knowledge, of course, and we are faced with the threat that "data" may turn into one of the most serious pollutants.

As a result of such technological growth on many fronts the rather old spaceship Earth is suddenly becoming an overcrowded global village. As with two people in a telephone booth, every action by one affects the other and cooperation as well as careful planning is essential if any useful objective is to be achieved. How can national leaders manage homeostatic or balanced development when we possess insufficient understanding of the global "system," its subsystems, and their interactions? Let us look at some of the difficulties.

Population Imbalance

Currently, there is a net increase of 69 million people a year. A world which accommodated 3.5 billion souls in 1968 may house 7.5 billion by the

---

[14] P. Armer, "The Individual, His Privacy, Self-Image, and Obsolescence," Panel on Science and Technology, Eleventh Meeting, Committee on Science and Astronautics, House of Representatives, January 28, 1970.

year 2000. Asia alone is likely to have a population equal to that of the entire world today. While seventy percent of today's population is poor and inadequately nourished, that fraction may rise to eighty percent by 2000. It is concentrated in Latin America, Asia, and Africa.

It is hoped that agricultural technology will prevent the great famines forecast for the next twenty-five years. Farm machinery, irrigation, pesticides, fertilizers, and "transplants" of high yield grains have combined to create the "green revolution." In 1967–68 the wheat harvest in Pakistan reached 7 million metric tons, compared to the previous all-time high of 4.6 million harvested in 1964–65.[15] World meat production has grown by 75 percent from 1950 to 1965, fish by 140 percent.[16] Even today, however, 25 percent of the world population uses 75 percent of its fertilizers. And the poor world has a physical environment which places it at a great disadvantage in food production. A fourfold increase in food production is required by the year 2000. Technology can:

1. Reduce waste and spoilage (currently estimated to cost 20 percent to 50 percent of the food produced).

2. Provide storage.

3. Increase yield per acre by production of more and lower cost fertilizers in poor areas.

4. Use salt water irrigation.

5. Produce more high yield varieties of plants (rice, wheat) and create new plants by genetic manipulation.

6. Produce food by direct chemical means (for example, synthetic foods, proteins from petroleum, and plant protein mixes).

Such changes require effective large-scale government implementation as well as new policies in the area of distribution and pricing of agricultural output.

Reduction of the birthrate is another approach. Contraceptives for fertility control are available but government-sponsored programs have not proven effective to date. India introduced a birth control program in 1951 when the nation's growth rate was 1.3 percent. After sixteen years, the growth rate rose to almost 3 percent. By 1967, only 2 percent of India's 95 million couples of reproductive age practiced contraception systematically.[17]

Social solutions offer another possible approach. For example, retirement benefits might be geared to the number of children produced by a couple, with income inversely proportional to the number of offspring.[18] The income tax structure can be altered to favor single persons and couples with few children

[15] N. E. Borlaug, O. Aresvik, I. Narvaez, and R. G. Anderson, "A Green Revolution Yields a Golden Harvest," *Columbia Journal of World Business* (September–October, 1969), p. 12.

[16] G. Borgstrom, "The World Food Crisis," *Futures* (June, 1969), p. 342.

[17] P. Ehrlich, "Paying the Piper," *New Scientist* (December 14, 1967).

[18] J. J. Spengler, "Population Problem: In Search of a Solution," *Science* (December 5, 1969), p. 1234.

rather than those with many. Foreign aid can be made dependent on effectiveness of population control in the recipient country. But the net effectiveness of such steps is not at all clear. Forrester's World Dynamics model indicates that neither population control nor increased food production suffice; they merely shift the global crisis from one type to another.[19]

Socioeconomic Imbalance

The population problem must be considered integrally with socioeconomic development. For example, capital formation increases as the birth rate declines.[20] In other words, reduction of the birth rate is far more critical for the future of the poor world than increase of the food supply. The combination of low fertility and low mortality is desired.

Even the most advanced countries, such as the United States, lack sufficient understanding of socioeconomic development. It has surprised many to learn that the Vietnam War is proving to be the most expensive in American history if the socioeconomic impact on this nation is taken into account.

In 1967, Litton Industries set out on a program of "nation building" in Greece. Over a twelve-year period, Litton's systems management approach was to attract more than $800 million in investments in the country. Two years later the investments amounted to only $3.35 million and the contract was canceled.[21]

In his book, *Maximum Feasible Misunderstanding: Community Action in the War on Poverty* (sic), Daniel Moynihan observed that the ultimate source of failure was the application of half-baked social science. The poverty warriors assumed they had a scientific answer to poverty when what they really had was merely an interesting hypothesis.

The combination of pressure for change and failure to achieve homeostasis in socioeconomic growth leads directly to violence. An analysis of eighty-four countries for the period 1948 to 1962 substantiates the hypothesis that the faster the rate of change in the modernization process within any given society, the higher the level of political instability in that society.[22] Another study[23] has shown that revolutions, whether French, Russian, or Cuban, tend to follow periods of significant improvement when expectations diverge from aspirations ("the revolutionary gap")[24] This helps to explain the marked increase in riots

[19] J. W. Forrester, Testimony for the Subcommittee on Urban Growth of the Committee on Banking and Currency, U.S. House of Representatives, October 7, 1970.

[20] H. Frederiksen, "Feedbacks in Economic and Demographic Transition," *Science,* (November 14, 1969), p. 841.

[21] *Forbes,* December 1, 1969, p. 38.

[22] I. K. Feierabend, R. L. Feierabend, and W. W. Conroe, "Aggressive Behaviors Within Polities 1948–62: A Cross-National Study," *Journal of Conflict Resolution* (September, 1966).

[23] R. Tanter and M. Midlarsky, "A Theory of Revolution," *Journal of Conflict Resolution* (September, 1967).

[24] The Cuban example is illustrated in Exhibit A–4. The reversal in GNP per capita is followed by Castro's success.

following the dramatic legislative gains in United States civil rights under President Johnson.

To say, then, as Drucker does, that all countries have capital for development and that the claim of a "lack of capital is euphemism for mismanagement" is to beg the question.[25] We cannot manage very well when we cannot effect balanced growth and this is not possible without an understanding of the "system." The dramatic increase in mobility and instant audiovisual communications throughout the world is likely to have the following result: The gap between similar levels in most countries will decrease, while the gap between different levels in any one country will widen. In other words, the rich will lead very similar lives in most countries, but the rich and poor in each country will become extremely dissimilar. The dangers inherent in this situation represent one of the most important lessons of history. Consider the Roman Republic:

> In an age when all the emphasis is upon wealth, great is the frustration of those forced to remain poor.... No wonder, therefore, that violence and street battles on a scale unheard of before began to disfigure life in Rome.... Was not [Cicero's] plaintive motto *concordia ordinum* or "cooperation between the social classes" a plea for the restoration of a vanished social harmony? No remedy indeed, but a description of the state of society which a remedy ought to produce, could it have been found.[26]

Ecological Imbalance

Man's impact on the environment was of little concern when his numbers were small and his technological rate of achievement modest. But in recent years the population explosion (Exhibit 1) and the technological explosion have created a situation where a threat to man's survival on the shrinking spaceship Earth is looming. A few examples suggest the seriousness of the threat.

Air Pollution. In the United States man-made pollutants are currently added to the air at a rate of 160 million tons annually. Smog is increasing the incidence of respiratory disease. Chlorinated hydrocarbons, for example, DDT, are increasingly present in the atmosphere and extrapolation of the present trend may lead to thousands of deaths before 1980. Pollution may significantly alter the weather. Atmospheric temperature changes during the next fifty years are likely to affect the ice caps and the climate over the entire earth. Air pollution changes the rainfall pattern[27] and may also lead to a reduction in absorbed solar radiation.

---

[25] P. F. Drucker, *The Age of Discontinuity* (New York: Harper and Row, 1968), p. 128.

[26] F. R. Cowell, *Cicero and the Roman Republic* (New York: Chanticleer Press, 1948), pp. 276–77.

[27] *Wall Street Journal,* December 31, 1969, p. 1.

Water Pollution. Chlorinated hydrocarbons also slow down photosynthesis in marine plant life. Certain phytoplankton types may prove resistant and displace other microorganisms. Tiny animals which feed on such organisms will in turn be affected by the change and the impact will be felt on higher ocean fish life. Diatoms (sometimes known as "red tides") may proliferate and destroy vast amounts of marine animal life.[28] Drinking water in many areas of the United States does not meet Public Health Service standards. A recent sampling indicates that 8 million Americans consume water which exceeds the bacteriological standards set up by the Federal Government.[29] Excessive amounts of iron, manganese, and arsenic have been observed and pesticides have turned up. Cadmium is another dangerous water pollutant. Another study indicates that the United States may face a fresh water shortage by 1983.

Resource Management. The recent fouling of the Southern California coast as a result of oil drilling operations is well publicized. The cost-benefit evaluation of the Glen Canyon Dam may ultimately prove that project to have been a colossal blunder. Leakage into the reservoir walls may approach fifteen percent of the total average annual flow of the entire Colorado River. Evaporation may cause another twelve percent annual loss.

In Egypt the waterweeds clogging the shoreline of Lake Nasser behind the Aswan Dam may accelerate evaporation through transpiration to such a degree that there will be insufficient water in the lake to drive the generators. Furthermore, the dam has stopped the flow of silt which in the past offset the natural land erosion. Thus, as much productive farmland may be washed away as is opened up by new irrigation systems around Lake Nasser. As the rich silt has disappeared below the dam, so has the Egyptian sardine catch in the Mediterranean (18,000 tons in 1965, 500 tons in 1968). Finally, the irrigation area has provided a home for a snail which has infected the local population with schistosomiasis, an agonizing liver and intestinal disease.[30] The refrain is familiar: we do not adequately understand the system and therefore cannot achieve balanced growth.

Military Imbalance

Conducting war has constituted a major function of nations. Recent research has brought to light some interesting results and shattered familiar slogans. In the past foreign conflict behavior has been related to various national attributes such as domestic instability, level of economic or technological development, totalitarianism, international communications or transactions, psychological motivations of its people, and power or military capabilities. For example, we commonly hear that "a strong military establishment assures

---

[28] P. Ehrlich, "Eco-Catastrophe!" *Ramparts* (September, 1969).
[29] *Los Angeles Times,* December 21, 1969, p. C–1.
[30] *Time,* February 2, 1970, p. 62.

peace." Rummel's analysis of all nations in the 1950s shows that none of these factors correlate significantly to a nation's foreign conflict behavior.[31] The one critical positive correlation in an individual country's foreign conflict behavior appears to be time. The generation cycle theory suggests that major formalized violence surges in a nation approximately every twenty-five years.[32] Exhibit 2 shows the striking correlation of this cycle with United States conflicts ever since this nation was created (except for Korea). The generation cycle constitutes a worldwide pattern, as seen in Denton's work. This suggests some degree of validity in the claim that man's basic drives are at work here. The male apparently has an instinctive need for adventure, for exhibiting courage in front of one's peers, and for aggression. For many, war seems to satisfy these needs.[33] A generation which has fully experienced war tends to develop a

[31] R. J. Rummel, "The Relationship Between National Attributes and Foreign Conflict Behavior," in *Quantitative International Politics: Insights and Evidence,* ed. J. D. Singer. (Reprint No. 6, *The Dimensionality of Nations Reprint Series,* Department of Political Science, University of Hawaii.)

[32] F. H. Denton and W. R. Phillips, "Some Patterns in the History of Violence," *Journal of Conflict Resolution,* XII, No. 2 (June, 1968).

[33] These needs also explain the lack of popularity of nuclear and nonlethal biological/chemical weapons. Pushing buttons in a silo and annihilating millions does not offer much opportunity for "exhibiting courage in front of one's peers." Neither does the use of nonlethal weapons.

As Dr P. G. Bourne points out: "What people don't like to believe is that there is a real thrill in killing people." ("Men and Stress in Vietnam," to be published, quoted in *Newsweek,* December 8, 1969, p .35.)

Exhibit 2. United States Wars.

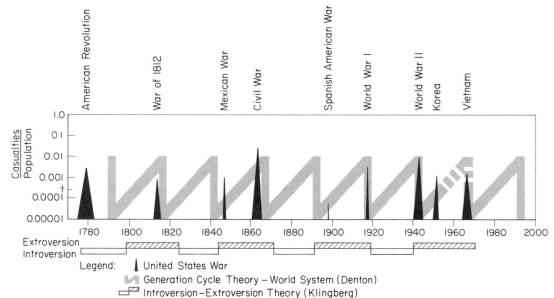

* Vietnam casualties to February 1970.
† Automobile fatalities in 1968.

revulsion against it, but slowly tolerance is built up for the next military orgasm. It is interesting to speculate whether or not the opposition to a large commitment in Vietnam is not the result of the Korean War. Following too quickly upon World War II, it did not leave time for the normal buildup, which would have led to much stronger support for the Vietnam War (Exhibit 2). Some advanced countries appear to have been successful in sublimating these human needs into alternatives more desirable than war (for example, Sweden, Switzerland). Have they paid a price?

If war occurs who are the likely opponents? In studying nation pairs in the 1950s, Rummel finds that "the more dissimilar two nations are in economic development and size and the greater their joint technological capability to span geographic distance is, the more overt conflict they have with each other."[34] Exhibits A–1 to A–3 measure size (population) as well as economic development (GNP per capita). Superimposing the ability to span geographic distance, some candidate conflict pairs are indicated by shaded double arrows.

Israel and its neighbors Jordan and the UAR provide an excellent illustration (Exhibit A–1). They are far apart in wealth and size and will be moving still further apart in the next fifteen years. At the same time they can easily reach each other. The widening gap between rich and poor nations, evident in Exhibits A–1 to A–3, therefore, portends increasing dangers of internation conflict between *nonequals*. Since the United States and Soviet conventional (general purpose) forces have been, and continue to be, based on the Western European tradition of war between *equals,* an imbalance in the military thinking of both superpowers becomes apparent. In Colonel W. F. Long's words, "the thrust of effort in Vietnam has been as countertradition as it has been counterinsurgent."[35]

Of the $200 billion spent each year on military forces in the world, about two thirds is expended by the United States and Soviet Union.[36] And they produce seventy-five percent of all the world's arms. Since World War II the United States alone has shipped $50 billion worth of conventional weapons to other countries.[37] In Vietnam we have even become the quartermaster for the enemy. Thus conventional forces have been forced to serve backward as well as advanced nations. The equipment provided may have very little relation to needs. The Soviet Union has shipped amphibious armor to Algeria. Most Latin American countries have combat jet aircraft for prestige purposes; the UAR has surface-to-air missiles and tanks which it cannot operate effectively. Our military advisors in Vietnam for years sought to pattern the RVN forces after the Korean forces which had, in turn, been patterned after the United States forces. Both superpowers train military personnel from poor as well as advanced

[34] R. J. Rummel, "Some Empirical Findings on Nations and Their Behavior," *World Politics* (January, 1969), p. 238.
[35] W. F. Long, Jr., "A Perspective of Counterinsurgency in Three Dimensions—Tradition, Legitimacy, Visibility," *Naval War College Review* (February, 1970).
[36] *Wall Street Journal,* May 5, 1969.
[37] Conventional is used here in the sense of traditional weapons for European style nonnuclear warfare.

countries to fight frontal wars with their conventional weapons. Although this suggests the continuing occurrence of conventional wars between nations, the pointlessness of supplying conventional weapons to poor countries is now widely recognized.

1. They do not assure balance and stability (Israel against UAR).
2. They do not obviate superpower involvement (543,000 United States troops in Vietnam in 1968).
3. They are ineffective against motivated guerrilla forces.

The unsuitability of conventional forces in this type of conflict is a glaring fact.[38] Even when military victory by such forces is attained, it is likely to be hollow. The English defeated the Mau Maus in Kenya, the French defeated the FLN in Algeria; in both cases the loser was the ultimate winner.

Violence in the underdeveloped world will be widespread in the next fifteen years. United States involvement is in no case inevitable but in all cases physically possible. No countries other than Mexico and Canada can reach the United States with significant ground forces. On the other hand, the United States can effect a large physical presence anywhere. A conscious United States decision to become militarily involved is less likely in the next decade in view of the Vietnam experience. However, the 1985–95 period is again one of danger if Exhibit 2 has significance. This suggests the possibility of a buildup in United States general purpose forces after 1980.

The primary concern for all nations in the 1970–85 period will be *internal upheaval*. The reasons are apparent from the imbalances described earlier. The following catalog exemplifies the kinds of conflict today's military planner in any country should consider of paramount concern:

1. Coup.
   Revolutionary: Iraq, 1958.
   Reform: Dominican Republic, 1963.
2. Majority uprising.
   Passive resistance: India.
   Guerrilla war: Cuba, 1953–59; Angola, present; Rhodesia, present.
3. Minority uprising.
   Urban ghetto: Warsaw, 1944; Watts, 1965; Detroit, 1967.
   Tribal region: Biafra, 1967–70.
4. Other types.
   Paramilitary persecution of minorities (by private organizations).
   Wanton vandalism (destruction of wealthy communities by mobs of poor and destruction of foreign property and personnel in poor nations).
   Antiestablishment disobedience (student sit-ins, assassination of establishment leaders, and physical attack on ecological degradation sources such as industrial plants).

[38] "Why We Didn't Win in Vietnam," *U.S. News and World Report,* February 9, 1970, pp. 44–45.

Such conflicts may rock rich and poor nations to their foundations. William Butler Yeats has given poetic expression to this drift:

Things fall apart; the centre cannot hold;
Mere anarchy is loosed upon the world;
The blood-dimmed tide is loosed, and everywhere
The ceremony of innocence is drowned;
The best lack all conviction, while the worst
Are full of passionate intensity.

And in their final report, Dr. Milton S. Eisenhower and the members of the National Commission on the Causes and Prevention of Violence warn the President that while serious external dangers remain, the graver threats today are internal:

> We solemnly declare our conviction that this nation is entering a period in which our people need to be as concerned by the internal dangers to our free society as by any probable combination of external threats.[39]

Organizational Imbalance

Man creates and innovates continually, yet he also resists change. In Hamlet's words, we would "rather bear those ills we have than fly to others that we know not of." While the pace of technology has accelerated dramatically, the governmental framework for creative action has not altered significantly. The result is increasing incompatibility. In nature, obsolete organisms die; in society they seem to flourish. When new organizations are created the old ones continue and even expand, regardless of need. In the United States Federal Government we still have the Rural Electrification Administration created in 1935 and the Subversive Control Board created in 1950, although their raison d'être has long since disappeared. Parkinson gives the example of the British Admiralty: between 1914 and 1928 Admiralty officials increased in number by seventy-eight percent, while capital ships in commission decreased by sixty-seven percent.[40] The whole organizational apparatus from planning to action becomes rigid in structure and operation rather than rational and creative. Individual security is sought in organizational constancy. The Civil Service in the executive branch of government, the seniority system in the United States Congress, the Curia in the Roman Catholic Church, and the Soviet Central Committee (average age of voting members nearly sixty) are all examples of this principle. When individuals at the top or bottom want to implement change they are frustrated. Pope John XXIII and the ghetto priest may have agreed on desirable actions but neither could force change through the organizational

---

[39] *Los Angeles Times*, December 13, 1969, p. 1, and December 14, 1969, Sec. A, p. 8.
[40] C. N. Parkinson, *Parkinson's Law*, (New York: Houghton-Mifflin, 1957).

hierarchy. For President Nixon the ratio of actual to potential power is very much smaller than it was for President Lincoln.

The United States military establishment is geared to large-scale conventional operations in the European tradition. The combination of age, size, and success is very apparent and it promotes organizational rigidity.[41] Creative planning for small, unorthodox actions in underdeveloped country situations poses virtually insuperable difficulties. In Vietnam, the novel Combined Action Platoons and "Sting Ray" patrols were pioneered by the small United States Marine Corps and Special Forces rather than the large United States Army. Partly as the result of public attitudes the military academies are staffed by conformists whose task it is to produce more conformists who will preserve the traditional establishment. Conformism in Vietnam meant that "the best defense is a good offense" and that the South Vietnamese should be trained for a frontal type war.[42]

Calcification of the organizational framework for action is not restricted to government. It is apparent in industry also.[43] We have seen the rapid and dynamic growth phase of railroads in the last century, automobiles and steel in the first third of this century, aircraft in the middle third of this century, and computers today. We already recognize the innovative failures of the railroad industry and the rigidity of the American steel and automobile industries. Innovation in offroad ground vehicles has come from aerospace companies, innovation in steel processing from small European companies. Often top management and the lowly engineer are both anxious to try new ideas and diversify but find themselves stymied by middle management. Labor unions once were vital instruments of change; today they are bastions of conservatism. Automation is resisted and craft unions force industry to operate with deadly inefficiency. The search for security has led to concepts (for example, seniority, retirement benefits) which assure obsolescence. The Peter Principle and the Paul Principle plague us everywhere.[44] The first states that capable individuals

41 The United States gained militarily 3.1 million square miles in the period 1776 to 1935, second only to Great Britain's 3.5 million square mile gain. ("Why an Army?," *Fortune* (September, 1935), p. 48.)

42 "Why We Didn't Win in Vietnam," *U.S. News and World Report,* February 9, 1970.

Col. W. F. Long, Jr., points to a key difficulty:

> U.S. tradition has not encouraged political development of its military officer corps. . . . Further, there is no coherent link between military behavior which is domestically acceptable and that which may be required for successful participation in a clandestine war. Without a political mandate or relevant psychological inclination, the military man is simply not professionally fitted to make substantive contributions to political, economic, or sociological problems as regards foreign involvements. He must understand these factors within the context of his profession, but an excessive commitment to them will only detract from his primary concern—military capability.

"A Perspective of Counterinsurgency in Three Dimensions—Tradition, Legitimacy, Visibility," *Naval War College Review* (February, 1970).

43 The difference is that a business enterprise may merge, be absorbed, or go bankrupt.

44 See footnote 14.

are promoted until they attain a position which they are incompetent to fill. The second tells us that individuals in a given position become more incompetent in that position as time goes on, since the job changes and the individual does not. The result: incompetence squared! One of the few bright spots today in the area of organization change is the university. Spurred by students and young faculty members, it has entered a period of flux to end centuries of organizational rigidity. There are numerous harbingers of change. Interdisciplinary courses, futures seminars, theme-oriented colleges, and other new concepts are sweeping through the halls of ivy.[45]

What conclusion do we draw from this "new" look at the future? The price tag of technology is becoming visible. The vital social units of our global society—represented by the family and the nation—are moving into a state of extreme crisis in the wake of decades of phenomenal technological change. Ferkiss insists that

Industrial society is not so much being transformed into a post-industrial, technological society as it is breaking down—economically, politically, and culturally.[46]

And Soviet writer Andrei Amalrik reminds us that:

In fifth century Rome, with its six-story houses and its steam driven mills, a planner might have predicted twenty-story houses and mechanized industry for the sixth century. However, in the sixth century goats were chewing at the grass of the Forum.[47]

CHALLENGE: NEW KIND OF PLANNING

Words Versus Deeds

In the last few years, much has been written and mouthed about the future. Reports by distinguished establishment members discuss possible new national strategies and their costs. Futurologists paint glowing pictures of a utopia in which technology overcomes poverty, war, environmental degradation, old age—and in the process probably also humanism. The missing element almost always is a prescription or road map outlining with some degree of realism the means to move from here to there. Government is looked upon for funding which will rapidly lead to answers to the unsolved problems.[48] And somehow

---

[45] H. A. Linstone, "A University for the Postindustrial Society," *Technological Forecasting*, I, No. 3 (1970).

[46] V. Ferkiss, *The Technological Man: The Myth and the Reality,* (New York: George Braziller, 1969).

[47] Andrei A. Amalrik, "The Fall of the Russian Empire", *Atlas,* February 1970, p. 24.

[48] Actually the government can effect the content of technology much more than the rate of advance.

catastrophic side effects will be avoided. On July 13, 1969, President Nixon created a National Goals Research Staff.[49] He noted that:

> It is time we addressed ourselves, consciously and systematically, to the question of what kind of a nation we want to be as we begin our third century. We can no longer afford to approach the longer range future haphazardly.

The Staff functions were to include:

> —forecasting future developments, and assessing the longer range consequences of present social trends,
>
> —measuring the probable future impact of alternative courses of action, including measuring the degree to which change in one area would be likely to affect another,
>
> —estimating the actual range of social choice—that is, alternative sets of goals that might be attainable, in light of the availability of resources and possible rates of progress,
>
> —developing and monitoring social indicators that can reflect the present and future quality of American life, and the direction and rate of its change,
>
> —summarizing, integrating and correlating the results of related research activities being carried on within the various Federal agencies, and by State and local governments and private organizations.

An annual report was to be issued setting forth the key choices and their consequences.

The most significant passage in the Presidential announcement is the following:

> There is an urgent need to establish a more direct link between the increasingly sophisticated forecasting now being done and the decision-making process. The practical importance of establishing such a link is emphasized by the fact that virtually all the critical national problems of today could have been anticipated well in advance of their reaching critical proportions. Even though some were, such anticipation was seldom translated into policy decisions which might have permitted progress to be made in such a way as to avoid—or at least minimize—undesirable longer range consequences.
>
> We have reached a state of technological and social development at which the future nature of our society can increasingly be shaped by our own conscious choices. At the same time, those choices are not simple. They require us to pick among alternatives which do not yield to easy, quantitative measurement.

[49] R. M. Nixon, "The Establishment of a National Goals Research Staff," The White House, July 13, 1969. Reproduced in *Technological Forecasting*, I, No. 2 (Fall, 1969).

A first—and last—report was issued on July 4, 1970.[50] Unhappily it falls far short of the hopes stimulated by the admirable announcement. There are expansive generalities such as the following:

> The central ingredient in the development of a growth policy will be for the American people to decide just what sort of country they want this to be.... We will have to develop better institutional arrangements for the people to relate to the leadership and better mechanisms of policy analysis to serve all parties.

There is no attempt to formulate alternative goals, policies, and plans in a sufficiently specific manner to permit meaningful public debate. A hundred years ago we had the Lincoln-Douglas debates. In an era of one-minute political campaign commercials and Madison Avenue packaging of major candidates one wonders whether such discussion would even find an audience.

The experience of the National Goals Research Staff lends support to Donald Michael's pessimism. He believes that there is no chance of society adapting quickly and thoroughly enough to cope with the problems it faces in the coming decades.

> Our technological ability to change our world will exceed our ability to anticipate whether we are using it wisely.
> We are almost certain to face disaster if we don't plan.... But we are also almost certain to be in deep trouble even with planning because our best plans will be developed and fostered by limited human beings picking and choosing among limited knowledge, very often ignorant of the extent of their own ignorance.[51]

The Bellagio Declaration on Planning warns that "the pursuance of orthodox planning is quite insufficient in that it seldom does more than touch a system through changes of the variables."[52] Our options are:

1. Resist change.
2. "Muddle through" with minimum planning, relying on luck and divine guidance to prevent uncontrolled, violent, and detrimental change.
3. Accept the challenge to develop our future society through rational creative action, to become *magister ludi,* master of the game.

The last alternative requires not only a true future orientation and a willingness to embrace change, but a recognition that:[53]

[50] "Toward Balanced Growth: Quantity with Quality," Report of the National Goals Research Staff, The White House, Washington, D.C., July 4, 1970.
[51] Donald N. Michael, *The Unprepared Society: Planning for a Precarious Future,* (New York: Basic Books, 1968).
[52] E. Jantsch, (ed.), *Perspectives of Planning,* (Paris: OECD, 1969).
[53] V. Ferkiss, op. cit.

1. Man is part of nature rather than apart from it.
2. The whole physical and social environment is a single dynamic system.
3. This system can be controlled and shaped from within.
4. Knowledge is the primary strategic resource if such control is to be exercised properly.
5. Change is not the sole responsibility of the Federal Government. It is the responsibility of all—the individual and the state and the communal group and the corporation.
6. Change must flow through values, policies, strategies, and tactics, from forecasting to planning to decision making to action.

Forrester's work[54] best illustrates the total systems approach required.

He sees the world as a non-linear feedback system, hence strongly counterintuitive in its behavior. Continued growth, as recent generations have known it, is out of the question in his highly aggregative model. At best the quality of life may be stabilized near current levels by *simultaneously reducing* the rates of change of a whole set of variables: natural resource usage, pollution generation, capital investment, birth rate, and food production! Whether his particular model is valid or not, it suggests an approach we must exploit. With the future of our advanced civilization at stake, the urgency of work in this direction transcends that of national defense.

[54] Jay W. Forrester, *World Dynamics,* Wright-Allen Press, Inc., Cambridge, Massachusetts, forthcoming.

Exhibit 3. Planning—The Realistic Setting.

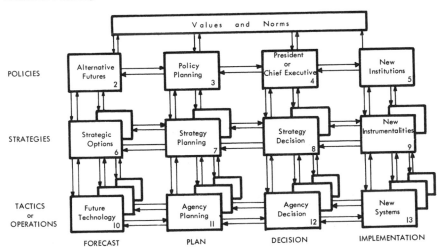

Adapted from: E. Jantsch, "From Forecasting and Planning to Policy Sciences," AAAS Paper (Boston: December, 1969).

The difference between the total system planning envisioned here and present practice is fundamental. Consider the current mechanism of change in defense systems. The focal point is the dialog between the technologist and the planner of the using organization (Exhibit 3, Boxes 10 and 11). At time t_0, when system S_0 with capability C_0 is operational, there is considerable communication between the planner and the technological community (initiated by either side) concerning feasible new capabilities relevant to the missions performed by system S_0. How accurately can a missile be delivered at a given range by 1980? How fast can submarines move at a given depth? Now the technological forecaster tends toward pessimism in very long-range projections and optimism in shorter-range forecasts. This phenomenon has been illustrated by Buschmann[55] (Exhibit 4a). The curved line (C_1) suggests that no serious technologist anticipated capability C_1 in 1900. By 1920, C_1 was expected to occur after the year 2000 and in 1935 optimism began. The hindsight line shows the actual occurrence date. A recent example, forecasting the fast reactor capability, is also shown. An interesting implication follows from this figure:

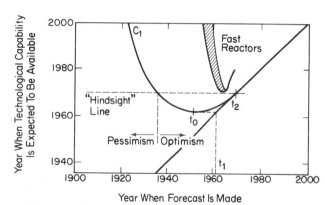

Exhibit 4. Forecasting-Planning Problem.

(a) Pessimism Versus Optimism in Forecasting.

Source: R. Buschmann, "Balanced Grand-Scale Forecasting," *Technological Forecasting*, Vol. 1, No. 2 (Fall 1969).

there is an optimum time to make a forecast, and it is probably in the ten- to twenty-year rather than five- to ten-year span.

Let us assume that our forecaster tells the planner (at time t_0) that a significant advance in technology (C_1) is feasible at time t_1. The planner then includes in his document (Box 11) a "requirement" for a new system S_1 at time t_1 which will take advantage of this anticipated component or subsystem technology. It is noted that we use the term "capability" and not "effectiveness."

[55] R. P. Buschmann, "A Research Problem: Balanced Grand-Scale Forecasting," *Technological Forecasting*, I, No. 2, (Fall, 1969).

At time t_1 it becomes apparent, however, that technology will only reach capability C_1' (Exhibit 4b). In most cases the planner's decision is to move forward with the new system according to plan rather than delay initiation until capability C_1 is attained at time t_2. The capability gap $(C_1 - C_1')$ is subjected to negotiation, and a compromise, C_1'', is reached. The technology forcing function $C_1'' - C_1$ then becomes the source of cost overruns and nonfulfillment of system "specs." Thus system S_1 is acquired and proves to have only marginally superior capability over the existing system S_0. The "plan" has become the tool for justifying an acquisition previously based on a significant, but unrealizable, technological improvement. We now return to Exhibit 3. On the basis of the plan alternative designs are produced and evaluated, a choice is made (Box 12), and the system is created (Box 13). Future environmental forecasts are produced independently (Box 2) as are national goals studies (Box 3). The National Security Council is the strategic decision-making body (Box 8). Henry Kissinger has tried to create a strategic planning staff (Box 7), and the National Goals Research Staff might have formed Box 3. Their relation to the other boxes is not clear. The driving force of Boxes 10 and 11

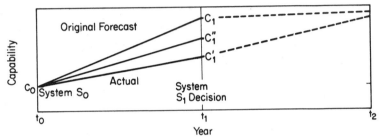

Exhibit 4 (cont.)

(b) Capability Gap Illustrated.

in the form of new hardware system capabilities and user system replacement planning as well as the dominance of the "tactical" level for the entire process are striking. We see a clear reflection of the technological explosion of recent decades. However, the weakness in effecting balanced and controlled change of the total system is also evident. There is no mechanism for shifting emphasis between technology and motivation or training, for reorganization of the Department of Defense, or for elimination of obsolete forces as a result of changing environments or values. Policy planning does not proceed from forecast to creative action in a rational process nor is there proper interaction between the policy, strategic, and tactical levels.

The desired framework is shown in Exhibit 5. Boxes 10 and 11 no longer provide the dominant impelling force to the entire process. New institutions and new instrumentalities as well as new systems are created (Boxes 5 and 9). There is feedback between all boxes. In other words, *innovation flows through the entire network,* and this is the challenge of the coming years for every member of our society.

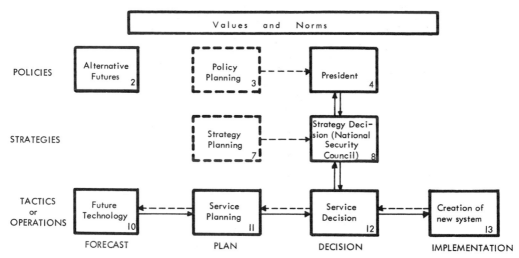

Adapted from: E. Jantsch, "From Forecasting and Planning to Policy Sciences," AAAS Paper (Boston: December, 1969)

Exhibit 5. Planning—The Ideal Setting.

IMPACT ON DEFENSE PLANNING

Our stated purpose was to provide a starting point for a defense planning study. We will confine ourselves here to a very brief glimpse at the next steps in relating this background material to defense planning. We must consider alternative United States (and world) futures and develop general defense guidelines for each. We then combine the effects of current force extrapolation and new technology with the implications of the preceding environmental discussion to arrive at alternative force structures for the 1980–85 period. Several iterations of this process are desirable. Finally, a weighting of the likelihood of these alternatives should uncover "good risk" recommendations for planning. The four alternative United States futures used in this work have the following political-military features:

A. Militaristic-Imperialistic
—Populist, anti-intellectual, reactionary
—Stability, old values highly prized
—Use of "Communist threat" to motivate and unite people
—Authoritarian, intolerance toward dissent, witch hunts
—Military forces overseas expanded
—Control of communications media and universities
—Military needs receive top priority
—Manned space program shifted to the Department of Defense

B. Surprise-Free
—Present trends continue (government generally follows rather than leads)
—"Responsible center," middle of the road administration
—Maintenance of current institutions

—Some new goals set to motivate public

—Dissident elements alternately accommodated and restrained

—Lip service paid to long range planning

—Some recognition of a shift in the "threat to survival" from military external to nonmilitary external and internal problems

—Reduction in military spending

—Increase in social and environmental control programs (approach conventional)

—Large scale action only when crisis and disaster imminent

C. Nationalistic-Inward Oriented

—Nationalistic, but not imperialistic

—Major shift in priorities to address critical internal problems

—Isolationist, return to Monroe Doctrine

—Strong defensive capability, but greatly reduced "general purpose" forces

—Major drive to provide internal stability through a combination of strong, modernized police forces and a large scale attack on underlying socioeconomic and ecological problems

—Space program severely curtailed

—New technical task forces formed to work on selected crisis problems

—Changes in U.S. institutions, although obsolete organizations still proliferate

—Supranational programs not supported effectively

D. A New Society

—Most profound departure from surprise-free world

—Fundamental reformation of U.S. structure to reduce or eliminate imbalances and prevent breakdown in our society

—Total creative planning emphasized to bring technological and social/behavioral change into harmony (requires mobilization of intellectual talent)

—"Future" orientation in education and government

—Drive to eliminate obsolete institutions

—Considerable decentralization and creation of new structures to encourage local experimentation and diversity

—Experimentation involving new social arrangements and groupings

—Massive shift of resources from national security to urban, educational, and ecological systems improvement programs

—Surrogates for war (i.e., peaceful competition) sought

—Supranational projects encouraged

—Military emphasis confined to strategic retaliatory forces.

Table 2 shows the relative emphasis in resource allocation to major defense areas as a function of these four alternative United States futures.

Exhibit 6 shows the result of such an analysis in terms of military expenditures as a function of GNP. A simplistic assumption is made in this figure. Real GNP growth and inflation are the same in all four cases (average of 4.4 percent for the former and 2.3 percent for the latter). The impact of alternative postures on the total economic growth is not yet well enough understood to permit a meaningful differentiation. One fact is known, however: a high level of defense spending is not a prerequisite for rapid GNP growth. Japan has experienced phenomenal growth while spending only about one

Table 2. Relative Emphasis in Four United States Futures.

	United States Future			
	A	B	C	D
	Militaristic-Imperialistic	Surprise Free	Nationalistic Inward Oriented	"New Society"
Strategic offense/defense	••••	••	•••	•
General purpose forces	••••	•••	••	•
Manned space programs (nonmilitary)		•		•
Other new programs (nonmilitary)		••	•••	•••••
Sample subjective estimate of likelihood	20%	40%	25%	15%

Exhibit 6. United States Military Expenditures as Percent of GNP.

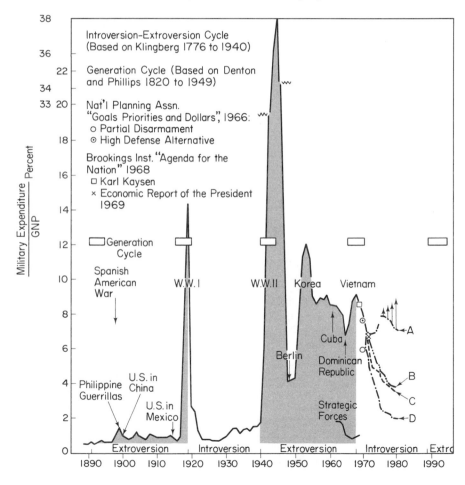

percent of its GNP on defense. For comparison, the figure also shows forecasts prepared by other studies; all of them feature a reduction in the fraction of the GNP given to military expenditures. We note that even the minimum of two percent in 1980 still exceeds the fraction spent in over half the years covered by the data. The budget line of Future A is probably a lower bound for this militaristic environment; as suggested by the vertical arrows, it might rise much higher. In constant dollars Futures B, C, and D lead to a continuing decline in defense budgets.

The figure sets the future into the context of the past and repeats the generation and introversion/extroversion cycles from Exhibit 2. We observe that United States external military activity has not always required vast military expenditures. In the 1890 to 1914 extroversion period this nation was involved with military operations in Latin America and Asia while spending only about one percent of its GNP on military expenditures. Conversely, a large military establishment has not assured peace. The data thus tend to underscore the findings of R. J. Rummel: the military budget does not correlate significantly with foreign conflict involvement.[56]

Any analysis of future forces must take full account of the present planning approach (Exhibit 3) and the hurdles which impede rational creative planning (Exhibit 5). Unless there is a dramatic change in modus operandi as well as in the structure of the Department of Defense (doubtful with the expectations of alternative futures used in Table 2), the gap between true defense needs and stated requirements will become a chasm in the next fifteen years. And this makes "needs analyses," when differentiated from market research, as barren an exercise for the defense supplier as it is for the automobile manufacturer.

APPENDIX A:
NATIONS OF THE WORLD

Sources:

F. M. FULTON AND D. S. RANDALL, "Crisis Control Environments 1975–85," Stanford Research Institute, Naval Warfare Center Memorandum 53, 1969.

H. KAHN AND A. WIENER, *The Year 2000* (New York: Macmillan and Co., 1967).

H. A. LINSTONE, ET AL. "Mirage 80," Lockheed Aircraft Corporation Report DPR/46, 1966.

---

[56] Rummel, *op. cit..*

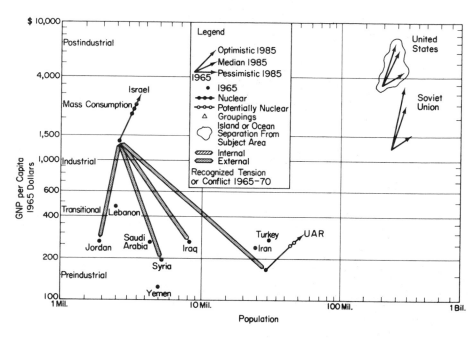

Exhibit A-1. Middle East, 1965–85.

Exhibit A-2. Europe, 1965–85.

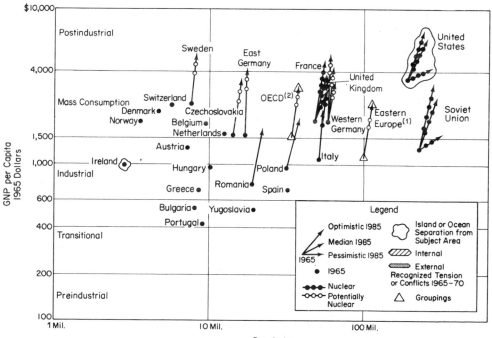

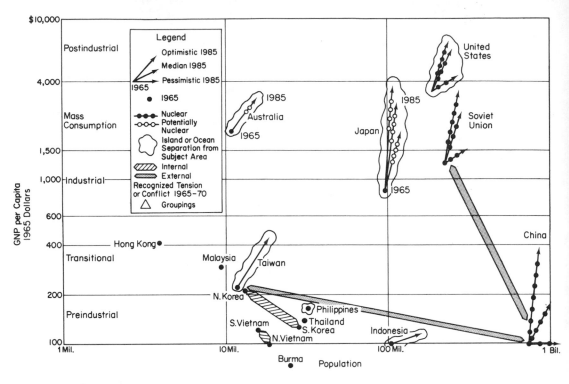

Exhibit A-3. Pacific Area, 1965–85.

Exhibit A-4. The Underdeveloped World.

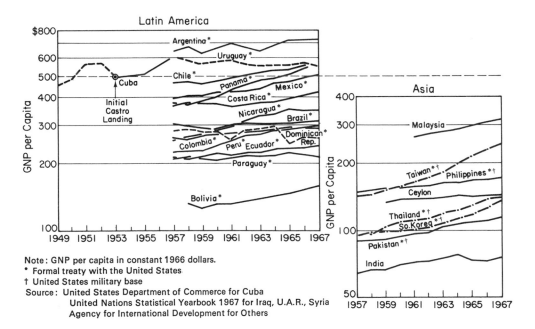

Note: GNP per capita in constant 1966 dollars.
* Formal treaty with the United States
† United States military base
Source: United States Department of Commerce for Cuba
        United Nations Statistical Yearbook 1967 for Iraq, U.A.R., Syria
        Agency for International Development for Others

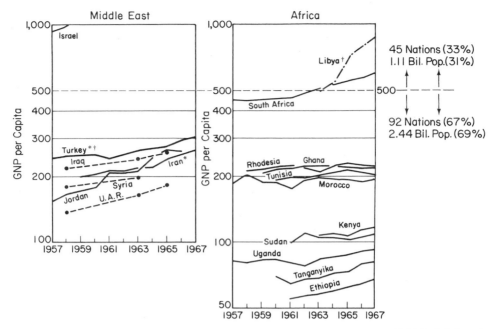

Input-Output Analysis and Business Forecasting

ELISABETH K. RABITSCH

Input-output analysis is a powerful and accepted tool for economic forecasting. The author gives the technology forecaster a brief overview of the concept, then stresses how basic economic changes shown through input-output studies could help to reveal the impact of new technologies. She reviews the data sources available to business forecasters and planners through the input-output study of the Office of Business Economics (redesignated the Bureau of Economic Analysis, effective January 1, 1972) and some of the input-output research conducted under the federal government's Interagency Growth Study Project in the United States, examines the use of input-output tables to assess the short-term impact of economic events and to estimate the effect of long-term structural and technological changes on economic growth, and discusses the application of the input-output concept to business forecasting and planning.

Late in 1969 the Department of Commerce, OBE (Office of Business Economics), published a summary report on the "Input-Output Structure of the U.S. Economy: 1963" [1]. The input-output transaction table shown in this report consisted of eighty-five industry sectors. A few weeks later the OBE released a considerably expanded input-output table for 1963 which contained detailed information for 367 separate industries [2]. The OBE's 1963 input-output study provides a data base to arrive at up-to-date estimates and forecasts of the input-output relationships of major industrial sectors in the United States.[1]

The OBE's detailed *1963 Input-Output Tables* are published in three volumes. All productive activities of the United States economy are grouped into 367 industry sectors. The *first volume* contains the transaction table which shows for 1963 the dollar value of the transactions among the various industries, the sales of each industry to final markets, and the value added of each industry. The *second volume* contains the direct requirements table (table of direct input-output coefficients) which relates for 1963 each of the inputs of an industry to its total output. The *third volume* contains the total requirements table (table of indirect or inverse input-output coefficients) which shows for 1963 the direct and indirect effects of final sales on the output of each industry. The industry grouping in these tables is based on the SIC (Standard Industrial Classification) four-digit code. In addition to the printed tables, the OBE released a computer tape containing 1963 input-output transaction data for 478 industry sectors. The time lag between the data base and the publication of the OBE's input-output tables reached almost six years. Various research projects are now in progress to update the OBE's input-output tables to more recent base years in order to make the information and the forecasts more relevant for business planning.

The OBE's research in input-output analysis is conducted as part of the Federal Government's Interagency Growth Study Project. Guidance for this research program is provided by an interagency steering committee consisting of representatives of the OBE, the BLS (Bureau of Labor Statistics of the Department of Labor), the Office of Management and Budget, and the Council of Economic Advisers. The committee is chaired by a member of the council. Various agencies and research groups perform research under its auspices.

The growth projections of the United States economy to 1980, prepared and published by *BLS* in 1970, have also been based on the input-output tables developed by OBE. [(3)–(5)]. The 1980 projections, which are used for a wide variety of planning and policy development purposes, are made in a series of distinct but closely interrelated steps. First, the potential gross national product is developed based on a projection of the labor force, assumptions regarding the rate of unemployment and the level of the Armed Forces, and by projecting trends in average hours and output per man-hour. Given the

---

[1] The 1963 data were obtained as part of the second full-scale input-output study by the OBE, the first having covered the year 1958. The 1958 table was updated to 1961 in: Office of Business Economics, U.S. Department of Commerce, "Input-Output Transactions: 1961," *Staff Working Paper in Economics and Statistics,* No. 16 (Washington, D.C.: Government Printing Office, 1968). For an explanation of the 1958 I–O tables prepared by OBE and a brief account of I–O techniques, see the November 1964 and September 1965 issues of the *Survey of Current Business.*

potential gross national product, projections are developed of the composition of GNP among demand components—government, consumption, business investment, and net foreign demand. Once the composition of GNP is determined, the detailed distribution of each of these final demand components is projected. In order to translate projections of industry demand into industry output requirements, input-output relationships are used.

The input-output tables providing the framework for the economic growth model are the 1958 tables published by OBE for 87 separate industry sectors of the United States economy. As these input-output tables incorporate the technology and product mix for 1958, they do not adequately reflect the technology and product mix for 1980 for most industries. A great amount of research has been conducted by BLS (and several other economic research groups) in order to project to 1980 the changes in the ratios of purchases to outputs. Variable input-output coefficients are thus built into the economic growth model. The calculation of industry growth rates leads finally to an estimate of man-power requirements in 1980 for each of 82 industry sectors. (Five sectors of OBE's table are not identifiable as industries in the ordinary sense.)

The information on *capital goods* contained in OBE's input-output table for 1963 was expanded by the recent publication of a "Capital Flow" table (6). This is the first such table prepared by OBE. While the input-output table records the interindustry transactions in goods and services purchased on current account (the output of new structures and equipment is shown in such a table as purchased by the final demand sectors "investors", "foreigners", and "government"), the capital flow table disaggregates the investors sector so as to show the flows of new structures and equipment from producing industries to using industries. The capital flow table can be put to several uses:

a. One obvious and important use of input-output tables by a business firm is for the purpose of measuring market penetration. Since the input-output table provides information on market transactions only for those products which are purchased on current account, capital goods producers have not been able to use the input-output table for detailed market analysis. The capital flow table fills this gap by providing a fairly detailed breakdown of sales for each type of capital good.

b. The capital flow table can also be used to measure the impact of changes in total investment spending on each of the capital goods producing industries. The table, therefore, serves as a basis to estimate the detailed requirements placed on the construction and equipment producing industries through a change in the level of investment spending.

c. The data in capital flow tables furthermore provide information needed in estimating capital stocks. A series of capital flow tables would make it possible to estimate capital stocks in fairly great detail, for example by type of asset for each user industry shown in the capital flow tables.

The OBE is continuing its research to study the *application of input-output methods to a variety of economic problems,* such as:

a. Measurement of the direct and indirect effects of stipulated changes in the output of one or more industries upon the outputs of all other industries.

b. Measurement of the effects on prices throughout the economy of changes in the costs or prices of one or more industries.

c. Assessment of markets for individual companies or industries, taking account of indirect demand that reaches the company or industry through a chain of interindustry repercussions.

d. Calculation of industry outputs consistent with specified levels of gross national product (GNP).

In recent publications, the OBE examined various of these applications, but in particular the use of input-output methods for long-range economic projections [(7)–(9)].

Research in input-output analysis is also conducted, outside government agencies, by a number of private research groups which are either connected with universities or management consulting firms. It is beyond the purpose and scope of this article to list or evaluate the various services which are being offered.

USE OF INPUT-OUTPUT TABLES IN SHORT-TERM AND LONG-TERM BUSINESS FORECASTING

The second part of this paper will illustrate the use of input-output analysis to determine the direct and indirect requirements of industrial chemicals, man-made fibers, plastics, and paints of the motor vehicle industry. The illustrative examples given here will be presented in a form to explain the basic concepts of input-output analysis.

Short-Term

Estimating the automobile industry's direct and indirect requirements of materials and supplies was a very timely subject in late 1970. The prolonged shutdown of all General Motors major production facilities, which started on September 15, 1970, and lasted for more than two months, had been affecting industry and service sectors throughout the economy. It was of particular importance to estimate the effects of a prolonged shutdown not only for those industries which are direct suppliers to the automobile industry but also for the intermediate processing and service industries.

The OBE's 1963 input-output tables provided a bench mark for estimating the impact of the automobile industry shutdown on the supplier industries. In the I-O tables the motor vehicles and parts industry is classified under the I-O industry number 59.03 (corresponding to the four-digit SIC industry 3717). The principal product lines of a chemical and fiber producing company, such as Celanese Corporation, are classified under five different I-O industry numbers. The I-O classification of the major product lines and the corresponding SIC codes are shown in Table 1.

Table 1. Classification of Celanese Principal Product Lines Four-Digit SIC Basis.

I–O Industry No.		SIC Classification
27.01	Industrial inorganic & organic chemicals	281, exc. 28195
28.03	Cellulosic man-made fibers	2823
28.04	Organic fibers, noncellulosic	2824
28.01	Plastics materials & resins	2821
30.00	Paints & allied products	2851

Sales of the chemical and fiber industries, in millions of dollars at 1963 producers' prices, are recorded in Volume I of the OBE's three-volume tabulation. Each horizontal row of the I-O table shows the distribution of sales (outputs) of one industry to each of the 366 other industries recorded in the table. Similarly, the entries in each vertical column show the distribution of purchases (inputs) of that industry from each of the 366 other industries.

Reading across row 27.01 (Table 2) one can see that in 1963 the sales of chemicals to the motor vehicle industry amounted to $15.7 million. Total sales of chemicals to other industries (total intermediate output) amounted to $10.6 billion. Gross output, which equals intermediate output plus sales to final markets, amounted to $12.6 billion.

It becomes apparent in this table that the *direct* sales of chemicals and man-made fibers to the motor vehicle industry are very small. The *direct* sales of surface coatings for motor vehicles are fairly large. When reading across the row of industry number 30.00, one can see that in 1963 the sales of paints and allied products to the motor vehicle industry amounted to $182 million or almost eight percent of total intermediate output.

Table 2. Input-Output Structure of United States Economy, 1963, Sales to Motor Vehicles Industry (in millions of dollars at producers' prices).

I–O No.	Title	Motor Vehicles & Parts (I–O No.: 59.03)	Total Intermediate Output	Gross Output
27.01	Industrial inorganic & organic chemicals	15.7	10,573.9	12,613.2
28.03	Cellulosic man-made fibers	—*	769.6	810.3
28.04	Organic fibers, noncellulosic	—*	1,359.8	1,495.3
28.01	Plastics materials & resins	10.7	2,939.4	3,239.0
30.00	Paints & allied products	181.9	2,363.0	2,462.1

* Less than $500,000.

The figures shown in the next illustration (Table 3) measure the direct requirements of materials per $1,000 of gross output in the motor vehicle industry. In other words, for each $1,000 of output by the motor vehicle industry, a *direct* input of forty cents worth of chemicals and $4.66 worth of paints and surface coatings is required. These figures are the *direct* input-output coefficients which are published in Volume II of the OBE's 1963 input-output table.

Table 3. Input-Output Structure of United States Economy, 1963, Sales to Motor Vehicles Industry. Direct requirements per $1,000 of gross output (producers' prices).

I-O No.	Title	Motor Vehicles & Parts (I-O No.: 59.03)	Gross Output: ⟶$39,067.5 Million
27.01	Industrial inorganic & organic chemicals	0.40	
28.03	Cellulosic man-made fibers	——*	
28.04	Organic fibers, noncellulosic	——*	
28.01	Plastics materials & resins	0.28	
30.00	Paints & allied products	4.66	

NOTE: Figures shown in the cells are direct I-O coefficients, or the direct requirements per $1,000 of gross output. Gross output of the motor vehicles industry in 1963 = $39,067.5 million.
* Amount negligible.

The input-output method makes it possible to estimate the changes in total (direct and indirect) input requirements when final markets of a particular industry are expected to increase or decline.

The *indirect* (or inverse) coefficients shown in Table 4 provide a measure of the total requirements of materials per $1,000 of final sales of the motor vehicle industry. The *total* requirements of chemicals per $1,000 of final sales in 1963 amounted to almost $20 and those for paints and surface coatings to more than $9. In 1963 when final sales of the motor vehicle industry reached $23.5 billion, the total requirements of chemicals were estimated at about $470 million and those of paints at over $200 million. The requirements of man-made fibers amounted to $74 million; these fibers are processed by the textile industry and the rubber industry, for example, to be consumed eventually in automobile production in the form of seat covers, wall linings, carpets, and tire fabrics.

Using the indirect coefficients of the 1963 I-O table, one can estimate the effects of the automobile industry shutdown as follows: A cutback in production by 800,000 units during the last quarter of 1970 would reduce the value of motor vehicle shipments by about $2 billion (assuming an average price per

Table 4. Input-Output Structure of United States Economy, 1963,
Sales to Motor Vehicles Industry.
Total requirements (direct and indirect) per $1,000 of delivery
to final demand (producers' prices).

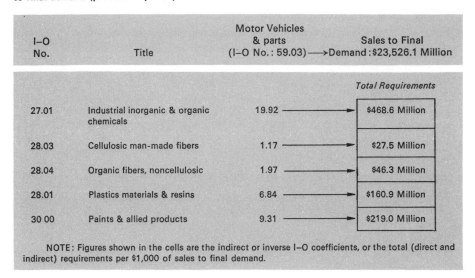

I–O No.	Title	Motor Vehicles & parts (I–O No.: 59.03) ──→	Sales to Final Demand: $23,526.1 Million
			Total Requirements
27.01	Industrial inorganic & organic chemicals	19.92 ──────────→	$468.6 Million
28.03	Cellulosic man-made fibers	1.17 ──────────→	$27.5 Million
28.04	Organic fibers, noncellulosic	1.97 ──────────→	$46.3 Million
28.01	Plastics materials & resins	6.84 ──────────→	$160.9 Million
30 00	Paints & allied products	9.31 ──────────→	$219.0 Million

NOTE: Figures shown in the cells are the indirect or inverse I–O coefficients, or the total (direct and indirect) requirements per $1,000 of sales to final demand.

vehicle of $2,500). A reduction in value of shipments by $2 billion would cut the total requirements of chemicals by almost $40 million, of man-made fibers by $6 million, of plastics materials and resins by almost $14 million, and of paints by $18 million.

This 1970 estimate is based on the 1963 indirect coefficients and, therefore, on the input-output relationships or technology which existed seven years earlier. The assumption of constant technical coefficients is likely to introduce a fairly large margin of error into our estimates.

Changes over time, in the relationships between producing and consuming industries and, therefore, changes in the input-output coefficients, can have various reasons. First of all there is the change in technology which adds new or modified materials and processes to the production stream. Second there is the change in product mix which may occur in any one of the industries; if the product mix of an industry changes over time, then the input-output coefficients may change for the entire industry sector. Third a shift in the relative prices of two industries producing competitive products will result in a coefficient change if the less expensive product is substituted for the more expensive one.

Obviously the input-output coefficients of the motor vehicle industry must have changed since 1963 for any one of the reasons stated above. Plastics materials have replaced metals; certain types of man-made fibers have been replaced by other types. Prices of many industrial materials, for example basic organic chemicals, have declined, whereas the cost of services and labor has rapidly increased over the past seven years. In order to arrive at a more reliable estimate of the automobile industry's input requirements in 1970, one would have to make adjustments in the input-output coefficients. The availability of more timely data from government sources would bring a major improvement in this area.

The problem is complicated when input-output coefficients are used to estimate the effect of a cutback in output of one industry, such as the automobile industry, on a particular manufacturing company supplying materials, either direct or indirect, for automobile production. In any input-output table (even at a 367 industry-sector detail), it is necessary to aggregate a number of industrial activities into a single sector. The problem of aggregation is very difficult, but it should be obvious that the technical coefficients are different for each industrial activity and that the level of aggregation used in the input-output model will have an important bearing on the input-output coefficients. To apply the coefficients of the input-output table to a particular manufacturing company, the production functions of the manufacturer and the I-O sector would have to be identical. However a rough estimate of the impact of an industry shutdown on an individual company can be made by calculating the potential loss of sales on the basis of market share.

Long-Term

The input-output table used as a bench mark in long-term forecasting should be as timely and as detailed as possible. The *static* input-output model does not provide information on advancement in technology, or in other factors, which would cause the technological coefficients to change over time. The Leontief type static input-output models, however, have been generalized to simulate the dynamic behavior of the economy by considering changes in technology, capital stocks, inventories, and other factors.

A change in technology affects both the industry in which the change takes place and the industry which produces the intermediate materials. In projecting the input-output relationship of consuming and producing industries for the coming decade, any foreseeable technological advance in an industry sector has to be translated into changes in demand for direct or indirect material requirements. Variable input-output coefficients are then introduced into the forecasting model.

The growth in automobile production over the next ten years will affect the demand for automobile tires, which in turn will affect the demand for fibers used in production of tire fabrics.[2] Change in automobile and tire technology will have to be considered in such ten-year projections. In the past, changes in tire technology have had a significant impact on several industry sectors. Before World War II, tire fabrics were made almost exclusively from

2 U.S. production of passenger cars is projected to reach 11.5 million units in 1980, up almost thirty percent from 8.8 million units produced in 1968.

cotton. In the 1940s and 1950s, rayon fibers made from wood cellulose were increasingly used. Nylon fibers gained in importance in the late 1950s, and polyester fibers in the late 1960s. Both nylon and polyester are made from petroleum derivatives. Thus, the basic raw material sources changed from agriculture to forestry and then to crude oil production.

What are the changes in tire technology expected during the next decade? Obviously, any changes will affect the demand for fibers, chemicals, petroleum, and many other materials. An input-output forecasting model projecting material requirements to 1980 would have to allow for such changes in technology.

INPUT-OUTPUT ANALYSIS AND BUSINESS PLANNING

The usefulness of input-output analysis to business management depends to a large degree on the timeliness, accuracy, and completeness of the basic data. Additional efforts and expenditures by government agencies, research institutions, and business corporations will be required in the future to provide up-to-date and detailed information. In coming years more and more industries and business firms will start to compile and tabulate statistics on sales and purchases in a form applicable to input-output analysis.

In a large corporation with several operating divisions and many plant locations, the collection of basic sales and purchasing statistics can become a monumental task which cannot be undertaken without strong systems and electronic data processing support. The collection of sales and purchasing data in all operating divisions and, if relevant also in all affiliate companies and subsidiaries, should be based on compatible systems, so that sales and purchase reports can be aggregated at a corporate data processing center.

A company's activities can be grouped by either operating divisions, principal product lines, profit centers, and so on. The principal product lines of Celanese Corporation are classified under five different four-digit SIC industries. (Table 1). Similarly all customer and vendor establishments should be classified by four-digit (or possibly five- or six-digit) SIC categories. The estimated number of Celanese customers and vendors by four-digit SIC industries is shown in Table 5. The plastics division, for example, sells its products to more than 150 different customer industries, the chemicals and coatings division each sells to more than 70 different industries. The cellulosic fibers division purchases materials, fuel, and services from 46 different supplier industries.

There are three basic methods of industrial classification: (1) the commodity basis, (2) the activity basis, and (3) the establishment basis. It was decided to follow the establishment classification rules prescribed in the SIC manual issued by the Bureau of the Budget. According to the SIC manual, an establishment is not a legal entity or a company but rather an economic unit which produces goods or services such as a farm, store, mine, or manufacturing

Table 5. Estimated Number of Celanese Vendors and Customers by Four-Digit SIC Industries.

I–O Industry No.	Product Lines	Purchases from or Sales to Estimated Number of Four-Digit Industries	
		Vendors	Customers
27.01	Chemicals	26	72
28.03	Cellulosic fibers	46	39
28 04	Organic fibers, noncellulosic	18	33
28.01	Plastics	28	157
30.00	Paints	27	76

plant. Each establishment is assigned to its primary product. Then all the establishments assigned to the same product lines are grouped into an industry. The definition of establishment becomes important as many large customers are in five, six, or more different product lines and have to be classified correspondingly.

Once all customer establishments have been classified by SIC categories, the SIC code numbers can be incorporated in a customer sales reporting system. This reporting system would provide corporate managment with the overall dollar volume of sales to major SIC industries on a quarterly and year-to-date basis. This report can be prepared at a corporate data processing center from inputs provided by operating divisions and affiliate companies at the end of each calendar quarter. The operating divisions and affiliate companies produce the data in the form of a punched card, a transaction card of a given format. The transaction card records the customer account number which is unique for each customer establishment; it also records the SIC code and the year-to-date sales amount. Sales to each SIC industry can then be totaled by operating division and for the corporation as a whole.

When relating the information on corporate sales to input-output tables of the United States economy, many important factors have to be considered: (1) the product mix of each SIC sector; (2) the corporation's market share; (3) the pricing of products (interindustry shipments are recorded in producers' prices but sales to final markets reflect transportation costs, trade markups, and indirect taxes); and (4) the reconciliation of the Census of Manufactures' shipments with the gross output of an industry sector shown in the I-O table.[3] There are other factors, of course, which cannot be discussed here in detail.

Once corporate sales and purchases are classified by SIC code, a dollar flow matrix of corporate transactions can be assembled for a given time period,

[3] The U.S. Department of Commerce, Office of Business Economics, plans to release in the near future background information which will facilitate the interpretation of the OBE's 1963 input-output table.

such as a calendar year. A computer program was developed in Celanese Corporation to compile the dollar flow matrix and to compute the direct and inverse coefficient matrices.

This program is written in FORTRAN IV for a GE-625 computer. The configuration of the GE-625 computer makes it possible to do the inversion on a 300 × 300 matrix in core. With a larger 256K core, the same program could handle a 500 × 500 matrix with only minor changes.

Due to the GE-625 multiprocessing system, the inversion of smaller matrices (up to 100 × 100) passes through the job stream virtually unnoticed by the system because of the extremely fast execution. For this reason the program is exceptionally well fitted to remote terminal execution. Using a remote terminal, it is possible to vary the entries in the square matrix or demand columns to get a feel for the sensitivity of the solution.

The inversion of a 300 × 300 matrix on the GE-625 computer requires approximately twelve minutes.

A corporate static input-output model, of course, serves as a bench mark only. The emphasis in input-output analysis will increasingly be placed on the design of more complex dynamic models and the use of computer simulation. Considerably more research work in this area will be required before complex dynamic input-output models will become available to management as a flexible planning tool.

In recent years, management of large and medium sized corporations has been gaining appreciation of potential uses of the input-output method in business planning. In many corporations, however, management has shown reluctance to experiment with input-output analysis as a planning tool because of (1) the lack of timeliness of the available data base (for example, the OBE's 1963 input-output table was released in late 1969), and (2) the complexity involved in collecting and setting up comparable company data.

The most important contribution that input-output analysis could make to corporate planning is the element of equilibrium it brings to the planning process. As research work continues, input-output analysis will eventually be developed into a useful link between the forward plans made by the operations people of a business firm and the strategic planning done at the corporate management level because it can provide a useful tool for evaluating the different long-range plans that come before corporate management.

REFERENCES

1. OFFICE OF BUSINESS ECONOMICS, U.S. DEPARTMENT OF COMMERCE, "Input-Output Structure of the U.S. Economy: 1963," *Survey of Current Business,* Vol. 49, No. 11, (November 1969), 16–47.

2. ——, *Input-Output Structure of the U.S. Economy: 1963:* Vol. 1, *Transactions Data for Detailed Industries;* Vol. 2, *Direct Requirements for Detailed Industries;* Vol. 3, *Total Requirements for Detailed Industries.* (Washington, D. C.: GPO, 1969).

3. BUREAU OF LABOR STATISTICS, U.S. DEPARTMENT OF LABOR, "The U.S.

Economy in 1980: A Preview of BLS Projections," Reprint 2666, from the *Monthly Labor Review,* (April 1970).

4. ———, "Patterns of U.S. Economic Growth; 1980 Projections of Final Demand, Interindustry Relationships, Output, Productivity, and Employment," *Bulletin 1672,* (1970).

5. ———, "The U.S. Economy in 1980; A Summary of BLS Projections," *Bulletin 1673,* (1970).

6. A. H. YOUNG, L. C. MALEY, JR., S. R. REED, and R. A. SEATON, II, "Interindustry Transactions in New Structures and Equipment, 1963," *Survey of Current Business,* Vol. 51, No. 8, (August 1971), 16–22, 44.

7. B. N. VACCARA, "An Input-Output Method for Long-Range Economic Projections," *Survey of Current Business,* Vol. 51, No. 7, (July 1971), 47–56.

8. ———, "Changes over Time in U.S. Input-Output Relationships," Office of Business Economics, U.S. Department of Commerce, (July 1969) (mimeographed).

9. ———, "Changes over Time in Input-Output Coefficients for the United States," *Applications of Input-Output Analysis,* Vol. 2, Amsterdam, (1970).

Value Considerations in Public Policy Issues of Year 2000

NICHOLAS RESCHER

The SST case is a dramatic current example showing how changes in societal values affect appraisals of the future economic worth imputed to new technology. Since this is so, should not the forecaster consider value changes when assessing the future? Yet the average person, including most forecasters, has had virtually no training in the nature of values and value changes.

Classifying key values developed by the Commission on the Year 2000 into three groups—individual right values, life-setting values, and personal-characteristic values— Rescher argues that each of them is an integral part of any trend and should, therefore, be included in any projection. He also provides some much needed insight on the nature of values and value changes. While this chapter provides only a brief introduction to the complex topic of value analysis, we are certain that the forecaster must begin to introduce something on value changes into his work.

ROLE OF VALUES[1]

Values are intangibles. They are, in the final analysis, things of the mind that have to do with the vision people have of "the good life" for themselves and their fellows. A person's values, such as loyalty, economic justice or self-

[1] This chapter was originally prepared by the author while he was a consultant to the Rand Corporation, Santa Monica, California. Any views expressed are those of the author. They should not be interpreted as reflecting the views of The Rand Corporation or the official opinion or policy of any of its governmental or private research sponsors.

aggrandizement, represent factors that play a role in his personal welfare function, the yardstick by which he assesses the extent of his satisfactions in and with life. Values of course manifest themselves concretely in the ways in which people talk and act and especially in the pattern of their expenditure of time and effort and in their choices in the marketplace. And it is primarily through these concrete manifestations that values secure their importance and relevance. This, however, does not alter the fact of their own abstract and mentalistic nature, nor does it mitigate their methodologically problematic character.

People's values function both as constraints and as stimuli. A man's adherence to a certain value, say patriotism, motivates him in doing some things (for example, resenting an aspersion upon his country) and in refraining from doing others (for example, any disloyal action). The prevalent values of a society, particularly of a democratic society, significantly condition the ways it conceives of and goes about discharging its business. The flat rejection of the prospect of preventive war by the United States and the no-first-strike policy maintained through several successive administrations reflect in significant part the appreciation by America's political leadership of certain fundamental value commitments of the American people. And somewhat similar value-inherent restraints were operative in circumscribing American actions during the Bay of Pigs adventure and the subsequent Russo-Cuban missile crisis. Values are thus an important element, not merely for the sociologists understanding of national traits and character but even for the appreciation of political realities. Values are important for politics, that art of the possible, because they play a significant role in the determination of what *is* possible.

When one talks of the values of American society, the question immediately arises: whose values? Few societies are homogeneous, and American society certainly is not. For present purposes we shall focus upon the prevalent values of the society, those that are sufficiently pervasive and prominent throughout its fabric to be invoked by its major value spokesmen (including politicians, newspaper editorialists, graduation exercise speakers, and sermonizers.) The existence of subsocieties with divergent value schedules must of course be recognized (for example the Amish, or the hip bohemians). Indeed, such subsocieties may well play an increasingly prominent role in the future.[2] But there unquestionably exists (and will continue to exist) a sufficiently prominent value consensus to underwrite a basis for proceeding with the discussion at the level of aggregation of a system of "generally accepted" values, without becoming enmeshed in the ramifications of social fragmentation.

---

2 The reasons for subsocieties may be different, however. For in the past the opportunities for interaction were circumscribed, while in the future relative desire for, or rather against, interaction may well be the key factor, as various groups "opt out" of the mainstream.

All writers on the subject, however much they may disagree in other respects, seem to concur in the thesis that substantial pressures for change will come to bear upon American values within the next two generations. And this conclusion is pretty much beyond question, since a rather obvious mechanism is at work in tending to produce such changes. Environmental changes affecting the circumstances of life (for example, demographic crowding) in turn affect the "cost" of realizing a certain value (for example, privacy) or else affect the "benefit" to be derived from its realization. In this way, the environmental change (economic, social, political, or demographic) exerts a pressure upon the extent to which a certain value (privacy or the like) can be realized with the result that this value comes to be reevaluated: downgraded or upgraded, as the case may be.[3]

Although the phenomenology of value change is readily understood at a very general level of consideration, the specific forecasting of specific value changes is quite another thing. There are great complexities in the interaction of values with one another (for example, in cases where only one mode of justice, say economic justice, may be involved, the linkages[4] with other parts of the value cluster may be close enough to create substantial side effects. Nor is it an easy matter to say how a value will react to a specific environmental change. (For example, when the realization of privacy becomes operationally more difficult, and we are in the position of having to settle for less, the reaction toward the value itself may be either to downgrade it as a value—the sour-grapes reaction—or else to upgrade it as something to be treasured all the more for the difficulty of its attainment. Also, there is considerable reciprocal interaction between values and changes in the social and economic environment, since our values condition the quality and especially the quantity (that is, rate) of such environmental changes. Such considerations, and kindred ones, point up the difficulties in the forecasting of specific value changes.

The intrinsic nature of value changes is intricate and complex. A value change is generally not an on-off matter of subscribing to or abandoning a certain value, but is a matter of the *extent* to which the value is held and of the *way* it is combined with others. For the values we have, both as individuals and as a society, tend to conflict with one another when taken in the aggregate, in just the way our life needs (for rest, say, and amusement and exercise and rewarding activity) contradict one another, not only in competing for time, attention, and resources but also in representing vectors that point in different

---

[3] Downgrading and upgrading are themselves complex, multidimensional phenomena. Without changing the place of a value in relation to others in the hierarchy of a value scale, we may still "settle for less" in respect of its realization. That itself is a mode of downgrading.

[4] Purely ideological linkages, that is, for there is no question of *logical* interdependence.

directions. Precisely because they are abstract, our values also exhibit a self-inconsistent character (as novelty can easily clash, for example, with convenience), just as we find this in the proverbial wisdom of the race: Haste makes waste, versus a stitch in time saves nine.

The practical implications of the problem of future values, however, is such that attempts to deal with it cannot be deferred on excuse of its difficulties. One must cope with the problem with the crude tools at hand because its importance is too pressing to brook delay until the predictive methodology for tackling such complexities is in a more satisfactory state than is currently the case. It is necessary to grapple, however imperfectly, with the question of possible and probable value changes with which the policy makers of the future may have to reckon.

LIST OF KEY VALUES FOR FUTURE-ORIENTED INQUIRIES

The subsequent considerations of future American values will focus upon those particular values that have featured explicitly and prominently in the first set of working papers of the working party on "Values and Rights" of the Commission on the Year 2000 of the American Academy of Arts and Sciences.[5] In the deliberations comprising this document, the following values hold a place of prominence:

Privacy.
Equality (legal, social, and economic).
Personal integrity (versus depersonalization).
Welfare (personal and social).
Freedom (of choice and action).
Law abidingness and public order.
Pleasantness of environment.
Social adjustment.
Efficiency and effectiveness in organizations.
Rationality (organizational and individual).
Education and intelligence.
Ability and talent.

[5] *Working Papers of the Commission on the Year 2000 of the American Academy of Arts and Sciences,* Vol. V (Boston, 1966). Over half of these papers were published in *Daedalus: J. of the Amer. Academy of Arts and Sciences,* Summer, 1967 issue.

These are the values which the authors of the document in question apparently regard as particularly affected by or involved in foreseeable future developments. This list will set the stage for our subsequent deliberations.

It is useful to observe that the values in this list can be sorted into three categories:

1. Individual rights values: privacy, equality, personal integrity, welfare, freedom.
2. Life-setting values: public order, pleasantness of environment.
3. Personal-characteristic values: efficiency, rationality, social adjustment, education, intelligence, ability, and talent.

These three groupings would seem to provide a natural classification for the values of our basic list.

It is not difficult to supply by conjecture the (tacit) rationale by which the specific values of our basic list come to be accorded their place of eminence. Each of them falls into one of two groups when considered against the background of the major projectable demographic, social, economic, and technological trends. For the working of these trends is shaping a future society in which the value under consideration is either (1) markedly threatened, or (2) badly needed. The first is the case in particular with the values of categories 1 and 2 (the indicated individual rights values and life-setting values), all of which are such that their realization is apt to be substantially more difficult in the America of the future. The second is the case in particular with the values of category 3, which are such that their espousal, maintenance, and realization will be especially important in the future society. The pair of indicated factors would thus appear to constitute a plausible rationale to explain why the values registered in the basic list figured in their prominent role in the document we have cited.

Two interestingly contrasting sets of considerations are operative here. In saying that a value is "threatened" by future developments, one takes one's footing on the *current* present-day value system. For it is to say that they (those future ones—and it is not, of course, to be precluded that we ourselves may be among them) will be under pressure to *downgrade* this value, that is, to be willing to settle for less in regard to it. Indeed, describing the value as threatened we raise the spectre of its abandonment. (There is, after all, the phenomenon value obsolescence, when what was once under certain conditions of life a genuine and authentic value becomes outmoded under changed circumstances: chivalry and noblesse oblige, for example.) By contrast, to say that a certain value will in the future be more badly needed and so to suggest that they of the future will be under pressure to *upgrade* this value is not necessarily to suggest that this value has its proper place in the current value system. There is an asymmetry here. Saying that a value is threatened is to reflect the posture of the present value system in a way in which saying that a value is underrated is not.

An important feature of the strategy of value-change prediction should be noted in this connection. One of the key objects of the enterprise is obviously to identify threats to values, for example, predictable developments that may exert a pressure upon the maintenance of a value in the society to the point where it becomes downgraded in the future. Such a prediction of a change regarding values can thus serve in an early-warning role to trigger or facilitate preventive action to assure that the untoward foreseen consequences do not occur. The very prediction of a value downgrading may thus (hopefully) have a self-defeating effect by being a causal factor in a chain of preventive developments.

FURTHER ASPECTS OF BASIC LIST

Several further aspects of the basic list of values that seem likely to be affected by future developments deserve to be noted. One of the most interesting of these is the fact that the majority of these items are characteristically *social values,* in that they represent arrangements that his social environment affords to an individual (for example, equality, privacy, freedom, public order) and of the rest the bulk consist in *virtues of ability* involving personal qualities that individuals prize but over whose *possession,* albeit not whose *cultivation,* they have relatively little control (for example, rationality, intelligence, ability, and talent). Thus a striking feature of the list is the absence of the traditional *virtues of character,* in contrast with the virtues of ability, including such values as:

Honesty.
Loyalty.
Idealism.
Friendliness.
Truthfulness.

What accounts for this gap? One is tempted to assume that the explanation of the absence of the character virtues from the basic list is implicit in the previously conjectured rationale of the list itself. For it was perhaps the view of the authors at issue that these specific values are neither particularly threatened by predictable trends shaping our future environment nor that the degree to which they will be requisite in the future society is significantly greater than their present-day desirability. If this was indeed the view of these authors, then I should like to enter a sharp dissent from their position.

There can be little question that there are at least some values within the category of the virtues of character for which the crowded, depersonalized, and complex society of the year 2000 will have a yet greater need, however highly these values may have been prized heretofore. For example, one thinks here specifically of those values that are oriented toward service to others: service to one's fellows, a sense of responsibility, and dedication to humanitarian ideals. American society has treasured these values throughout its history, but in the future they will very likely be markedly upgraded as a social environment takes shape in which they are increasingly indispensable.

Over the past generation there has been a marked tendency in American life—one whose continuation and intensification can confidently be looked for —to shift from the Protestant ethic of getting ahead in the world to the social ethic of service to one's fellows—from the gospel of profit and devotion to mammon to the gospel of service and devotion to man. Many forces have produced this phenomenon, ranging from spiritual causes such as a decline of traditional Protestant ethic on the one hand to material causes such as the rise of the welfare state on the other (God helps those who help themselves, perhaps, but the state is not so exclusive about it, and so it is less urgent in our affluent society to look out for oneself). There has thus been in many sectors of our national life, including the industrial, a distinct elevation of the historic American value of public service. And this is a trend that will certainly continue and possibly intensify. However, this will presumably happen in such a way that the present tendency to the institutionalization of these social service values (for example, the Peace Corps, Vista, and so on) will continue and perhaps intensify.

This contention that certain of the virtues of character will be upgraded in the America of the year 2000 must not be inflated beyond intended limits, however. Some of these traditional values seem very definitely on the decline. A trend toward the welfare state, for example, will very likely add charity to the roster of outmoded values. Again, compassion may well become downgraded in an environment in which violence is an increasingly familiar phenomenon. (One thinks with dismay of incidents like the murder of Miss Genovese in the midst of some twoscore unbudging witnesses.)

Yet another negative feature of the basic list deserves to be noted: the absence of aesthetic values. Increasing affluence and leisure, and the growing worldliness and sophistication that come in their wake, promise to give a definitely more prominent place to the ornamental aspects of life. Specifically, there is every reason to think that the aesthetic element will play a vastly more significant role in American life in the year 2000. Moreover, one need not dwell upon the unpleasing aspects of a graphic picture of the degradation in the quality of life that will come to realization if our response to the needs of the future for housing, transportation, and so on are dictated by considerations of economy and efficiency alone, without due heed of human considerations, prominently including aesthetic factors. The probable upgrading of the aesthetic element is something that an examination of American values of the year 2000 cannot afford to overlook.

The crowding of the avenues of action in modern life increasingly puts the individual into a position not so much of interacting with others as individual agents, but of reacting to them as a mass comprising a complex system.

Many of the things that go wrong are best looked at, at any rate from the standpoint of their victims, as system malfunctions. It seems probable in this context that we will less and less treat such failures as matters of individual accountability. If X cheats, burgles, or inflicts motor vehicle damage upon Y, the view will increasingly prevail that Y should not have to look to X for recovery from loss but to a depersonalized source, that is, some agency of the society. We will not improbably move increasingly towards the concept of a "veteran's administration" for the victims of the ordinary hazards of life in our society. Individual responsibility and personal accountability has suffered some depreciation in American life over the past two decades, but it seems likely that in the years ahead social accountability will become an increasingly prominent ideal.

This illustrates a rather more general tendency with respect to future values: it would appear on the basis of present trends and predictable developments that we can look for a significant upgrading of institutional values.

Although America has in the past been able to achieve prodigies of progress on the basis of a large measure of individual effort and individual initiative, the value of organization in common effort for the public good has been recognized since the days of Benjamin Franklin as a natural channel for constructive individual effort. The value of common action through association and institutionalization has always been accepted, despite a strong strain of independence of the "No thanks, I'll go my own way" type. But in days ahead with an increasing dependence on and need for (and thus augmented although at first no doubt ambivalent, respect for) institutions we may reasonably expect an increased respect for institutions and an upgrading of institutional values and the social ethic generally. The decline of independence remarked on above may be looked upon to undermine the traditional individualistic (and even anarchic) strain in American life with its concomitant duality towards laws, rules, and law-abidingness. Our increasing reliance on institutions will increasingly emphasize and strengthen institutional values. The present-day revolt of the Negro, for example, is not so much a rebellion against American institutions as such as a protest on the part of those who look on themselves as outsiders excluded from them. Note that the current remedies to the exclusion problem (for example, the war on poverty and the Job Corps) are of a strictly institutional character, so are various of the means to solve the problem of youth anomie by creating institutional means for helping the young to find the meaning of life in socially useful service (for example, the Peace Corps). (Such ventures represent an institutionalization of idealism; they involve a scaling down in individual ambition, a shift from playing a part on the big stage oneself to being a small cog in a big venture.)

Such a predictable upsurge of institutional values and the social ethic is an important trend in future American values whose working would not be suspected on the basis of even a careful scrutiny of the basic list.

Considerations such as these underline the fact that it is critically important for the viability of our future society that people *have values of a socially beneficent sort,* that they be persons of the type one characterizes as being "dedicated to sound values" and as having a "good sense of values." The need of society for such people raises the question of where they are to come from —of how people are to be trained in this way.

This question brings vast difficulties to the fore. With the decline of religious committment in American life, the churches are losing their traditional role as value inculcators. The weakening of the family undermines many of its roles, that of a center for value teaching among them. Nor, in this "post-ideological" age are politicoeconomic doctrines a potent force for value propagation. The question of the teaching of values, especially of their transmission to the younger generation, in the operative environment of the future would seem to constitute one of the large problem issues in the public policy domain of the (nonremote) future.

This problem has been rendered the more difficult to grapple with by a significant social development of relatively recent vintage. This relates to the importance not merely of *having* socially desirable values in the weak sense of according them verbal recognition and adherence but in the significantly stronger sense of actually *acting* upon them. The present "revolt" of a sector of American youth against the mainstream of adult thought and behavior is in fact largely based on an *acceptance* (rather than rejection) of mainstream values. The disaffected young do not advocate an abandonment or transvaluation of the traditional values, rather, they accuse those who have succeeded in the mainstream of American life of a *betrayal* of these values (a sort of *trahison des arrivistes*). In the eyes of disaffected youth, adult society is not *misguided* in holding the wrong values but *hypocritical* in not acting on the correct values it overtly holds and professes. American society is sick in the eyes of its youthful critics because it *fails to implement its own values* in an appropriate way: the society's actions distort its own value scale (for example, in giving low priority to personal aggrandizement, success, and material welfare in the value teaching of the middle class, in contrast with their intensive pursuit in practice, or at the national level in propagandizing the ideals of a great society at home and a good neighbor abroad while expending a vastly greater share of the national budget on the space race than on foreign aid, on military preparedness than on domestic welfare). The cynicism about our values created by such developments will greatly complicate the intrinsic difficulties of value teaching in the days ahead.

And just here lies one of the great gaps in our knowledge in this sphere: how to generate adherence to values. Precious little is known about many key aspects of values, despite the monumental labors of Clyde Kluckhohm and his school in the area of value studies. And no part of this knowledge gap is more notable than that relating to the *teaching* of values, nor more acute in an era when the historic media of value teaching are losing much of their traditional effectiveness.

BIBLIOGRAPHY

1. Baier, K. and N. Rescher, eds., *Values and the Future* (New York: The Free Press, 1969).

2. Heinrichs, H. H. and G. M. Taylor, eds., *Program Budgeting and Cost-Benefit Analysis* (Pacific Palisades: Good Year Publishing Co., 1969).

3. Kluckhohm, C. M., "Have There Been Discernible Shifts in American Values During the Past Generation?" in E. Morison, ed., *The American Style* (New York, 1958), pp. 145–217.

4. Prest, A. R. and R. Turvey, "Cost-Benefit Analysis: A Survey," *The Economic Journal,* (December, 1965), pp. 683–735.

5. Rescher, N., *Introduction to Value Theory* (Englewood Cliffs, N.J.: Prentice-Hall, Inc., 1969). Contains an exhaustive bibliography.

Experiment in Simulation Gaming for Social Policy Studies

THEODORE J. GORDON
SELWYN ENZER
RICHARD ROCHBERG

Single dimension forecasts (for example, technical, social, or political), tend to be inaccurate, particularly as the scope of the system being forecasted is enlarged. One of the thorniest and most complex problems is to forecast the reactions of society to future policy alternatives.

The Institute for the Future conducted a unique experiment for the State of Connecticut. They attempted to identify or forecast likely reactions of various groups in the State to different policies that the State might choose to employ. The basic concept was to establish fourteen measures of social and economic conditions, then to have informed persons representing various elements in the state society react to suggested policies, events, and technical choices.

Whether such simulation gaming can properly predict future reactions is open to argument, of course. Yet, we believe that this novel concept deserves the reader's imaginative consideration. One of the dilemmas facing society today is to anticipate how people will feel in years to come about the policies and actions we are establishing now.

This paper[1] describes the objectives, design, and play of a simulation game developed by the IFF (Institute for the Future) under a grant from the Connecticut Research Commission. The general subject of the research conducted under this grant was the design of methodologies which promised to be useful in forecasting societal change and the application of these methodologies to derive societal forecasts for the state. The research included preparing forecasts of technological change, social change, and issues and opportunities in prospect for Connecticut; these forecasts came to focus in a simulation game conducted toward the conclusion of the overall project. In this game, forecasts of the future economic status and "quality of life" were derived in view of certain of the expedited changes and alternative legislative policies. The game was a methodological experiment designed to identify techniques for forecasting conditions of society, in light of potential societal and technological changes and the reactions and values of individuals and groups these changes would be likely to affect.

TYPES OF SIMULATION

Simulation, in the sense used in this paper, is the approximation of complex systems by dynamic models. These models may exist in several forms: they may be mechanical analogs, for example, a wind tunnel model of an SST; mathematical analogs, a set of equations depicting the economic situation of a country; metaphorical analogs, the growth of a bacteria colony depicting human population growth; or game analogs, in which interactions between "players" are taken to represent social interactions. These models all have the ability to change with time and imposed conditions, and are used where experimentation with an actual system is too costly, is morally impossible, or involves the study of problems which are so complex that analytic solution appears impractical.

OBJECTIVES OF SIMULATION

The exact natural sciences have "laws" which link cause immutably to effect and provide the means of forecasting the impact of one parameter on others which are relevant. The less social sciences are almost devoid of generally applicable laws which permit the forecasting of social behavior. Such laws would be most useful in assessing the likely impact of anticipated policy actions.

[1] Much of the material in this chapter is drawn from a study performed by the Institute for the Future, Middletown, Connecticut, for the Connecticut Research Commission. As such it represents the work of many members of the staff in addition to the authors, particularly Dr. Olaf Helmer, Mr. Raul deBrigard, and Mr. Robert Buchele. IFF Report R–9, "A Simulation Game for the Study of State Policies," describes this work in more detail.

Simulation methods (and other techniques in development) may eventually serve as surrogates for these laws which can be used in forecasting in the same way that natural laws permit forecasting some interactions between physical systems.

Before these methods can seriously serve in this application, however, they must meet the following criteria:

1. The results produced must be independent of the model user and his institutional setting.
2. The results produced must match the performance of the real system being simulated in historical as well as future situations.

There are already many simulations which meet these criteria. Scaling rules for mechanical analogs exist which permit aeronautical engineers to use models to understand the performance of full-scale aircraft. Mathematical analogs gain independence from their users and precision by using the past as a basis for modeling. Econometric models are based on regression analyses of past economic data. They can successfully duplicate the historical performance of economic systems and, at least for the short term, produce economic forecasts which appear to anticipate reality. (However, when the forces that shaped the regression constants change, the model is apt to fail in prediction.) Mathematical models of physical systems have also been successful; for example, computer simulations of smog conditions in the Los Angeles basin have led to insight as to the relative contributions of automobiles and power stations to atmospheric pollution and the expected effect of varying either input. Metaphorical models are important semantically, and perhaps even in reaching an understanding of underlying phenomena, but the precision of such a tool no doubt depends on the "stretch" of the metaphor.

The fourth category of simulation, simulation gaming, fails in both criteria. It can as yet offer neither repeatability nor precision in forecasting. Yet the technique has promise and may develop into a useful method. It appears that the effort in developing the tool is warranted because it alone among the four techniques mentioned offers the most immediate promise of including human values, which, after all, are the most important ingredient of social policy studies.

If simulation gaming has not yet satisfied the criteria of repeatability and forecasting precision, it does serve other ends, including:

1. It teaches players and experimenters about the issue.
   A. A well-designed simulation game includes descriptive elements which acquaint the players with definitions of and interrelationships among some of the variables, at least as perceived by the experimenters.
   B. The experimenters, in designing the game, learn to define the parameters thought to be relevant to the issues being simulated and to describe their mode of interaction.

C. Tolerance (or at least sensitivity) to various points of view is usually a part of the players' game experience.

2. It provides a framework for systematic review.

A. In constructing a game it is usually necessary to define the constituent elements of the issues being simulated and consider how these elements are likely to interact. Because games usually require a structure of sorts, these considerations are usually orderly and may thus replace haphazard and sporadic probes into issues.

B. The game structure often forces the role players to consider the issues surrounding the simulation from attitudes unfamiliar to them, usually in a systematic fashion.

3. It allows role players to exhibit emotions and make decisions normally excluded in real life but permitted in the game environment. In effect the game may become a psychological device, a means of experimenting with and exerting power in circumstances not likely to be detrimental to society at large.

4. It improves communications between players. The cooperation or competition engendered by the roles of a game may promote relationships between the players which last long after the game is over. By exposing concerns in the game environment, players may build new sensitivities to each other's personalities and behavior patterns in real life.

These ancillary potentialities of simulation gaming have led to its use in applications as diverse as the testing of military strategies and the teaching of political science on the secondary level and above. They speak for its application to conflict resolution between individuals, groups separated by ethnic or socioeconomic differences, or nations. They speak for its use in arriving at democratic decisions. These uses are practical; such practical applications can provide the experience and data necessary to develop the tool itself into a more precise forecasting instrument.

DESIGN CONSIDERATIONS

A simulation game generally has the following elements:

1. Rules of play govern the "moves" and interactions between players, and other variables of the game.

2. Objectives or goals, allow players to work cooperatively toward a joint goal or competitively toward goals which cannot be shared. (A single game may incorporate both modes, as in Monopoly when players decide to "gang up" on the "apartment house owner.")

3. A method of translating the moves of the players into indicators which measure the degree of attainment of goals may be a game board, a mathematical model, or another group of participants charged with responsibility of judging the effect of the players' moves.

4. A display system illustrates the progress of the game.

5. A set of exogenous variables is used to introduce "outside events" into the play.

More generally, games should be designed so that goals can be achieved or a winner determined within the time span available to the players; the time of the players should be well occupied; and the rules should permit freedom of action (at least to the level available in society) but without introducing frustration. And all of this should be done in a manner which approximates real social intercourse.

DESIGN OF THE CONNECTICUT GAME

It was intended that IFF's Connecticut game explore methods for assessing the likely effects of various governmental legislative policies in view of potential external events and developments. The game had several objectives:

1. Analyzing the effects of external (that is, world and national) technological and societal developments and of alternative courses of action (policies) on the State.
2. Providing a better understanding of the roles of governmental decision makers and the elements of society affected by their decisions.
3. Generating discussion which would lead to better understanding of the State's future needs and opportunities and wider recognition of the alternative courses of action available to the State.

The participants were divided into three groups. Two of them were called "players," representing legislators and developed programs of action for the state. The third group, called evaluators, simulated society at large and (using a voting machine) assessed the status of the indicators before and after each round of play. Examples of the displays used for recording indicator values are presented in Exhibit 1. One team of players was instructed to maximize

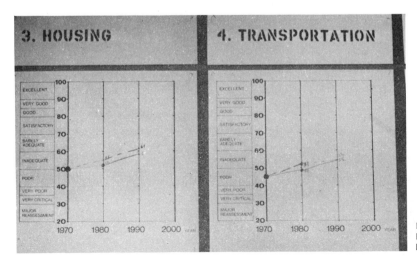

Exhibit 1. Indicator Display Boards as They Appeared at the End of the Game.

Exhibit 2. The Gross State Product (GSP) Team.

Exhibit 3. The QOL (Quality of Life) Team.

the per capita GSP (gross state product) and the other to optimize the Connecticut QOL (quality of life). The GSP team is shown in Exhibit 2, the QOL team in Exhibit 3, and the evaluators seated at the voting machine appear in Exhibit 4. The basic task of the evaluators was to record society's degree of satisfaction with affairs in Connecticut. Each member of the evaluating team represented a certain segment of society: the urban poor, the cultural elite, the middle class, the older citizens, youth, city management, the financial community, and the federal government. Role playing was used to encourage explicit consideration of the interests of these sectors of society.

The procedural flow of the game is shown in Exhibit 5. The game began with the evaluators discussing their satisfaction with aspects of life in the state in terms of fourteen economic and social indicators:

Exhibit 4. The Evaluators Seated at the Voting Machine.

1. Government effectiveness
2. Business climate
3. Housing
4. Transportation
5. Satisfaction with tax structure
6. Employment
7. Standard of living

8. Social climate
9. Control of crime
10. Physical environment
11. Health
12. Education
13. Recreation
14. Personal liberties

Two aggregate indicators, per capita GSP and Connecticut QOL, were also assessed. As shown in Exhibit 1, the scale ranged from 20 (situation

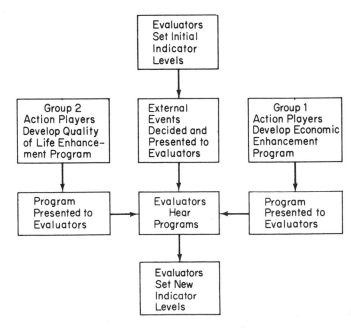

Exhibit 5. General Game Flow.

requiring major reassessment) to 100 (excellent situation) except for the per capita GSP which was in absolute units.

The players heard (and were permitted to contribute to) the discussion of the current rating of each indicator and so were aware of the criteria used by the evaluating team in its assessment. The evaluators discussed each issue, voted on a preliminary rating, discussed the issue again to seek a consensus, and then cast a final vote. The median of each rating was recorded on the indicator charts. Each round of the simulation represented a ten-year period.

Although group competition was not intended as an element of the game, the two teams of players competed in the sense that the impacts of their decisions on the indicators were compared, allowing an evaluation of the relative success of different governmental policies.

Each team was provided with candidate action cards derived from the IFF Delphi study of future issues and opportunities in Connecticut, described in IFF Report R–8.[2] The cards described potential actions and presented short background discussions of the concepts. In addition they indicated a minimum cost and impact of each action on the fourteen economic and social indicators. A typical action card is presented in Exhibit 6, and a complete list of the titles of the actions is presented in Exhibit 7.

[2] S. Enzer and R. deBrigard, *Issues and Opportunities in the State of Connecticut: 1970–2000*, IFF Report R–8, September, 1969.

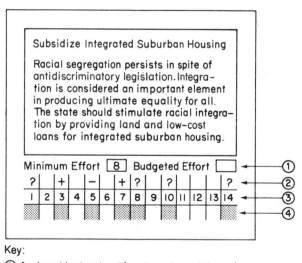

Key:
① Assigned budget level (not less than minimum).
Suspected indicator impacts economic and social indicators:
② { + = probable improvement
− = probable degradation
? = likely to vary with societal sectors
③ Indicator number
④ Color code for easy identification

Exhibit 6. Typical Action Card.

Exhibit 7. List of Candidate Actions.

1. Implement community development programs on a regional basis.
2. Establish an ombudsman's office.
3. Institute a statewide government data management system.
4. Centralize public service and facility districts.
5. Establish statewide regional governments.
6. Subsidize on-the-job retraining.
7. Offer incentives to high-technology industries for migrating to this state.
8. Establish a business climate planning agency.
9. Subsidize moderate-income housing.
10. Subsidize housing for senior citizens.
11. Rezone urban areas to permit greater population density.
12. Make multiple use of highway space.
13. Enact statewide flood controls.
14. Rezone to decentralize urban areas.
15. Place building code control at the state level
16 Establish minimum quality standards for new housing.
17. Reform building codes to accept technological innovations.
18. Establish minimal maintenance requirements for housing appearance.
19. Subsidize low-income housing.
20. Increase taxes on poorly maintained property.
21. Enforce health and safety standards in housing vigorously.
22. Establish new communities in rural areas.
23. Reform building codes to accept standardization.
24. Place zoning controls at the state level.
25. Expand bus services in urban areas.
26. Establish quasi-public companies for interstate mass transportation systems.
27. Provide advanced mass transit systems in urban areas.
28. Acquire land for future transportation needs.
29. Build a bridge to Long Island.
30. Provide mass parking facilities around urban perimeters.
31 Construct "floating" airport in Long Island Sound.
32. Enlarge Bradley Field.
33. Construct a new airport in western Connecticut.
34. Charge a toll for personal cars entering urban centers.
35. Acquire rural land for future airports.
36. Establish quasi-public companies for intrastate mass transportation systems.
37. Ban personal cars from urban centers.
38. Provide land for mass transportation systems.
39 Provide for future expansion in new highway systems.
40. Acquire land adjoining existing roads for future expansion.
41. Provide incentives for urban mass transportation systems.
42. Provide mass parking facilities in urban centers.
43. Raise state sales taxes.
44. Remove all highway and bridge tolls.
45. Initiate a state income tax.
46. Use special-purpose taxes for special purposes only.
47. Limit property taxes to one percent of market value
48. Equalize tolls on all limited-access highways.
49. Offer tax incentives to encourage profit sharing.
50. Provide low-cost business loans for welfare recipients.
51. Offer welfare bonuses for family planning.
52. Offer bonuses to welfare recipients who become self-supporting.
53. Subsidize new businesses for locating in ghettos.
54. Subsidize integrated suburban housing
55. Strengthen racial anti-discrimination legislation.

Exhibit 7 (cont.)

56. Teach birth control in public high schools.
57. Limit prices charged to ghetto residents.
58. Provide free day care for children up to age fourteen.
59. Initiate penal system reforms.
60. Upgrade educational requirements for police.
61. Make drug use a noncriminal act.
62. Increase penalties for selling drugs.
63. Require registration of firearms.
64. Control the sale of ammunition.
65. Enact and enforce antinoise pollution standards.
66. Increase taxes on property use obstructive to regional development.
67. Enact and enforce atmospheric antipollution standards.
68. Enact and enforce antipollution standards for water.
69. Enact and enforce solid waste disposal standards.
70. Encourage health insurance programs which maximize efficiency in medical services.
71. Offer incentives for development of a system of echelon hospitals.
72. Provide computer services for medical diagnostics.
73. Make nutritional education mandatory for welfare recipients.
74. Provide free medical insurance for all.
75. Provide free medical insurance for poor people.
76. Discourage use of "heroic" medical measures.
77. Improve educational TV programs.
78. Require teacher conferences with parents of failing primary school students.
79. Offer adult educational retraining.
80. Require mandatory periods of nonteaching employment for educators.
81. Extend the educational period.
82. Institute high school work-study programs.
83. Improve program of general scientific understanding in secondary schools.
84. Acquire and use teaching machines.
85. Provide incentives for school district unification.
86. Eliminate all split sessions in public schools.
87. Raise teachers' salaries.
88. Limit individual public school classes to thirty pupils.
89. Enlarge vocational training programs.
90. Offer compensatory payment for scholastic achievement.
91. Employ professional managers to administer institutions of higher education.
92. Reform college administrative procedures to include formal student participation.
93. Reduce college admission and scholastic standards for minority group students.
94. Provide free college for all students.
95. Collect all taxes at the state level.
96. Provide free college for qualified poor students.
97. Eliminate the tenure system in schools.
98. Provide low-cost loans for College education.
99. Provide more state parks.
100. Provide wildlife reserves.
101. Provide community recreation facilities.
102. Make state parks self-sufficient through use of fees.
103. Perform research on innovative techniques for creating new recreational facilities.
104. Provide public pools in urban areas.
105. Provide free transportation to recreational facilities for ghetto residents.
106. Keep school recreation areas open during off-school hours.
107. Legalize homosexuality.
108. Legalize abortions.
109. Legalize marijuana-type drugs.
110. Legalize abortions in special situations.
111. Require courses in family relations for all high school students.

The minimum price of an action was taken as an intuitive measure of the combined political effort and monetary cost required to establish a program at a level that would have a noticeable impact on the state. The least expensive actions were assigned a cost of one point.

A pseudoeconomic system was superimposed on the play to force consideration of priorities and limit the number of actions the players could choose. The players had sixty points to spend in the first decade; in the later decades the available budget varied as a function of the action taken and the effect of certain outside events. The two budgetary constraints were that each action had to be allocated, at least its minimum price, and that total expenditures had to be within the budget.

The players were urged not only to consider the prepared actions but to propose new ones, which were approved and given minimum levels of effort by the evaluators.

Both teams were told that the programs they were preparing were in addition to the normal operation of the state. For example, if no action had been taken in the area of education, this would not have meant that schools would be closed or even that the normal rate of school construction would be slowed; it meant rather that no innovative improvements would be made in education.

Since future events and developments external to Connecticut will also have an important impact on the economic and social well-being of the citizens of the state, such externalities had to be introduced into the game. Accordingly, fifty potential external developments were selected from technological and societal Delphi studies conducted at IFF. The events were displayed on cards and in the form of a so-called cross-impact matrix, which indicated their probability of occurrence with the expected linkages of interdependencies among events. The outcome of the external events for each simulation period was determined by a probabilistic procedure. Items were chosen at random from the list of fifty external events, and their occurrence in the interval being simulated was "decided" by comparing their probabilities of occurrence as determined from the Delphi study with a random number obtained by casting dice. If an item "happened," the probabilities of the other items were adjusted according to instructions contained in the matrix. The process was continued until every item was decided. A typical event card is shown in Exhibit 8. Exhibit 9 presents a complete list of the titles of the fifty events.

Events determined to have happened were described to the evaluators while the players were choosing their programs. The evaluators were instructed to consider these events later when they assessed new values for the societal indicators.

After the players had selected their actions, the evaluators were briefed on these programs and the indicators were reassessed for 1980. The evaluators considered how the indicators might be affected from the point of view of the societal sector they were simulating. This assessment again involved debate which centered around recognition of pluralistic societal interests and the

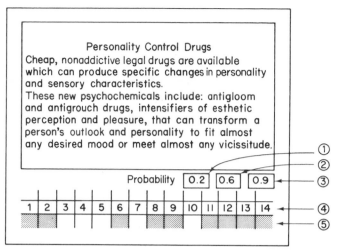

Key:
Probability of occurrence as determined
by Delphi study during decade of:
① 1970—1980
② 1980—1990
③ 1990—2000
④ Indicator number (economic and social Indicators)
⑤ Suspected (nondirectional) indicator impact.

Exhibit 8. Typical Event Card.

likely effects of external technological and social change. Once this had been accomplished, the cycle was repeated for the next decade.

DYNAMIC SIMILARITY
BETWEEN SIMULATED AND REAL ENVIRONMENTS

The game design presented in the previous section contains elements of all basic societal forces: government, the governed, and external events. To assess how closely this simulation reflected real societal functions, three separate aspects must be considered:

1. Game concept: whether or not it reflected the variables and dynamics of the State of Connecticut.
2. Rules: whether or not they imposed unrealistic constraints or offered excessive freedom to the participants.
3. Simulation: whether or not the concepts and rules provided a framework in which the participants would perform effectively.

Game Concept

The game concept assumed that the participants were sufficiently knowledgeable of state functions to portray them accurately. The external events

Exhibit 9. List of External Events.

Event	Probability 70s	Probability 80s	Probability 90s
1. Multistate regional authorities	80	90	90
2. Average retirement age: 55	10	30	50
3. Advanced building codes	60	80	90
4. Nuclear disarmament	20	40	40
5. Biological weapons	30	30	30
6. Wars not directly involving the United States	80	80	80
7. Limited wars directly involving the United States	30	40	40
8. Credit card economy	70	80	90
9. Business recession	20	20	20
10. Continued inflation	60	60	60
11. Educational network	30	50	70
12. Data banks	60	70	80
13. Household robots	20	30	40
14. Thirty five hours worked per week	30	50	60
15. High I. Q. computers	30	60	70
16. Behavior control	20	40	50
17. Cheap power from controlled thermonuclear power	10	50	70
18. Limited weather control	30	50	70
19. Antismog devices	20	60	90
20. High-speed mass transportation	10	50	80
21. Hypersonic transport	00	20	30
22. Aging control	10	15	15
23. Sex choice	20	50	60
24. Personality control drugs	20	60	90
25. Relative decrease in defense budget	10	20	20
26. Mass produced housing	40	80	90
27. Urban riots	20	20	10
28. European SST	30	20	20
29. NASA budget cut	20	20	10
30. Oceanography program	20	30	40
31. Increased crime	30	50	20
32. New cities programs	20	30	40
33. Negative income tax	40	60	70
34. Extended vacations	00	20	30
35. Multihome families	10	30	50
36. Women in the labor force	30	40	50
37. Negro political control of cities	50	40	20
38. New family arrangements (for example, communal marriages)	05	10	20
39. Expanded federal welfare programs	20	30	40
40. Leisure expenditures double	40	60	90
41. General immunization	20	60	90
42. More autos per family (doubling of ratio)	20	40	50
43. Automated purchasing	30	50	80
44. Low-cost portable telephones	00	10	50
45. Growth of air traffic	10	40	60
46. Abandonment of low status jobs	40	40	40
47. Decrease in tobacco consumption	30	50	60
48. Limited genetic control	00	10	50
49. Automated plebescite	00	10	20
50. Growth of public service professions	20	20	20

derived from other portions of this effort were analytically evaluated and introduced on a probabilistic basis. As in actual societal activities, such events are often foreseeable and can be assigned subjective likelihoods of occurrence; and in the simulation, as in life, the final word is left to the random operation of chance.

The dynamic similarity to external events was modified to the extent that the set of events was limited to fifty, a limitation that was introduced because practical reasons as well as the impossibility of accurately listing all possible developments. On balance, it appears that while the basic game concept contains all the elements necessary for reasonable dynamic similarity, details of the rules under which the participants had to operate often compromised this similarity.

Rules

The players and evaluators were exposed to differing degrees of similitude. For the players, the simulation was considered good because:

1. The choice of possible actions was not limited to the prepared set. In fact, the players were encouraged to create additional actions, which were assessed for credibility and minimum level of effort by the evaluators. This freedom facilitated communication between society and government regarding possible social actions that were not explicit in the simulation.
2. The players, like their real-world counterparts, were exposed to trends and prospects within and outside the state in selecting their action programs.
3. The players had budgetary limitations.
4. Budgetary savings from complementary actions were added to the budget in the following period. This allowed the players freedom to break an action into incremental steps and spread the cost over more than one simulation period.

The rules imposed certain constraints on the players that are not imposed on government in real life. These were:

1. No mechanism was provided to allow the players to raise their budgetary levels in subsequent periods by increasing taxes.
2. An extremely long planning interval of a decade was used for each simulated action program, compared with the biennium actually used in Connecticut.
3. No distinction was provided in the budget between the obstacles of cost and social or political difficulty.

The environmental realism of the evaluators was also affected by the rules. Similarities were:

1. The evaluators played societal sectors and therefore represented more closely the distribution of real social forces.
2. The evaluators, like the players, were exposed to trends and prospects within and outside the state throughout the simulation.

Dissimilarities were:

1. The evaluators did not reflect the actual mix and relative weight of the individual societal sectors of the state. It is quite possible, however, that an actual mix could never be determined.
2. The indicators were not rigorously specified, and questions were raised as to their content, mutual exclusivity, relative importance, and the uncertainty of whether each was to be evaluated in absolute terms or relatively to other states.
3. The communication between players and evaluators was in some ways better than between actual government and the public, and in some ways worse. In real life there is generally a lengthy period during which issues and prospective actions are publicly aired and opportunities for feedback and reconsideration exist. The simulation rules gave the evaluators little exposure to the players' program before it was enacted. On the other hand, the players listened to and engaged in the evaluators' discussions and assessments of the programs and events, and of their relation to the setting of the indicators. This is probably more direct and precise than real communication in Connecticut.
4. The evaluators could not vote the players out of office.
5. The evaluators had no mechanism for confrontation.

Simulation

The remaining difficulties in the game can be associated with the simulation itself. Among the more notable are the following:

1. The evaluators were unable to cope with all of the events and actions in the time available. This led to many apparent inconsistencies in subjective assessments and some bewilderment on the part of the players.
2. The evaluators in many cases were portraying roles too removed from their own backgrounds for them to be completely effective. This may be improved in subsequent simulations by use of participants who are better acquainted with the values of the groups whose roles they are to play.
3. Competition between the two player teams, which were being evaluated against the same indicators even though each had different policy instructions, tended to turn the simulation into a competitive game. The players also sensed the overburdening of the evaluators and resorted to oral presentations of their action packages. As the simulation proceeded, their presentations sounded more and more like election campaign oratory.

SIMULATION RESULTS

The game was played on June 6–7, 1969, at IFF in Middletown, Connecticut. In the sense that the exercise was designed to provide an experimenta test of the simulation methodology, it was successful. Twenty-two representatives of state agencies, universities within the state, and other concerned groups participated. Many shortcomings were evident, and some were serious.

After the objectives and procedures of the game were presented to the participants, the evaluators began the conference with an initial setting of the economic and social indicators. They described their constituencies' satisfaction and dissatisfaction with each indicator, and the players were invited to ask questions.

The player teams then set about developing their action programs for the 1970–80 period in accordance with their different policy instructions. In the initial deliberations they decided which indicators needed improvement in view of their goals and the remarks of the evaluators when the indicator values were set. The deliberations were then reviewed in light of the probabilities of the external events and their likely impacts, and as a result the players formulated an initial action program that included the formulation of new actions.

After "pricing" the new actions contemplated by the players, the evaluators were presented with the results of the probabilistic assessments of certain external events occurring during the 1970–80 period. During this time the player teams pared their action programs down to budget. The review of the external events by the evaluators and the finalization of the action programs were completed at the same time so that program review and reappraisal of the indicators for 1980 could begin without any group of participants being idle. The results of the appraisal were two possible futures for Connecticut, one for the GSP team and another for the QOL team, both of which would be embedded in the same external environment. These futures represented the two different starting points for the second round, 1980–90, in which the process just described was repeated.

The indicator satisfaction levels, which became the point of departure for the simulation conference, are presented in Table 1. All but the per capita GSP were established by the evaluation as described above and in accordance with the game rules presented earlier; the gross state product indicator started with the projected 1970 value for Connecticut. In interpreting these indicator values it must be remembered that the evaluators, in role playing, had to produce ratings based upon a potentially highly biased and unrepresentative sampling of the Connecticut population.

Several observations can be made from these values. First, they are low. While there was general agreement that Connecticut compares very favorably with the rest of the United States, it was felt that much more could· and should be done. The actual ratings apparently stemmed from the differences between the achievements of the state and the expectations and aspirations of the evaluators. Second, the QOL indicator is higher than the average of the fourteen others, showing that the evaluators considered some indicators more important than others, and gave their contribution to the quality of life greater weight. Interestingly, among the indicators rated below average are *housing, transportation,* and *education,* all regarded by the Connecticut Delphi panel

Table 1. Indicator Satisfaction Levels.

Indicator	Initial Value (1970)	GSP Team				QOL Team			
		Changes in 1970s	1980 Value	Changes in 1980s	1990 Value	Changes in 1970s	1980 Value	Changes in 1980s	1990 Value
1. Government effectiveness	50	6	56	4	60	4	54	5	59
2. Business climate	70	5	75	3	78	4	74	4	78
3. Housing	50	5	55	6	61	2	52	5	57
4. Transportation	45	6	51	3	54	4	49	5	54
5. Satisfaction with tax structure	40	−1	39	0	39	−2	38	0	38
6. Employment	70	4	74	4	78	2	72	3	75
7. Standard of living	65	4	69	4	73	2	67	4	71
8. Social climate	50	4	54	2	56	1	51	4	55
9. Control of crime	40	2	42	2	44	2	42	6	48
10. Physical environment	65	4	69	4	73	2	67	5	72
11. Health	60	2	62	6	68	2	62	5	67
12. Education	50	6	56	6	62	4	54	4	58
13. Recreation	60	1	61	4	65	0	60	8	68
14. Personal liberties	60	0	60	−2	58	0	60	1	61
Quality of life	60	6	66	3	69	2	62	6	68
Gross state product per capita	$5,000	$500	$5,500	$500	$6,000	$200	$5,200	$600	$5,800

as containing the most important issues likely to occur in this state during the remainder of the century.

Table 1 also shows the substantive progress of the teams during the two simulation rounds, noting the evaluators' assessment of each team's actions in light of the external events. The actions selected by each team in each of the two rounds are shown in Table 2. The external events and the results of the probabilistic assessments of these events are shown in Table 3. The initial probabilities of these events occurring within the 1970–80 time period are noted

Table 2. Actions Selected in Two Rounds of Play.

No.* GSP Team	Round One: 1970–80 Action Title	Minimum† Effort
1	Implement community development program on a regional basis	3
7	Offer incentives to high technology industry for migrating to Connecticut	4
17	Reform building codes to accept technological innovation	4
19	Subsidies for low-income housing	7
22	Establish new communities in rural areas	5
41	Provide incentives for urban mass transportation systems	4
45	Initiate a state income tax	2
53	Subsidize new businesses for locating in the ghettoes	2
79	Offer adult educational retraining	3
82	Institute high school work-study programs	3
89	Enlarge vocational training programs	4
90	Offer compensatory payment for scholastic achievement	3
201	Construct a new airport in eastern Connecticut	5
203	Provide free post-secondary education for all students	5
209	Keep schools open evenings and weekends	4
210	Offer subsidies for the creation of new categories of service employment	3
		61

QOL Team

4	Centralize public service and facility districts	3
5	Establish statewide regional governments	8
7	Offer incentives to high technology industry for migration to Connecticut	4
15	Place building code control at the state level	5
45	Initiate a state income tax‡	2
81	Extend the educational period	8
89	Enlarge vocational training programs	4
204	Establish a state new community development corporation	8
205	Collect all taxes at the state level	3
210	Offer subsidies for the creation of new categories of service employment	3
212	Reconstruct public educational system	8
214	Enlarge educational opportunities for adults	3
		61

* Actions numbered 1–111 were provided in the prepared deck. Actions numbered in the 200's are new actions proposed by the teams in the first period.

† Each team had a budget of 60 points for the first decade. Both the Green and the Buff programs sum to 61 effort points due to an accounting error which was not discovered until after the programs were evaluated.

‡ This is the only action of the entire game which was budgeted at more than minimum cost; it was budgeted at a level of 4.

Table 2 (cont.)

No.§ GSP Team	Round Two: 1980–90 Action Title	Minimum Effort
3	Institute a statewide government data management system	3
4	Centralize public service and facility districts	3
12	Make multiple use of highway space	1
30	Provide mass parking facilities around urban perimeters	4
52	Offer bonuses to welfare recipients who become self-supporting	3
58	Provide free day care for children up to age fourteen	3
66	Increase taxes on property use obstructive to regional development	3
71	Offer incentives for development of a system of echelon hospitals	3
77	Improve educational TV programs	2
79	Offer adult educational retraining	2
84	Acquire and use teaching machines	4
90	Offer compensatory payment for scholastic achievement	3
99	Provide more state parks	3
103	Perform research on innovative technology for creating recreational facilities	1
105	Provide free transport to recreational facilities for ghetto residents	1
302	Construct a deep-water port in Long Island Sound near New Haven	3
303	Encourage establishment of new rural and urban communities	7
304	Provide incentive for development of computerized at-home instruction	2
305	Establish quasi-public corporation to encourage new manufacturing businesses based on technological innovation	1
306	Restructure public education to make program more relevant	6
308	Subsidize groups or corporations which develop leisure time products	2
309	Initiate system of fines which would make air and water pollution uneconomical for polluter	3
		63

§ Actions numbered in the 300s are new actions proposed by the teams in the decade of the 1930s.

in Column 2; the events which the analysis concluded had occurred in this decade are in Column 3; the new probabilities of these events occurring within the 1980–90 time period are in Column 4; and the events deemed to have occurred by 1990 are noted in Column 5.

The action programs and the evaluation of these programs in terms of fourteen indicators provide a very rich source of insight into the goals and methods of the participants. A review of the actions for the 1970–80 time period shows that four of the actions were common to both teams. Both teams faced the housing issue squarely. The GSP team used the prepared action addressing new communities, and the QOL team formulated an even more costly program in this area. Both teams also dealt strongly with the issue of education. The GSP team addressed the transportation issue at the expense of somewhat less ambitious housing and educational programs than were presented by the QOL team.

In total, the 1970–80 programs for the two teams must be regarded as quite similar, with the QOL team electing to concentrate on fewer areas than the GSP team. This concentration and its attempt at greater thoroughness was not appreciated by the evaluators as much as the broader package offered by the GSP team, which can be seen to have scored higher in almost every indicator, including the QOL indicator itself.

Table 2 (cont.)

QOL Team

1	Implement community development programs on a regional basis	3
2	Establish an ombudsman's office	1
14	Rezone to decentralize urban areas	5
60	Upgrade educational requirements for police	2
62	Increase penalties for selling drugs	1
63	Require registration of firearms	4
70	Encourage health insurance programs which maximize efficiency in medical services	2
72	Provide computer services for medical diagnostics	3
99	Provide more state parks	3
101	Provide community recreation facilities	3
105	Provide free transportation to recreational facilities for ghetto residents	1
304	Provide incentive for development of computerized at-home instruction	2
305	Establish quasi-public corporation to encourage new manufacturing businesses based on technological innovation	1
308	Subsidize groups or corporations which develop leisure time products	2
311	Lower voting age to 18	1
312	Construct and *lease* housing (lease to cover cost of services and maintenance)	3
318	Develop transportation system to enable young, poor, old to get to leisure areas	4
320	Expand State highway system	5
322	Develop two new major airports with complete new communities (including employment opportunities)	8
323	Eliminate substandard housing through acquisition and demolition or restoration	3
328	Develop new underground drilling techniques	2
331	Intern field training programs for government personnel	2
332	Make Connecticut the national data-bank center	4
		65

The player action programs in the second round are much more diffuse than the first round programs and probably cannot be characterized clearly as having particular areas of emphasis. Both programs for round two have more low-priced actions and more new actions. Perhaps the most striking feature of the 1980–90 programs is the noticeable emphasis both teams put on *recreation* and *physical environment*, two indicators that had been largely ignored on the first round. As noted in Table 1, five of the second round actions were the same. Each team also adopted one other action that had been selected by the other team for the previous decade. Other actions, concerning airports, restructuring the educational system, new highway systems and so on, had generally common elements that varied in approach and level of effort. The relatively small differences in total programs seem to be reflected in the relatively close assessments of the indicators at the end of the 1980–90 decade.

The task of the evaluators in assessing these actions was far more complex than the players' task of preparing the programs. The evaluators had to assimilate the external events and mentally project the impact into the scenario that led to the initial indicator settings. Consideration also had to be given to whether the external events made certain actions feasible, more desirable, and so on. Evaluation of the action programs required comparing the relevance of

Table 3. Probabilities of Occurrence of External Events.

External Events	Probability of Occurrence in the 1970s	Occurred in the 1970s	Probability of Occurrence in the 1980s	Occurred in the 1980s
1. Multistate regional authorities	.80	yes	.90	yes
2. Average retirement age: 55	.10	yes	1.00	
3. Advanced building codes	.60		.80	yes
4. Nuclear disarmament	.20		.55	
5. Biological weapons	.30		.20	
6. Wars not directly involving the United States	.80	yes	.80	
7. Limited wars directly involving the United States	.30		.35	
8. Credit card economy	.70		.85	yes
9. Business recession	.20		.30	
10. Continued inflation	.60	yes	.45	yes
11. Educational network	.30		.45	
12. Data banks	.60		.70	yes
13. Household robots	.20		.40	
14. Thirty-five hours worked per week	.30		.55	
15. High I. Q. computers	.30	yes	1.00	▭
16. Behavior control	.20	yes	1.00	▭
17. Cheap power from controlled thermonuclear power	.10	yes	1.00	▭
18. Limited weather control	.30		.50	yes
19. Antismog devices	.20		.65	yes
20. High-speed mass transportation	.10		.65	
21. Hypersonic transport	.00		.20	
22. Aging control	.10		.15	
23. Sex choice	.20		.45	
24. Personality control drugs	.20		.55	yes
25. Relative decrease in the defense budget	.10	yes	.65	yes
26. Mass produced housing	.40		.75	yes
27. Urban riots	.20		.20	
28. European SST	.30		.30	
29. Nasa budget cut	.20	yes	.20	
30. Oceanography program	.20		.50	
31. Increased crime	.30		.50	
32. New cities programs	.20	yes	.60	yes
33. Negative income tax	.40		.55	
34. Extended vacations	.00		.20	yes
35. Multihome families	.10		.25	yes
36. Women in the labor force	.30		.30	
37. Negro political control of cities	.50	yes	.40	yes
38. New family arrangements	.05		.10	
39. Expanded federal welfare programs	.20	yes	1.00	▭
40. Leisure expenditures double	.40	yes	.60	yes
41. General immunization	.20		.50	yes
42. More autos per family	.20	yes	1.00	▭
43. Automated purchasing	.30	yes	1.00	▭
44. Low cost portable telephones	.00		.10	
45. Growth of air traffic	.10		.30	yes
46. Abandonment of low status jobs	.40		.40	
47. Decrease in tobacco consumption	.30		.50	yes
48. Limited genetic control	.00		.10	
49. Automated plebiscite	.00		.10	
50. Growth of public service professions	.20		.20	

Key: "yes" = Event occurred in decade shown.
▭ = Event occurred in the previous decade and had sustained effects.

the action to the need, comparing the magnitude of the action to the need, assessing whether the action would improve or be harmful to the situation, and determining a confidence level associated with practical implementation of the action program. To make matters worse, the evaluation had to be made using fourteen societal indicators which were not rigorously defined, and two different team programs had to be assessed.

One prerequisite of a successful simulation is to have a clearly defined area to be simulated. Both the simulation designers and the participants had difficulty with the always complex and occasionally ambiguous relationship between federal, state, and local governments. This kind of ambiguity may result from the difficulty in thinking of the state as a sharply defined, independent, social and economic entity. This problem was made worse by awareness of the ambivalence of the people living in the state, some regarding themselves primarily as citizens of the state, and others regarding themselves as primarily citizens of the country and only secondarily citizens of the state. These and other factors make a state a very difficult entity to simulate.

After the conference, the participants were invited to offer suggestions, comments, and criticisms in which defects of the game were listed and reviewed. There was a strong consensus that the conference has been useful and that this type of research should be continued. With the results of this simulation and these promising innovations in design and play, it may be possible to improve future activities of this kind. Among the suggestions that seem to offer possibilities for improvement in the realism of the game and the utility of its results are the following:

1. Including on the panel actual members of the sectors represented.
2. Defining the list of indicators more precisely, and, where possible, making them quantitative, not subjective (substituting, for example, *crime rate* for *control of crime*). In any event, making the numerical and verbal scales more consistent possibly and using fewer indicators. Determining aggregate indicators, *per capita GSP* and *Connecticut QOL* in the present game, as functions of the levels and changes in the other indicators.
3. Allowing participants to propose new indicators at the start of the game before the initial indicator settings are made.
4. Introducing statistical background material on the recent history and current status of the indicators.
5. Changing the roles of the evaluators. Sharpening the definitions of the roles and introducing new roles.
6. Giving the evaluators more time, more interpretive information, or distributing their work so they could consider the more subtle interrelations and implications of some of the events and actions.
7. Defining the actions to include quantitative descriptions of the programs.
8. Integrating the external events and their cross-impacts into the game in a better way.
9. Facilitating the mental leap into the future, possibly with audiovisual displays and written or verbal scenarios.

10. Introducing a number of political realities into the simulation, which might include allowing the teams to raise taxes to produce greater revenue for the next round, introducing budget cuts just before the programs are presented, using a format in which the evaluators would elect a team to office on the basis of their platform, using two-year planning periods, and forcing the two teams to produce a compromise program that incorporates features from the proposals of both.

11. Eliminating the sense of competition between the two teams or incorporating it into the game in a productive way.

12. Including the opinions, insights, and suggestions of people who are not represented by current consensus views to avoid a framework of reinforced consensuses, since the backgrounds of the people who produced the initial list of suggested actions and of those who participated in the game were too similar to allow serious study of the effect of innovation on the future.

A number of useful suggestions were made concerning technical details of the game.

In summary, the game did demonstrate that the format could serve as a teaching tool, promote communication between players, provide an orderly framework for the consideration of alternative potential political actions, and spark debate on issues of concern to various sectors of society. The game in its present form did not provide a significant forecast of the future of Connecticut or provide detailed insight into the effects of alternative policies. In spite of this, and perhaps most importantly, the game is a forum in which planners and forecasters can learn a great deal about how they can support each other in pursuing a brighter society.

ORGANIZING FOR TECHNOLOGICAL FORECASTING

PART FIVE

Although many industrialists and government officials have been inclined to apply technological forecasting to their concerns, their dilemma has been to organize the effort. Because widespread attention to this decision-making aid is so new, there has been little experience to draw upon. This section describes forecasting applications to functions and organizations on an increasing scale of effort.

Lt. Colonel Joseph Martino describes how the R&D manager can apply technological forecasting to make his plans more pertinent to the technological advancement needed by his organization. Two Honeywell authors, Dempsey and McGlauchlin, then show how they used a technological forecasting exercise to sharpen the relevance of central R&D laboratory work to the needs perceived in the various divisions of their corporation.

A most encouraging chapter has been provided by Richard Davis of Whirlpool Corporation. He explains how this consumer products firm has achieved a solid record of product and policy successes, based on its organization of a technological forecasting program. Examples and techniques are involved. For the large-scale government agency, Cetron's description of the U.S. Navy's first technological forecast will be equally useful.

This book closes, appropriately, with the far-reaching views of Eric Jantsch. This pioneer in the field has gradually shifted his studies to forecasting on the highest policy levels of national and multinational programs. To make the most of the future, new organizational forms for forecasting will be required.

Technological Forecasting for Planning Research and Development

JOSEPH P. MARTINO

The R&D manager, faced with planning an R&D program, must reckon with three types of uncertainty: technical uncertainty, target uncertainty, and process uncertainty. The primary role of the technological forecast as an input to planning is to help convert uncertainty to risk. The author shows that by providing the R&D manager with quantitative estimates of reasonable goals and the probability of reaching them, technological forecasts can make the whole planning process more quantitative and less intuitive.

The proper role of the technological forecast in the planning process is as an *input* to a *plan*. It is not in itself a plan. The technological forecast is a statement of possibilities, of capabilities which could be achieved. The forecast contains no implication that a decision has been made to realize these possibilities. The forecast does not imply a commitment to allocate resources. The forecast may include a statement of the resources which would be needed to achieve the capabilities projected, but the decision to achieve the forecast capabilities and the decision to make the needed allocation of resources represent a later stage in the planning process. The role of the forecast is to state what could be done, and possibly to state the resources which would be needed to do it. As such, it is only one of the many inputs needed before decisions can be made and resources allocated. Other inputs required include estimates of the resources available and the benefits from the achievement of the forecast capabilities. Discussion of these would take us too far afield.

The R&D manager must plan for the total package of elements which make up his R&D program. The elements included in this package are the facilities, funding, technical program itself, and manpower to carry out the technical program. None of these are static in nature. On the contrary, each element in the package, as well as the interrelationships among the elements, is dynamic. The plan must take into account this dynamic nature and assure that the changes in these elements over time are in accord with the goals and interests of the organization within which he works.

The facilities planning problems of the R&D manager are unique within the industrial setting. The usual facilities problems in industry are concerned with the efficient conduct of standardized and repetitive operations. Problems such as plant layout, work flow, materials handling, and storage of raw and semifinished products are dominant. The R&D manager, on the contrary, is concerned with one-of-a-kind facilities for the conduct of one-of-a-kind operations. He must assure that his facilities are updated when needed, that the existing facilities still serve a useful purpose in terms of his R&D goals, and that requirements for new facilities are recognized far enough in advance that they can be ready when they will be needed.

In the area of funding the R&D manager must concern himself not only with the flow of funding to continue his R&D program but also with the availability of funds to exploit the results of his program. A deficiency in either area could severely cripple his effectiveness. In particular he must be aware of the tendency to cut the R&D budget when his organization must reduce its expenditures and begins searching for "controllable" items. He must be in a position to show that a proposed cut in R&D is really being made at the expense of a specific organizational goal and that postponement of the R&D will result in postponing achievement of that goal.

Regarding the technical program, the R&D manager must very carefully work out a long-range plan and a set of short-range plans. For the long-range plan he must select major areas of science and technology in which his R&D organization is to build up a special competence. This choice must be based on the relationship between expected results in these areas and the goals of his organization. For the short-range plans he must select a menu of projects within these areas of competence. These projects must be selected to achieve R&D goals in an efficient and expeditious manner.

Finally, manpower planning involves a twofold aspect. First, the R&D manager must assure the needed flow of scientists and engineers with the specialties required by his long-range plans. This means recruiting the right mix of people. It also means planning the career development of those people once they are recruited. In addition, it means purging the R&D staff of those people who will no longer be required because of planned phase-downs in effort in certain areas. Second, the R&D manager must plan for motivation of his R&D staff and the incentives required to increase the likelihood of achievement of his R&D goals.

In planning for all the elements of his program, the R&D manager is plagued by the problem of uncertainty. This is true of every planner, of course. But the uncertainties facing the R&D planner somehow seem more acute. He has not only the uncertainty about demand, which faces every production planner, but also uncertainty about supply, since he often is not certain just what will come out of his program. In the face of this uncertainty, many people tend to throw up their hands. Some even go so far as to claim that R&D is inherently unplannable; the uncertainties simply preclude rational planning.

This argument is logically inconsistent, of course. Even the condition of "no plan" represents a plan of sorts. However, those who argue this way have missed the connection between uncertainty and planning. If there were no uncertainty, if the future were absolutely certain and fixed, there would be no need for planning. The fact that the future is neither certain nor fixed, however, opens up choices to us. Exercising these choices inherently involves planning. Hence, uncertainty is the basic reason for planning. It should never be allowed to become an excuse for not planning.

Most of the uncertainty facing the R&D planner involves, in one way or another, the question of the future progress of technology. It is because of this fact that technological forecasting plays an important role in R&D planning. While it may be of value to any planner, it is of particular value to the R&D planner, since it deals with the very stuff of his future problems and product.

Exhibit 1 indicates schematically the way in which a technological forecast can be used as an input to R&D planning. This schematic represents the manner in which a technological forecast is used in the U.S. Army. It is taken from Army Materiel Command Pamphlet 705–1. There is nothing unique to the Army in this process; the manner of use would be the same in almost any large organization. The Army, however, has a formally structured planning system, in which the technological forecast plays an integral part, which has been in operation for many years. Hence it serves as an excellent example of the role of technological forecasting in R&D planning.

We start with the technological forecast in the upper left of the figure. In the Army, this is a formal and official document produced and updated periodically by the staff office charged with this continuing responsibility. The forecast is a statement of potential capabilities in the field of weapons, equipment, vehicles, and other areas of concern to the Army. In the process which leads to development of national objectives, DoD (Department of Defense) objectives, and Army objectives, the Army makes use of its technological forecast. Hence the forecast is shown as an input to the formulation of these objectives. However, the objectives, once formed, in turn impact on the forecast

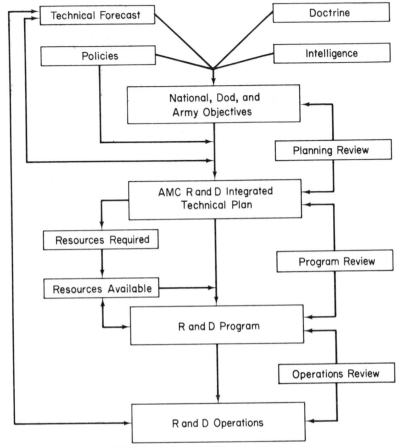

Source: AMC Pamphlet 705–1, Vol. 1, January 1966.

Exhibit 1. R & D Planning Interrelationships in the Army.

itself. National, DoD, and Army objectives will result in the emphasis on certain types of activities and lesser emphasis on others. Thus the technological requirements may be altered. A technology which does not appear to support an important objective is unlikely to get support and, hence, may not advance as rapidly as it could with higher support. Thus the forecast for that technology may need to be modified downward. Similarly, a technology which appears to support an important objective may move faster than it has in the past, calling for an upward revision of the forecast.

On the basis of national, DoD, and Army objectives and the current technological forecast, the Army Materiel Command prepares its integrated technical plan for R&D. This includes a general plan for all the elements in the package making up the R&D program. When the decision is made to allocate resources to these elements of the package, the result is the R&D program of the Army Materiel Command. This program in turn leads to specific R&D

operations, covering the entire spectrum from mission-oriented basic research to weapon system development. These specific operations are often influenced by the projections made in the technological forecast. But the forecast itself, in turn, may be altered on the basis of the results of the R&D operations. Technologies may be found to progress at a rate different from that assumed in preparing the forecast; extreme difficulties may be uncovered and some lines of approach abandoned entirely, or completely new and unexpected possibilities may be opened up. All of these results must be reflected in the technological forecast.

As indicated by this illustration, the technological forecast is not a static document. It influences the goals of the organization and, in turn, may be influenced by them. More particularly, it not only is used in formulating an R&D plan but, in turn, may be altered on the basis of that plan or on the basis of the results of the R&D activities finally carried out.

The schematic in Exhibit 1 gives only an overall picture of the use of a technological forecast as an input to an R&D plan. We will now take up in more detail the manner in which a technological forecast can be used in specific types of R&D plans. To do this, we need to examine in more detail the types of uncertainty facing the R&D planner.

The R&D planner is faced with three types of uncertainty: technical, target, and process uncertainty.

Technical uncertainty deals with the question of whether the technical goals of a project can in fact be met. A technical development project can be viewed as a search for information, information about how to build a device to perform some function. At the outset it is never completely certain whether a given set of goals can be achieved. If achievement were certain, there would be no need for the development project.

Target uncertainty deals with the question of whether or not, even if the technical goals were met, the device would function satisfactorily in the operational environment. Expressed another way the planner is never certain whether or not the future needs of the ultimate user have been translated accurately into a set of technical goals. It may be that some factor omitted from the specification through some oversight will turn out to be more important than all those included.

Process uncertainty deals with the question of whether or not the R&D organization is properly structured to carry out a specific R&D program.

The importance of these types of uncertainty will differ with the type of R&D activity engaged in. A technology advancement program, for instance, is most concerned with technical uncertainty. Reduction of this uncertainty is, in fact, the reason for undertaking a technological advancement program. Target uncertainty is a lesser question, and process uncertainty is relegated to third priority. In product development, however, the order of importance of the priorities is reversed. Getting a product on the market on time involves the coordination of many people and many interrelated activities. Hence the question of proper organization looms large, and uncertainty is never entirely eliminated before the project is completed. Target uncertainty is of lesser importance, and technical uncertainty should largely have been resolved before commitment to product development. At a stage between these two, where emphasis is on a working prototype for field testing, target uncertainty becomes the top priority item, with technical uncertainty second and process uncertainty still lowest. If a prototype development program has been preceded by an adequate technology advancement program, then the most important question to be answered by a prototype is that of whether or not a device meeting the given set of specifications will, in fact, turn out to serve the user's needs in the environment in which the user will have to operate it.

Having identified the types of uncertainty which the R&D manager faces, we can now examine the role of the technological forecast. We can define uncertainty as an unquantified lack of knowledge about an outcome or result. Uncertainty might be characterized as large or small, but for our purposes that will be the extent of the possible quantification of uncertainty. We will use the term *risk* to describe a quantitative measure of the degree of probability of risk or failure. It may be argued by the purist that risk is a subset of uncertainty. However, for our purposes it will be more useful to distinguish between them.

We can now express more clearly the role of the technological forecast in R&D planning. Its function is to convert uncertainty to risk. The role it plays is to give the R&D manager a quantitative estimate of the likelihood of an event as opposed to the unquantified uncertainty he faces without the technological forecast.

Technological forecasting is of greatest value in helping convert technical uncertainty to risk. It is of somewhat lesser value in helping convert target uncertainty to risk and of very little utility in helping convert process uncertainty to risk. Hence, its role will be different in planning for different types of R&D activity. The technological forecast plays a much bigger role in planning a technology advancement program than it does in planning a product development program. However, its use is still important in the latter case as we will see below.

In the remainder of this paper, we will examine the application of technological forecasting to technology advancement programs and to product development programs. We will use some specific examples to illustrate the approaches involved.

PLANNING FOR TECHNOLOGY ADVANCEMENT

The R&D manager concerned with a technology advancement program has two basic questions: (1) What general areas should we be working in and (2) what specific projects, with what specific goals, should be undertaken in those areas? The R&D manager does not want to expend resources exploring an area which is already nearly worked out and which does not have much potential for additional advancement. When he sees that an area will no longer be fruitful, he wants to initiate work in another area, building it up as the first phases down. However, his choice of a new area should be based, not on a fad of the moment, but on the needs of the larger organization of which he is a part. Thus he must identify the requirements for new technology which this larger organization will have at the time his program is reaching fruition. Technological forecasting can be very helpful in making this identification. Once the areas of effort are decided, whether they are new or a continuation of areas previously worked, it is necessary to set goals for projects. These goals must neither be so ambitious that failure is virtually certain nor so near at hand that an aggressive competitor can exceed them easily. Technological forecasting can also play a major role in setting goals for individual projects.

We will now consider an example of the choice of an area in which to pursue technology advancement on the basis of a forecast. We will consider the currently very popular topic of air pollution. One aspect of this problem is shown in the following table, which is extracted from Bureau of Mines Circular 8384, *An Energy Model for the United States.*

U.S. Consumption of Gasoline for Transportation

Year	Consumption (Millions of Barrels)
1947	787.2
1950	984.3
1955	1323.2
1960	1496.6
1965	1720.2
1980*	2677.0

* Forecast.

The forecast calls for an increase between 1965 and 1980 which is approximately equal to the increase already seen between 1950 and 1965. That is, by 1980, the amount of gasoline consumed for transportation will be more than half again as large as today's consumption. The implications of this forecast for air pollution in urban areas, as well as densely populated areas such as the stretch of Atlantic seaboard between Boston and Washington, D.C., are readily apparent. An already barely tolerable situation will become intolerable.

Suppose now the manager of an industrial laboratory wonders what is in it for his company from the standpoint of reducing pollution. Does the forecast existence of this problem have any implications for new technology from which his company might profit and on which his laboratory should therefore get busy long before 1980?

Starting from the existence of a problem, that of pollution from automobile engines, the R&D manager might proceed as shown in Exhibit 2. Here we

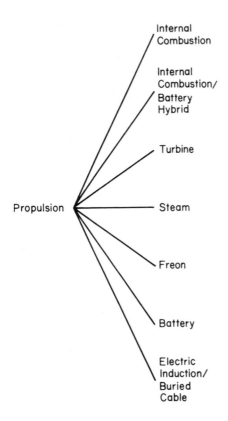

Internal
Combustion

Internal
Combustion/
Battery
Hybrid

Turbine

Propulsion — Steam

Freon

Battery

Electric
Induction/
Buried
Cable

Exhibit 2. Possible Propulsion Systems for Low-Pollution Cars.

have the beginnings of a relevance tree which lists possible approaches to a low-pollution automobile. To start with, the internal combustion engine might be improved so that pollution can be minimized while the current advantages of the internal combustion (high power-to-weight ratio, good acceleration characteristics) can be retained. An alternative possibility is the use of a hybrid propulsion system wherein a small internal combustion engine charges a battery. Since the engine could operate at constant load and would not be subject to acceleration and deceleration, the exhaust pollutants would be reduced considerably. Another possibility is a turbine engine. An external combustion engine using either steam or Freon as a working fluid represents another possibility. Or one might consider a purely electric car operating on batteries. Finally, a possibility is dispensing completely with energy storage in the car and obtaining power by induction from a buried cable. Each one of these possibilities delineates a major area of work in which technology advancement would be beneficial from the standpoint of reducing pollution. The R&D manager must now select one or more of these areas to initiate a technology advancement program.

It must be kept in mind that an attractive technological possibility is not alone a sufficient reason for the R&D manager to initiate a technology advancement program. In fact, many of the factors he must take into account are nontechnical in nature, and he must seek help from other officials of the company before making a selection. Let us suppose, for instance, that he feels there are good technical reasons for working on steam engines. Prospects of technological advance are high; he already has some good people with the right skills; he can draw on existing test facilities. Before making a positive decision regarding this technological area, however, he should also determine whether his company can produce steam engines or even wants to produce steam engines. Can his company market them and provide service for them? Is the company willing to acquire a marketing and servicing capability if it does not now have one? Discussing this in more detail would take us too far afield. The essential point is that if the company is to make a worthwhile profit on the results of its R&D program, that program must come up with products the company can make, sell, and service. If the products do not fit the company's operations, then its only return on the investment in R&D may be licensing. While licensing of by-products and incidental inventions is a valuable source of revenue, the primary interest of an industrial R&D laboratory is products which the company can exploit itself, not merely license.

Let us assume further, then, that the company is in the electrical products business and that should an electrically powered car become practical, the company could make and sell a variety of components for such a vehicle. Hence, it appears that the battery powered automobile would be a fruitful area for a technology advancement program. This is still a big area and needs further refining.

Exhibit 3 illustrates a refinement of this area of technology. The propulsion system of an electric car contains three major electrical subsystems. First is the battery itself, which must be cheap, have a high energy density (kilowatt hours per pound), and be capable of accepting a rapid recharge a large number of times. Next is the electronic controls which control flow of energy from the battery to the drive motors. These must be cheap, reliable, and simple to operate and maintain. Finally, there will be one or more drive

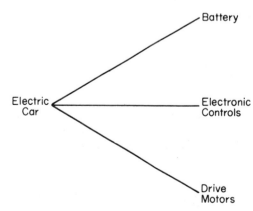

Exhibit 3. Major Electrical Subsystems of an Electric Car.

motors in the car which must be lightweight, have adequate torque and speed, and be reliable and long-lived. We will assume that the company could manufacture and market any or all of these items should the laboratory manage to make a significant contribution to the state of the art.

This illustrates the process by which an R&D manager can utilize a technological forecast as input information for selecting major areas for a technology advancement program. In this case we took a forecast of gasoline consumption and recognized that with current technology an intolerable problem would exist. Then by turning to a normative method (in this case a relevance tree), we derived several areas in which technology advancement programs might be worthwhile. Some of these could be eliminated on the grounds that the company was not set up to exploit them even if the programs were to be successful. Our hypothetical R&D manager was then left with a number of areas in which he could initiate technology advancement programs, any of which would be beneficial to the company if it succeeded. The R&D manager must now make choices on the basis of the total resources at his disposal, the existing laboratory staff and its facilities, and so on. He still has many decisions to make in selecting among possible areas of effort. Technological forecasting would be useful in making many of these decisions also. However, we will terminate the illustration of choice of areas here and take the question of projects.

Suppose now our R&D manager decided that batteries might be one good area for his laboratory to work in. He selects a small team from his professional staff and asks them to look into the possibility of initiating a project to develop an improved battery which might be suitable for an electric car. This team looks into the area and reports back that they favor the initiation of a battery project. The laboratory already has most of the needed skills available so additional hires would be minimized. Floor space and instrumentation can be made available from a project being phased out. Finally, some recent findings reported in the basic research literature indicate that new approaches to batteries are feasible, hence the project has a good chance of success.

Assuming the R&D manager decides to approve the recommendation of his team and initiate a development project for a new battery, he still has a problem. What goals should be set for the project? What should be the objective for energy density in watt hours per pound? How many charge and discharge cycles should the battery be capable of? What should be the maximum allowable charging time after a deep discharge? It is not sufficient to tell the development project team: Do as well as you can in each of these areas. The R&D manager must establish goals for the project, goals against which he can measure progress. He must set these goals high enough that accomplishment is worthwhile but not so high that success is out of reach. How can he decide what the goals should be?

To keep the example simple, we will restrict our attention to the energy density in watt hours per pounds. The approach to the other parameters would be similar to this, and examining them would not add appreciably to the utility of the example as a vehicle for explaining the use of technological forecasts.

Figure 12 of Ayres' paper in Bright, *Technological Forecasting for Industry and Government,* reproduced here as Exhibit 4, gives a forecast of energy density for secondary batteries through 1975. For 1975 the forecast appears to call for a level of about 450 watt hours per pound. Our hypothetical R&D manager then would be well advised to set his project goal at this level. If he sets it significantly lower than this level, he runs considerable risk that his development project will actually do no better than the goal set while his competitors, responding to higher goals, exceed the achievements of his laboratory. Conversely, if he sets the goals too high, it is likely that his development team will respond on an "all or nothing" basis, trying only those approaches which, if successful, will come near achieving the goal but which have a low probability of success. He thus runs an extremely high risk of having no useful results come out of the project.

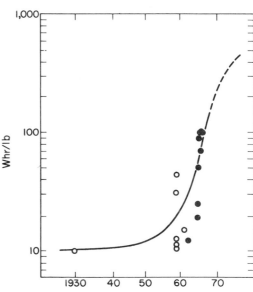

Exhibit 4. Energy Density of Secondary Batteries.

Source: R. Ayres, "Envelope Curve Forecasting," in *Technological Forecasting for Industry and Government,* J. R. Bright, ed. (Englewood Cliffs, N.J.: Prentice-Hall, Inc., 1968), p. 91.

Thus again we see that a technological forecast can be a valuable input to the decision on selection of project goals. A forecast of what is a reasonable possibility provides a useful goal to shoot for. Setting a goal which deviates too much from what is forecast to be possible increases the risks of getting nothing at all or getting an achievement which will be surpassed easily by the competition.

These examples have illustrated the use of technological forecasts in the planning of technology advancement programs. The primary use of technological forecasts in technology advancement programs is to help decide what areas should be pursued and what the specific goals of projects in those areas should be.

PLANNING FOR PRODUCT DEVELOPMENT

When a device reaches the stage where it is ready to be introduced to the market, most of the technical uncertainties should have been resolved by prior technology advancement programs. The target uncertainties should have been resolved through field trial of one or more prototypes. As indicated above, the remaining problems are usually of an organizational nature. Hence technological forecasts play a much lesser role in planning for the development of a product to be placed on a market than they did when the same device was still in the technology advancement stage.

There is, however, still an important role for the technological forecast as an input to the planning for a product to be marketed. A decision must be made regarding the level of functional capability which the product should have. There is usually some range of levels of capability which the product can be designed to meet. The higher the design level of capability, the more competitive the product will be and the longer it will be before the product is rendered obsolete by a competing product. The higher the design level of capability, however, the greater the risk that the development program will be unsuccessful or that introduction on the market will be delayed. This relationship between risk of early obsolescence and risk of technical failure is portrayed graphically in Exhibit 5.

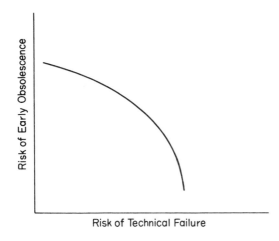

Exhibit 5. Relationship Between Risk of Obsolescence and Risk of Technical Failure.

Here again the role of the technological forecast is to convert some or all of the uncertainty to risk. The R&D manager faced with a decision about the level of capability for which a new product should be designed needs some quantitative estimates of the risk of obsolescence and the risk of failure.

Exhibit 6 illustrates one approach to estimating the risk of obsolescence. Here, for illustrative purposes, we assume that in the past the level of functional capability can be described by a linear trend (this may be an exponential trend if the level of functional capability is plotted on a logarithmic scale). The forecast is simply an extension of this past trend. Using standard statistical techniques, we then plot confidence limits about the forecast. However, we label these in a manner somewhat different from the usual method. Usually, these confidence limits are labeled in terms of the probability that a device introduced in a particular year will have a level of functional capability falling between the limits. Here, instead, we have plotted the probability that the level of functional capability will fall above the confidence limit (for example, the upper ninety percent limit is relabeled five percent; the lower ninety percent limit is relabeled ninety-five percent).

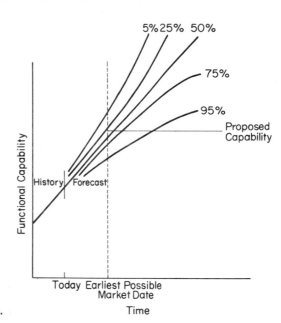

Exhibit 6. Estimating the Risk of Early Obsolescence.

By then drawing a horizontal line showing the level of capability proposed for the product, we can obtain a more quantitative estimate of the risk of early obsolescence. For the year in which the "proposed capability" line crosses the ninety-five percent confidence limit, the probability is ninety-five percent that a competing device introduced then will exceed the level of capability of our product. (Note that this is *not* the cumulative probability that a superior product will be introduced by that date. Calculating that would take us too far afield.) Conversely, for the years near the "earliest possible market date," we can also determine the probability that competing devices introduced then will be superior. This plot can give us an estimate of the risk of early obsolescence associated with any proposed level of capability.

Exhibit 7 shows the other half of the story. This plots the probability of achievement versus possible market dates for a given proposed capability level. The basis for this plot, of course, must be historical data from previous products. If this is unavailable, subjective estimates obtained from the R&D staff, possibly through use of Delphi, can be helpful. This plot can give us an estimate of the likelihood of meeting a specified market date for a product with any proposed level of capability.

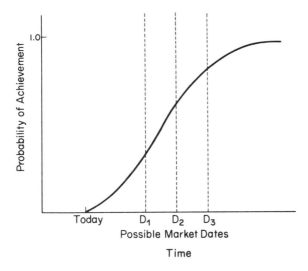

Exhibit 7. Estimating the Technical Risk in Meeting Possible Market Dates.

It must be recognized that these forecasts do not in any way force the decision of the R&D manager regarding the design level of capability for the product. It is still his task, along with managers from Marketing, Production, and other areas involved to make this decision. The point is that with the information from these technological forecasts, he can make a more precise estimate of the risks he is taking with any given design level of capability. It is still his job to make the appropriate trade-off between the risk of early obsolescence and the risk of technical failure. The technological forecast helps him through providing a better estimate of just what these risks are.

Since the primary uncertainties differ with different types of R&D programs, the role of the technological forecast will differ also. In a technology advancement program, the technological forecast can provide information about major areas of technology which should be investigated and can help in setting goals for specific projects within those areas. In a product development program, the technological forecast can provide quantitative estimates of the risk of technical failure and the risk of early obsolescence associated with any proposed design level of capability for the product.

In no case does the technological forecast force the decision of the R&D manager. It is still his responsibility to make the various trade-offs involved in setting objectives and making plans for his R&D program. By giving him a quantitative estimate of reasonable goals and prospective risks, however, the technological forecast makes his task considerably easier.

Technological Audits:
An Aid to
Research Planning

LAURENCE D. McGLAUCHLIN

How can the corporate research center of a large multiproduct corporation develop and maintain a research program that serves its divisions effectively? Honeywell's approach includes an unusual "customer" (meaning company division) evaluation weighing of major research projects, based on a corporate-wide technological forecasting effort, supplemented by inputs from managers, accountants, and planners. Central to this procedure is the utilization of the Delphi method for technological prediction. Delphi results were circulated for executive and division management assessment and reaction. The whole procedure forms an interesting way of improving the relevance of research efforts and improving communications about technical efforts.

Almost every industrial laboratory would like to have an effective method for planning and evaluating its research program. Planning the best use of our scientific talents is an important task. It is important for individual industrial firms. It is important for our nation, and the improved utilization of our scientific talent could benefit all of society. Better research planning may be the key to survival for many organizations known as technology based companies. Technological forecasting is the key to better research planning.

We will discuss four major topics. First, we will indicate why research planning is inherently difficult. Second, we will describe the method we have evolved for planning our research at Honeywell. Third, we will discuss how we have made technological forecasting an integral part of this planning activity. Finally, we will indicate some of the benefits we have derived by doing a technological forecast and then using it to prepare a research plan.

Now turning to our first topic, as we see in Exhibit 1 research planning has certain difficulties which arise because of the intrinsic nature of research. Research is a quest into the unknown and, therefore, highly uncertain. This is not to say that the purpose of the search is unknown. People everywhere are beginning to accept the notion that research can have, should have, *must have* a clear, explicit goal. The direction and the manner of the search should also be known. What is unknown is the redirection which may become necessary because of what lurks beyond the first corner in the road. For this reason schedules and other plans must be tentative, subject to periodic modification.

Exhibit 1. Difficulties of Research Planning.

A. Feedback is long delayed

B. Violates tradition

C. Identification with present products

D. Influence of the external world

Professor Donald Marquis at the Sloan School of Management has stated that research planning is difficult because research is not repeated. This is in marked distinction to production activities. He further points out that feedback is delayed for a long time. Financial planners and forecasters often get feedback once each quarter. Sales data are available monthly in many cases. Production data are available so that adjustments can be made each day and may be available hourly. Research results usually require a period of the order of ten years before economic or social benefits can be measured and fed back to planners.

Also, there has been the feeling that research could not be planned. We refer to this by saying that it violates tradition. Some scientists resist planning. They feel that it is a corruption of their dedication to science. There was a feeling that the most valuable results from scientific research came about as a result of synergism and *not* as planned work toward a preannounced goal.

One of the other reasons that planning is difficult in an industrial laboratory is that there is a strong tendency for most operating divisions to identify their future business with their present products. Research is expected to outdo engineering in making present products achieve higher performance. For example, if a company makes gyroscopes there is a tendency to urge the research activity to make the rotors stronger, make the bearings better, reduce the angular drift rate, and so on.

Finally, research is hard to plan because any single laboratory does only a small fraction of all the research done in the world no matter what special field one considers. Thus, a planned program of research on lasers needs review and redirection more often because of the influence of discoveries made elsewhere than because of progress within the laboratory where the research was originally planned.

Despite these difficulties, we at Honeywell have been attempting to plan research for some time, and some of our experiences may be useful to others. In order to tell you how we planned, we should begin by enumerating the types of inputs utilized. First we will describe the procedure we used initially. Then we will tell you how we have changed our planning method so as to utilize some of the more recent techniques of technological forecasting.

Exhibit 2 shows that originally we solicited inputs from three groups of people: the engineering and marketing people at Honeywell's eighteen divisions, the scientists at our Corporate Research Center, and the corporation's general management at corporate headquarters.

Exhibit 2. Factors of Importance for Each Input.

A. Divisional input

1. Consensus
2. Possess temporal stability
3. Be suitable for research

B. Scientific input

1. Spotlight new development
2. Tolerate success
3. Recognize that $P_S V_R$ = a constant

C. Corporate input

1. Identify research goals
2. Specify size and growth rate
3. Provide economic and legal guidance

We addressed a letter to each of our operating divisions asking them what research they would like to have us do. A number of years ago we asked for such information on a random basis throughout the year. As a result, each division's requests or recommendations, when they were received, tended to become the focus for attention which caused the research activities either to be a succession of "fire fighting" activities *or* the division's inputs had to be given scant attention. By asking for recommendations from all our research users, all the divisions *simultaneously,* we could apportion our resources of scientific talent, equipment, and money among all of the most important needs. The task was then analogous to slicing a cake, or an even better simile would be "carving a turkey," because we had a number of different types of talent we could serve to the divisions. By knowing all of the needs, all of the applications for metallurgical research, we could see to it that those kinds of experts were spread over the most important problems.

In the old days we might have learned about a metallurgical problem at Division A in February and assigned our best talent to work on it. Then in

June when they were getting their teeth into the problem, find out that Division B had a much more important need for that same talent. Furthermore, hearing about all of the research needs at once enables us to determine regions of overlap. In other words, we are more likely to recognize several needs within the company for the same type of research. When several divisions need the same type of project, we have probably identified a topic for which the research results are more likely to benefit the company.

Now when we asked our customers, our operating divisions, for research recommendations, we gave them certain ground rules. We asked them to develop a consensus. We had sometimes found that a division's engineering department favored one type of research; their marketing group saw a different set of needs as being the most important; and the divisional manager often had some special topic of interest which was slightly different from those of either group. So, we asked for a resolution of these differences.

Next we asked that their recommendations possess temporal stability. By that we meant that they should choose needs that were going to last. It had to be recognized that the research had to be initiated; equipment modified, or in some cases, designed and constructed; and results obtained and communicated to the appropriate division. They then had to accept the results; all of which takes time. The results had to be incorporated into a development phase which might mean using them to develop a new product. A decision was needed as to where this would be done and how the work would be funded. The development then had to be started and the product developed, tested, and accepted before the design and production engineering phases could be started. Eventually quality control and field sales and service need to be brought into the act. Unless the need is quite a basic, long-standing need, fundamental to the business of the division requesting it, there is not much use to be made of starting a research program on it. During the years that must elapse before the research can lead to profit, the problem must remain important, or, as we said a moment ago, the input has to possess temporal stability.

Finally, the recommendation had to be suitable for a research type of organization. We do not have expert design engineers. Our people are not process or production engineers. Therefore, they were unlikely to be successful on a design type of problem. So much for divisional input. We also asked our scientists for research recommendations. Most companies expect their scientists to spotlight or recognize new scientific trends or new engineering developments. We expect the same thing. In order to recognize those unusual applications, they must know the company's technical problems, and they must know the company's sales force and its markets.

As we queried our scientists, we required that the projects they recommended be able to tolerate success. Perhaps we can clarify this. A research project, if successful, should not simply generate a need for more research. And yet, some research projects are like that. It is vital to ask the question: Where would we be if we were successful? What would we do next? If the answer is, a lot more research, the question should be posed for that new research. In any event, this type of thinking may run up a warning flag.

Next, we required our scientists to realize that the product $P_s V_r$ is equal to a constant. That is, that P_s, the probability of technical success, times V_r, the economic value of the result, is a constant. To see the plausibility of this, consider the following. If a program is begun on a problem so straightforward that technical success can virtually be guaranteed, that is, if P_s is essentially unity, then almost any other laboratory could achieve the same result. Many probably will. Thus, the economic value of that technical success is likely to be small because the market for resulting products will be supplied by all of the companies whose laboratories chose to begin similar research. They will all succeed.

Conversely, if a technically difficult project is selected, for example, if P_s is very small, but the research team is very talented and/or persistent, the results are likely to have substantial economic value. If any other company, seeing the result, attempts to duplicate it, they may well have a long and arduous path to follow. The technical success of the original firm provides a monopoly for a time and during this lead time margin and profits will probably be high.

The third source of guidance and recommendations for our research planning was our general management group. We asked them to indicate how much growth we might anticipate in our research budget. Knowing approximately what increase in funds and what growth rate in research staff is likely to be approved is vital. If such data are available, planning a sound program is much easier. The general managers of the company should also provide information as to which segments of the business they would prefer to expand most rapidly. In order to offer guidance of that kind, they need to take into account economic, political, and legal factors influencing overall operations. The man in charge of all research activities within Honeywell is a corporate vice president who reports to the chief executive officer of Honeywell, the chairman of the board.

In addition to actively soliciting these kinds of input, we used another procedure. We used an evaluation of our present research program as a guide for planning the future. To accomplish the evaluation we relied upon people outside the research organization. We thought it best to rely upon the judgment of selected individuals in the "user" group, that is, individuals who would be expected, some day, to make use of the research results to provide new products and services.

We therefore selected three or four people from each division who were assigned responsibility at their location for new product development or new product planning and asked them to rate our currently active research programs. If the division was large we asked four different people to rate our program. If the division was small we asked only two or three people to do the rating. We sent each voter a description of the active research program and asked him to rate it in terms of its potential value to his division. We instructed him to rely on any and all knowledge he had about each category of research. In other words, he was not to rely *only* on the program descriptions which accompanied the request for his evaluation.

There were fourteen topics, research areas, or categories of projects which were rated by individuals from eighteen different operating divisions. The ratings for any individual program could be displayed in the form of a bar graph as shown in Exhibit 3. We expected to see that programs were rated

Exhibit 3. Divisional Evaluation of Research Center Programs.

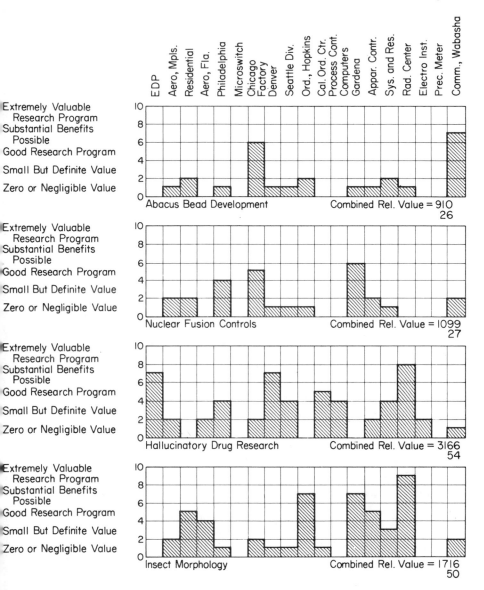

very high by one or two key divisions or were rated moderately well by a broad spectrum of divisions. In either case it would seem easy to justify research on the topic in question.

There is inherently greater risk in pursuing a program of intense interest to one division but of little interest to others. If that division's plans change, or its markets change due to the activities of a competitor, the research may be useless. On the other hand, research results which would have broad application extending over many divisions within the corporation has a markedly greater probability of providing benefits someplace within the company's range of activities and interests.

The exhibit shows two different ways of integrating the ratings given a program by all of the operating divisions. One obvious measure is to simply add the shaded squares. This procedure assigns equal weights to the judgments of all the divisions. Another method is to compute what we have called the combined relative value. For each research program the columns are arranged from left to right so that the evaluations of the largest divisions appear at the left. Each rating is multiplied by the annual sales of that division in millions of dollars. We then sum these products to obtain a ranking for that research program. The large divisions quite obviously have their preference count more heavily in such a summation.

One might argue that the small divisions need research help more desperately. Perhaps so, but it can also be argued that Honeywell has a larger stake in preserving the strength of those divisions which are already such a significant segment of its entire business. It is also true that the small divisions may have only recently been placed in operation as separate entities because they seek to exploit some newly acquired technology acquired by purchase of a small company or from research. Therefore, the small divisions may not be in a good position to assimilate still more new data and new research results. In any event both sums are available and the ratings can undoubtedly be combined in many additional ways.

Thus far we have talked about how we planned before we began to incorporate technological forecasting techniques in an explicit way. Obviously by the very nature of the research planning task, we were implicitly attempting to forecast technological change. However, attendance at conferences, such as those presented by the Industrial Management Center, made us aware of tools being used by other organizations. We will describe how we attempted to incorporate some of these additional forecasting techniques.

We were particularly impressed by the descriptions of the Delphi techniques by Helmer and Gordon.[1] We were also interested in Thompson Ramo Woolridge's use of Delphi in the paper given by North.[2]

---

[1] Olaf Helmer, "Analysis of the Future: The Delphi Method," and Theodore J Gordon, "New Approaches to Delphi," *Technological Forecasting for Industry and Government,* ed. J. R. Bright (Englewood Cliffs, N.J.: Prentice-Hall, Inc., 1968) p. 110 and p. 134.
[2] Harper Q. North, "Technology, the Chicken—Corporate Goals, the Egg," *Ibid.* p. 412.

In October of 1967 we selected a Delphi panel at Honeywell. We chose eight scientists at our Corporate Research Center and eight engineers, one from each of the larger divisions of the company. Then we picked sixteen marketing people. Most of our operations, scattered over eighteen geographical locations, were represented in the group of sixteen by one of their marketing people whose past contacts at the Research Center seemed to indicate that he was imaginative and interested in new technology. To round out a panel of forty, we selected eight more people whose functions were varied. Several were divisional new product department managers; one served as an advisor to the chairman of the board in evaluating candidates for mergers, one is a professor of physical chemistry at a major university and a consultant to Honeywell at our Research Center. These together with accountants and planners rounded out the panel.

Three of us at the Research Center planned and coordinated the exercise. After selecting the panel and before proceeding further, we traveled around the country as necessary to interview and instruct each of the panelists in his role. They had each received a letter from one of the authors of this paper asking them if they would help us to forecast Honeywell's future by participating in a Delphi exercise. All accepted. Nevertheless each was asked again if he was willing to devote the time (in most cases his personal time) to complete the series of responses.

In Round A we asked each of the forty panelists to list three events, occurrences, or developments which would have a substantial effect on Honeywell's business ten years hence. These events could be expected to divide neatly into two categories, threats to established areas of business and opportunities for new business. In either case, the individual making a prediction was asked to:

1. Describe the event in a short paragraph which would indicate what was expected, so that an average reader would be able to form an opinion as to its importance.
2. Supply data which argue for the plausibility of the predicted occurrence, so that the reader could make an estimate as to the likelihood of occurrence.

Thus, our use of the Delphi technique was somewhat different from the usual procedure in that the panelists did not merely respond to events predicted by outsiders. They first generated the data anonymously then exchanged arguments about the predictions.

The forty panelists generated sixty-four clearly separate and distinct ideas. We three processors at the Research Center had examined approximately 120 predictions and found that there were many overlaps. Many panelists were forecasting the same occurrence under different captions, supplying descriptions in different words and with variations in their supporting data. We tried to be cautious in our editing. If all three of us felt confident that separate panelists were really talking about the same event we melded them into a single write-up attempting to select the panelist's own words while condensing the data and changing things as little as possible.

In Round B we mailed a thirty-five page booklet back to the panelists. It contained the sixty-four predictions *without* identifying their authors and a few pages of instructions as shown in Exhibit 4. As instructions for Round C we sent instructions as found in Exhibit 5.

We then made up an additional set of booklets so that in our solicitation for Round D inputs we could provide forty copies of the original predictions and directly across from each of the sixty-four predicted developments the arguments both pro and con.

In Round D the panelists were asked to submit their final vote on each item together with their rebuttals of any arguments unacceptable to them.

We were able to make up a chart showing the final vote on each predicted event. There was some small but apparently significant convergence during the three rounds of voting. There was a large spread in the value assigned to the "most important events" compared to the "least important events" by the panelists. The most important events were voted twenty times more important than the items at the bottom of the list.

We prepared a summary of the results for our Executive Research Committee. It contained the five events which ranked at the top of our Delphi

Exhibit 4. Sample Instructions for Round B of Delphi.

1. Read everyone's predictions
2. Assign priority rankings:

   i. Extremely important (dire threat or big opportunity)
   ii. Important
   iii. Slight importance
   iv. Insignificant

   When you assign priorities think of Honeywell, not your division. Think of the future (approximately ten years from now). When ranking events use the following criteria:

a. Is the event (development, prediction) likely?
b. Will it affect Honeywell?
c. Can we do anything about it to protect our business or to take advantage of an opportunity?
d. Consider interactions among the predictions: if event E_1 occurs, will it raise or lower the importance of some other event, say for example, E_6?

   In other words, please return this booklet (Round A predictions) with your priority rankings indicated. Your Round B input will then consist simply of a circle around one of the Roman numerals opposite each of the predictions. Please send this to reach any one of us, airmail-special delivery.

   S A M P L E of your Round B input:
   USE OF CHIMPANZEES AS DOMESTIC HELP   I   (II)   III   IV

   As you know, your ideas have been and will be completely anonymous with respect to the other panel members. However, we must know who you are as you periodically submit information to us, so that we can properly record your participation. Therefore, please leave this page attached when you mail back this booklet because it contains your name, which is:

Exhibit 5. Instructions for Round C of Delphi.

This booklet is the one you sent as your Round B input. However, it now contains a feedback of information representing a consensus of the panel. For each item, the red arrow represents the panel consensus (simple average), while the circle you had made remains. In those cases where your rating was markedly different from the consensus, the entire block of Roman numerals is circled in red meaning that you owe the panel an argument about why you voted as you did on that item.

This, as (fictitious) examples:

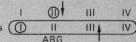

A. Electronic control of personality   I   (II)   III   IV

B. Use of telepathy in communications   (I   II   III   IV)
                                           ARG

In example A your initial rating was II, whereas the panel average was slightly above II, so that you essentially agree with the panel consensus and, therefore, need not submit an argument. In example B, your rating was far from the average so that the large circle marked "ARG" represents our request for an argument indicating why you voted that way. Please tell the others about factors they may not have considered. Cite facts and figures where possible.

Thus, for your Round C input please supply the following to us:

1. Reconsider each of the sixty-four items and mark your new rating of that item using the green pencil provided. It may be the same as before or different depending upon whether or not and how much you are influenced by the panel consensus.

2. For each of the items marked ARG state in not more than fifty words why you are above (or below) the panel's average rating.

According to your count, you _____ John Doe _____ owe the remainder of the pane sixteen   (16)   such statements.

---

forecast. The arguments pro and con for each and the rebuttals supplied by the panelists were combined into a brief discussion recommending certain actions by the company. This summary was forwarded to Honeywell's board chairman who is the chief executive officer to the president and to several group vice presidents. Thus concluded our first use of the Delphi method.

The year after this first Delphi exercise we originated a technological survey of the various divisions. A questionnaire was prepared and sent to the eighteen divisions over the signature of Honeywell's board chairman. In brief, this questionnaire asked each division to supply the following information:

1. A list of technologies that would have the greatest impact on the division's business during the coming five years.

2. A comment on whether the technology listed would threaten a present part of our business or offer an opportunity for expansion into a new field.

3. An indication of the action to be taken and the magnitude of the business affected (in either case threat or opportunity).

Each of the divisions responded, and responses varied greatly. Some divisions mentioned technological trends or developments, such as microelectronics or fluidics, while many others concentrated instead on new product needs.

Their inputs were processed by the author of this chapter.

1. Those who had named new products were singled out.
2. For each such product the key component was identified.
3. The new technologies which seemed most likely to provide a substantial improvement in that key component were then listed.
4. The technological developments listed by those other divisions who responded more appropriately were then added to the list.
5. Duplications were eliminated.
6. Technologies were ranked in accordance with the breadth of their application, for example, in accordance with the amount of business they affected.
7. Those technologies well-developed within Honeywell were temporarily set aside. Thus technologies which were important but needed strengthening within our company were identified.

A summary was prepared and presented to the board of directors.

BENEFITS OF PROCEDURES

The divisions have become much better informed about our research program. They have given more realistic thought to their own research needs. We believe that respect for the company's research capabilities seems greater.

Our research managers have come to a realization that they needed to improve communication of research plans and results to our divisions. Some were shocked to find that the intended beneficiary did not know what we were trying to do. They now recognize their responsibility to explain and interpret our programs to the rest of the company.

We now have a better balanced research program. Honeywell's corporate officers have indicated that they wanted part of our research to be exploratory, part directed to support of present products, and part directed toward the creation of new kinds of business.

Finally, we have placed a part of the burden for liaison on our scientists. Everyone likes to feel that his work is useful. Scientists' discussions with the divisions who might use their results seem to increase motivation.

Organizing and Conducting Technological Forecasting in a Consumer Goods Firm

RICHARD C. DAVIS

Many people assume that technological forecasting is only for the high science firm. This is not so. Indeed, one of the best technological forecasting efforts in industry is done by Whirlpool Corporation and relates to consumer appliances. With two solid successes and one major "assist" since 1963, Whirlpool's TF (technological forecasting) performance has been remarkable.

Davis shows how his firm organizes for TF and gives examples of their work. Basically, he uses a very strong monitoring effort supported by Delphi and trend extrapolation.

INTRODUCTION

Within the confines of this paper it is the intent of the author to present *one* method which has been used by a multiproduct corporation to introduce technological forecasting into its product planning procedure. It is not the belief of the author that the system presented here will work for every company, there being as many quirks to corporate operations as exist. It is presented here for the use of anyone who has just been handed the assignment as technological forecaster in the hope that some of it will be useful to him as he attempts to

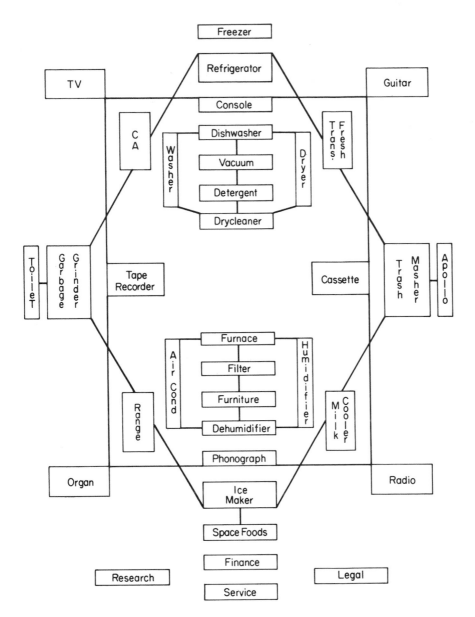

Exhibit 1. Products Made and Marketed by Whirlpool Corporation.

get a program underway. The reader is urged to use any parts which may fit his particular circumstance and to reject the remainder.

We define the general area of Whirlpool Corporation's business as "life support." Our intent is to manufacture machines, appliances, and other products which contribute to the general health, comfort, and well-being of the family within the home. At the present time we manufacture approximately forty products to achieve this goal. We categorize these products into four groupings:

1. Food: preparation, preservation and waste disposal.
2. Home comfort.
3. Home entertainment.
4. Fabric maintenance.

Exhibit 1 shows these products diagrammatically.

At Whirlpool, we divide our forecasting into two separate types. The first of these is what we call "defensive forecasts." This type is exemplified by Exhibit 2. We simply do not want our officers to make statements like this without having benefit of the latest knowledge on his subject. "Offensive forecasting," our major endeavor, I shall leave and discuss later. Generally, offensive forecasting is more product oriented.

Thus, to answer the question, Why forecast? our answers are: to assure that the latest in knowledge is available to all levels of management for any decisions to be made in regard to public position on an issue or in regard to the development of a new product or feature.

Exhibit 2. Examples of Defensive Forecasts.

| Early May 1970 | President of one of the large lead producers told his stockholders that a ban on the use of lead in gasoline "would be unwarranted and would be inconsistent with the economic facts." He assured them that such would not occur. |
| Late May 1970 | President Nixon suggested a tax on lead used in gasoline. Tax would vary from about 2.3 cents/gallon to 4 cents/gallon. |

NOTE: To date there has been no tax on lead in gasoline, but all of the major oil companies are either out with or are hard at work on lead-free gasoline.

MECHANICS OF FORECASTING

TF (technological forecasting) begins and ends with information. The objective of the TF exercise is to gather as much information as possible. This information may be either factual or, to a certain degree, speculative. Data are sorted, stored in a convenient device, retrieved, and finally acted upon. In these departments we have evolved a system which we, and others, have found very convenient.

Gathering of Information

The precursor group to our TF effort at Whirlpool was called our Technical Liaison and Intelligence Department. The author was given this assignment back in 1963. The challenge given was to find out what other noncompetitive corporations were doing which we might use in our own research to shorten or even alter the course, of some of our projects.

Personal visitations still constitute a vital part of our information gathering process. Members of our TF staff each travel an estimated 75,000 miles per year. Telephone, not even the picture telephone, can ever replace the quid pro quo potentialities which exist in a face to face conversation. Much is said of the reticence of the researcher. We have never found this to be the case. We have found the typical researcher more than eager to discuss his work, especially with someone relatively knowledgeable about the field. We do, on some occasions, encounter the point of corporate secrecy. At this point, our people have instructions to back away. When this point is reached, we usually have found out most of what we sought.

Information gathered in this manner is the best. It is far better than reading the technical journals. While reading of technical journals is encouraged, it must be remembered that the data reported here is, in reality, "stale mail." It took six months to a year to gather the reported data. It took three months to put it into a readable paper and another month to get corporate approvals to publish. After waiting a year for publication in some of our journals, the data finally appears in print two to three years after derivation. In face to face talks, this is the amount of advanced notice you have over simply monitoring the printed word. In a fast moving industry such as ours, the year or more advanced notice is frequently the difference between being there first and having to say "me, too."

It would appear that the reading of technical literature has been maligned in the previous paragraph. This was not meant to be. It is necessary to the forecaster that he be fully up to date on at least one basic science. It is preferable that he have an in-depth, though not necessarily perfect, knowledge of at least one more science or technology. Reading the technical literature is the one way to become sufficiently knowledgeable.

But the technical literature is not the sole source of good information. It is well to know that some sequestered solon has theorized that there might be oil beneath the North Sea. It is also necessary to know that the Ekofisk fields are already yielding several thousands of barrels per day. This type of information is available in, for example, *The Wall Street Journal*. The *Journal* and many other "obscure" journals are prime sources of information to the forecaster, and they should be studied diligently. Exhibit 3 shows a few of these "obscure" journals we use in our efforts.

Personal interview and perusal of general information sources are but two of the methods we use to gather information. There are a number of others. In this case, the individual forecaster, knowing the peculiar situations which exist in every major corporation, should improvise other means of getting data, as have we.

Exhibit 3. Obscure Journals.

Journal	Area of Use
Atlas	International politics
Daedalus	Population problems
Soviet Life	Politics, science technology
Center Magazine	Domestic politics
Technocrat	Japan, Inc.
Environment	Conservationist thoughts
New Leader	Sociology

NOTE: These are referred to as "obscure" journals not because they are not well-known but because technologists are as guilty of not reading these as non-technologists are of not reading technical journals.

Information Storage and Retrieval

"A little knowledge is a dangerous thing. . . ." The same can be said about a lot of information. The forecaster will gather, in a very short time, reams of information which is so varied in content that it will become a confused hodgepodge in his mind. The answer is, of course, do not trust to memory. Evolve a system which allows storage of vast quantities of material and which allows rapid access to the stored data.

In our TF effort we began by attempting to store all of our data in loose-leaf notebooks. Exhibit 4 shows a typical page from one of our books. This system involved simply typing full bibliographic information and a short abstract onto the pages in a random sequence. It is not hard to visualize that this system rapidly became too extensive and too cumbersome for practical usage. It very soon presented too many books and too many pages. Further, access was limited to knowing the approximate date of an event. The system was quickly abandoned.

Our next effort toward a workable retrieval system involved a commercially available method called Termatrex. This, too, is a random access system in which large index cards are punched according to an established thesaurus of terms. A typical Termatrex set allows storage of up to about 10,000 bits of data. It allows a multiple search methodology; that is, it allows one to subdivide a search through several thesaurus terms.

We have found the Termatrex system a useful one for a newly evolving TF department. It is recommended over the notebooks discussed above. It is, however, limited to relatively few bits of data which you shall soon discover as your TF effort expands.

The obvious answer to data storage and retrieval is the use of a computer.

Exhibit 4. Example Page from Notebook.

Textiles

1. *Chicago Trib.*, June 17, 1969. Chicago has converted Audy Home, juvenile prison, to disposable bed sheets. Previously shifted Cook Co. Hosp. at savings announced at $40,000/year.

2. *DNR*, June 12, 1969, pp. 16. Disposables are growing at rate of 20–25%/yr. according to Robt. Alpert, Pres. Blessings, Inc.

3. *DNR*, June 11, 1969, p. 1. Qiana appearing in neckties. About dozen major manufacturers. Cost retail about $10 down to $7.50.

4. *DNR*, June 11, 1969, p. 31. July 3 has been designated "Sock it to Sears Day" by NWRO, Natl. Welfare Rights Orgn. Will use out-of-store demonstrations, "militant" sit-ins and shop-ins. Demanding special credit considerations.

5. *DNR*, June 13, 1969, p. 2. Establishment of quotas on imports coming. Dirksen supports. Textile imports up 300% in last 8 years. Will probably be quoted at 1968 levels.

6. *DNR*, June 13, 1969, p. 2. D-M offering SR shirting at prices which do not reflect 2c/yd. for such treatment. D-M might be attempting to stir sales of SR. Shirt manufacturers have shown no interest. Even work shirt makers are adding to only about 10% of line. Oily soil on collars of dark shade a real problem. 80 PE/20% cotton worst offender in collar stain problem. Manufacturers are adding hang tags or printing on shirttails to pretreat with liquid detergents before washing.

7. *DNR*, June 13, 1969, p. 2. Kendall Mills major endeavor no longer textiles. Disposables. Last year growth 25%. Expect 1969—50% growth in this area.

8. *DNR*, June 10, 1969, p. 1. Ford Draper, Div. of Marketing, DuPont Fibers Div. announced total annual capacity for Dacron now 600 million lbs. DuPont will expand worldwide divisions to achieve single company capacity of 1 billion lbs.

At Whirlpool we progressed to this conclusion by trial and error; a route not recommended at this point in time. From all of our trial and error has evolved what we call our WIN (Whirlpool Information Network) system. Exhibit 5 shows a typical sheet which is the beginning point in WIN.

As the scientist or engineer reads, he is encouraged to fill out the information on this WIN sheet for each bit of information which he feels is germane to Whirlpool's business. Thus, the first advantage to the TF department is that it makes the readings of all the technical people in the corporation available for use in the TF department. The person submitting information is asked to give full bibliographic data, prepare a short abstract or include the entire article if it is not too long, and, finally, to suggest a list of key words which fit the content of the article. This done, the WIN sheet is sent to the central library where a clerk takes the key words suggested and fits these words into a uniform syllabus of words prepared especially for WIN. In such a transposition, the suggested key word *paint* might be changed to *coating, organic* which appears in the WIN syllabus. These translations made, the clerk next assigns a general "class category" number to the bit of information. Class categories simply define the general areas of science or technology implicit in the data bit. Examples here would be such broad areas as: *textiles, energy,*

WIN *WHIRLPOOL INFORMATION NETWORK* **SUMMARY**

1. TITLE (OR CITATION):	3. *Class category:
	4. *WIN summary no.:
	5. Division:
	6. Department:
2. AUTHOR(S) (OR SUBMITTOR):	7. Report date:
	8. Report file no.:
	9. No. of pages:

10. SUMMARY:

Columbia Gas will import $1.4 billion in LNG from Sonstrack, Algeria Oil Agency. Will buy from El Paso. Price $\$0.50/10^6$ BTU. Gas runs 1030-1200, BTU/ft^3, ergo priced 52.5¢--60¢/1000 ft^3. Pipeline gas Pittsburg--today--38.64¢/1000 ft^3--Boston--69.21¢/1000 ft^3.

11. SUPPLEMENTARY NOTES:

12. DISTRIBUTION:

13. KEY WORDS: (Suggested by Author)

14. *WIN INDEX TERMS:

PROPRIETARY INFORMATION. This information is *CONFIDENTIAL* to the extent it is original, and is intended for the exclusive use of authorized personnel of Whirlpool Corporation and its subsidiaries, Warwick Electronics, Inc. and Heil-Quaker Corporation.

S8W001020

*NOTE: ITEMS 3, 4 AND 14 WILL BE COMPLETED BY INFORMATION CENTER.

Exhibit 5. Information Sheet for WIN.

or *political science*. Each category is assigned a number; for example *textiles* fall into category 17. Next, the clerk assigns a serial number to the article. This number, for example 00188, means this article is the 188th article filed under a given category. A full WIN number thus appears: 17–00188. The bit of information is now advanced to the computer section.

To make the entry onto the computer memory, the full WIN number, the bibliographic data, name of person submitting information, the WIN index terms, and finally the abstract are keypunched. The original WIN sheet is returned to the library and filed under the name of the person submitting information.

Once the data is entered into the computer, it becomes a part of the total corporate knowledge bank. By proper manipulation, one can extract the full corporate knowledge on any given subject. This one can do to any depth of inquiry designed by simply increasing the number of WIN key words for which he requests search by the computer. It is possible to make a given search as cursory or as exhaustive as desired.

Once each month the TF department requests a printout of the title, bibliography, and the name of the person submitting information on every article it has had submitted. We do this simply to refresh our memories of the input. Exhibit 6 shows a part of this type printout. Just prior to general forecast preparation, we request exhaustive printouts of all data within a general category which is to be covered in our forecast.

This is our stance today. The WIN system has served us well, and we intend to continue to use it and expand it.

Utilization of Information (Forecast Preparation)

Too often perfect systems evolve, yet they produce nothing. This has not been the case with the various systems of information gathering, storage, and retrieval we have been discussing. A very pragmatic usage is made of the information we gather and store. It is used to produce the various types of forecasts required of us.

Technological forecasts at Whirlpool fall into three distinct categories.

1. Newsletters (actually month to month forecasts).
2. Interim forecasts.
3. Annual general technological forecast.

Certain fast moving industries which impinge upon the activities of Whirlpool, and whose developments are of critical importance to us, merit almost day to day scrutiny. Officers who make decisions which require knowledge about these areas must be constantly informed about developments. To assure that this knowledge is always up to date, we circulate a series of newsletters. Exhibit 7 shows the front page of a typical forecast of this type: Textile Topics. This newsletter is of vital importance to officers, supervisors,

```
        WIN ALERT        09/09/71

15   RESEARCH, TESTING, METHODS AND EQUIPMENT . . . . . . . . .  PAGE  01
INCLUDES - NEW PRODUCT DEVELOPMENT, LABORATORY PROCEDURES,
STANDARDS AND SPECIFICATIONS.

 -  -  -  -  -  -  -  -  -  -  -  -  -  -  -  -  -  -  -  -  -  -  -  -


TESTING METHODS AND TECHNIQUES (DOC) BY NASA
   15 71 00932 P                 INFO. CENTER                  RES

A GUIDE FOR FATIGUE TESTING AND THE STATISTICAL ANALYSIS OF
   FATIGUE DATA (DOC) BY ASTM
   15 71 00937 P                 INFO. CENTER                  RES

PERFORMANCE CURVES OF AN AIR CONDITIONER FOR A COMPUTERIZED
   ON-LINE TEST
   15 71 00957             U.   REMBOLD                        RES

LABORATORY TEST TO DETERMINE THE FEASIBILITY OF TESTING AIR
   CONDITIONERS IN A SHORT PULL DOWN TEST
   15 71 00958             U.   REMBOLD                        RES

FEASIBILITY OF 100% ON-LINE TESTING OF WINDOW AIR CONDITIONERS
   15 71 00976             P   CHEN              AND OTHERS     RES

ON-LINE AIR CONDITIONER PERFORMANCE TESTING OF SAMPLED UNITS
   15 71 00977             P   CHEN                             RES

ON-LINE MEASUREMENT OF WINDOW AIR CONDITIONER CAPACITIES
   15 71 00978             P   CHEN                             RES

CONTINUOUS SHORT PULL DOWN TEST TO MEASURE THE PERFORMANCE OF AIR
   CONDITIONERS ON THE PRODUCTION LINE
   15 71 00993             U   REMBOLD         P   CHEN         RES
```

WIN ALERT

Whirlpool INFORMATION NETWORK

NOTE: This *Win Alert* cites all information entering the network during the month in the subject category indicated above. The number in the citation identifies a WIN Summary. Summaries of the above unrestricted citations are available from *WIN Files* located at most divisions, subsidiaries, and from the Information Center. The WIN Program is intended for the exclusive use of authorized personnel of Whirlpool Corp. and its subsidiaries.

MEDIA CODE: (Designated after the WIN File number) internal report = no letter; published literature = P; vendor data = V; information alert = A.

INFORMATION CENTER   RESEARCH & ENGINEERING CENTER
                        EXTENSION 7272

S8R042051

Exhibit 6. Sample Printout of WIN Entries.

and engineers in our plants making and designing laundry machinery. This newsletter is circulated approximately one issue per month to about 125 people intimately involved with washers, dryers, water softeners, and so on. We also publish, at varying intervals, Materials Newsletter, Power Picture, Noise News, Air and Water Pollution, and others.

In our newsletters we attempt to accomplish two things at the same time. First we offer brief statements of new developments in the area of the newsletter, and second, we include, under the title What This Means to Whirlpool, *one* interpretation of that news and any potential impact on the corporation. We do not care particularly whether or not the reader agrees with our analysis of the news item. The important thing is that, for a moment, we stopped him and made him *think* about the significance of the item.

The question arises: Do the recipients of these newsletters stop and read them? The answer is: Yes, they do. How do we know this? We have built-in means of determining our "rating." First, the telephone starts ringing within two hours after an issue hits the desks. They want more information. Second, we occasionally insert a sleeper sentence. In our latest Textile Topics, for example, we said: "We have prepared an assessment chart showing the effects the new carbonate-silicate detergents are expected to have on washers. If you want a copy of this chart, call us." Within four hours, we had requests for thirteen copies.

Interim forecasts are special forecasts made to cover relatively short-term events effecting specific products. By short-term, we mean events from the present to a maximum of about four years hence. The specific products for which the interim forecast is made might be a new product, or it might be one of our present products which is under a threat. In the last part of this paper the interim forecast and its use will be discussed more fully.

The *magnum opus* of our department is our annual GTF. This work we publish about August 1 each year and distribute to the top 126 people in Whirlpool and our subsidiaries. The timing is such that the information presented is useful in budget planning which gets underway in August and in the economic forecasts which issue in November.

The GTF is a massive document, composed of:

1. An executive summary containing individual forecasts.
2. A summary of pertinent data supporting the forecasts.
3. An in-depth study of developments in a number of sciences and technologies of interest to the company and the implications of these developments.

In toto, a typical GTF will be composed of about fifteen technology surveys, the executive summary, and so on and will reach 300 to 350 pages in length. It is not expected that every officer will read the entire document; he is expected to study closely only the summary and those surveys pertinent to his portion of the business. That the document is read is evidenced by replies and

Whirlpool CORPORATION

TEXTILE TOPICS

I.   Cotton

1.   The Cotton Producers Institute has allocated more than one million dollars in research funds to try to develop Permanent Press cotton. Some of the best known organizations (Stanford Research, Gillette, Battelle, Gagliardi and Southern Research) have been brought into the picture with grants ranging from $7,500 to $120,000.

2.   U.S. needs of cotton from the 1969 crop will be down by 700,000 bales from last years usage of 8.9 million.

3.   The three big U.S. shirt manufacturers have begun production of a line of pure finish cotton shirts. Prices for these garments will range from $8 to $20 per garment.

WHAT THESE MEAN TO WHIRLPOOL: A part of the cotton industry realizes that their market has shrunk and is trying to do something about it. Unfortunately, the projects they are funding are the same projects which have not been successful in the past. The projected usage of 700,000 fewer bales in 1969 reflects the textile industry's attitude toward cotton. The entry of high priced cotton shirts should not be construed as a change in philosophy. There will be few of them made from those people who have been slow in shifting to synthetic/cotton blends.

4.   Cotton is witnessing a big intrafamily feud these days. It seems that growers expected high 1967 prices to carry into 1968, so they planted low bearing, high quality varieties. Prices fell and their incomes tumbled. Now they blame textile mills, ginners, Cotton Council, and just about everyone for their troubles. They say they will revert to high yield, lower quality varieties in 1969.

WHAT THIS MEANS TO WHIRLPOOL: Just more fuel on the fire that is burning out cotton and promoting synthetics. It seems cotton will remain controlled by individualists and will never become an industry.

II.   Synthetics

1.   Monsanto announced recently that they will start production soon on a new polyester fiber differing in chemistry from any now on the market. The fiber will be expensive and is not expected to gain mass usage for a number of years. It is a specialty.

Exhibit 7. Typical Newsletter.

questions we receive from them. Exhibit 8 lists the technologies covered in our 1969 GTF (general technological forecast).

Exhibit 8. Sciences and Technologies in General Technological Forecast of 1969.

Technological-political	Detergents
Electronics	Fuel and energy
Food	Engineering research
Chemistry	Manufacturing engineering
Physics	Informational science
Life sciences	Industrial design
Materials	Product safety
Computer technology	Engineering home economics
Textiles	

METHODOLOGIES

As stated earlier, we are the technologists in the TF field. We do not create new methodologies. Our task is to forecast, not to be particularly innovative.

What methodologies do we use? Listed in order of the relative frequency of usage by our department, they would be:

1. Literature monitoring.
2. Trend extrapolation.
3. Delphi or modifications of Delphi.
4. Relevance trees.
5. Impact analysis.

Exhibit 9 shows a typical example of trend extrapolation we have used, and Exhibit 10, a typical Delphi study.

Exhibit 9. Uses of Selected Textile Fibers, 1965–90.

1. When will cotton constitute less than 20% of the fiber used, either alone or blended, in washable fabrics purchased for use in the American home?
Answer _____

2. When will more than 50% of the apparel bought by the American consumer be discarded after one use?
Answer _____

3. When will sheets and pillowcases purchased for home use become disposable after one normal use?
Answer _____

4. When will knitted fabrics surpass woven fabrics in usage for outer wear designed for the American male?
Answer _____

5. When will bonded structures, designed for casual wear by both men and women, be made fully washable in existing home laundry equipment?
Answer _____

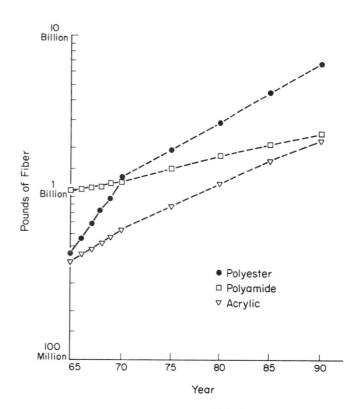

Exhibit 10. Delphi Questionnaire-Textiles
February 28, 1969.

RESULTS OBTAINED

No TF department can expect to be allowed to continue to exist in a profit making operation unless it produces results which contribute to the profit position of the company. This statement has become more or less the modus vivendi with us. We constantly assess our effort to keep it in synchronization with overall corporate performance.

The question naturally arises: What have you done to merit continuance?; or, expressed more succinctly, What have you done for me lately? We are prepared to answer this question at all times.

Permanent Press Cycles for Laundry Machines

When we were still in the technical liaison and intelligence phase of our effort, an opportunity arose to do some forecasting which could produce meaningful results (profit). Exhibit 11 shows a brief record of events that led to our first success.

While no complex methodology was involved in arriving at our first success at forecasting, it becomes obvious as you progress down the list of events that monitoring the environment can produce profitable results. Observe also that these events occurred early in the 1960s when the industry was not aware of the advantages of TF.

Exhibit 11. Chronology of Major Events Involving Permanent Press Products and Their Impact on Whirlpool.

Winter 63–64	Rumors in textile industry of new "delayed cure" process for resin applications.
April 64	Confirmation of rumor—Spartanburg, S. C.
May 64	First glimpse of process—San Francisco.
May-June 64	Education of Whirlpool personnel about permanent press garments. Forecast of doubling of dryer sales in a year.
August 64	"Development Conference" arranged for VP, laundry at large fiber manufactory. Permanent press garments shown at Whirlpool for first time.
September 64	Textile industry introduces permanent press.
January 65	Permanent press cycles on washers and dryers. First by appliance industry.
March 65	Research project for a "new concept ironer" dropped. Monies diverted to other uses.
May 65	Dryer sales reported more than double May 64.
November 65	Introduction of permanent press sheets.
February 66	Volume of dryer increased to app. 6 ft.³ without change in exterior dimensions. This to accommodate large permanent press sheets and other items.
February 67	Forecast that knitting would grow to challenge weaving for control of garment textiles by midyear 1970.
May 67	Work begun on knit cycles for washers and dryers.
June 69	Textile industry forecast that knits would control 30% of trouser market in 1970.
January 70	Introduction of knit cycles on washers and dryers.
May 70	"No-iron" concept of textiles postulated.
August 70	Forecast that by 1975, knit structures would constitute 50% of men's wear market.
February 71	Five textile experts forecast that by 1975 knits would constitute 50% of men's outerwear.

There is another salient point demonstrated here. It is apparent that, as one's expertise increases, the time advantage can become more effective. Note the February, 1967, forecast that by midyear 1970, knitted structures would challenge wovens for preeminence in the garment trade. Then note the June, 1969, textile industry forecast that knits would control thirty percent of the trousers market in 1970. Both of these forecasts were fulfilled. Success of this order led us to forecast in our 1970 GTF that by 1975, fifty percent of the men's outerwear market would be controlled by knitted structures. This forecast was repeated by textile industry spokesmen in February, 1971. A modified Delphi study led to our 1970 GTF forecast.

In a pragmatic sense, such forecasting is useless unless acted upon with dispatch. Observe the January, 1970, introduction of "knit cycles" on our laundry machines.

Evolution of "Nonpolluting" Detergents

This subject is one of great current interest. Beginning as far back as 1963 (refer to Exhibit 12), a logical chain of events began to evolve which has not completely unfolded today. The attack on use of phosphates began,

Exhibit 12. Chronology of Phosphate Replacement in Detergents.

1963–64	Numerous photographs of suds on streams. Dams and waterfalls hidden by mounds of foam.
1964	Threat of legislation to force elimination of suds from surface waters.
1964–65	Voluntary shift to "biodegradable" surfactants by primary and secondary detergent manufacturers.
1965	Swedish experimentation with NirtiloTriAcetate as a detergent builder to replace phosphates.
1965–66	Serious drought in Eastern U.S. Heavy growth of green algae in Delaware and Susquehanna Rivers noted and commented upon by conservation groups.
1966	Polyphosphates used in detergents openly blamed for algae growth. Contributions to eutrophication by phosphates claimed. Appearance of algae growths in "sterile" waters of northern Canada noted. Said to be caused by water fowl bringing algae and phosphate from U.S.
Aug. 1967	Secretary of Interior Udall warns Soap and Detergent Association about use of phosphates.
Nov. 1967	Whirlpool forecast that replacement of phosphates would be required by law by late 1972.
1967	First U.S. detergent using NTA as partial phosphate replacement.
1967	Scientific evidence published showing carbon, not phosphorous, responsible for algae blooms.
1968	Strengthening of carbon cause theory by other scientific evidence.
1969	Swedish experience shows NTA substitution fails to halt eutrophication.
1969	Introduction of several bills into Congress which would regulate phosphate usage in detergents.
Feb. 1970	Passage of phosphate control legislation by Canadian Parliament.
Summer 1970	Several small detergent makers announce phosphate free detergents. Chicago passes city ordinance.
Nov. 1970	Suffolk County, N.Y. bans sale of detergents after late 1971. Reason: pollution of *ground* waters.

not as a frontal assault, but as a flanking maneuver against sudsy streams. Once a dent in the armor of the detergent companies had been made, the subsequent events assumed a definite and logical ordering until today the detergent makers are under violent attack by conservationists and environmentalists. These early attacks led us to conclude in November, 1967, that phosphates would be outlawed as detergent ingredients. While this forecast has not yet been fully accomplished, I am sure that any detergent maker will agree with me that inevitable is the word for the situation.

As a logical phosphate replacement, the industry had waiting in the wings a product called NTA (sodium nitrilotriacetate). In looking at the reasons for the demise of phosphates, we came to the conclusion in February, 1970, and so stated publically in May, 1970, that NTA would not be permitted for use. Our forecast called for the banning of NTA in mid-1971; the actual occurrence was in November, 1970.

A portion of our public forecast of May, 1970, called the period mid-1971 to 1975 an era of chaos in the detergent industry. Anyone familiar with the industry knows that we are already in that era.

There are a number of other areas of technology in which we have had meaningful (profitable) results. In the realm of fuel and energy, we have caused each of our plants to evaluate their energy needs for the next five to ten years. This action has led to the purchase of new auxiliary generation equipment and installation of LPG (Liquid Propane Gas) tanks.

TECHNOLOGY ASSESSMENT

A new assignment which we have just undertaken is to do complete technological assessments of all of our products. This is in keeping with the latest thinking in Washington that a company is responsible for every aspect of the performance of its products. As a consequence, we will undertake our product assessments from three directions:

1. Whirlpool oriented.
2. Consumer oriented.
3. Ecology oriented.

FROM FORECAST TO PRODUCT

As stated earlier, offensive forecasting is one of our goals. Offensive forecasting has, at least in part, the creation of ideas for new products or new features. We have already discussed some of these ideas. Now the question arises: How do we progress from the idea concept to a product or feature?

Exhibit 13 shows how we do this. We have created what we call an "opportunity group." This group has the function of screening ideas of products and defining quickly corporate interest or lack of same. Represented in that group are all the corporate activities which will ultimately have anything to do with the product. This includes research, manufacturing, engineering, sales and marketing, and finance. The group is headed by the vice president of engineering and is responsible directly to the president.

The group operates in this manner. An idea comes before the group. That idea might originate anywhere, inside or outside the company. The idea is discussed fully. It might be dropped immediately, as many are. Reasons for dropping are, typically: we have no way to market it; it would sell at too low a price to make it profitable; and would require too much research time which, by the time it is ready to produce, will lead to the solution of the problem by other means. If the idea is accepted for further study, an assessment of its present state of development is made and the idea is assigned to whatever department most closely matches the state of development. For example, if more research is needed, the idea goes to the director of research who gives the

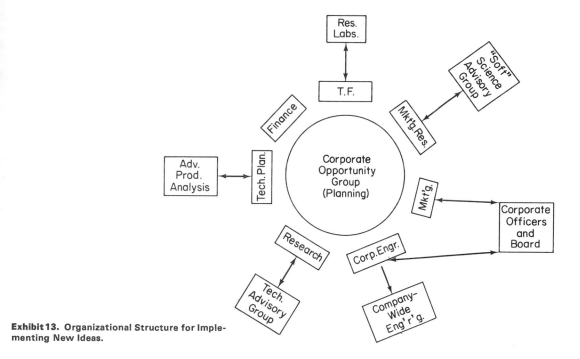

Exhibit 13. Organizational Structure for Implementing New Ideas.

assignment to someone on his staff for study. A date for report back is set, and the recipient of the assignment is expected to meet that date. The idea is passed back and forth across the table many times before a product finally evolves. Each time it changes hands is in essence a vote of confidence for the idea to progress further in the system. It can also be killed along this line at any point.

The TF department is a constant participant in this process. If it takes more than a year for an idea to reach prototype state, it is entirely probable that at least two interim forecasts of the continuing efficacy of the idea have been made.

When all segments of the opportunity group have finally given a green light to the idea, and prototypes are desired, the vice president of engineering appoints an APA (Advanced Product Analysis) task force. These people are charged with building the first models. Generally the people assigned to APA are of several disciplines. The APA group is usually composed of five to seven people.

From this point forward our procedures are quite normal. We progress through field testing, correction of difficulties found in field tests, preproduction models, more field testing, industrial design, and finally to pilot plant production. At this point the opportunity group steps aside and lets the product move forward at its own pace. Usually, members of the APA group follow their baby directly into production.

EPILOGUE

As was stated at the outset, this is *one* way that TF has been used by a major, multiproduct corporation. It is not, by any means, the only possible way. The reader is encouraged to use whatever he may find applicable in this paper for his own uses. If the experience of the author or his staff would be of interest, you are invited to call or visit us at any time.

Producing the First Navy Technological Forecast

MARVIN J. CETRON
DONALD N. DICK

In 1968, Marvin Cetron and his colleagues in some twenty-four Navy laboratory installations and commands prepared and published the U.S. Navy's first technological forecast. Cetron is widely known as a pioneer and a most energetic and preceptive author on forecasting here and abroad. The Navy's approach was developed only after a very careful study of Army, Air Force, and other industrial forecasting efforts. The organization and implementation of this forecasting effort provides a thought provoking guide for other large organizations.

During 1968 the Navy prepared and published its first NTF (Navy Technological Forecast). The Navy effort involved sixteen major laboratory/centers and eight system commands. The forecasts range from functional technologies to system options and the overall effort is comprised of approximately five hundred individual forecasts. It was implemented, prepared, and published by Cetron in seven months and it is estimated the overall forecasting task cost approximately $1.9 million. The implementation of such a task, requiring the efforts of a large number of activities in a relatively new field over a short period of time represented a challenge to all involved.

We should start by acknowledging the assistance gained from those who formally prepared forecasts some time ago, notably the Air Force and Army.[1] Although the Navy waited until 1968 to prepare a Navy-wide forecast, three years were spent studying other forecasts for the considerations of methodology, structures, and overall approaches. The end result of this study is contained in *A Proposal For a Navy Technological Forecast, Part II, Back Up Report*[2] which has, and still does, serve as the bible for the NTF as well as a good introduction to the subject of technological forecasting.

ORGANIZATION OF NTF

A review of the above report will indicate the Navy chose an approach similar in part to both the Air Force and the Army. The result is a program with a very broad scope. The basic NTF is composed of three major parts. These are: *scientific opportunities*, Part I; *technological capabilities*, Part II; and *probable system options*, Part III. As indicated in Exhibit 1, the three major areas constitute the basic forecast. In addition to these planned-for forecasts, provision has been made for specific technological projections for paramount needs. These are called TENIS (technological needs identification studies). The categories shown in Exhibit 1 indicate the complete program as it is envisioned at present.

1 See John R. Bird and Halvor T. Darracott, "Developing and Using The Army Long Range Technological Forecast," *Technological Forecasting For Industry and Government*, ed. J. R. Bright (Englewood Cliffs, N.J.: Prentice-Hall, Inc. 1968), p. 385–411, for the history and operation of the Army forecasting system.

2 M. J. Cetron, et al., *A Proposal For a Navy Technological Forecast, Part II, Back Up Report*, Hdq. NMC, May, 1966. An unclassified version is available from DDC or Clearinghouse, AD 659 200.

Exhibit 1. Diagram of Navy Technological Forecast Program.

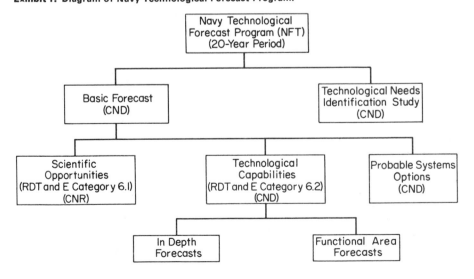

During 1968 the technological capabilities and probable systems options were prepared under the direction of the CND (Chief of Naval Development). The first forecast of scientific opportunities was prepared by the ONR (Office of Naval Research) during fiscal year 1969 (FY 69).

The probable systems options section of the present Navy forecast suggests examples of systems or subsystems which could be developed to satisfy broad mission requirements such as: strategic, amphibious, antisubmarine, and so on. The forecast identifies known technological barriers and indicates fund-time requirements for system development. It differs from an unconstrained wish list in that concept feasibility has been investigated by analytic studies and/or experimentation to indicate gross operational capabilities. The Navy System Commands (Air, Ordnance, Ships) are responsible for preparation of the systems options forecasts. The 1968 NTF contains submissions from Ship Air and Ordnance System Commands. The 1969 NTF contained submissions from all commands.

The technological capabilities portion, in terms of volume, is the largest contribution of the 1968 NTF. The technological capabilities forecast contains projections of applied research and development activities of the Navy. The area covered is that normally associated with the RDT&E 6.2 planning series. In the system acquisition process, it is the area of exploratory development. As such, it represents the technologies that convert basic science concepts into the fundamental building blocks of Navy systems.

The remainder of our discussion will be confined to the Navy experience in the technological capabilities forecast. The forecasts cover a broad spectrum of technologies, from nominal scientific approaches to specific Navy needs to the broad functional technologies necessary to support general Navy missions. Because of the nature of this area of "technology," the problems of planning and providing a technological forecast are intensified. Any specific procedures or actions have broad implications.

Of the many factors considered in the initial stages of planning the Navy forecast, two were considered fundamental. These are: (1) the forecast format, and (2) the technological structure. The determination of these two considerations will crystalize many aspects of forecasting such as: who will use the forecast, who will prepare the forecast, the overall utility, and the definition of technology.

FORMAT

A major factor influencing the forecast format is the group of potential users. In general, these include groups from corporate executives through to the technical staff working at the bench level. The Navy group having the greatest need, and therefore the highest potential use, are the military planners in the CNO (Chief of Navy Operations) offices. This dictated that the Navy forecast be presented in a *format meaningful to the operating Navy*. At the same time, a comprehensive technical projection was considered necessary to support any projections of Navy operational capabilities. Further, it was assumed that individual readers might not be familiar with Navy problem areas and be in doubt concerning the relativeness of technology areas, giving rise to the desirability of general background information.

Attempting to provide a forecast that satisfies all may provide a forecast that satisfies none. However, the format used by the Navy is designed to accomodate a diverse and wide audience. Table 1 indicates the major sections of a technological capabilities forecast. The amplifying statements under each section are the abbreviated directions given to each forecaster for the 1968 preparation.

As indicated in Table 1, the technological capabilities forecast begins with a brief discussion of past applications and significance to the Navy, continues with a discussion of the present state of the art, gives a technical prognostication, relates the technical projection to operational considerations, and ends with credibility references and a listing of where, in the Navy, to look for such expertise.[3]

One question immediately arising from the specification of this "total assessment of technology" type of forecast is: Why such an extensive treatment,

[3] This same format is used extensively in industry today. Illustrations may be found in *Industrial Applications of Technological Forecasting*, M. J. Cetron and C. A. Ralph (New York: John Wiley and Sons, 1971).

Table 1. Guidance for Technological Forecast Preparation.

1. *Background:* Identify, if possible, the exploratory development goals or other objectives to which the forecast will contribute, for example, statements indicating why the forecast is being made. Describe briefly the significant factors which influenced past developments and those which will tend to emphasize or de-emphasize further developments.

2. *Present status:* Describe briefly the present state of the art. For hardware items cite, in vertically-parallel columns, advantages and disadvantages of present items. Include, where appropriate, a description of limitations (technological barriers or gaps in technology) which are (or may become) troublesome.

3. *Forecast:* Utilize the functional parameters (that is specific weight, shaft horsepower/unit weight, shaft horsepower/unit volume, and so on) which are the most meaningful in your technological area. Describe anticipated changes in complexity, cost, and, where appropriate, physical characteristics and performance which have the potential for alleviating the limitations. (Charts and/or graphs should be employed where possible.)

4. *Operational implications:* Utilize the appropriate operational parameters, (that is, cruising range, CEP's, speed, operating depth or altitude, and so on). Describe the effect of the forecast changes on cost/effectiveness, manpower requirements, and any other factors affecting operational efficiency. (Charts and/or graphic techniques should be employed, where possible.)

5. *References:* Cite the publications from which authoritative direction has been elicited and list technical documents in this function subarea which tend to add credibility to the forecast.

6. *Associated activities:* List the organizations who have contributed in this forecast.

why not just a technical projection? Hopefully, it will suffice for now to say that in addition to any logical discussion of the utility of serving a wide audience, there is one economic fact of life that must be faced concerning technological forecasting. The cost for a technology projection (the forecast section of the NTF) is a very large percentage of the cost for an NTF type production, approximately eighty percent. There are additional beneficial side effects gained by requiring the NTF broad assessment type of forecast which will become apparent only after acceptance and utilization by Navy planning and analysis groups.

DEFINITION OF TECHNOLOGY

With a definite outline of the information desired about each technology (as indicated in Table 1) for each user, we can turn to the second fundamental issue: which technologies? Better yet, one can ask: What do we really mean by technology? To answer this question, it should be recognized that "technology" is a widely used and abused magic word of the day. It has recently seen expanded use in political circles as concern over the "technology base." We hear statements such as "technology is the solution to many of man's social ills," "technology made the moon shots possible," "technology is making possible the wide application of integrated circuits to everyone's packaging needs." The word is used at many levels of abstraction.

Through this haze there does seem to be a gross consensus of what technology "means." This meaning is summarized by Galbraith[4] as, "technology means the systematic application of scientific or other organized knowledge to practical tasks." In this sense, technology is a relative term and the full meaning requires association of scientific knowledge *with* a distinct problem or goal. This may seem somewhat picayune concerning definitions, but it has been our experience that communication barriers become visible when one tries to explain what one means by technology. To many, technology is used synonymously with areas of scientific activity, say superconductivity. The distinction we wish to make is: to say "superconductivity technology" is to *imply* the superconductivity activities are directed to the solution of a problem, as opposed to activities oriented toward research. This coupling between organized knowledge and problem areas is a critical issue to long-range forecasting.

There are strong implications in how one defines technology for technological forecasting. The longer the range of the forecast, the more important the determination of technology becomes. A good forecast of the wrong technology is like the right solution to the wrong problem. It could be misleading and it could represent an incomplete assessment.

4 J. K. Galbraith, *The New Industrial State* (Boston: Houghton Mifflin, 1967).

The Navy's interpretation of technology (for technological forecasting) is the application of scientific procedures and findings to Navy problem areas. The Navy forecast was originally planned to be a twenty-year projection. The definition of Navy problem areas becomes critical to the definition and conduct of the NTF. For the NTF, the Navy specified a set of problem areas and required the submission of forecasts for the areas. This constituted a major portion of the technological capabilities forecast.

Exhibit 2. Exploratory Development Planning Structure.

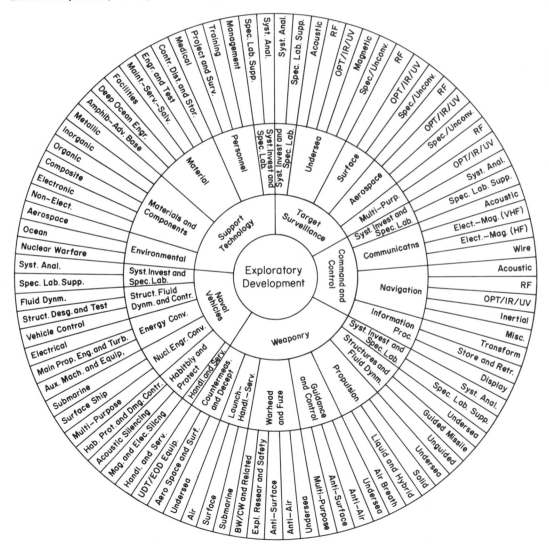

Over the years, the Navy has evolved a structure of problem areas in the exploratory development phase of the system acquisition process. The structure, the exploratory development planning structure, has two very desirable features for long-range technological forecasting. First, the problem areas have evolved to describe the unique needs of the Navy, giving a communicative base with the technical community. Second, the categorization of the structure is a functional description of Navy problem areas, independent of specific system concepts. The structure describes the generic, continuing problem areas of the Navy. It served as the base for the structure of the twenty-year NTF.

Exhibit 2 shows the Navy Exploratory Development Planning Structure. It is a three level hierarchal structure, beginning with the military responsive functions of command and control, target surveillance, weaponry, vehicles, and support. These accurately reflect problem areas of the Navy; however, they are too broad for a technology assessment of the type envisioned in the NTF. The second level is identified as the subareas of the exploratory development planning structure. It maintains the functional aspects and overall completeness required for long-range forecasting. However, the subareas are still quite broad for overall NTF needs. At the third level of the structure, the problem areas begin to take on specific orientation for platforms and technical approach or continue with a further breakdown of functions. Although the breakdown does reflect the uniqueness of each area, the third level does not represent a single specific orientation. There is no reason the present structure should; however, for technological forecasting it would be desirable to be comprehensive with all categories having the same orientation.

SPECIFYING FORECASTS

For the first NTF, the exploratory development planning structure was modified slightly to serve as the definition of Navy functional technology. The second level was maintained for completeness, with the third level modified to give specific overall orientation. The orientation selected was platform/target problems: air, surface, undersea.

A technological forecast was requested for each third level category. For example, a forecast was requested in weaponry (first level), propulsion (second level), and undersea (third level). These forecasts are called broad area forecasts of functional areas, indicating the intent to have a broad assessment of a general problem area.

The broad area forecasts represent the continuing problem areas in which the Navy required forecasts. Provision was made in implementing the NTF to encourage the submission of forecasts in areas the laboratory/centers felt were important. These forecasts are identified as in-depth forecasts. One of the first acts in implementing the NTF was the solicitation of in-depth forecasts. The laboratory/centers submitted approximately eighty proposals covering probable system options and technological capabilities. The CND approved and funded eighteen in-depth forecasts of technological capabilities.

The broad area forecasts were prepared by the laboratory/centers through overhead funding. A total of seventy-five broad area forecasts were requested. The assignment of activities to prepare the forecasts was based upon laboratory missions. Each laboratory was assigned a group of forecasts to prepare. Each laboratory was instructed to prepare the forecast with the entire Navy in mind, and to contact other laboratories for inputs, assistance, and review. The laboratories selected the individuals to prepare the forecasts. In general the forecasters were technical experts, as opposed to system analysts or operations researchers.

RESPONSE

In a period of approximately three months, the laboratory/centers prepared technological capabilities forecasts covering the assigned broad areas. The technological capabilities forecasts were submitted directly to CND from the laboratories. In response to the request for 75 broad area and 18 in-depth forecasts, CND received approximately 600 forecasts. Of the 600 forecasts received, 300 were accepted and 300 were returned to the preparing activity for modification. Of the 300 returned, 190 were resubmitted and were used in the published NTF.

The submission of 600 forecasts in response to a request for 75 forecasts was not due to a zealous desire on the part of the laboratories to forecast. The large number of forecasts indicated an incompatibility or misinterpretation of the intent of broad area forecasts. In those broad areas where a great number of forecasts were submitted, the forecasts appeared to be prepared under the following interpretations:

1. Preparation of a forecast for each project in progress at the time of the forecast.
2. Preparation of a forecast for each known technical approach to the broad functional area.

RELATIVE QUALITY

Each forecast received by CND was rated for forecast quality. *The rating made no attempt to determine technical validity*. Rather, the rating attempted to determine the extent to which the forecast met stated objectives. The objectives were that the forecast be:

1. A function of time, resources, and confidence level.
2. In compliance with the overall format.
3. Quantitative in discussion in all forecast sections.

A section-by-section indication of relative quality is shown in Exhibit 3. Also indicated on Exhibit 3 are the relative needs of one user, CNO. The background and present status sections, in general, more than meet minimum levels of required information. The forecast section drops off slightly, and the operational implications section is by far the lowest in relative quality. This should not be interpreted as poor or invalid information in the operational implications section. It reflects an assessment of the *type* of information contained in the section which appears to reflect the type of individuals who prepared the forecast. As was indicated previously, the majority of forecasters were technical experts, people required to dig deeply into technological areas, having little experience in Navy operations. The best balance across the forecast sections was achieved by activities that used a technical expert for the forecast section and a systems analyst (or operations researcher) for the operational implications section.

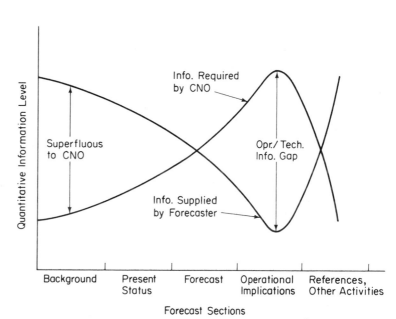

Exhibit 3. Indications of Relative Quality.

Note: This comparison shows one user's needs (Chief of Naval Operations); other users, such as systems analysts, would not indicate the same gaps.

METHODOLOGY

The format of the technological capabilities forecasts required each forecast to be both exploratory and normative. The forecast section was primarily exploratory while the operational implications section was need oriented. Several forecast sections were oriented toward specific goals[5]; however, most forecasts are exploratory.

The techniques used for forecasting were primarily trend extrapolation. The techniques used and the approximate percentage of forecasts are shown in Table 2. There was one major Delphi effort.[6]

Table 2. Usage of Forecasting Techniques

Trend Extrapolation	80%
Trend Correlation	10%
Growth Analogies	5%
Intuitive	5%

GENERAL ASSESSMENT

As an overall assessment of the first NTF, as represented by Part II, technological capabilities, the verdict is good. The actual utility in the end depends upon the specific information required by a user; however, the NTF contains some excellent examples of forecasting. Exhibit 4 is an example (edited) of a projection that satisfies the requirements of the NTF. These are:

1. A pacing parameter has been identified and is used as the basis of projection.
2. The projection is a function of time (twenty years).
3. The projection indicates expected values and confidence limits.
4. The variation to funding changes is indicated.
5. The known practical limit is identified.

As with most things in life, it is easier to identify the discrepancies of the NTF than the good points. The NTF shortcomings were of the following types:

1. Technology areas: The response of several hundred forecasts for broad functional areas indicates the necessity to develop a technology structure compatible with expert working areas as well as general Navy needs.
2. Pacing parameter: In many forecasts a pacing parameter was not identified. This was expected, since the task of identifying pacing parameters is generally very difficult.

---

[5] The Navy has recently published a set of specific quantitative needs for technology. These are called the "Exploratory Development Goals," published by CND, and were the specific goals used by a few forecasters in the forecast section.

[6] An extensive survey of what forecasting techniques are being used in the U.S., based on some 1,114 respondents to a National Science Foundation sponsored survey, may be found in Cetron and Ralph, *op. cit.*, Chapter 12.

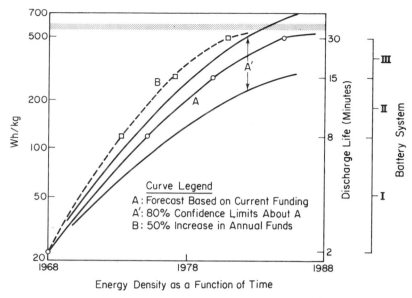

Exhibit 4. Technological Forecast: High-Power Density Batteries.

3. Quantification: The quantification of information is the key to forecast utility. Many forecasters are reluctant to be precise about a future where they doubt the overall accuracy of the projection. Both forecaster and user must recognize the need to think in terms of confidence levels of projections.

4. Relative quality: It appears there could be a considerable increase in relative quality across forecast sections by utilizing different types of people to prepare different sections.

5. Data bank: Many forecasts did not indicate specific data points in the projections. This could indicate the absence of any organized data bank for some of the technology areas.

UTILITY AND FUTURE PLANS

A summary of the characteristics of the Navy Technological Forecast is given in Exhibit 5. Ultimately, the value of the NTF will be shown by the use of the document. Although we cannot indicate known specific uses here because of the unclassified nature of this chapter, the NTF, not yet one year old, is being used extensively by some very influential (Navy) groups. As could be expected the first groups are those who must cut across technological lines or who must gain in indication of future operational implications.

Currently the laboratories are reviewing, and updating where necessary, the forecasts submitted for the second technological capabilities forecast. We anticipate a relatively small percentage of significant changes, most will be updating parameter selection. The Office of Naval Research has completed the second draft of Part I, scientific opportunities. The Navy System Com-

Exhibit 5. NTF Fact Sheet

A. *Size and scope of forecast:*
   75 broad functional areas covered
   16 major laboratory/centers participated
    8 systems commands and bureaus
   Published 460 technical area forecasts
              30 probable system forecasts
          100 percent in-house with some of the laboratory/centers requesting information from
   industry, the academic community and other government agencies

B. *Cost information:*
   Funded—$375,000 to laboratories/centers
              45,000, publication expenses
            $420,000 total
   Actual including overhead $1.9 million
            $400 per technical area
            $25,000 per broad functional area
   About two-thirds of one percent of the exploratory development budget

C. *Administrative information*
   Time : Six months to produce the first NTF
           CNM instructions were accepted with militant pessimism
           All forecasts submitted on time
           Update periodically, loose leaf arrangement
           CND edited and revised the basic volume (operational implications)

mands are greatly expanding the systems contained in Part III, probable systems options. In addition, the systems commands are referencing the options to the projections contained in the Part II technological capabilities forecast for additional continuity. In October, 1969, the Navy published a complete assessment of technology, from scientific opportunities through the functional technologies to a multitude of system options. Although there are no formal plans at present, this common experience across the entire Navy technical community will provide an opportunity in the future for a comprehensive in-house review of all aspects of forecasting and planning for the Navy. The NTF, along with the exploratory development goals, are fast becoming a formal part of the Navy planning system. The probable systems options are presently recognized by CNO as a predecessor to the preparation of PTA's (planned technical approaches). The NTF is being used by people who live by the golden rule: "He who has the gold, makes the rules." As such, it it becoming a valuable aid to planning and funding technology.

New Organizational Forms
for Forecasting

ERICH JANTSCH

Erich Jantsch authored, while consultant to Organization for Economic Co-operation and Development (Paris, France), the first major study on technology forecasting Since then his international teaching and consulting work has led him to reflect forecasting as one element of policy.

This chapter* suggests that new organizational forms for forecasting must be designed consistent with the new roles and interrelationships of the social institutions for which the forecasting is performed. Corporations, cities, nations all are moving towards integrated activities that make them more and more a part of a single system, rather than individual, segregated, autonomous units. Thus, the author proposes some interesting considerations in terms of coordinated activity if forecasting, and the subsequent planning based on the forecasts, are to be meaningful and realistic.

* This paper was originally presented at the conference on Technological Forecasting: Concepts and Methods, June 8–12, 1969, held on Hilton Head Island, South Carolina, organized by the Industrial Management Center under the direction of Professor James R. Bright.

Forecasting takes a look ahead at potential dynamic developments. There are obvious interdependencies among such developments in different domains—economic, social, political, technological, and psychological, for example. Thus the integrative approach[1], emphasizing concern with the dynamics of multidimensional systems, is becoming increasingly accepted as a conceptual and methodological basis for long-range forecasting and planning. Both the long-range view and the hierarchical structure of normative forecasting and planning enforce a dynamic approach embracing many dimensions of change.

The inherent difficulties of integrative forecasting become apparent in the naive attempts to obtain a synopsis by combining forecasts obtained more or less independently and along one-dimensional lines of development. The objective is frequently to derive a scenario, a cross section at some given moment in the future. Most technological forecasts are still derived individually from relatively autonomous criteria and subsequently merged with the other ingredients of a dynamic scenario.[2]

This would have been more appropriate for the recent past when the thrust of technological development was less guided by social, ecological, psychological, and anthropological criteria than it probably will be in the future.

If we reverse—as, indeed, we should—the currently dominating principle of "adapting society to technological innovation," to make it "adapting technology to the evolution of society," we make technological forecasting more difficult in two ways: First, forecasted technological feasibilities will mean little in themselves if the normative elements are to be derived primarily from social criteria—we will need fully-developed social forecasts against which we then may match these technical feasibilities. And, second, the entire institutional framework, as it exists within society, must itself be taken as a variable in any social forecast.

This implies that forecasting for the purposes of a specific institution—an industry, a corporation, a government agency, and so on—can no longer assume that institution to be static. The forecast must include the evolution of the institution itself: its varying structures; its evolving pattern of resources; its dynamic interrelationships with other existing, emerging, or decaying institutions; and the adaptation of its responses to different sets of challenges. Technological forecasting, geared as it is to the purposes of institutions acting as agents in technological change, is to a particularly high extent confronted with this problem of institutional change; for technological change takes

---

[1] Frank P. Davidson, Macro-Engineering—A Capability in Search of a Methodology, *Futures* Vol. I, No. 2, December 1968. Erich Jantsch, "Integrative Planning of Technology" in *Perspectives of Planning* (Paris: OECD, 1969).

[2] Herman Kahn and Anthony Wiener, *Toward the Year 2000: A Framework for Speculation* (New York, 1967). It is also astonishing to see the authors extrapolate up to the year 2000 the concepts of power and economic strength in the rigid framework of a multiplicity of nation-states as they are defined today.

place at a very high rate and is very closely coupled to social change. From the above it follows that useful technological forecasts must have two attributes:

1. Forecasts made for a specific institution must be fully integrated with forecasts for the entire relevant social system, beyond the specific environment with which the institution directly interacts.
2. Forecasts must emphasize the potential for continuous self-renewal of the institution for which they are made.

Looking at technology, and the way it interacts with society, we may distinguish three levels of organizational complexity. These may also be understood as levels of hierarchical dependence:

1. "Joint systems" of society and technology (for example, systems of urban living, communication, transportation, health, food production, and so on).[3]
2. Missions of technology within society (for example, shelter, personal transportation, cure of diseases, agriculture, and so on).
3. Specific technologies (for example, prefabricated construction elements, internal combustion engine cars, specific drugs, hybrid plant species, and so on).

Today, forecasting activity focuses primarily on the lowest level, and many of its techniques encourage staying there. In industry this type of forecasting favors the development of product lines and specific processes to their ultimate possibilities. It also conveys the comforting—but misleading—feeling that institutions will not have to change.

Some institutions, government agencies as well as industrial corporations, are at present more or less successfully trying to come to grips with the complex forecasting problems at the middle level. The United States Government's Planning-Programming-Budgeting System (PPBS) is designed to encourage and facilitate forecasting and planning at this level, frequently referred to as the "strategic" level. In industry, the introduction of corporate long-range planning has gradually led some corporations to reformulate their goals in terms of broad missions, or functions of technology vis-à-vis society.[4] Here, the institution which gives rise to technological change, accepts for itself a dynamically changing role while retaining its basic identity. An example would be those petroleum companies who conceive of their future role as "energy for transporation." This might involve, for example, switching from the production and distribution of gasoline to that of stored electricity produced in giant base-load operated nuclear reactors.

[3] For further elaboration of the concept of "joint systems," see Erich Jantsch, "Integrative Planning of Technology" in *Perspectives of Planning* (Paris: OECD, 1969).
[4] Erich Jantsch, "Integrating Forecasting and Planning through a Function-Oriented Approach" in James R. Bright (ed.), *Technological Forecasting for Industry and Government* (Englewood Cliffs, N.J., 1968) and Erich Jantsch, Technological Forecasting in Corporate Planning, *Long-Range Planning*, Vol. I, No. 1, September, 1968.

Today most institutions hesitate to acknowledge any responsibility for integrating their own pattern of actions into the wider concepts of the "joint systems" constituted by society and technology. However, this issue will soon become a matter of life or death for the existing institutions, especially for industry. If companies choose to continue acting as independent entities pursuing the goals inherent in their own (present) structure and interacting with other independent units, new institutions will be forced to assume responsibility. Instead of industry, giant publicly-owned monopolistic enterprises of regional or world scope may take over the planning and possibly also the production and distribution tasks in entire areas defined by social needs. However, the existing institutions are still free to participate actively in the present move towards a more fully integrated society. They can even assume a leading role in this movement.[5]

The task of technological, more appropriately integrative, forecasting for any institution may be divided in two. The two subtasks determine the ways one may properly organize the forecasting function:

1. Forecasting the role of the institution in a broad societal context beyond the environment with which the institution interacts directly.
2. Forecasting the deployment of the resources available to the institution to fulfill the forecasted roles.

The types of organization suited to these two subtasks will be briefly discussed in the following sections.

First, a short note on technological forecasting methodology: most exploratory techniques in use at present (perhaps with the exception of envelope curve techniques) tend to enhance a static and rigid position of the institution. They facilitate linear and sequential modes of forecasting and planning. On the other side, normative techniques, particularly relevance tree techniques, contribute significantly to the clarification of alternative technological options falling under specific missions. These techniques thus favor a flexible and adaptive position of the institution itself.

As Roberts points out, exploratory techniques are about to enter a stage of development comparable to complex multivariate econometric forecasting in the economic area.[6] In the process they will climb one step in their usefulness for integrative forecasting. Concurrently, feedback models linking exploratory and normative aspects may soon come into the stage of large-scale application. It is evident that forecasting institutional change will depend a great deal on feedback thinking and simulation.

[5] Erich Jantsch, *Technological Planning and Social Futures* (London and New York: Business International, 1972).
[6] Edward B. Roberts, Exploratory and Normative Technological Forecasting: A Critical Appraisal, *Technological Forecasting*, Vol. I, No. 2, November, 1969.

The trend away from the present fragmented social systems—corporations, cities, nations—and towards a more fully integrated system of world society is inherent in the psychosocial evolution of mankind itself. We are standing at a new "threshold" of this evolution, comparable in importance to those thresholds crossed by mankind when it developed primitive agriculture or trade links between villages and cities.

The unity of mankind and the indivisible responsibility for the planet Earth constitute the challenge at this new threshold, which must be crossed more rapidly than was ever required before. The response has to be formulated and implemented in the decades left of this century. If it turns out to be unsatisfactory, not only will mankind fail to achieve the necessary degree of integration, but severe dynamic instabilities will throw us back to lower levels and earlier phases of evolution.[7]

For many problems at the macrolevel of the "joint systems" formed by society and technology—food, environment, development, and so on—nothing short of such a global view will be required in the horizontal integration of forecasting and planning.

To industry the integration of its processing activities will become of particular concern. Product and service lines, developed—and frequently pushed to their extremes—in a piecemeal way, interact within larger systems, such as the systems of urban living and of the human environment. Linear and sequential modes of action—in other words, the indiscriminate pursuit of feasibilities in a fragmented way—inhibit the healthy development of such systems and the coordination of the *outcomes* of lines of action.

Inevitably, industry will face the paramount task of inventing, planning, designing, building, and possibly also operating, such joint systems of society and technology. Inherent in technological innovation is the task of social innovation. Technology must be forecasted and planned in this wider connotation.

This implies that forecasting and planning also have to be vertically integrated. They become part of inventing, planning, and creating the future of society, not just of the particular institution. Thus far, implementation and goal-setting have been pursued separately, the former by independent institutions such as industry and the latter by government. But now integration of the goal-setting and implementing activities is becoming necessary to deal appropriately with the "joint systems" of society and technology.

Competition will increasingly assume the form of competition between ideas and plans, inventions and designs of systems—in other words, between social software alternatives rather than industrial hardware. This has led some to anticipate a shift in the roles of industry and university. Daniel Bell, for

---

[7] For the most comprehensive statement on this issue, see the following book written by an industrialist, Aurelio Peccei, *The Chasm Ahead* (New York, 1969) as well as R. Buckminster Fuller, *Operating Manual for Spaceship Earth* (Carbondale, Ill., 1969) and Jay W. Forrester, *World Dynamics* (Cambridge, Mass., 1971).

example, writes: "Perhaps it is not too much to say that if the business firm was the key institution of the past hundred years, because of its role in organizing production for the mass creation of products, the university will become the central institution of the next hundred years because of its role as the new source of innovation and knowledge."[8]

There may be a fallacy here in assuming that industry will not change in its institutional character with this new challenge. Historically, industry has taken over from the university the bulk of applied scientific research and technical development and is now becoming deeply involved even in fundamental research. There is no reason why industry should not acquire the capability to deal imaginatively with the new challenge presented by the "joint systems" of society and technology. However, one may agree more readily with Daniel Bell when he continues: "To say that the major institutions of the new society will be intellectual is to say that production and business decisions will be subordinated to, or will derive from, other forces in society; that the crucial decisions regarding the growth of the economy and its balance will come from government. . . ."[9]

The pressures emanating from this development may well lead to the emergence of three of the existing types of institutions as "joint planners for society": government (which may gradually change as the principle of the nation-state increasingly gives way), industry, and the university. Both industry and the university ought to become *political institutions* in the broadest sense, contributing to an integrated planning process for society at large instead of pursuing independently conceived subgoals.[10]

The central problem here may be seen in analogy to the problem of organizing forecasting and planning within a corporation. But the degree of complexity, and therefore the danger of failure, in getting this subtly integrated process well under way is much higher at the level of society. But, in analogy to the corporation, the principle of *decentralized initiative and centralized synthesis* provides the only effective way of soliciting creative contributions from all parts of society. The outstanding role of industry and the university merely derives from their effectiveness in organizing the intellectual capacity of society. If industry should prefer to stagnate and concentrate on optimizing mass production techniques, think tanks in some form may emerge as competitors with ideas and plans. Industry could become an appendage.

At present we may assume that both industry and the university—after a period of resisting change and clinging to the old principle inevitably evoked in pertinent discussions today—will accept their new roles as political institu-

[8] Daniel Bell, Notes on the Post-Industrial Society, *The Public Interest,* No. 6, Winter 1967.

[9] *Ibid.*

[10] Erich Jantsch, Technological Forecasting for Planning and Institutional Implications, in Proceedings of the Symposium on *National R&D for the 1970's,* National Security Industrial Association, Washington, D.C. (1968).

tions and develop systems laboratories focusing on the "joint systems" of society and technology. These systems laboratories would then provide the organizational framework for the bulk of technological forecasting. Technological feasibilities would be tested in the context of the dynamic behavior of complex social systems. This would finally result in the purposeful pursuit of technological options. One would build new systems structures and no longer simply adapt social systems to new technologies.[11]

While the university may then compete with industry on equal terms in selling its ideas and systems designs, it may also be called upon by government to provide prototypes by developing experimental designs in strategically selected sectors. It may study, or contribute to the study of systems too large to be dealt with by individual corporations. And it may organize and back up the study of global problems which will be so crucial for the future of mankind.

For a transitory period, industry may utilize to an increasing extent the services of nonprofit research institutes which will offer a considerable capability in forecasting and systems planning.[12] More and more of the large industrial corporations are recognizing the desirability of projecting their future roles against a common background while retaining the principle of competition in planning and action. They recognize the necessity to harmonize their policies at the societal level, while competing with each other at the level of strategies. If, for example, a thorough background study would establish beyond doubt the necessity to develop before the end of the century nonagricultural food production technologies, and these findings would also be accepted at the political level, decisions leading to the development of alternative technological strategies would be much easier for the individual corporation.

Harmonization of basic policies within a broad societal context would open the way to an important gradual shift in human values, easing up on the overstressed value "competition" which governs relations between nations, institutions, corporations, and individuals today in Western civilization and supplementing it by the value "responsibility."

Ultimately, a "common policy background" may be developed in many areas of concern by concerted actions of governments—and through them of universities and research institutes—in a joint effort of the advanced countries. This may be a way to approach the problem of planning on a planetary scale and fighting the disruptive forces developing in our society and our planet.

---

[11] On the principles of purposefully designing social systems, see Jay W. Forrester, "Planning Under the Dynamic Influences of Complex Social Systems" in *Perspectives of Planning* (Paris: OECD, 1969) and Jay W. Forrester, *Urban Dynamics* (Cambridge, Mass., 1969).

[12] The prototype of a "pure" institute of that sort is the Institute for the Future which has begun operation in Middletown, Connecticut, and has recently moved to Menlo Park, California. Other research institutes, such as the Battelle Memorial Institute, Stanford Research Institute, and A. D. Little, have shifted from concentration on hardware research to developing capabilities in these new areas.

If the mobilization of industries and universities may be expected to ensure the creative inputs, the decentralized initiative, new types of organizations will be required to synthesize them. Synthesis will become necessary for any scale of system—communities, nations, regions, continents, and the entire planet Earth. There can be little doubt that this function of synthesis will come to dominate the form of "systems government"; however, this will be defined, then, in similar ways as it already impresses itself on the organization of industrial management and starts to influence the structures and functions of national government. To deal with the most pressing and threatening problems of the future, those of planetary dimensions, a World Forum has already been proposed, which would be set up jointly by the advanced countries of the world.[13] Its activities would focus on forecasting and simulating the consequences of changes in policy (systems structure) and various courses of action impinging on the present. The World Forum would also serve in the role of synthesizing. It would have only a small permanent staff, but could elicit a maximum of creative inputs from all participating countries and bodies. It would involve them in a feedback loop of translating objective consequences of alternative courses of action and potential implications of recognized feasibilities into each other within a dynamic systems context.

This same principle of feedback and translation is already becoming applied in the organization of forecasting within an institution, as will be briefly discussed in the following section.

ORGANIZING FORECASTING WITHIN AN INSTITUTION

The same principles that should guide forecasting and planning in the broad framework of society—leading to a clear statement of the individual roles of interdependent institutions—also apply to the organization of forecasting and planning within an institution. Only their difficulties are reversed: The "translation mechanism" may be established more effectively within an institution, but it will encounter considerable difficulties in embracing and channeling through the overall systems view obtained at the level of society. The temptation to short-circuit the feedback between creative individuals working in an organization and the concerns of society at large will be great. Institutional, piecemeal objectives may replace societal objectives as they have replaced them in institutional planning so far. In industry, for example, this led to the crude "maximization of profit" motive which can no longer provide a real challenge to the creative professional. Another danger has been evoked by Galbraith with the notion that the "technostructure," the creative technical and midlevel management cadres, may pursue false objectives inherent in technology itself.[14] These can become a challenge to the professional if he is lacking the framework of reference to work for a real purpose.

13 Aurelio Peccei, *The Chasm Ahead* (New York, 1969).
14 John K. Galbraith, *The New Industrial State* (Boston, 1968).

What is needed, and more and more explicitly understood in advanced-thinking industry today, is coupling between social and institutional objectives. Institutions, forecasting their role in a changing environment, will reformulate their own objectives in the light of the anticipated societal context. All forecasting and planning within the institution will be aligned to this dynamic view of the institution's policy.

The first requirement for the organization of forecasting within an institution is therefore the establishment of an *"environmental radar" and a policy planning forum.* Clearly, since the transformation of the institution itself is subject to forecasting and planning at this scale, this must be the job of top management and the Board of Directors. Few industrial corporations or universities have yet set up such a function. Typical organizational forms include the appointment of "Officers of the Board" working directly for the Board and the President of a company and high-level "Policy Committees" meeting regularly and perhaps supported by a small staff.

It is also the top management's task to keep the vertical translation process between objectives and opportunities running in both directions, upwards and downwards. This vertical stream of information implements the feedback process between exploratory and normative technological forecasting involving the entire institution. In some advanced-thinking industrial corporations, vice presidents devote more than half of their time to this translation process. They thereby create self-motivation at all levels and elicit creative ideas which become parts of the institution's forecasting and planning process.

Whenever top management fails to recognize its role in this process, the feedback between normative and exploratory forecasting will be severely curtailed. This is generally the case today. Imaginative people at lower hierarchical levels will be uncertain of the dynamic changes anticipated for the institution itself, or will have to assume it as rigid, and their inadequate attempts to link institutional and social futures will sooner or later lead to frustration. Top management sits at the top of an institution not to control but to guide the human fabric of the institution. It should attempt to synthesize any responses to a challenge in form of an explicitly spelled out policy in a societal context.

The next level down is that of *forecasting alternative technological strategies to contribute to recognized technological missions* which the institution has adopted under its role. There, the "vested interest" in specific technologies has to be overcome, as this prejudice is embodied not only in capital investment and special human skill but also in the general thrust of specific technological development lines. People identify strongly with specific technical objectives. The foremost task at this level is to educate corporate personnel to look beyond specific projects to the missions of technology and to their implications in the context of society.

Two organizational principles help to overcome the difficulties in obtaining a perspective of the future that is not obstructed by barriers inherent in particular technologies:

1. Build divisions of the institution around technological missions rather than around specific technologies or product and service lines.
2. Create a strong "corporate development" function, in which alternative strategies for missions are synthesized and assessed, newly developed (for example, in corporate-level research laboratories), and viewed in the context of dynamically evolving corporate policies; this leads to a high-level organizational split between the "present" and the "future" of the institution.

Several industrial corporations have adopted both principles and organized themselves to bring the long-range future into sharper focus. Exhibit 1 sketches the example of Westinghouse, the first multibillion-dollar corporation to adopt such a structure.

The same principles may be applied to nonindustrial organizations, in

Exhibit 1. High-level split between the "present" (operating companies) and "future" (corporate development) as exemplified by recent reorganization of Westinghouse Electric Co.

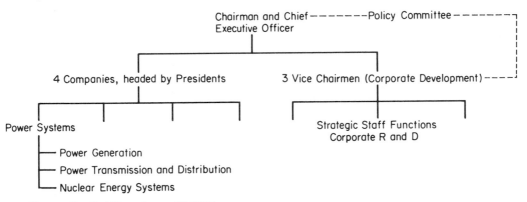

(Source: *New York Times*, January 12, 1969).

Note: In particular, the structural subunits built around missions such as "power generation" and "power transmission and distribution" replacing the traditional units focusing on specific technologies, such as steam turbines, gas turbines, generators, transformers, and so on.

particular to government[15] and to the university.[16] In the future it will no longer be the functions of forecasting and planning looking for a place in the current organization, it will be the long-range perspective on the social and technological future determining the organization of the institution. The trend is already visible.

[15] Erich Jantsch, Technological Forecasting for Planning and Institutional Implications, Proceedings of the Symposium on *National R&D for the 1970's,* National Security Industrial Association, Washington, D.C. (1968).

[16] Erich Jantsch, *Technological Planning and Social Futures* (London and New York, 1972).

The flexibility required by advanced corporate planning cannot be attained in current administrative and operating structures, though higher degrees of automation may change this in the future. Today, a common, satisfactory solution is to superimpose a flexible "innovation emphasis structure" over the more rigid administrative and operating structures. Exhibit 2 shows such a simple "innovation emphasis structure" worked out and maintained by the "corporate development" side of an American electronics company. Ten-year strategies are updated annually through a process involving the entire company, including the operating divisions. As forecasted developments become adopted and mature, they give rise to the corresponding organizational changes. The important point is that current organizational structures are kept completely outside this framework of long-range forecasting and planning.

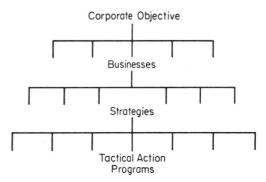

Exhibit 2. "Innovation emphasis structure," (example from American electronics company) superimposed over the administrative structure, resembles relevance tree and enforces normative thinking at all hierarchic levels.

Technological forecasting, as part of planning, depends on decentralized initiative and centralized synthesis. Exhibit 3 proposes a scheme of involvement in this process, viewed for an industrial corporation, by successive stages of a typical technological innovation.[17] The responsibility for synthesizing technological forecasting may thus be shared as follows:

1. Technological forecasting for policy making at the Board and top management level.
2. Technological forecasting for strategic decision making at the corporate level, for example, by horizontal staff groups (usually incorporating mixed scientific, technical, and marketing expertise), working for top management and specifically within the framework of corporate long-range planning.
3. Technological forecasting for tactical decision making at divisional management level, also preferably by staff groups.

Exhibit 3 also demonstrates the decisive role of horizontal, corporate level staff groups in synthesizing technological forecasting. For three or four of the

[17] Erich Jantsch, Technological Forecasting in Corporate Planning, *Long-Range Planning*, Vol. I, No. 1, September, 1968.

Exhibit 3. Organization of Technological Forecasting Within Corporate Structure, Each Forecasting Stage involving Interaction Between Different Levels, and Synthesis Always Made at the Highest Level Involved. (Schematic Representation Does Not Show Full Feedback Taking Place at Each Forecasting Stage.)

	Forecasting Stages							Evaluation
	Prediscovery	Discovery	Creation	Substantiation	Development	Advanced Engineering	Applied and Service Engineering	
Board level								
Board officers	X	(X)						
Corporate level								
Horizontal staff groups	O	X	X	X	X	X	X	
Research laboratories	O	O	O	O	O	O	O	
Divisional level								
Management (staff groups & division laboratories)								
Project engineering groups		O	O	O	O	O	O	
RDT & E phases								
Environment	Corporate environment	Technological missions	Technological options	Implications for corporate objectives	Market implications	Repercussions in market	Feedback	
Science and technology	Basic potentials and limitations	Attainment of technological capability	Systems performance	Systems elements and dev. requirement	Systems design	Production and operating requirements	Feedback	
Forecasting objectives								

X Synthesis and inputs
O Inputs

six forecasting stages, the synthesis is best entrusted to such staff groups.[18]

The complete organizational framework of technological forecasting may thus also be viewed as unfolding by the interaction of synthesis at three different levels of an institution. But the involvement of creative people at all hierarchical levels and the stimulation of decentralized initiative are decisive. A particularly effective way of stimulating such institution-wide involvement has been found in comprehensive joint forecasting "rounds" involving management as well as key technical people, for example on the basis of Delphi technique panels.

This integral process involves the whole institution and reaches beyond its boundaries. With the participation in broad environmental forecasting and policy planning it reaches beyond the boundaries of the environment with which the institution directly interacts. Through this process the organizational problems deriving from the requirements and opportunities of forecasting and planning may now be seen as becoming *identical with the problem of self-renewal of the institution*. If planning is understood as the task of designing a system (a corporate policy), flexibly changing subsystems (corporate strategies) and the ways to operate them (operational plans), technological forecasting is the principal building material for this system.

The emphasis on institutional outputs moves more and more to ideas and plans dealing with the "joint systems" constituted by society and technology. Individual entrepreneurship within the institution will thus find increasing freedom to interact with explicitly stated policies. Instead of a split between corporate development and operating units, a new type of "dynamic systems laboratory" may develop, both in industry and in universities. This new organization would form units which focus neither exclusively on the future nor on the present but rather on dynamic behavior of large systems. Instead of deploying in a flexible and dynamic way relatively static "modules" of skill and equipment, as in current long-range planning of product development, emphasis on the "software" output of institutions—on ideas, plans, and designs of systems—will permit the organizational reunification of planning and action.

The "dynamic systems laboratories" will then be the natural place for technological forecasting, an integral part of its activities, involving all its personnel. If, at present, tactical planning may be entrusted to the individual carrying out the action, provided that a policy is stated explicitly, the inventor and planner of systems will in the future become responsible for strategic planning, too. Since competition between institutions will take place as competition between strategies, the principal interaction within an institution

[18] The OECD survey found such corporate level staff groups in 20 out of 23 American, and in 26 out of 39 European companies considered. However, in all but one American, and in 16 European companies, they were supplemented by forecasting groups in the research laboratories and the operating divisions. See Erich Jantsch, *Technological Forecasting in Perspective* (Paris: OECD, 1967).

will be the dialogue between policies and strategies. With this simplification inside the institution, the way may hopefully be free to acquire a new level of planning in the interaction between institutions: the planning of society's policies.

The people animating the institutional activities may then come close to Forrester's vision: "To deal with the system of which engineering is a part and thereby to make engineering more effective within that system, the engineer must understand the system components and structure. He needs an insight into the nature of the corporation, the tasks of top management, human motivation, the relationship of the individual to the organization, and the psychology and processes of innovation and change. The engineer could become a 'change agent' to precipitate improvements in our social system. . . . He would try to clarify the enduring goals and objectives for his organization and the people within it. . . . He would give more attention to the surrounding social system as a whole rather than as an array of isolated parts. . . . He would consider the transient and steady-state behavior of his organization from the same system viewpoint that he approaches complex physical systems. He would strive to perfect models of social processes that permit simulation studies leading to a better understanding of organization, information links, and policy. And he would bring to actual human organizations the courage to experiment with promising new approaches based on a foundation of design."[19]

Technological forecasting is a means to achieve this foundation of design.

[19] Jay W. Forrester, Common Foundations Underlying Engineering and Management, *IEEE Spectrum,* Vol. I, No. 9, pp. 66–77, September, 1964.

Index